European Consortium for
Mathematics in Industry 5

Manley, McKee and Owens (Eds.)

Proceedings of the Third European Conference
on Mathematics in Industry

European Consortium for Mathematics in Industry

Edited by
Michiel Hazewinkel, Amsterdam
Helmut Neunzert, Kaiserslautern
Alan Tayler, Oxford
Hansjörg Wacker, Linz

ECMI Vol. 5

Within Europe a number of academic groups have accepted their responsibility towards European industry and have proposed to found a European Consortium for Mathematics in Industry (ECMI) as an expression of this responsibility.

One of the activities of ECMI is the publication of books, which reflect its general philosophy; the texts of the series will help in promoting the use of mathematics in industry and in educating mathematicians for industry. They will consider different fields of applications, present casestudies, introduce new mathematical concepts in their relation to practical applications. They shall also represent the variety of the European mathematical traditions, for example practical asymptotics and differential equations in Britain, sophisticated numerical analysis from France, powerful computation in Germany, novel discrete mathematics in Holland, elegant real analysis from Italy. They will demonstrate that all these branches of mathematics are applicable to real problems, and industry and universities in any country can clearly benefit from the skills of the complete range of European applied mathematics.

Proceedings of the

Third European Conference on Mathematics in Industry

August 28-31, 1988 Glasgow

Edited by

John Manley

Lecturer in Industrial Mathematics,
University of Strathclyde, Glasgow

Sean McKee

Professor of Industrial Mathematics,
University of Strathclyde, Glasgow

and

David Owens

Professor of Dynamics and Control,
University of Strathclyde, Glasgow

 B. G. Teubner Stuttgart

KLUWER ACADEMIC PUBLISHERS

DORDRECHT / BOSTON / LONDON

Library of Congress Cataloging in Publication Data
CIP data available from publisher (Kluwer)

CIP-Titelaufnahme der Deutschen Bibliothek
CIP-data available from publisher (Teubner)

ISBN 978-94-010-6770-6   ISBN 978-94-009-0629-7 (eBook)
DOI 10.1007/978-94-009-0629-7

Sold and distributed in Continental Europe (excluding U.K.)
by B. G. Teubner GmbH, P.O. Box 801069, D-7000 Stuttgart-80

Sold and distributed in the U.S.A. and Canada
by Kluwer Academic Publishers,
101 Philip Drive, Norwell, MA 02061, U.S.A.

Kluwer Academic Publishers incorporates
the publishing programmes of
D. Reidel, Martinus Nijhoff, Dr W. Junk and MTP Press.

In all other countries (including U.K.), sold and distributed
by Kluwer Academic Publishers Group,
P.O. Box 322, 3300 AH Dordrecht, The Netherlands.

Printed on acid-free paper

Preface

The European Consortium for Mathematics in Industry (ECMI) was founded,
largely due to the driving energy of Michiel Hazewinkel on the 14th April, 1986
in Neustadt-Mussbach in West Germany. The founder signatories were
A. Bensoussan (INRIA, Paris), A. Fasano (University of Florence), M. Hazewinkel
(CWI, Amsterdam), M. Heilio (Lappeenranta University, Finland), F. Hodnett
(University of Limerick, Ireland), H. Martens (Norwegian Institute of
Technology, Trondheim), S. McKee (University of Strathclyde, Scotland),
H. Neuazert (University of Kaiserslautern, Germany), D. Sundstrom (The Swedish
Institute of Applied Mathematics, Stockholm), A. Tayler (University of Oxford,
England) and Hj. Wacker (University of Linz, Austria). The European Consortium
for Mathematics in Industry is dedicated to:

(a) promote the use of mathematical models in Industry

(b) educate industrial mathematicians to meet the growing demand for such
     experts

(c) operate on a European scale.

ECMI is still a young organisation but its membership is growing fast. Although
it has still to persuade more industrialists to join, ECMI certainly operates on
a European scale and a flourishing postgraduate programme with student exchange
has been underway for some time.

It is perhaps fitting that the first open meeting of ECMI was held at the
University of Strathclyde in Glasgow. Glasgow is and was the industrial capital
of Scotland and was, and arguably still is, Britain's second city after London;
when this volume appears it will have rightly donned the mantle of the cultural
capital of Europe. It is also fitting that such a meeting should be held at the
University of Strathclyde. Founded as Anderson's Institute in 1796, it emerged
from the industrial revolution and the so-called period of Scottish
Enlightenment during which the economist, Adam Smith, the moral philosopher,

David Hume, the chemist, Joseph Black, and the geologist, James Hutton, and many
others were at their height. Anderson, a Professor of Natural Philosophy at
Glasgow University, was perturbed that the older universities were not meeting
the new disciplines and skills demanded by the industrial revolution, and so in
his will he left money to found an 'Institute of useful learning'; this concept
is still strongly in force today.

Applied mathematics is characterised by the dual demand it requires of its
practitioners: competence in mathematical and numerical methods and a reasonably
deep understanding of some field of experimental results. While the pure
mathematician limits himself largely to the former, and the experimental
scientist is principally involved with the latter, the applied mathematician has
to use both deductive and inductive inference.

But what is Industrial Mathematics? It is simply applied mathematics which
is problem-driven rather than area-led. By this we mean that solving the
problem is tantamount. If in solving a problem mathematics is not required, or
indeed is not cost-effective then so be it, although the industrial
mathematician would probably, but not necessarily, leave it for a more
appropriate specialist to solve. Further, it is pointless claiming to be an
industrial mathematician and waiting for a problem to appear in, let us say,
Pade approximation. Area-led research has little to do with industrial
mathematics: it is the essence of industrial mathematics that the practitioner
be a generalist, who tackles the problem with an open mind using whatever
mathematics is necessary for that particular problem. Of course, all
mathematicians want to develop attractive and novel mathematics. One of the
motivations for the applied mathematician's involvement with industry is to
tackle difficult problems that occasionally, but we must emphasise this,
occasionally give rise to new ideas and theories. But it is crucial that the
problem itself, the raison d'etre, is not lost sight of in the quest for elegant
mathematics: we stress this again - it is the problem that is tantamount.

The industrial mathematician, often working within an interdisciplinary team, is presented with (more likely has to seek out) a problem or, more commonly, an ill-defined problem area. His first step is to define the precise physical, biological or economic problem. A model is then built and a solution is sought, usually by a combination of analytic and numerical techniques.

The results often in graphical form must then be interpreted and the model refined until it is validated by whatever experimental data is available or obtainable.

This volume contains many examples of this general strategy. It is divided up into three sections, one containing the invited speakers, one containing the three mini-symposia and a final section containing contributed papers. The invited speakers cover nonlinear magnetostatics, field-effect transistors, supercomputers, simulated annealing, domain decomposition techniques and the optimisation of chemical plants, while the mini-symposia concentrate on three areas: numerically intensive computing; vibrations in cables; and some novel partial differential equation problems arising in industry. The contributed papers cover many areas of science and engineering ranging from the freezing of meat to polymer crystallisation and from many aspects of control applications to gel electrophoresis.

Finally we would like to express our gratitude to the Scottish Development Agency, the Central Electricity Generating Board, IBM, Rutherford Appleton Laboratory and Shell for their financial sponsorship. However, we would particularly like to thank the conference secretary, Mary Doherty without whose tireless energy, efficiency and enthusiasm this meeting would have been very much less successful.

Glasgow, Summer 1989   Sean McKee

TABLE OF CONTENTS

INVITED SPEAKERS

On non-linear magnetostatics:
dual-complementary models and "mixed" numerical methods
A. BOSSAVIT   3

Modeling for field-effect transistors
E. CUMBERBATCH   17

Supercomputers – 1988
I. S. DUFF   31

Simulated annealing: theory of the past, practice of the future?
P. J. M. van LAARHOVEN, E. H. L. AARTS   45

Domain decomposition for a generalized Stokes problem
A. QUARTERONI, A. VALLI   59

Mathematical simulation and optimization of chemical plants
F. KOKERT, L. PEER, Hj. WACKER   75

MINI-SYMPOSIUM 1
(organiser: P. W. Gaffney)

Numerically intensive computing
P. W. GAFFNEY   92

Fomulation of a sea model with continuous density stratification in the vertical
G. K. FURNES   93

Electroplating simulation
L. J. GRAY, G. E. GILES, J. S. BULLOCK, P. W. McKENZIE   95

Elastic modelling on the IBM 3090 vector multiprocessor
A. H. KAMEL   107

Solving a model interface problem for the Laplace operator by boundary
collocation and applications
L. REICHEL   119

MINI-SYMPOSIUM 2
(organiser: R. Mattheij)

On flow-induced vibrations of overhead transmission lines
R. MATTHEIJ   132

Non-linear free vibrations of coupled spans of suspended cables
S. W. REINSTRA   135

Vibrations of overhead transmission lines: computations and experiments
P. HAGEDORN, M. KRAUS   145

On the modeling of a continuous oscillator by oscillators with a finite number of
degrees of freedom
A. H. P. VAN DER BURGH   159

MINI-SYMPOSIUM 3
(organiser: J. Ockendon)

Some novel partial differential equations problems arising in industry
J. R. OCKENDON   172

Modeling coronas and space charge phenomena
C. BUDD, A. WHEELER   173

Temperature surges in thermistors
A. C. FOWLER, S. D. HOWISON   197

Mixed hyperbolic-elliptic systems in industrial problems
A. D. FITT   205

CONTRIBUTED PAPERS

Linear and non-linear approximation of power density spectra with linear
dynamical filter systems
D. AMMON   217

Modelling software reliability from run-time data
A. W. ANDREW, R. J. COLE, J. GOMATAM   225

Software simulation of model reference adaptive control systems
M. BAKR, D. BELL   233

Effective length of an ultrafiltration device
N. G. BARTON   241

Truncated sequential tests for material control problems
R. BEEDGEN   249

On the approximation of free vibration modes of a general thin shell application to
turbine blades
M. BERNADOU, B. LALANNE   257

Integration algorithms for the dynamic simulation of production processes
M. BERZINS, P. M. DEW, A. J. PRESTON   265

Multicomponent flow computation with application to steam condensers
A. W. BUSH. G. S. MARSHALL. T. S. WILKINSON   273

Natural convection within a droplet as a result of a chemical reaction on its surface
E. H. DE GROOT — 283

Unwanted compartment fires: zone modelling the onset of hazardous conditions
H. A. DONEGAN, T. J. SHIELDS, G. W. SILCOCK — 293

A tree search approach based on an assignment relaxation for the solution of set
covering problems
E. EL-DARZI, G. MITRA — 303

Making a workpiece with spiral turns by means of forming cutters
H. W. ENGL, T. LANGTHALER — 313

State variables feedback control of stepping motors with flexible shaft
R. FIROOZIAN, J. G. BAJIR — 333

Application of mathematics to heat processing in the meat industry
G. S. FULTON, D. BURFOOT, S. J. JAMES, C. BAILEY — 343

Numerical modelling of conjugate heat transfer in an advanced gas-cooled reactor
fuelstandpipe
M. T. R. FUNG, R. P. HORNBY — 353

Restoration of NMR images
F. GODTLIEBSEN — 363

An on-line augmented price correction technique for hierarchical control of
interconnected industrial processes
Z. M. HENDAWY, P. D. ROBERTS — 369

A least-squares fitting technique for use with large non-linear plant models
J. HOPE — 377

Spline approximation of offset curves and offset surfaces
J. HOSCHEK, F-J. SCHNEIDER — 383

Diffusion flame ignition by a recirculating flow
M. KONCZALLA — 391

On the real-time simulation and control of the continuous casting process
E. LAITINEN, P. NEITTAANMÄKI, T. MANNIKKÖ — 401

Simulation of VLSI circuits: relaxation techniques
P. LORY — 409

Development of models for flashing two-phase jet releases from pressurised
containment
K. McFARLANE — 415

Robust recursive estimation: the Lp approach
D. W. McMICHAEL — 429

Numerical approximation of free boundary problems in polymer crystallization
S. MAZZULLO, M. PAOLINI, C. VERDI — 437

Ignition/extinction phenomena: an investigation of parametric sensitivity for a
strongly non-linear reaction-diffusion system
D. MEINKÖHN
445

On the methods for optimal shape design
P. NEITTAANMÄKI
453

Steady-state optimization of large gas networks
A. J. OSIADACZ, D. J. BELL
461

Modelling and control of acid-base blending systems
U. PALMQUIST
469

The prediction of cyclic plastic strain growth behaviour for severe thermal loading
problems in structural engineering using upper bound methods and linear
programming
A. R. S. PONTER, K. F. CARTER
479

Three problems in the integration of electric circuits by ROW-type methods
P. RENTROP
487

Controller design for industrial multipass processes
E. ROGERS, D. H. OWENS
495

Computer-integrated production-planning and inventory-control at an automobile-
engine producer
W. SCHNEIDER
503

Numerical solution of a liquid crystal problem by optimization
S. SINGER
513

LACTEO: A dairy management and forecasting system
D. SPREVAK, R. S. FERGUSON
523

Identification of amplitude and phase discontinuities in the intensity signal
of a Nd-YAG solid state laser
H-G. STARK
531

Gel electrophoresis and graph matching
D. SWAILES, S. McKEE
539

Mathematical models of silicon chip fabrication
A. B. TAYLER
547

On a partial integro-differential equation related to the dynaliser concept for
industrial rubber materials
R. VAN KEER, H. SERRAS
557

INVITED SPEAKERS

ON NON-LINEAR MAGNETOSTATICS:
DUAL-COMPLEMENTARY MODELS
AND "MIXED" NUMERICAL METHODS

A. BOSSAVIT, EdF, 92141 Clamart, France

Abstract: A rationale for the use of "mixed" methods and "mixed" elements in non-linear magnetostatics.

Introduction

We are interested in the numerical computation of eddy-currents in three dimensions in pieces of steel. The motivation is induction heating: a detailed description of how the magnetic field diffuses in steel, taking into account the variations of electrical characteristics of steel with the local temperature, would be of help in designing induction furnaces and their associated controlling devices. Our long-term commitment is to the design of numerical methods and codes for such computations.

Over the course of several years of investigations about this subject, we became convinced of the advantage of using special finite elements for this, unconventional inasmuch their degrees of freedom are associated with the *edges* of the mesh. Such elements belong to a family known to numerical analysts as "mixed" elements. They are actively studied. But electrical engineers tended to frown upon them, and to question the necessity of their use, when our computational method for eddy-curents was first presented /2/. So it became a challenge to us present a rationale for the use of such elements. This resulted in a relatively simple presentation of the whole topic of mixed elements, which we give here in the hope that it may be useful in other areas of mathematical modelling where the computation of fields via finite elements is necessary: in particular, Elasticity and semi-conductor modelling.

In order to focus on the essentials of this topic, we mainly deal here with a very particular model, namely Magnetostatics, with a non-linear behaviour law. The relationship with eddy-currents is as follows.

The general mathematical model of eddy-currents consists in the following set of equations in three-dimensional space:

$$\operatorname{curl} h = j, \ \partial_t b + \operatorname{curl} e = 0, j = \sigma e + j^s, \quad b = B(h),$$

3

J. Manley et al. (eds.), Proceedings of the Third European Conference on Mathematics in Industry, 3–16.
© 1990 *Kluwer Academic Publishers and B. G. Teubner Stuttgart.*

where b is the induction, h the magnetic field, j the current density, e the electric field. The two first equations are "Ampères's theorem" and "Faraday's law" respectively. The third relation is Ohm's law: the conductivity σ is 0 only in a bounded region C, the "conductor", and j^s is a given current density, somewhere outside C. The last relation is the non-linear magnetization characteristic (in the air, it reduces to the linear relation $b = \mu h$).

Magnetostatics obtains when all fields are time-independent, and $e = 0$ everywhere. So the first and last equations are kept (with now $j = j^s$, given) and div $b = 0$, which is consequence of Faraday's law, is substituted for it. It is probably not necessary to explain in detail how this problem is actually treated (see /2/) in order to understand that knowing how to treat the magnetostatics case is the key to the whole question: For each step of, let's say, a Crank-Nicolson scheme for the above problem consists in solving a static problem (only with non-zero source terms where magnetostatics has null terms, a minor matter). So we plan to concentrate on magnetostatics, paying special attention to the non-linearity $B(h)$, and to derive a rationale for the use of mixed methods and mixed elements.

This rationale will be found, as we shall see, in the mathematical structure of the question, so we pay more attention to this than to algorithms which could possibly be used in order to solve the proposed numerical schemes. On the other hand, we strongly emphasize the *duality* and the *complementarity* which exists in the magnetostatics model. A good part of this is standard convex analysis. But another part is novel: it consists in the realization that, thanks to classical concepts of differential geometry, known to mathematicians as "Whitney differential forms", which we propose to use as finite elements, such complementarity can be kept at the discretized, finite-dimensional, level. This is the real point of the present paper.

These duality facts stem from well-known duality theorems in algebraic topology and differential geometry. If differential forms were more familiar to engineers than they currently are, we would make use of them as the right tool to deal with the subject. But in dimension 3, fortunately, one can do without them: everything which can be said about differential forms in dimension three can be said in the more classical language of vector fields and functions. We shall adhere to this discipline in the present paper.

1. Non-linear Magnetostatics

Let D be a bounded region of space, S its boundary, partitioned as indicated in Fig. 1, n the outward going field of unit normals. We address the following problem: find a pair $\{h, b\}$ of vector fields in D, linked by some "behaviour law", and such that

$$\text{(1)} \qquad \text{curl } h = 0 \quad \text{in} \quad D, \qquad n \times h = 0 \quad \text{on} \quad S^h,$$
$$\text{(2)} \qquad \text{div } b = 0 \quad \text{in} \quad D, \qquad n \cdot b = 0 \quad \text{on} \quad S^b,$$

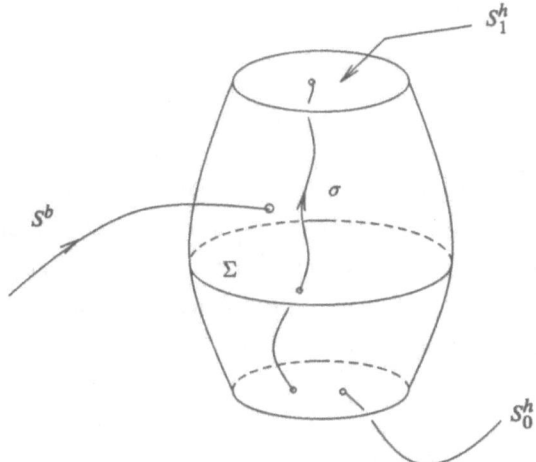

Figure 1 : Here, $S = S^b \cup S^h$ and $S^h = S^h_0 \cup S^h_1$; σ is a path from the bottom (S^h_0) to the top of the box (S^h_1). Σ is a "cutting surface", with its boundary in S^b.

plus some "source condition". Before discussing (1) (2), let us clarify the meaning of the two empha-sized expressions.

Let us assume that D is partially filled with some ferromagnetic material. Therefore, the relation

$$(3) \qquad\qquad b = \mu_0\,h,$$

valid in the vacuum, should be replaced by a more general one. For this, we assume that a pair of lower semi-continuous convex functions in duality, Φ and Ψ, is given on $\mathbf{L}^2(D)$, and we request, instead of (3),

$$(4) \qquad\qquad \Psi(b) + \Phi(h) = \int_D b \cdot h.$$

When $\Psi(b) = \int_D b^2/2\mu_0$ and $\Phi(h) = \int_D \mu_0/2|h|^2$, (4) and (3) are equivalent. Condition (4) is the law $b = B(h)$ of the Introduction, in a more symmetrical form.

As for the source conditions, let, as in Fig. 1, σ be a path which links the two components of S^h. If h is the magnetic field, we note

$$(5) \qquad\qquad J = \int_\sigma \tau \cdot h$$

(τ is the unit tangent vector) its circulation along σ. The flux across Σ ot the corresponding induc-tion is

$$(6) \qquad\qquad F = \int_\Sigma n \cdot b$$

(with n oriented in the same direction as τ). We shall consider that, as part of the problem specification, some relation

(7)
$$c(F, J) = 0$$

has been given, which accounts for the sources of the field. The idea is that D is part of an environment, for instance a magnetic circuit, as in Fig. 2, which creates a flux of magnetic field, and forces it through D. The magnitude of this flux, for a given energizing current, may be sensitive to what is put in the experimental region D, so we just assume, with (7), a definite *characteristics* for the external system. We want b and h to be such that (7) holds.

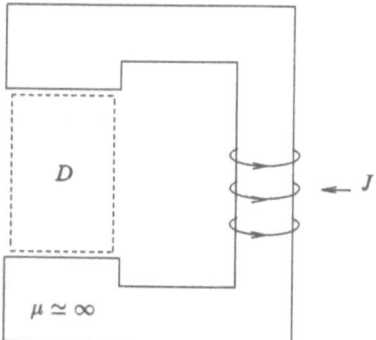

Figure 2 : In this case, J of (5) is the total intensity ("Ampère-turns") of the energizing current.

Let us now discuss problem (1) (2) (4) (7) as a whole. Equations curl $h = 0$ and div $b = 0$ stem from Maxwell equations in the static case, in the absence of imposed currents in D. The boundary condition $n \times h = 0$ means that S^h bounds an outer region of high permeability, for instance the poles of the magnetic circuit of Fig. 2, and the other boundary condition $n \cdot b = 0$ means that S^b is a tube of flux lines. (In most cases, region D may be defined in accordance to this requirement, at the price of a negligible error.)

To show that (7), the source condition, will determine the field, we may reason this way. Assume J in (5) is given. Now (as we soon shall see), (1) (2) (4) (7) determine b and h, so we get the flux F as an output. By doing this for all values of J (in practice, enough of them), we get a characteristic curve, but now of the inner region. (In the case of a linear $b - h$ law, this curve is straight, and the ratio J/F is the *reluctance* of region D.) The intersection with the outer characteristic now determines the actual value of F and J. So we shall find convenient to replace (7) by *either* (5) (circulation of h as data) *or* (6) (flux of b as data), and if necessary iterate by some Newton-like procedure. This is breaking, but in a controlled and harmless way, the symmetry between b and h present in the problem specification.

2. On the mathematical structure of the magnetostatics problem

We shall describe a mathematical structure which is, so to speak, "home" to the equations given above (and also to many of their extensions, which cannot find place in this paper). This underlying structure explains the facts of duality and complementarity that we may have already noticed, and which will become more and more apparent in the sequel.

Let us start from the following sequence of four Hilbert spaces and three differential operators (L^2 and $\mathbf{L}^2$ refer to functions and vector-fields respectively):

$$
(8) \qquad L^2(D) \xrightarrow{\;\text{grad}_h\;} \mathbf{L}^2(D) \xrightarrow{\;\text{curl}_h\;} \mathbf{L}^2(D) \xrightarrow{\;\text{div}_h\;} L^2(D)
$$

But beware: grad_h, curl_h, div_h, are not to be taken in the sense of distributions, as usual. They are *restrictions* of these, bounded operators whose domains are defined in a restrictive way, in some definite relation with the partition of S into S^h and S^b (the subscript, h or b, is intended to remind one of that). This is done according to the following table (Fig. 3):

operator	functions or vector fields which belong to its domain
grad_h	$\varphi \in L^2(D)$: $\text{grad } \varphi \in \mathbf{L}^2(D)$, $\quad \varphi = 0$ on S^h
curl_h	$h \in \mathbf{L}^2(D)$: $\text{curl } h \in \mathbf{L}^2(D)$, $\quad n \times h = 0$ on S^h
div_h	$b \in \mathbf{L}^2(D)$: $\text{div } b \in L^2(D)$, $\quad n \cdot b = 0$ on S^b

Figure 3: Here, grad, curl, div, take on their usual meaning. The conditions $\varphi = 0$, $n \times h = 0$, $n \cdot b = 0$ make sense, according to well-known trace theorems.

Sequence (8) is almost "exact": the range of curl_h coincides with the kernel of div_h, and the range of grad_h almost fills up $\ker(\text{curl}_h)$. The complement is a subspace of dimension one, made of fields of the form $h = \text{grad } \psi$, with $\psi = 0$, $\partial\psi/\partial n = 0$ on S^b, and ψ constant on S_0^h and S_1^h. We call it H_h^1 (*not* a Sobolev space!).

Now consider the operators grad_b, curl_b, div_b, obtained by substituting S^b for S^h in Fig. 3. They are (up to sign) adjoints to the previous ones: $-\text{div}_b$ to grad_h, curl_b to curl_h, $-\text{grad}_b$ to div_h. Thence an *adjoint sequence*:

$$
(9) \qquad L^2(D) \xleftarrow{\;\text{div}_b\;} \mathbf{L}^2(D) \xleftarrow{\;\text{curl}_b\;} \mathbf{L}^2(D) \xleftarrow{\;\text{grad}_b\;} L^2(D)
$$

Non-exactness occurs again at the same position: the complement of the image of curl_b in $\ker(\text{div}_b)$ is a one-dimensional space H_b^2 (isomorphic to H_h^1).

Let us display the two sequences, now columnwise, and watch how the equations take their place in this structure (this is a "Tonti diagram", see Ref. /13/):

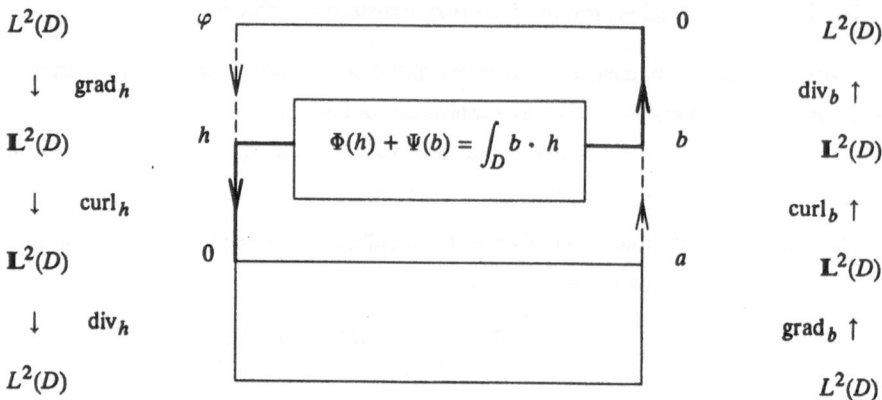

$$
\begin{array}{ccccc}
L^2(D) & \varphi & & 0 & L^2(D) \\
\downarrow \quad \mathrm{grad}_h & & & & \mathrm{div}_b \uparrow \\
\mathbf{L}^2(D) & h & \boxed{\Phi(h) + \Psi(b) = \int_D b \cdot h} & b & \mathbf{L}^2(D) \\
\downarrow \quad \mathrm{curl}_h & & & & \mathrm{curl}_b \uparrow \\
\mathbf{L}^2(D) & 0 & & a & \mathbf{L}^2(D) \\
\downarrow \quad \mathrm{div}_h & & & & \mathrm{grad}_b \uparrow \\
L^2(D) & & & & L^2(D)
\end{array}
$$

Figure 4 : Tonti diagram of the equations of magnetostatics.

For later convenience, let us name a few functional spaces:

(10) $H = \mathrm{dom}(\mathrm{curl}_h), \quad B = \mathrm{dom}(\mathrm{div}_b), \quad \Phi = \mathrm{dom}(\mathrm{grad}_h), \quad A = \mathrm{dom}(\mathrm{curl}_b)$

and (here again, s stands for "source")

(11) $H^s = \{h \in H : \int_\sigma \tau \cdot h = J\}, \quad B^s = \{b \in B : \int_\Sigma n \cdot b = F\}.$

These are affine, not vector subspaces. The corresponding parallel vector subspaces are

(12) $H^0 = \{h \in H : \int_\sigma \tau \cdot h = 0\} \equiv \mathrm{grad}_h \Phi,$

(13) $B^0 = \{b \in B : \int_\Sigma n \cdot b = 0\} \equiv \mathrm{curl}_b A.$

One has

(14) $H^s = h^s + H^0, \qquad B^s = b^s + B^0,$

where h^s and b^s are particular fields ("source fields") of H and B, built as follows. They should satisfy

(15) $\mathrm{curl}\, h^s = 0, \quad \int_\sigma \tau \cdot h^s = J,$

(16) $\mathrm{div}\, b^s = 0, \quad \int_\Sigma n \cdot b^s = F,$

but no other conditions ($\mathrm{div}\, h^s = 0$ and $\mathrm{curl}\, b^s = 0$ are *not* assumed). Such fields are easy to get, so we shall consider they are given, and may stand in place of the original data J and F. Note that, now, Eqs. (1) (5) are equivalent to $h = h^s - \mathrm{grad}\, \varphi$, for some φ in Φ, and (2) (6) to $b = b^s + \mathrm{curl}\, a$, for a in A. One calls φ and a *potentials*.

With these notations, we may restate the problem as follows:

Magnetostatics problem *Find h in H and b in B such that*

(17) $$\Psi(b) + \Phi(h) = \int_D b \cdot h,$$

(18) $$c(F, J) = 0,$$

where J and F are as defined in (5) and (6).

The main result is:

Theorem 1: *If* $\{h, b\}$ *is solution to* (17) (18) *then*

(19) $$h \in \mathrm{arginf}\,\{\Phi(h') : h' \in H^s, \quad \mathrm{curl}\,h' = 0\},$$

(20) $$b \in \mathrm{arginf}\{\Psi(b') : b' \in B^s, \quad \mathrm{div}\,b' = 0\}.$$

There exists potentials $\Psi \in \Phi$ *and* $a \in A$ *(non unique) such that* $h = h^s - \mathrm{grad}\,\varphi$ *and* $b = b^s + \mathrm{curl}\,a$. *Therefore, the constrained minimization problems* (19) (20) *can be replaced by the unconstrained ones*

(21) $$\varphi \in \mathrm{arginf}\,\{\Phi(h^s - \mathrm{grad}\,\varphi') : \varphi' \in \Phi\},$$

(22) $$a \in \mathrm{arginf}\,\{\Psi(b^s + \mathrm{curl}\,a') \in A\}.$$

Moreover, the pairs field-plus-potential are solutions to the following saddle-point problems:

(23) $$\{b, \varphi\} \in \mathrm{arg}\sup_{\varphi' \in \Phi}\ \inf_{b' \in \mathrm{dom}\,(\Psi)}\ \{\Psi(b') - \int_D h^s \cdot b' + \int_D b' \cdot \mathrm{grad}\varphi'\}$$

(24) $$\{h, a\} \in \mathrm{arg}\sup_{a' \in A}\ \inf_{h' \in \mathrm{dom}\,(\Phi)}\ \{\Phi(h') - \int_D b^s \cdot h' - \int_D h' \cdot \mathrm{curl}\,a'\}$$

All of this is almost direct consequence of the definitions. A bit more delicate is the converse statement:

Theorem 2: *If h is a solution of* (19) *or* (24) *and b a solution of* (20) *or* (23), *and if* (18) *holds, then the pair* $\{h, b\}$ *is solution to* (17).

Note that Th. 1 suggests *four* different methods to solve the problem: (21) and (23), which give φ, and (22) and (24), which give a. (For practical purposes, (19) and (21) are the same, as are (20) and (22).) Each of the four methods gives an element of the pair $\{b, h\}$, and the other one can be obtained by a direct application of the behaviour law.

All what precedes is standard convex analysis. We just took care to emphasize the symmetry of the structure, which is not always stressed in texts on the subject. Most often, problem (21) is considered as the "primal" one, and one proceeds to its dualization, in order to get the corresponding "mixed" formulation (24), the so-called "dual" problem. Such a procedure is not reversible, contrary to what the vocabulary ("primal", "dual") seems to suggest. But this is not the main point. Much more interesting is the fact that a discrete structure, completely analogous to that of Fig. 4, exists. By imbedding the magnetostatics equations into this structure, we shall obtain, almost automatically, discrete versions of the variational problems (19-24), and thus workable numerical schemes.

10

3. Whitney elements, a discrete structure analogous to the previous one

3.1 Whitney elements

First consider a tesselation of D by tetrahedra (curved images of straight reference-tetrahedra, if necessary). Two distinct tetrahedra share either a face, or an edge, or a vertex, or nothing. The sets of vertices (or *nodes*), of edges, faces, and tetrahedra (or *volumes*) are called $\underline{N}$, $\underline{E}$, $\underline{F}$, $\underline{V}$ respectively. Nodes are indexed by their numbers, i, j, k, etc., edges are denoted by $\{i, j\}$ or just by e, faces by f or by $\{i, j, k\}$, etc.

Let us assign to node number i the function λ_i so defined: if x is in one of the tetrahedra which have vertex number i in common, $\lambda_i(x)$ is the barycentric weight of x with respect to this vertex. Otherwise, $\lambda_i(x) = 0$. Now define the following vector fields (∇ is short for grad):

$$(25) \qquad w_e = \lambda_i \nabla\lambda_j - \lambda_j \nabla\lambda_i$$

if $e = \{i, j\}$, and, for face $\{i, j, k\}$,

$$(26) \qquad w_f = 2(\lambda_i \nabla\lambda_j \times \nabla\lambda_k + \lambda_j \nabla\lambda_k \times \nabla\lambda_i + \lambda_k \nabla\lambda_i \times \nabla\lambda_j).$$

These are piecewise linear vector fields associated with edge e and face f respectively. To volume $v = \{i, j, k, l\}$, we assign the function (supported on v, and constant there)

$$(27) \qquad w_v = 6(\lambda_i(\nabla\lambda_j \times \nabla\lambda_k) \cdot \nabla\lambda_l + \cdots \text{ [all similar combinations]}).$$

For consistency, λ_i is also noted w_n, i being the number of node n. Note that w_n is just the basis function of node n in the classical P^1 finite element approximation.

The field w_e also behaves as a finite element basis function (vector valued), but in a weird way. First, w_e has a curl which is in $\mathbf{L}^2(D)$, thus its tangential component is continuous across faces of the mesh, but its normal component is not: w_e is a "non-conforming" element (later, we shall challenge this terminology). Next, it is easy to check that the *circulation* of field w_e along all edges but e is equal to 0, and equal to 1 along e itself. So if h is a linear combination of the w_es:

$$(28) \qquad h = \sum_{e \in \underline{E}} \overline{h}_e w_e,$$

the real coefficients $\overline{h}_e$ (i.e. the *degrees of freedom* of h) are the circulations of h along the edges of the mesh, and not, as we are used to, nodal values of components of h. Remark that curl w_e is a linear combination of w_fs, grad w_n a linear combination of w_es, etc.

Similarly, the w_fs are such that if $b = \sum_f \overline{b}_f w_f$, the degree of freedom $\overline{b}_f$ is the *flux* of b across face f. The sum of w_v over volume v is similarly equal to 1.

We propose to call the w_n, w_e, etc., "Whitney elements". They first appeared in 1957 /15/, as differential forms, and of course for entirely different purposes.

Now call W^p, with $p = 0, 1, 2, 3$, the linear subspaces of L^2, $\mathbf{L}^2$, $\mathbf{L}^2$, L^2, spanned by the $w's$ of all degrees. From definitions (25) to (27), we check that

$$(29) \qquad \text{grad } W^0 \subset W^1, \quad \text{curl } W^1 \subset W^2, \quad \text{div } W^2 = W^3.$$

3.2 Approximation schemes

Whitney spaces will now serve as approximation spaces for H, B, Φ, A. By substituting Whitney spaces for them, we shall get numerical schemes. Why the W^p, and not others, like for example spaces of piecewise linear continuous vector fields ? A rigorous answer demands estimates of the truncation error (for this, see /6/), but a rationale can be given as follows. The tangential continuity of fields belonging to W^1 (that is, linear combination of edge-elements) is exactly what is requested for them to qualify as approximants of the magnetic field h, which has precisely this kind of continuity across material interfaces. On the other hand, the normal component of h should not be forced, so to speak, by the approximation scheme, to be continuous, since it may not be (for instance, again, across interfaces between materials with different magnetic properties). Otherwise, convergence will not occur. The same point can be made about b and W^2, where now normal continuity is desirable, and built in the structure of W^2.

To take care of boundary conditions, we introduce subspaces of the W^p. The W_h^p [resp. W_b^p] are obtained by equating to zero all degrees of freedom which correspond to simplices included in S^h [resp., in S^b]. By doing this, we obtain two sequences which are discrete counterparts (finite-dimensional) to (8) and (9):

$$(30) \qquad W_h^0 \xrightarrow{\text{grad}_h} W_h^1 \xrightarrow{\text{curl}_h} W_h^2 \xrightarrow{\text{div}_h} W_h^3$$

$$(31) \qquad W_b^3 \xleftarrow{\text{div}_b} W_b^2 \xleftarrow{\text{curl}_b} W_b^1 \xleftarrow{\text{div}_b} W_b^0$$

But the adjointness property of sequence (9) with respect to (8) is lost, for spaces W_h^1 and W_b^2, for instance, are not in duality (they have different dimensions). So what about dual sequences ? When restricted to the W_h^p, grad_h, curl_h, div_h do have adjoints, but these are no longer differential operators. They are *non-conformal approximations* of differential operators (to remind one of the difference, we shall use inverted commas to denote them: "curl" a, etc.). For instance, "curl" a is defined as the unique field in W_h^1 such that

$$(32) \qquad \int_D \text{"curl" } a \cdot h' = \int_D a \cdot \text{curl } h' \quad \forall h' \in W_h^1.$$

12

So "curl" maps W_h^2 into W_h^1. We have just uncovered the basic fact about the dual sequence in finite dimension: the operators in the direct one were differential operators, those of the dual sequence are not. We say that the dual sequence lacks *conformity*, a property which is enjoyed by the direct one. Conformity, understood this way, is a property of the sequences, not of the finite elements.

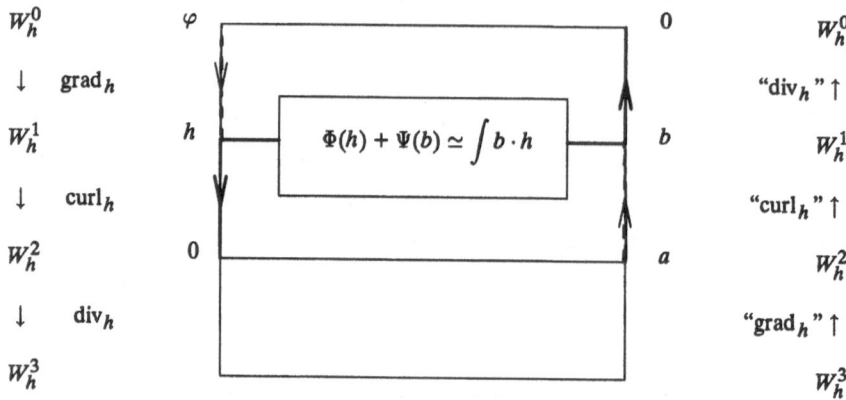

Figure 5 : Tonti diagram for the "h-conformal" methods. Functions Φ and Ψ are not in duality over W_h^1 any more, only approximately so. This is a source of approximation error which is not present if μ is constant.

As a result, the symmetry between the four methods is lost. As it often happens, this symmetry is recovered at a higher level, because we have now *two* discrete Tonti diagrams, instead of one, and therefore, *eight* different methods, instead of four. To see this, first consider the Tonti diagram relative to the W_h^p (Fig. 5). Let us substitute Whitney spaces for the original ones, as already explained, and see the outcome. The counterparts of methods (21) and (23) are:

$$(33) \qquad \varphi \in \arg\inf\{\Phi(h^s - \operatorname{grad} \varphi') : \varphi' \in W_h^0\},$$

$$(34) \qquad \{b, \varphi\} \in \arg \sup_{\varphi' \in W_h^0} \inf_{b' \in W_h^1} \{\Psi(b') - \int h^s \cdot b' + \int b' \cdot \operatorname{grad} \varphi'\}.$$

The analogs of (22) and (24) are

$$(35) \qquad a \in \arg\inf\{\Phi(b^s + \text{"curl" } a) : a' \in W_h^2\},$$

$$\{h, a\} \in \arg \sup_{a' \in W_h^2} \inf_{h' \in W_h^1} \{\Phi(h') - \int b^s \cdot h' - \int h' \cdot \text{"curl" } a'\},$$

which can (and should) be rewritten, according to (22), as

$$(36) \qquad \{h, a\} \in \arg \sup_{a' \in W_h^2} \inf_{h' \in W_h^1} \{\Phi(h') - \int b^s \cdot h' - \int a' \cdot \operatorname{curl} h'\}.$$

This last scheme is what is known as a "mixed", or "two fields", method.

A common feature of all these methods is that they will yield (either directly or through differentiation from the potential φ) a field h which is in W^1, thus "curl-conformal". But the corresponding induction $b = \partial\Phi(h)$ (or the one given directly by (34)) lacks the corresponding desirable property of "div-conformity": its divergence is not, in general, in $L^2(D)$. We see there the effect of the loss of symmetry introduced by discretization. The fact we observe here is commonplace in finite element studies: it is well known for instance that the standard P^1 approximation of the stationary heat equation yields a continuous temperature (and therefore its gradient, analogous to our h, is curl-conformal), but not a normal-continuous heat flux. So the calculated heat flux fails to be conservative, as it should be on physical grounds. This defect is especially obnoxious in some applications where this conservation property is essential (for instance, some studies in semi-conductor simulation, where the analog of the temperature is an electric potential, and the analog of the heat-flux, a current density), and mixed methods were apparently invented precisely in order to correct such defects.

So what if we insist on div-conformity on b? We choose the other discrete Tonti diagram (Fig. 6), and get the following schemes:

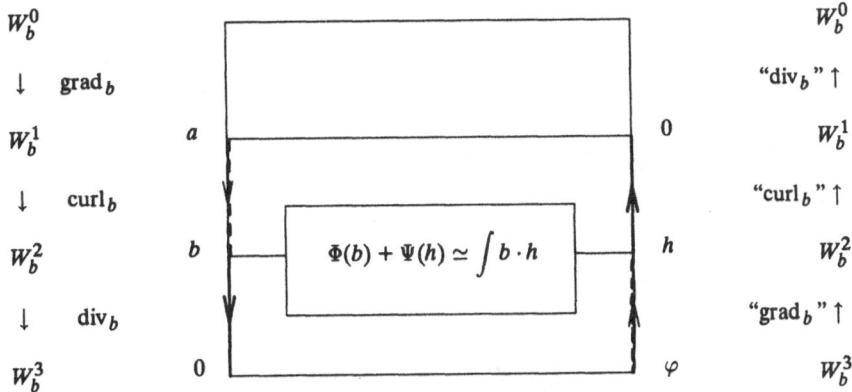

Figure 6 : Tonti diagram for the "div-conformal" methods.

(37) $$a \in \arginf \{\Psi(b^s + \operatorname{curl} a') : a' \in W_b^1\}$$

(38) $$\{h, a\} \in \arg \sup_{a' \in W_b^1} \ \inf_{h' \in W_b^2} \ \{\Phi(h') - \int b^s \cdot h' - \int h' \cdot \operatorname{curl} a'\}$$

(to be compared with (36): note especially the transposition of curl),

(39) $$\varphi \in \arginf \{\Phi(h^s + \text{"grad"}\varphi') : \varphi' \in W_b^3\}$$

$$\{b, \varphi\} \in \arg \sup_{\varphi' \in W_b^3} \ \inf_{b' \in W_b^2} \ \{\Phi(b') - \int h^s \cdot b' + \int b' \cdot \text{"grad"}\varphi'\},$$

which rewrites as the mixed method:

(40)
$$\{b, \ \varphi\} \in \arg \sup_{\varphi' \in W_b^3} \ \inf_{b' \in W_b^2} \ \{\Phi(b') - \int h^s \cdot b' - \int \varphi' \text{div} \ b'\}.$$

3.3 The linear case

The connection we claim with the common wisdom about mixed methods is easier to show in the linear case (Φ and Ψ quadratic), where the behaviour law is

(41)
$$b = \mu h, \quad h = \nu b \quad (\nu = \mu^{-1}).$$

Let us rewrite the two mixed schemes (36) and (40) in this case. Finding a saddle-point results in two Euler equations, which are, in the case of (36),

(42)
$$\int_D \mu h \cdot h' - \int_D a \cdot \text{curl} \ h' = \int_D b^s \cdot h' \quad \forall h' \in W_h^1,$$

(43)
$$\int_D \text{curl} \ h \cdot a' = 0 \quad \forall a' \in W_h^2$$

(a in W_h^2 and h in W_h^1), and in the case of (40),

(44)
$$\int_D \nu b \cdot b' - \int_D \varphi \ \text{div} \ b' = \int_D h^s \cdot b' \quad \forall b' \in W_b^2,$$

(45)
$$\int_D \text{div} \ b \ \varphi' = 0 \quad \forall \varphi' \in W_b^3,$$

with φ in W_b^3 (which, by the way, is no different from W^3) and b in W_b^2.

These are indeed the classical mixed schemes. To show their convergence, one usually appeals to the "Ladyzhenskaya-Brezzi-Babuska condition". In the linear case, our approach offers a simpler way, which we hint at on the example of (42) (43). Since curl $h^s = 0$, eq. (47), *due to the properties of the sequence of Whitney spaces,* implies $h = \underline{h}^s - \text{grad} \ \varphi$, for some φ in W_h^0, where $\underline{h}^s$ is the projection of h^s on W_h^1. Now, take $h' = \text{grad} \ \varphi'$ in (42): what comes out is the classical P^1-scheme for the potential φ, whose error analysis is standard. *As far as h is concerned, the "mixed" scheme* (42) (43) *and the conventional variational one in* φ *are equivalent,* so no particular analysis of the mixed scheme is necessary, and we may this way bypass the *LBB* condition. (This of course is valid only because the duality between Φ and Ψ is preserved on W_h^1 in the linear case, and the general case requires more study.)

At this stage, we hope we have made our point about mixed schemes: they come from imbedding the equations in the Whitney structure, just as naturally as conventional variational formulations do. Mixed schemes come out when the potential is located on the non-conformal side of the diagram. Putting them on the conformal side gives standard schemes.

Well, but so what ? Why mixed schemes in the first place ? Let us briefly answer this legitimate question by discussing (44) (45), the one which deserves attention since, as we have seen, mixed scheme (42) (43) is equivalent to the standard variational P^1 approximation in terms of φ. Suppose the degrees of freedom $\overline{b}$ and $\overline{\varphi}$ have been found.   Now $b = \sum\{\overline{b}_f w_f : f \in F\}$ is in W^2, therefore is normal continuous. As div $b = 0$ in all tetrahedra (this results from (45)), the conservation property div $b = 0$ is true *in strong form*. One will remember that this is not true of the classical P^1 scheme. (The following analogy may help in this respect: if we were dealing with the heat equation, φ would be the temperature, b the flux of heat; this flux should be conservative, in the absence of heat sources, but is not, with P^1 elements.) This strong conservation of b is what makes the mixed scheme interesting, and the reason why such schemes were developped in the Seventies.

Conclusion

We have presented a family of numerical schemes for the magnetostatics equations, which rely on "Whitney elements", a family of unconventional finite elements, with degrees of freedom associated with simplices of the mesh of all dimensions, and not only with the nodes. The schemes are intended to be used in the case of non-linear behaviour laws (though no rigorous proof of their convergence in this case has been proposed), and it is felt they extend in a straightforward way to the problem of eddy-currents with a non-linear $b - h$ law. Emphasis has been put on the structure of the equations, and on the fact that this structure can be preserved, in part, at the discrete level, thanks to Whitney elements.

One could perhaps venture from this point into a discussion on the philosophy of modelling, for there seems to be something there of general interest. Modelling is simulation, but only after a process of reduction of complexity, which filters out non-essential elements in the real situation, and keeps those which are deemed meaningful. Techniques for such reductions of complexity are numerous. Finite elements is just one of them, applicable in the case of systems ruled by partial differential equations. It is common sense that such a reduction of complexity should tend to preserve as much as possible of the structure of the system under study. We have analysed the structure of the equations of magnetostatics and looked for finite elements with the potential to reproduce this structure at the discrete level. We were fortunate to find, in a classical chapter of Mathematics, geometrical objects which could play this role: Whitney forms.

Bibliography and comments

Analyses of Maxwell equations in the language of differential forms can be found in many textbooks. See e.g. /12/, /14/. A systematic treatment of Magnetostatics in this way can be found in Kotiuga's thesis /7/. He also should be given credit for guessing about the role of Whitney forms in the subject.

Elements of convex analysis can be found in /1/. The representation of behaviour laws by using convex functionals in duality seems to come from Moreau /10/.

Mixed elements and the *LBB* condition are described, e.g., in /5/. For box-elements, see /9/. The importance of conservation properties in semi-conductor modelling is stressed in /8/.

Tonti diagrams have been around for a while /13/, and were often used in connection with "complementary" formulations. See e.g. /11/, among many references.

/1/ G. ALLEN: "Variational Inequalities, Complementary Problems and Duality Theory", *JMAA 58* (1979), pp. 1-10.

/2/ A. BOSSAVIT, J.C. VÉRITÉ: "A Mixed FEM-BIEM Method to Solve 3-D Eddy-Current Problems", *IEEE Trans.*, MAG-18, 2 (1982), pp. 431-35.

/3/ A. BOSSAVIT: "Two dual formulations of the 3-D eddy-currents problem", *COMPEL, 4*, 2 (1984), pp. 103-16.

/4/ A. BOSSAVIT: "Le calcul des courants de Foucault, en trois dimensions, en présence de corps à haute perméabilité magnétique", *Revue Phys. Appl., 23*, 6 (1988), pp. 1147-59.

/5/ V. GIRAULT, P.A. RAVIART: *Finite-Element Approximation of the Navier-Stokes Equations*, Springer V. (Berlin), 1986.

/6/ J. KOMOROWSKI: "On Finite-Dimensional Approximations of the Exterior Differential, Differential and Laplacian on a Riemannian Manifold", *Bull. Acad. Pol. Sc., 23*, 9 (1975), pp. 999-1004.

/7/ P.R. KOTIUGA: *Hodge Decompositions and Computational Electromagnetics*, PhD Thesis (MacGill University, Montréal), 1984.

/8/ J.J.H. MILLER, S. WANG, C.H. WU: "Mixed Finite Element Methods for Semiconductor Device Equations", in *IMACS Symposium* (Paris, 1988), pp. 197-99.

/9/ M.S. MOCK: "Analysis of a Discretization Algorithm for Stationary Continuity Equations in Semiconductor Device Models", *COMPEL, 3*, 3 (1984), pp. 117-49.

/10/ J.J. MOREAU: *Fonctionnelles convexes* (Séminaire EDP, Collège de France), Paris, 1966.

/11/ J. PENMAN, J.R. FRASER: "Dual and Complementary Energy Methods in Electromagnetism", *IEEE Trans., MAG-19*, 6 (1983), pp. 2311-16.

/12/ B. SCHUTZ: *Geometrical Methods of Mathematical Physics*, Cambridge U.P. (Cambridge), 1980.

/13/ E. TONTI: *La struttura formale delle teorie fisiche*, CLUP (Milano), 1976.

/14/ C. Von WESTENHOLZ: Differential forms in mathematical physics, North-Holland (Amsterdam), 1981.

/15/ H. WHITNEY: *Geometric Integration Theory*, Princeton U.P. (Princeton), 1957.

MODELING FOR

FIELD-EFFECT TRANSISTORS

by

Ellis Cumberbatch

1. Introduction

The transistor industry is vast, in its manufacturing aspect, in the wide use of its products, and in the research it generates both applied and fundamental. In universities most of this work is done in physics and electrical engineering departments and little of its mathematical requirements have been taken up by math faculty. Yet there has been extensive work on modeling, analysis and computer algorithms, and there remain many open and significant problems. In this talk I shall describe some of the modeling, give a brief introduction to the equations governing current flow in a device, review some of the analytic and numerical approaches to the solution of these equations, and refer to some specialized problems where analysis is useful.[1] In so doing I hope to pique the curiosity of some of my audience into taking a longer look at the rich phenomena in this field.

Transistor research splits into three main areas:

Process Modeling This is concerned with the fabrication of the silicon wafer and the various physical and chemical processes it undergoes on its way to becoming a layer of electronic circuits. These processes include doping (the scattering of impurity atoms in the pure silicon crystal) by diffusion, ion implantation or annealing, and lithography and etching by which the surface can be treated differentially. There are some challenging non-

---

[1]The readers will be directed to source material for some lengthy equations and formulae which are omitted from the written version of this talk.

J. Manley et al. (eds.), Proceedings of the Third European Conference on Mathematics in Industry, 17–30.
© 1990 Kluwer Academic Publishers and B. G. Teubner Stuttgart.

linear diffusion and free-boundary problems here, which Alan Tayler will describe in an accompanying article.

Device Modeling The current/voltage characteristics of a single device depend on current flow in a complicated geometry and in a complicated physical environment. The applications engineer requires good predictive information in order to design circuits, and preferably this should be in terms of formulae relating current to input voltages and device parameters.

Circuit Modeling This concerns the arrangement of the transistors and their interconnections on the surface of the wafer, so that the circuitry is correctly modeled. In addition, the design should give optimal advantage in terms of manufacturing technicalities and in circuit use, especially in reducing the timing constants of current flow.

This talk describes topics from device modeling only.

1. Basic Physics

There are excellent treatments, [1]-[5], of the structure of doped silicon so the discussion here is brief. Each silicon atom has 4 valence electrons shared with its nearest neighbors. When an impurity atom, say of phosphorus which has 5 valence electrons, is introduced at a site the extra electron is easily displaced and is available to contribute to electrical conduction. When displaced it leaves a positively charged ion. Such impurity atoms are called donors and silicon doped in this way is called n-type. Silicon of p-type has acceptor impurities, such as boron atoms, which have only three valence electrons, and each readily accepts a new electron, becoming negatively ionized. Shockley [6] modeled the "absence" of an electron as the presence of a hole. The number density of electrons, holes, ionized donor atoms and ionized acceptor atoms is denoted by n, p, N_D^+ and N_A^-, respectively, so that in regions where there is charge neutrality

(1.1) $$n + N_A^- = p + N_D^+ .$$

This equation, and the subsequent ones, identify the model as a "two-fluid" one, where electrons and holes flow in a background of static ionic charge and under the influence of population and voltage gradients. That is the electron and hole current densities are

(1.2) $$\underline{j}_n = q(n\mu_n \underline{E} + D_n \nabla n) \quad \text{and} \quad \underline{j}_p = q(p\mu_p \underline{E} - D_p \nabla p)$$

respectively, where q is the electronic charge, $\underline{E}$ is the local electric field, μ_n , μ_p are the electron and hole mobilities, and D_n , D_p are the electron and hole diffusion coefficients. The first terms on the right-hand side of (1.2) are called the drift currents, and the second terms are the diffusion currents. The mobilities depend on doping levels, electric field, temperature, etc., [1], but for the theory that gives rise to formulae they are approximated by constants. The diffusion constants are given by the Einstein relations

(1.3) $$D = \frac{kT}{q}\mu$$

where T is the temperature and k is Boltzmann's constant.

The equations governing current flow derive from conservation laws for electrons, holes and electric charge. They are

$$\frac{\partial n}{\partial t} = S_n + \frac{1}{q}\nabla \cdot \underline{j}_n,$$

(1.4) $$\frac{\partial p}{\partial t} = S_p - \frac{1}{q}\nabla \cdot \underline{j}_p,$$

$$\text{and } \nabla \cdot (K_s \underline{E}) = q(p - n + N_D^+ - N_A^-),$$

where K_s is the silicon permittivity and where S_n, S_p are source terms for electrons and holes. The latter have positive components when there is generation, e.g. under optical excitation, and there are negative terms when electrons and holes recombine. There are models providing formulae for these source terms, [1].

Introducing the electric potential, ψ, via

(1.5)
$$E = -\nabla \psi$$

and use of the electro-chemical or quasi-Fermi potentials ϕ_n and ϕ_p for electrons and holes, where

(1.6)
$$n = n_i e^{\frac{q}{kT}(\psi - \phi_n)} \ , \quad p = n_i e^{\frac{q}{kT}(\phi_p - \psi)}$$

allows (1.4) to be written as

$$\frac{\partial n}{\partial t} = S_n + \nabla \cdot (n\mu_n \nabla \phi_n) \,,$$

(1.7)
$$\frac{\partial p}{\partial t} = S_p - \nabla \cdot (p\mu_p \nabla \phi_p) \,,$$

$$\text{and} \quad \frac{K_s}{q}\nabla^2 \psi = n_i \{e^{\frac{q}{kT}(\psi - \phi_n)} - e^{\frac{q}{kT}(\phi_p - \psi)}\} + N_A^- - N_D^+.$$

In the above n_i is a constant, called the intrinsic carrier density (in the intrinsic state $p = n = n_i$). Equations (1.7) are called the Van Roosbroeck equation, [7]. In the time-independent case they are 3 non-linear Poisson equations.

2. The Gummel Algorithm

There has been much work on general properties of the solutions of (1.7) ([2], [3], [5], [10]) and on numerical schemes to generate solutions ([8], [9], [11]) (here we reference just one publication per author). Many of the techniques are based on the Gummel algorithm [12], which assumes a first approximation $\tilde{\phi}_n, \tilde{\phi}_p$ to ϕ_n, ϕ_p. Equation (1.7 c) is solved for ψ. Then (1.6) provide n, p and (1.7 a,b) are solved for ϕ_n, ϕ_p. This can be viewed as a map

(2.1)
$$(\varphi_n, \varphi_p) = T(\tilde{\varphi}_n, \tilde{\varphi}_p)$$

and T, and various discrete versions of T, have been shown to have a fixed point.

3. The P-N Junction: Devices

We now consider several simple configurations in order to build to the description of the more complicated transistors in use. These simple geometries allow explicit and precise solutions.

Consider a slab geometry where acceptor impurities (p) have been lodged in $x < 0$ and donor impurities (n) lodged in $x > 0$. Electrons are relatively plentiful in $x > 0$ and will diffuse to the left, and vice versa for holes. There may be some recombination as the holes and electrons intermingle. The diffusion process does not continue indefinitely due to the electric field generated: as the electrons diffuse to the left, and holes to the right, they each leave uncompensated charged ions whose fields inhibit the flow of charge across the boundary $x = 0$. There results a steady state configuration with further net diffusion restricted by the field (the barrier field) set up in the regions depleted of electrons and holes. (See Figure 1). The width of the depleted region can be changed by attaching a battery, say with leads at $x = \pm 1$; the battery helps or opposes the barrier field and current flows. The device resembles a diode in its current/voltage characteristics, see Figure 2.

P-N junctions may be adjoined to form a P-N-P transistor, see Figure 3. The forward biased junction on the left gives a low resistance element, whilst the reverse biased junction on the right gives high resistance. Hence a signal fed in at (1) can be extracted at (2) with a power gain. Such devices are useful and they form the basic building block of manufactured transistors. However, the slab goemetry indicated in Figure 3 is replaced by surface generated regions which are more easily mass-manufactured.

P-N Junction: Equations

The equations governing current flow across a device, such as that shown in Figure 1, are a special case of (1.2), (1.4), that is

(3.1)
$$\frac{K_*}{q} \frac{d^2\psi}{dx^2} = n - p + N(x),$$

$$\frac{1}{q} \frac{dj_n}{dx} = D_n \frac{dn}{dx} - n\mu_n \frac{d\psi}{dx},$$

$$\frac{1}{q} \frac{dj_p}{dx} = -D_p \frac{dp}{dx} - p\mu_p \frac{d\psi}{dx},$$

together with a specific model for hole/electron recombination viz.:

(3.2)
$$\frac{1}{q} \frac{dj_n}{dx} = \frac{-1}{q} \frac{dj_p}{dx} = \frac{np - n_i^2}{\tau_1 n + \tau_2 p + \tau_3}$$

where τ_1, τ_2, τ_3 are constants.

Please, [13], has results for the equilibrium case, where the re-combination is zero, the total current $j_n + j_p$ is zero, and where the doping is constant in the regions $x > 0$ and $x < 0$. In scaled variables this gives the boundary value problem

(3.3)
$$\frac{d^2\psi}{dx^2} = e^\psi - d^{-\psi} + \lambda N \quad \text{where } N = \begin{cases} -1 \text{ in } -1 < x < 0 \\ \\ 1 \text{ in } 0 < x < 1 \end{cases}$$

with $e^\psi - e^{-\psi} \pm \lambda = 0$ at $x = \pm 1$. Here λ is the non-dimensional doping level, the ratio of doping to n_i, for most devices a large parameter of the order of 10^5 to 10^{10}. Please obtains solutions in the limit $\lambda \to \infty$ by the method of matched asymptotic expansions. The outer and inner regions correspond to the neutral regions and depletion layer, respectively, but solutions in transition layers need to be found before the expansions can be properly matched. Please also obtains some results for $j_n + j_p \neq 0$.

Markowich and Ringhofer, [14], prove the validity of such asymptotic results and show their usefulness in numerical schemes. Ringhofer, [15], extends this work to the transient case. Cimatti, [16], has considered a two-dimensional problem where the boundary value problem (3.3) is restricted to a finite y-range and the differential operator becomes the Laplacian. Properties of the shape of the depletion region are found by variational inequality techniques. There are some interesting problems with multiple steady states, see Rubinstein, [17], for the case of a P-N-P-N junction.

4. MOS Diode

Figure 2 shows a small current flow (microamps) under reverse bias. This is due to the presence of a small number of acceptor ions in n-type materials (and a small number of donor ions in p-type). The current flow is generated by minority carriers. Under these circumstances what is nominally an n-type region becomes locally p-type (or vice versa) and this is called inversion.

The MOS diode shown in Figure 4 is designed to create an inversion region in the silicon (S) by raising the voltage in the metal layer (M) separated from it by an insulator $(O :$ for silicon oxide). Figure 5 shows the charge per unit area present under the oxide due to the potential there. The strong inversion at the steep part of the curve implies a large availability of electrons. The availability of these electrons by manipulation of voltage in the metal (the gate) becomes the key to the design of the transistor.

The equations governing the field in an MOS diode derive from (3.1). In the silicon

$$n = n_0 e^\psi \ , \quad p = p_0 e^{-\psi} \ , \text{where, at large } x, \ n_0 - p_0 = N_D^+ - N_A^- \ .$$

Hence the equation for the electric potential is

(4.1) $$K_s \frac{d^2\psi}{dx^2} = q \left\{ n_0(e^\psi - 1) - p_0(e^{-\psi} - 1) \right\}.$$

This integrates to

(4.2) $$x = \sqrt{\frac{K_s}{2}} \int^\psi \frac{de}{F(e)},$$

where $F^2(\psi)$ is the integral with respect to ψ of the right-hand-side of (4.1). The behavior shown in Figure 5 is found by relating the $x = 0$ values of ψ and $\frac{d\psi}{dx}$ obtained from this solution.

5. The MOSFET Device

Figure 6 shows a schematic for a field-effect transistor based on the MOS behavior just described. There may be $10^5 - 10^6$ of these devices on a single chip and so their length and width are of micron size; their depth is submicron size. The region labeled as the channel is where gate voltage influences the formation of an inversion region. Its length is being reduced to submicron size in current technology. As shown in Figure 6 the substrate is p-type (doped with acceptor atoms of density, say , of $10^{14} cm.^{-3}$) into which has been diffused/implanted higher concentrations of donor atoms (of density $10^{19} cm.^{-3}$) to make two n-type regions called the source and drain. Above are insulator (silicon oxide) and metal contact layers, all of the above being laid down by surface processes. When a voltage difference, V_D, is imposed across the source to drain contacts a current, I_D, flows, and this current can be controlled by the gate voltage, V_G, and the substrate voltage, V_B. In addition, this current depends on the geometry of the device and its physical characteristics (levels of doping, mobilities, etc.) That is

$$(5.1) \qquad\qquad I_D = G(V_D \, , \, V_G \, , \, V_B \, ; \, p_i)$$

where p_i is a set of parameters. Various solutions to the governing equations (1.7) have been obtained; routinely there are 9-10 parameters p_i. A short description of the steps involved in obtaining the function G in (5.1) is now presented.

The basic assumption used in deriving formulae for G is that the channel length is large compared with the depth over which changes take place in the x-direction below the gate, so that a slowly varying approximation is appropriate. The dependence of ψ on x is given by (4.2), and its dependence on y is through $\varphi_n(y)$, where

$$(5.2) \qquad\qquad n = n_i e^{\psi - \varphi_n}$$

is the electron density replacing $n_0 e^\psi$ on the right-hand side of (4.1). The current density in the y-direction from (1.2) is

$$j = -q\mu_n n \frac{d\varphi_n}{dy} \tag{5.3}$$

and when this is integrated, first in the x direction and then in the y-direction, there results

$$I_D = -\frac{q\mu_n}{L} \int_0^{V_D} \int_{\varphi_s}^{\varphi_F} \frac{1}{F} e^{\psi - \varphi_n} d\psi d\varphi_n \tag{5.4}$$

where φ_F is the bulk Fermi potential. This so-called double integral formula was established by Pao and Sah, [18], and numerical integration has shown it to be quite accurate when compared with experimental data. Various approximations to (5.4) have been obtained, resulting in algebraic representations for I_D, see [19]-[22]; some are based on approximations of the physics, some on expansions of F in different ranges.

There are numerous deficiencies in the model just presented, which have negligible effect at longer channel lengths. As L is reduced, the source/drain regions become more important because they are resistance elements in the circuit, and also because their P-N junction interfaces alter the fields under the gate. Ward, et al, [23], has obtained an asymptotic expansion combining both the large doping parameter expansion to obtain the MOS behavior and a slowly varying expansion for y variations. In particular the model includes boundary layers at the source/drain interfaces, and corner layers where these boundary layers meet the channel.

Current flow in the source/drain regions was first modeled by Berger, [24], as a one-dimensional transmission line. Subsequently there have been many planar simulations (in the plane $x = 0$) to see the effect of contact shape and misalignment. (The source/drain contacts have a more complicated configuration than that shown in Figure 6. See Loh et al, [25].) Cumberbatch and Fang have constructed a three-dimensional model, [26].

6. Parameter Extraction

The geometrical and physical parameters entering the formula (5.1) cannot be manufactured ab initio to close pre-set tolerances, and their direct measurement is difficult due to the small size of the device. In the design process extensive sets of current/voltage measurements are made so that the parameters can be identified from this data using a model formula. The extracted parameter values then can be used to predict device performance over all voltage ranges, and allow circuits to be designed using the manufacturing process under scrutiny. Subsequently, during production, test sections on each chip are subjected to the same process to check that the parameters remain within set tolerances. A Claremont Mathematics Clinic project for the VLSI group at JPL was concerned with the numerical optimization techniques used in parameter extraction, see [27], [28]. More recently, prompted by the cost of data collection and computation, our attention has turned to the experimental design problem of reducing the number of measurement points taken whilst retaining accuracy of extracted values, see [29]. Parameter extraction at long distance will be achieved when current/voltage measurements of hardware devices on deep space probes will be transmitted back to JPL, where changes in physical characteristics due to the space environment can be evaluated.

The technology of using a model formula (5.1) is not uniform across the industry. Two alternatives are in use. One substitutes (5.1) by a numerical solution of the full equations (1.7). That is, iterative estimates of parameter values are used in (1.7), whose solutions are then compared with measurements. This technique requires substantial computing power and efficient numerical PDE solvers. The other technique uses extensive measurements, which are stored as look-up tables with interpolation completing the spectrum of data.

FIGURE 1

P-N JUNCTION

DEPLETION
REGION
P-TYPE
REGION
N-TYPE
REGION

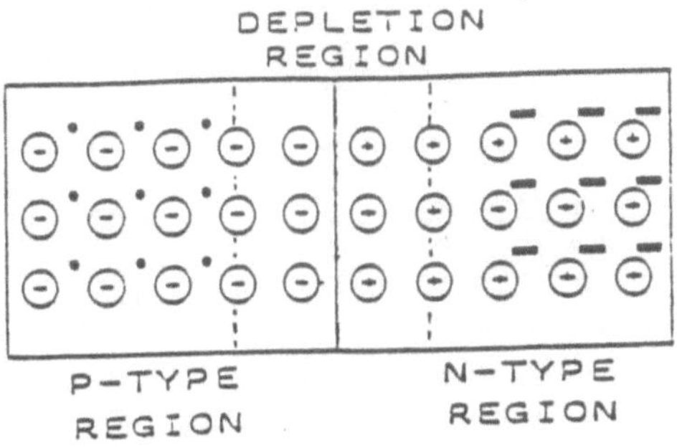

• Hole
▬ Electron
⊕ Donor Ion
⊖ Acceptor Ion

FIGURE 2

Current in P-N Junction with forward and reverse bias

CURRENT
MAJORITY
CARRIERS
← REVERSE BIAS
FORWARD BIAS →
MINORITY
CARRIERS

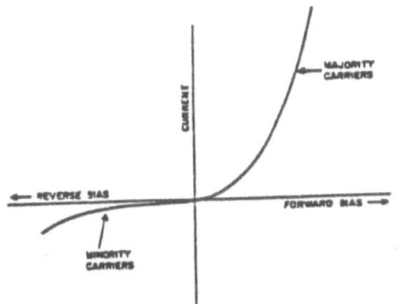

DEPLETION REGIONS
(1)
(2)

FIGURE 3

P-N-P JUNCTION

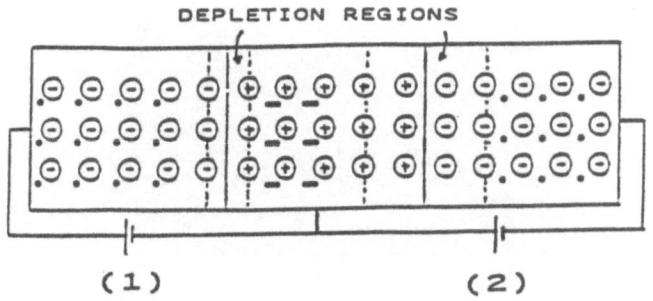

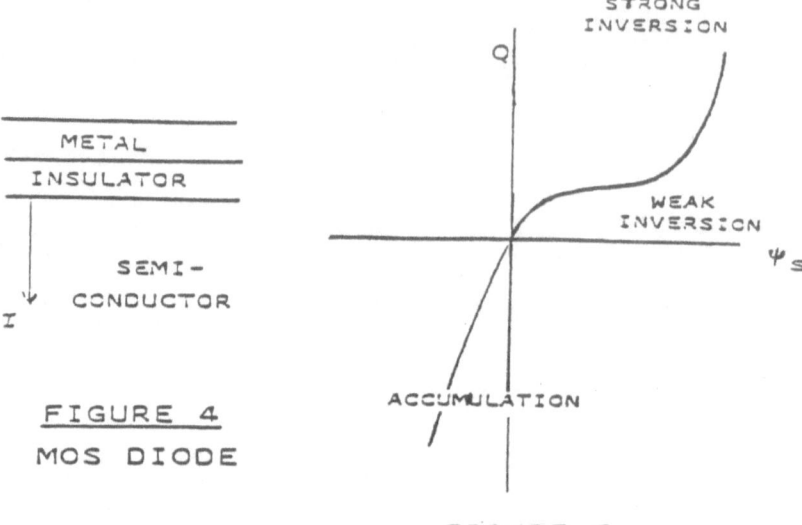

METAL

INSULATOR

SEMI-
CONDUCTOR

I

FIGURE 4

MOS DIODE

STRONG
INVERSION

Q

WEAK
INVERSION

ψ_S

ACCUMULATION

FIGURE 5

Space charge density Q
versus surface potential
ψ_S for an MOS diode

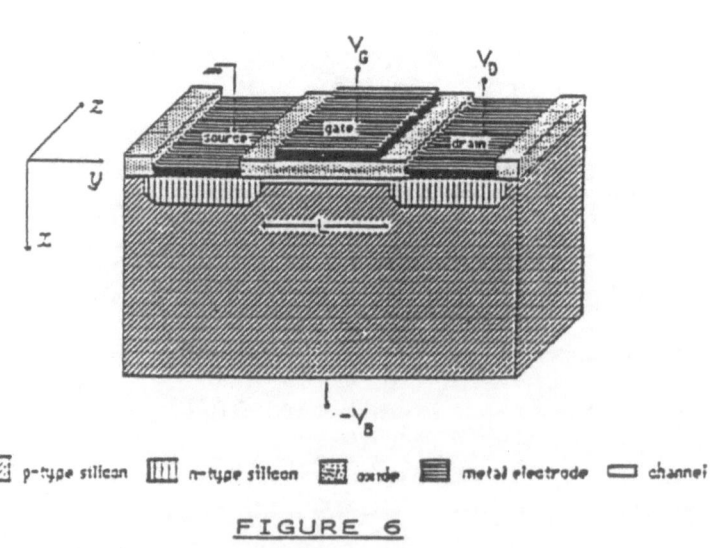

V_G V_D

source   gate   drain

z

y

x

$-V_B$

▨ p-type silicon ⊪ n-type silicon ▩ oxide ▤ metal electrode ▭ channel

FIGURE 6

MOSFET Schematic

References

[1] Sze, S. M.: Physics of Semiconductor Devices. John Wiley & Sons, New York, Second Edition, 1981.

[2] Mock, M.S.: Analysis of Mathematical Models of Semiconductor Devices. Dublin: Boole Press 1983.

[3] Selberherr, Siegfried: Analysis and Simulation of Semiconductor Devices. Springer-Verlag Wien, New York 1984.

[4] Smith, R. A.: Semiconductors. Cambridge: Cambridge University Press 1978.

[5] Markowich, P.A.: The Stationary Semiconductor Device Equations. Springer-Verlag, Wien-New York 1986.

[6] Shockley, W.: Electrons and Holes in Semiconductors. Van Nostrand, New York 1950.

[7] Van Roosbroeck, W.: Theory of flow of electrons and holes in germanium and other semiconductors. Bell System Tech. J., 20 (1950) pp.560-607.

[8] Bank, R., Jerome, J., and Rose, D.J.: Analytical and numerical aspects of semiconductor device modeling. Computing Methods in Applied Sciences and Engineering V (R. Glowinski and J. Lions, Eds.), North Holland Publishing, Amsterdam (1982), pp. 593-597.

[9] Jerome, J.: Consistency of Semiconductor Modeling: an Existence/Stability Analysis for the Stationary Van Roosbroeck System. SIAM J. Appl. Math., Vol. 45, No. 4, 1985 pp. 565-590.

[10] Seidman, T.: Steady state solutions of diffusion reaction systems with electrostatic convection. Nonlinear Anal., 4 (1980), pp. 623-637.

[11] Bank, R., Rose, D.J., and Fichtner, W.: Numerical methods for semiconductor simulation. SIAM J. Stat. Sci. Comp., 4 (1983), pp. 416-435.
See also: Special Issue on Numerical Simulation of VLSI Devices. IEEE Trans. Electron Devices, Vol. ED-32 (1985).

[12] Gummel, H.K.: A self-consistent iterative scheme for one-dimensional steady state transistor calculations. IEEE Trans. Electron Devices, Vol. ED-11, (1964), pp. 455-465.

[13] Please, C.P.: An Analysis of Semiconductor P-N Junctions. IMA Jour. Appl. Math. (1982) 28, pp. 301-318.

[14] Markowich, P.A., Ringhofer, C.A.: A Singularly Perturbed Boundary Value Problem Modelling a Semiconductor Device. SIAM J. Appl. Math. Vol. 44, No. 2, (1984), pp. 231-256.

[15] Ringhofer, C.: An Asymptotic Analysis of a Transient p-n Junction Model. SIAM J. Appl. Math., Vol. 47, No. 3, (1987), pp. 624-642.

[16] Cimatti, G.: On the Shape of the Region of Depletion in a P-N Junction. Bollettino U.M.I. (5) 18-B (1981), pp. 393-409.

30

[17] Rubinstein, I.: Multiple Steady States in One-Dimensional Electrodiffusion with Local Electroneutrality. SIAM J. Appl. Math., Vol. 47, No. 5, (1987), pp. 1076-1093.

[18] Pao, H.C., Sah, C.T.: Effects of Diffusion Current on Characteristics of Metal-Oxide (insulator)-Semiconductor Transistors. Solid-State Electronics, Vol. 9 (1966) pp. 927-937.

[19] Brews, J.: A Charge-Sheet Model of the MOSFET. Solid-State Electronics, 21, (1978) pp. 345-355.

[20] Ihantola, H.K., Moll, J.L.: Design Theory of a Surface Field-Effect Transistor. Solid-State Electronics, 7, (1964) pp. 423-430.

[21] Pierret, R.F., Shields, J.A.: Simplified Long-Channel MOSFET Theory. Solid-State Electronics, Vol. 26 (1983) pp. 143-147.

[22] Van De Wiele, F.: A Long-Channel MOSFET Model. Solid State Electronics, Vol. 22, (1979), pp. 991-997.

[23] Ward, M., Odeh, F., Cohen, D.S.: Asymptotic Methods for MOSFET Modeling. Accepted by SIAM J. Appl. Math.

[24] Berger, H.H.: Models for Contacts to Planar Devices. Solid-State Electronics, Vol. 15, (1972), p. 145.

[25] Loh, W.M., Swirhern, S.E., Schoeyer, T.A., Swandon, K.C. Saraswat: Modeling and Measurement of Contact Resistances. IEEE Trans. Electron Devices, Vol. ED-34, No. 3, (1987), pp. 512-524.

[26] Cumberbatch, E., Fang, W.: Three-Dimensional Modelling for Contact Resistance of Current Flow into a Source/Drain Region. Claremont Graduate School, Math Department, Preprint. 1988.

[27] Gribben, R.J., Martelli, M., Rykken, C., Meiser, V., Turner, G., Wang. Q.: Parameter Extraction and Transistor Model. Mathematics Clinic, Claremont Graduate School, Final Report, 1985.

[28] Gribben, R.J., Martelli, M.: Optimal parameter extraction of the Brews charge-sheet MOSFET model. Math. Engng. Ind., Vol 1, No. 2, 1987, pp. 155-168.

[29] Andersson, G., Allen, D., Fleishman, R., Hamza, H., Lacey, S., Larsson, K., Panagiotacopulos, D., Velasco-Hernandez, J.: Parameter Extraction from a Nonlinear MOSFET Model. Mathematics Clinic, Claremont Graduate School, Final Report, 1988.

Mathematics Department
The Claremont Graduate School
Claremont, California 91711

SUPERCOMPUTERS – 1988

Iain S. Duff, Harwell Laboratory, Oxfordshire, England.

Summary: This talk is concerned with the what, where, why, and how of supercomputers. We begin by defining what we mean by a supercomputer and how we differentiate it from other large computers. Naturally, the fine detail is very time-dependent but we can establish general guidelines for such an identification. By way of example, we examine some of the supercomputers available today. We then indicate the environments housing the supercomputers, the "where" of our talk. We look at the distribution within Europe both by discipline and geographical region. The reasons "why" supercomputers and supercomputing power is required and "how" they are used in scientific and industrial research will be considered. The influence that they are having in this area will be discussed. We shall illustrate our comments with some examples of computations performed on the CRAY-2 (and earlier vector processors) at Harwell.

1   Introduction

We consider the whys and wherefores of supercomputers in an industrial research and development environment. It is our contention that supercomputers are becoming an increasingly essential tool both in the search for greater truths, the so-called grand challenges, and in the more mundane world of designing aircraft or developing new drugs. We work on the premise that while you are all expert mathematicians and most of you have written the odd line or two of Fortran, there is not a universal awareness of what constitutes a supercomputer and why it is important in scientific and engineering research.

We first define what we mean by a supercomputer in Section 2 and explain what features differentiate it from a regular mainframe machine. In particular, we describe the principal architectural features that establish the prefix "super". It is often thought that such beasts are a rare breed, but we show by studying the distribution of supercomputers in Europe in Section 3 that this is indeed not the case. This represents an almost *de facto* proof of the importance of supercomputers in the modern scientific and engineering world. We then discuss in broad terms the current uses of supercomputers, drawing heavily on the principal activities of the sites that we have just listed. In Section 4 we examine why one might want the power of the supercomputers just discussed and how such a large investment can be justified. In Section 5, we give more detailed examples of the benefits of supercomputers and supercomputing. Finally, we comment in Section 6 about the future of supercomputing and reflect on new breeds which are emerging.

31

J. Manley et al. (eds.), Proceedings of the Third European Conference on Mathematics in Industry, 31–44.
© 1990 *Kluwer Academic Publishers and B. G. Teubner Stuttgart.*

2   What?

There are many good conference jokes about the definition of a supercomputer, and indeed a particular manufacturer has alternatively claimed that one of its product line is both a supercomputer and not a supercomputer, depending on whether they are trying to attract a customer or obtain a US export licence. In this presentation we adopt the conventional definition that supercomputers are the fastest and most powerful general-purpose scientific computing systems available at any given time. They offer speed and capacity significantly greater than mainframe computers, the workhorses of the commercial world and mixed computing environments.

To see how supercomputers get their speed, we make the assumption that the computer is governed by a clock and it is only at each tick, or clock cycle, that anything can happen. Furthermore, if we are concerned with floating-point computations, then it needs several clock cycles (or clocks) for each such calculation. Clearly one way of designing ever faster machines is to reduce the clock cycle time, introducing problems of scale, material purity, and cooling *inter alia*. Modern high-speed mainframes have clock cycles of around 20 nsecs; the fastest supercomputers have cycle times of around 4 nsecs and it becomes increasingly difficult to reduce it much more, the 1nsec barrier as elusive as the four-minute mile in the late forties. So, although it is important, a fast clock neither gives us the speed advantage desired nor looks likely to help us much in the future, as speed of light limitations become an increasingly important issue (light moves only about one foot in a nsec).

One of the principal ways of turbocharging our machine is to utilize the fact that a floating-point calculation, an addition say, is not atomic but has distinct stages, for example exponent match, shift, add, normalize, and to capitalize on these stages by organizing them as segments of a pipe and allowing different operands to be in different segments at the same time. The normal way to use this facility is to compute with vectors so that consecutive components are in adjacent segments. At each clock, the two components move to the next segment thus enabling the output (when the pipe is full) of one result every clock cycle. We illustrate this vector pipelining in Figure 2.1.

This means that a machine with a 10 nsec cycle would have a peak performance of 100 Megaflops (or 100 million floating-point operations per second), the common unit for rating computational speed. However, as we shall see shortly when we look at actual performance figures, that is still not sufficient. There are two further common routes to increased speed and both involve a simple form of parallelism. One can increase the

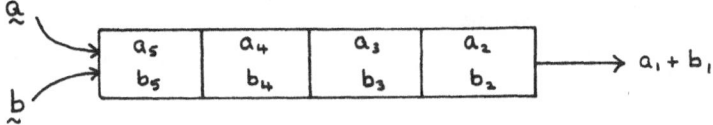

Figure 2.1. Illustration of vector pipelining.

number of pipes and allow different vectors or segments of vectors to be processed simultaneously. This is the route taken by the Japanese manufacturers who now have machines with 16 vector floating-point pipes. The American route is to go for only a few (sometimes effectively two) pipes but have several (presently up to 8) processors all capable of acting independently either on one job or several, and all with access to the same common memory. More recently, manufacturers have designed machines with larger degrees of parallelism and local instead of, or in addition to, shared memory. We defer discussion of such potential supercomputers to Section 6.

The first machine to widely receive the appellation "supercomputer" was the CDC 6600, first marketed in 1966. It had a performance of 1 Megaflop! Since then the peak performance of supercomputers has increased dramatically, and we illustrate this growth in Table 2.1, taken from Dongarra and Duff (1988). The projection for 1990-1995 is expected to have a maximum speed of at least 200 Gigaflops, more than 200,000 times that of the CDC 6600.

So now we have the thorny problem of where to draw the line for determing when a computer is powerful enough to be classified as a supercomputer in 1988. We certainly have an intuitive feel for what is currently a supercomputer and what is not and, by good chance, if we draw our line at a peak rate of 500 Mflops, our intuition is satisfied. Several companies manufacture supercomputers under this definition; we summarize those currently available in Table 2.2. Note that we have included the CRAY-1 computer rated at 160 Megaflops. We include it not only out of sentimentality but also largely as a benchmark since it could not now be considered a supercomputer in terms of

Year	Machine	Speed	Speed Increase	
			10 years	20 years
1966	CDC 6600	1 Mflops	–	–
1975	CDC 7600	4 Mflops	4	–
1979	CRAY-1	160 Mflops	100	–
1983	CYBER 205	400 Mflops	100	400
1986	CRAY-2	2 Gflops	500	2000
1990-1995	?*?	200 – 1000 Gflops	1000	250,000

Table 2.1. Performance trends in supercomputers.

performance and is no longer manufactured by Cray. We also include both the IBM 3090/VF even though it meets our cut-off only in maximum or next to maximum configuration, and the CYBER 205 although it makes the grade in only 32-bit precision. Remember that we are only including machines deemed to be fully general-purpose scientific machines. We consider novel architectures whose nominal peak rate puts them in the supercomputer class in Section 6.

Machine	Performance in Mflops (Max configuration)	Memory in Mbytes	Number of Processors
CRAY-1	160	32	1
CRAY X-MP	940	512	1,2,4
CRAY Y-MP	2664	256	8
CRAY-2	1952	4096	2,4
CRAY-3	16000	16384	16
CYBER 205	400 [a]	128	1
ETA-10	10826	2048 [b]	1,2,4,6,8
Fujitsu VP-400	1714	1024	1
Hitachi S-820/80	3000	512 [c]	1
IBM 3090/VF	686	256 [d]	1,2,3,4,5,6
NEC SX-2A	1300	1024 [e]	1

(a) 800 Mflops for 32-bit arithmetic
(b) Also 4 Mwords of local memory with each processor
(c) Also a 12-Gbyte extended memory
(d) Also a 2-Gbyte extended memory
(e) Also a 8-Gbyte extended memory

Table 2.2. Current supercomputers.

Cray and ETA are presently the only supercomputer manufacturers to offer multiple-processor machines although it is likely that some other vendors will announce multiprocessors in the near future. Both Fujitsu and Hitachi systems are IBM System 370 compatible. All the Crays offer a UNIX operating system called UNICOS, and ETA hope to mount a UNIX system shortly on their machines. All the machines and rates are for 64-bit arithmetic implemented in hardware. The CDC and ETA machines also allow 32-bit working at double rates. The Fujitsu machines are marketed in the West by Amdahl (the 1100 to 1400 range) and by Siemens (the VP-50 to 400 range).

Such power does not come cheap. The typical cost for a top-end machine is remarkably time independent and is between \$15M and \$20M. The entry level is somewhat more debatable since it depends at what level you draw the defining line. At over \$5M you should be starting to look at something respectable although some manufacturers, notably ETA systems, offer a range of machines starting at around \$1M.

To put things in perspective, think of solving a full system of linear equations with a 100×100 coefficient matrix using the SGEFA/SGESL combination from LINPACK (Gaussian elimination with partial pivoting). We quote some ratios from Dongarra (1988) in Table 2.3 where it should be added that the last four machines are only using 32-bit arithmetic and the particular Fortran implementation exploits the capabilities of the supercomputers rather poorly. Nevertheless the difference in actual computational rates is quite sobering and illustrates that maybe there are some problems for which a workstation is not adequate! The figures presented depend quite a lot on the actual hardware/compiler/operating system combination. Details of these are given in the tables in Dongarra (1988).

Computer	Time (secs)	Ratio
CRAY Y-MP	0.008	1
IBM 4381	0.43	50
VAX 11/780	1.94	250
SUN 3/50	7.40	1000
IBM PC-XT	17.0	2000
Apple Macintosh	96.5	12000

Table 2.3. Illustration of supercomputer speed.

Another illustration of the growth in supercomputing performance and technology over the last 12 years can be seen by looking at the products of the leading supercomputer manufacturer, Cray Research Incorporated. Clearly, as we see in Table 2.4, the trends of reduced cycle times, larger memories, and greater parallelism are very evident.

	1976 CRAY-1	1982 CRAY-XMP-2	1984 CRAY-XMP-4	1985 CRAY-2	1988 CRAY-YMP	1990 CRAY-3
CPU	1	2	4	4	8	16
Clock (nsec)	12.5	9.5	8.5	4.1	6.0	2
Memory	1 Mword (1 port/CPU) bipolar (55 nsec)	4 Mword (4 port/CPU) bipolar (35 nsec)	16 Mword (4 port/CPU) bipolar (25 nsec)	256 Mword (1 port/CPU) DRAM (130 nsec)	32 Mword (4 port/CPU) bipolar (15 nsec)	512 Mword (2 port/CPU) SRAM (25 nsec)
Total access	138 nsec	133 nsec	119 nsec	230 nsec	102 nsec	50 nsec
Banks	16	32	64	128	256	512
Issue rate	80	220	468	480	1328	8000
Peak Mflops	160	440	936	1800	2656	16000
Gate time	850 psec	650 psec	650 psec	650 psec	350 psec	200 psec
Gate	5/4	16	16	16	2500	300-500
Ratio	1	3-4	6-8	5-12	30	100

Table 2.4. Development of Cray supercomputers.

It is perhaps salutary before we get carried away in a state of *super*euphoria, to reflect on the status of supercomputing vis-a-vis the computing world in general. Cray Research are by far the current market leaders in supercomputing with an installed base of over 200 machines out of a total market of just over 300. However, if we look at the following table taken from a recent issue of Datamation, we see that Cray comes quite a long way down the league table of computer manufacturers and suppliers.

1987 Rank	Company	1987 Information Systems Revenue ($million)	1987 Rank	Company	1987 Information Systems Revenue ($million)
1	IBM	50,485.7	51	Northern Telecom	900.0
2	Digital Equipment	10,391.3	52	C. Itoh	829.2
3	Unisys	8,742.0	53	General Electric	800.0
4	Fujitsu	8,740.0	54	Telex	788.9
5	NEC	8,230.5	55	Commodore International	785.0
6	Hitachi	6,273.7	56	Sun Microsystems	755.9
7	Siemens	5,703.0	57	Storage Technology	750.0
8	NCR	5,075.7	58	Arthur Andersen	748.9
9	Hewlett-Packard	5,000.0	59	Texas Instruments	740.0
10	Olivetti	4,637.2	60	Motorola	724.9
11	Toshiba	3,441.3	61	Cray Research	687.3
12	Wang Laboratories	3,045.7	62	Mannesmann	686.0
13	Apple Computer	3,041.2	63	Martin Marietta	685.7
14	Groupe Bull	3,007.5	64	Cap Gemini Sogeti	682.3
15	Control Data	3,000.9	65	Econocom International	674.3
16	Nixdorf Computer	2,821.5	66	Computer Associates International	648.8
17	Matsushita	2,628.5	67	Intergraph	641.1
18	Philips	2,601.6	68	Bell Atlantic	634.9
19	Xerox	2,415.0	69	Alps Electric	632.5
20	STC	2,123.9	70	Samsung Electronics	569.5
21	Honeywell Bull	2,059.0	71	Computervision	564.0
22	Alcatel	2,052.1	72	Apollo Computer	553.6
23	AT&T	2,000.0	73	Lockheed	553.0
24	TRW	1,960.0	74	Racal Electronics	549.1
25	Tandy	1,682.4	75	Amstrad	533.0
26	Mitsubishi Electric	1,673.9	76	Comparex Information Systems	530.8
27	Canon	1,673.4	77	CSK Group	486.9
28	Ericsson	1,511.6	78	Microsoft	456.7
29	Amdahl	1,505.2	79	Xidex	455.7
30	Automatic Data Processing	1,467.0	80	3M	455.0
31	Electronic Data Systems	1,440.5	81	Harris	446.0
32	Data General	1,303.9	82	Finsiel	424.1
33	Nippon Univac Kaisha	1,294.6	83	Norsk Data	422.6
34	Ricoh	1,275.7	84	Continental Information Systems	405.4
35	McDonnell Douglas	1,241.8	85	Emhart	404.8
36	Inspectorate International	1,225.0	86	Ferranti	398.8
37	Compaq Computer	1,224.1	87	Wyse Technology	396.2
38	Seiko Epson	1,198.4	88	Lotus Development	395.6
39	Comdisco	1,153.0	89	Shared Medical Systems	390.7
40	Oki Electric Industry	1,137.3	90	Convergent Technologies	384.8
41	Computer Sciences	1,133.8	91	Nokia	375.3
42	Nippon Telegraph & Telephone	1,128.5	92	Tandon	374.0
43	Tandem Computers	1,099.6	93	Sony	365.5
44	Seagate Technology	1,075.7	94	Miniscribe	362.5
45	Memorex International	1,041.1	95	Diebold	345.5
46	Zenith Electronics	1,040.0	96	MAI Basic Four	334.2
47	National Semiconductor	985.0	96	Dataproducts	334.2
48	Société Générale	970.1	98	Datapoint	320.7
49	Prime Computer	960.9	99	Tektronix	320.0
50	Atlantic Computers	959.7	100	Gould	299.3

Table 2.5. Datamation list of information systems companies.

3   Where?

Who then buys and uses such machines and where are they located? We feel it is instructive to study this a little for two main reasons. The first is to emphasize that their use is indeed very widespread, and the second indicates the main uses of supercomputers through the principal characteristics of the sites. Since this is an ECMI meeting, we will consider only the use and distribution of supercomputers within Europe. This can be considered an update to previous reports on the same (Duff 1984, 1988).

The European market accounts for between one quarter and one third of the supercomputers installed to date. It is the second largest market to the US, although recently the Japanese market has been expanding at a fast rate. We present the data in Tables 3.1 to 3.4, where we have excluded sensitive sites and IBM installations. IBM sites have been omitted for two reasons. The first is that, as a company, they do not disclose information on their sites, thus making the information both hard to obtain and possibly unreliable. The second is that only the full system qualifies as a supercomputer. It is perhaps worth mentioning that there are reputed to be about fifty sites with vector facilities in Europe, and that IBM is encouraging the growth of vector computing by supporting about five centres in the public sector, all of which will probably get the six processor machine eventually. The centres currently in operation or shortly to be opened include RAL (England), CERN (Switzerland), Aachen (Germany), and CNUSC (France). In addition, IBM are sponsoring a competition for single vector facilities to be given to institutions with 3090s already installed. About twenty five such sites in Europe will be so favoured. There are also two major IBM research laboratories in Europe: the Bergen Scientific Center currently with an IBM 3090/200 VF and the ECSEC Centre in Rome with a full IBM 3090/600 VF. Perversely, we have included members of the ETA range whose peak performance does not merit the supercomputer label. The natural upgrade path for these (still quite powerful) systems leads us to include them in the tables.

Institute	Location	Supercomputer
AWRE	Aldermaston	CRAY X-MP/28
		CRAY 1S
BP Exploration	London	CRAY X-MP/24
Cray Research (UK)	Bracknell	CRAY X-MP/28
ECMWF	Reading	CRAY X-MP/416
GECO (UK)	London	Amdahl VP 1200
Harwell Laboratory	Didcot	CRAY 2
Merlin Profilers Ltd	Woking	CRAY 1S
RAE	Farnborough	CRAY 1S
RAL	Chilton	CRAY X-MP/48
RARDE	Fort Halstead	CRAY 1S
Shell	Wythenshawe	CRAY X-MP/14
UK MET	Bracknell	ETA-10 (4 proc)
ULCC	London	CRAY 1S
		CRAY 1S
		CRAY X-MP/28 (on order)
UMRCC	Manchester	CYBER 205
		Amdahl VP 1100 (on order)
Western Geophysical	London	Amdahl VP 1200

Table 3.1. Summary of supercomputers in Britain.

Institute	Location	Supercomputer
Aerospatiale	Toulouse	CRAY X-MPE/18se
CCVR (Ecole Polytechnique)	Palaiseau	CRAY 2
CEA	Cadarache	CRAY X-MP/14se
	Limeil	CRAY X-MP/416
		CRAY 1S
	Saclay	CRAY X-MP/28
	Vaujours	CRAY 1S
CERFACS	Toulouse	ETA-10P
CIRCE	Orsay	Siemens VP-200
CISI-Framatome (CIFRAM)	Gif-sur-Yvette	CRAY X-MP/416
Citroen/Puegeot PSA	Neuilly-sur-Seine	CRAY X-MP/14
CGG	Massy	CRAY X-MP/24
		CRAY 1S
EDF	Clamart	CRAY X-MP/216
		CRAY X-MP/24
Michelin	Clermont Ferrand	CRAY X-MP/14se
ONERA	Chatillon	CRAY X-MP/116
SNEA (Elf Aquitaine)	Pau	CRAY X-MP/12
Total CFP	Paris	CRAY X-MP/24
		ETA-10P

Table 3.2 Summary of supercomputers in France.

Institute	Location	Supercomputer
TU Aachen	Aachen	ETA-10 (8 Proc) on order
Adam Opel	Russelsheim	CRAY X-MP/14
BMW	Munich	CRAY X-MP/28
Ruhr-Universitat Bochum	Bochum	CYBER 205
Daimler-Benz	Stuttgart	CRAY X-MP/24
DFVLR	Oberpfaffenhofen	CRAY X-MP/216
DKRZ	Hamburg	CRAY 2
HLRZ	Julich	CRAY X-MP/416
IABG	Ottobrunn	Siemens VP-200
Universitat Kaiserslautern	Kaiserslautern	Siemens VP-100
Universitat Karlsruhe	Karlsruhe	CYBER 205
KFA-Julich	Julich	CRAY X-MP/22
KFK-Karlsruhe	Karlsruhe	Siemens VP-50
Universitat Kiel	Kiel	CRAY X-MP/18
Liebniz Rechenzentrum	Munich	CRAY X-MP/24
Max Planck Institut	Garching	CRAY X-MP/216
	Hamburg	CYBER 205
Prakla-Seismos	Hannover	CYBER 205
Siemens AG	Munich	Siemens VP-200
Universitat Stuttgart	Stuttgart	CRAY 2
VW	Wolfsburg	CRAY X-MP/24
Wetter Dienst	Offenbach	ETA-10 (8 proc)

Table 3.3. Summary of supercomputers in Federal Republic of Germany.

Country	Institute	Location	Supercomputer
Belgium	University Brussels	Brussels	CRAY X-MP/14se
	UIA	Antwerp	ETA-10P
Denmark	UNI*C	Lyngby	Amdahl VP 1100
Finland	VTKK	Espoo	CRAY X-MP/14 (on order)
Italy	CINECA	Bologna	CRAY X-MP/48
			CRAY X-MP/24
Netherlands	ENR	Petten	ETA-10Q
	KSEPL	Rijswijk	CRAY X-MP/164
	NLR	Petten	NEC SX2
	SARA	Amsterdam	CYBER 205
Norway	GECO	Stavanger	Amdahl VP 1100
	Norcomp	Nortodden	Amdahl VP 1100
	RUNIT	Trondheim	CRAY X-MP/14
Spain	CASA	Madrid	CRAY 1S
Sweden	SAAB-Scania	Linkoping	CRAY 1S
			CRAY X-MP/48 (on order)
Switzerland	CERN	Geneva	CRAY X-MP/48
	EPFL	Lausanne	CRAY 2 (2 proc)
	ETH	Zurich	CRAY X-MP/28

Table 3.4. Summary of supercomputers in other European countries.

The distribution of supercomputers in Europe is similar to the situation over the last few years with most of them in Britain, France, and Germany and about the same number in each of these countries. About one quarter of the installed base is shared between all the other countries. One should add that it is difficult to be very precise about figures, since new orders are being made all the time and the delays between initial order, licence export approval, installation, and acceptance can be quite substantial. Although the centralization on Paris and London (discussed in Duff 1988) is much less than formerly, about half of the supercomputers are at sites within 50 km of Paris, London, or Munich.

At one time, most supercomputers were either used in defence or weather forecasting applications, and perhaps the most significant change in recent years has been the diversification to many other areas. Regional and University computing centres still account for about one third of the installations, although it is important to recognise that many of these centres are used by industrial as well as academic users, and the applications supported are very diverse indeed. In Table 3.5, we show a rough breakdown of the European supercomputer base. When we compare it with the worldwide figures from Cray, we see a similar pattern, although the university/regional centre sector is more heavily favoured in the European figures, which may partly be an artifact of the subdivision we have adopted.

Application	Percentage of total installed base	
	European sites	Worldwide Cray
Universities	36	11
Government	18	33
Petroleum/seismic	16	14
Aerospace	9	14
Automotive	8	5
Weather	5	3
Service bureau	4	5
Energy	4	9

Table 3.5. Distribution of supercomputers by application.

4   Why?

Why one might ask do the activities mentioned in the previous section require the power of a supercomputer and why are people prepared to fund, or can persuade the necessary funding, for the same? The essential nature of a supercomputer as an instrument of large-scale scientific computing is now widely recognized and indeed many large simulations, including any realistic 3-D simulation, would be impractical without such equipment. Although some manufacturers can make a convincing case that supercomputing is "cost effective", the strongest argument is in the effectiveness, an argument so powerful that an excellent case can be made for the high level of funding required. Indeed, although supercomputers may at first seem expensive, their cost is often considerably less than the experiment they are used to simulate; oil reservoir modelling and computational wind tunnels being two outstanding examples. Perhaps the true yardstick should be to compare the cost of a supercomputer with the cost of not doing the calculation for which it is intended. A good example of this is given by a single run of a simulation of enhanced oil recovery from wells in Prudhoe Bay, on a CRAY 1 (Cray 1981). This simulation gave a strategy that was estimated to yield a 7% increase in oil recovery. When one speaks of quantities of 20 billion barrels at the (then) oil price of over $20/barrel, the saving is considerably more than the lifetime costs of the supercomputer.

It could be argued that a fast machine is somewhat of a luxury and that any calculation could be done on a workstation given sufficient patience and time. I think the ratios in Table 2.3 should make one feel a little nervous about this line of argument over and above any problems with memory size. In general then, the elapsed time must be feasible. There are few calculations that merit running so that one's children or grandchildren will reap the results. Indeed, the underlying physical theory may well have changed during the computation to render the results meaningless. In some applications it is important to obtain results in a fixed relatively short time. Geerd Hoffmann of ECMWF (private communication 1986) gives, as an example, the run of a fine-mesh weather forecasting model that he estimates would require 10^{16} floating-point operations and must be run in about 4 hours (10^4 seconds) for the results to be of use to the forecasters. The Megaflop rate required is thus 10^6 Megaflops, the upper figure in the 1990-1995 projection in Table 2.1!

In general, supercomputers are necessary in numerically intensive real time applications (witness their use in film making) and in time-critical situations, such as the weather forecasting model just mentioned. In many instances, they are important in keeping the elapsed time manageable and we quote an example in the next section.

Increasingly memory requirements, for example for three-dimensional simulations, also necessitate the use of a supercomputer.

5   How?

How does the availability of a supercomputer affect our ability to to do science? The main response to this is that it gives us answers to questions that might otherwise remain unanswered. The vast number of scientists doing lattice gauge computations are a case in point although closer to home at Harwell, we had a complex parameter fitting problem involving the repeated solution of a set of nonlinear partial differential equations whose discretization had around 30,000 degrees of freedom. The elapsed time for a single run on the CRAY was around 12 minutes while it was a full 24 hours on the IBM 3084 mainframe. Admittedly, this was partly due to scheduling issues but at some point these constraints must be met. The result was that calculations that could be completed in an overall time of a couple of days on the CRAY would have required over one month on the IBM, and since several such numerical experiments were performed, the infeasibility of doing them on our reasonably powerful mainframe was quite apparent.

A more recent example is given in the work of Jones, Simcox, and Wilkes (personal communication 1988) in modelling aspects of the disastrous fire at King's Cross Underground station on 18th November, 1987. The calculations for the Health and Safety Executive, performed using the HARWELL-FLOW 3D computational fluid dynamics package, illustrated features of fire spread in inclined trenches that were not at all intuitive and required subsequent costly test rig experiments to verify. Typical run times for a single simulation on the CRAY 2 were over 40 hours using a code with a highly vectorized kernel. A summary of the numerical experiments can be found in the transcript of the formal investigation (Jones 1988).

6   Next?

So what comes next in the world of supercomputing? We indicated trends for conventional supercomputing in Table 2.1, but we look in this section at potentially more dramatic advances through the use of massive parallelism. A major problem and limitation of the architectures that we have previously discussed is that memory, bus, or switch contention become considerably worse as we increase the number of processors accessing a common memory. This is one of the principal reasons why the parallelism of shared memory multiprocessors tends to be low, for example 8 on the CRAY Y-MP, 6 on the IBM 3090/VF, 8 on the Alliant FX Series, and 30 on the Sequent machines, admittedly with a slower CPU so that the bus overhead is masked.

Several solutions have been proposed to resolve this, the most radical being to use many processors with memory local to the processors, passing data as messages between processors. The idea is to use reasonably cheap off-the-shelf chips for the individual nodes and connect them in some configuration, the current most popular one being a hypercube. The crucial aspect of such a machine is the communication time for message passing which, on the first generation of hypercubes, effectively ruled out any calculations at high computational rates.

More recently, encouraging results are being obtained for fairly complex calculations in physics, chemistry, pattern recognition, and image processing *inter alia*. It is quite exciting to contemplate tens or hundreds of thousands of processors, each with a computational power of a few Megaflops working in unison on one problem, avoiding or masking message passing overheads.

We list some of the current offerings in this area in Table 6.1. The machines in this table are all in the marketplace just now. One area which is on the threshold of emerging from research project to the marketplace is that of dataflow machines (for example, Gurd, Kirkham, and Böhm 1987), where, as the name suggests, the computation is driven by data availability and can be described using directed graphs with nodes corresponding to machine instructions and arcs representing the flow of data. Other promising research projects include IBM's RP3 project where memory is held locally but can be efficiently accessed globally, the hierarchical CEDAR design with shared memory clusters (Alliant FX series machines), and the SUPRENUM project — again using a hierarchical structure.

Machine	Chip	Parallelism	Connection
Active Memory (DAP)	CMOS	1024(SIMD)	near-neighbour
Ametek Series 2010	68020/68882	1024	hypercube
BBN Butterfly GP1000	68020/68882	256	Banyon network
CYBERPLUS	Own	256	Ring
FPS T-Series	Transputer	16384	hypercube
Goodyear MPP	VLSI	16384(SIMD)	near-neighbour
Intel iPSC/2	80386/80387	128	hypercube
IP-1	Own	33	cross-bar
Meiko	Transputer	No limit [a]	user-configurable
Myrias 4000	68020/68881	512	hierarchical bus
NCUBE	VLSI	1024	hypercube
Parsys SN 1000 [b]	Transputer	256	reconfigurable
TMC CM-2	VLSI	65536(SIMD)	hypercube
Saxpy MATRIX 1	VLSI	32(SIMD)	bus/systolic

(a) Maximum system delivered to date has 300 processors
(b) From ESPRIT supernode project, manufactured by Thorn EMI

Table 6.1. Potential supercomputers.

References

Cray (1981). CRAY-1 pays for itself in one simulation. User News. CRAY CHANNELS 3(2), 20.

Dongarra, J.J. (1988). Performance of various computers using standard linear equations software in a Fortran environment. Report TM 23, April 26, 1988 edition. Mathematics and Computer Science Division, Argonne National Laboratory.

Dongarra, J.J. and Duff, I.S. (1988). Advanced architecture computers. Report AERE R 12415 (Revision 1), HMSO, London.

Duff, I.S. (1984). The use of advanced large computers in Europe. Report AERE R 11432, HMSO, London.

Duff, I.S. (1988). Supercomputing in Europe – 1987. In *Supercomputing*. E.N. Houstis, T.S. Papatheodorou, and C.D. Polychronopoulos (editors). Springer-Verlag, Berlin, 1031- 1041.

Gurd, J., Kirkham, C., and Böhm, W. (1987). The Manchester dataflow computing system. In *Experimental Parallel Computing Architectures*. J.J. Dongarra (editor). North-Holland, Amsterdam, New York, and London, 33-43.

Jones, I.P. (1988). Transcript of evidence in *Formal investigation into the King's Cross Underground fire*. Day 82, Part 2 and Day 83, Part 1. Department of Transport, London.

Simulated Annealing: Theory of the Past, Practice of the Future?

PETER J.M. VAN LAARHOVEN
and
EMILE H.L. AARTS

Abstract

Simulated annealing is a general approach for approximately solving large combinatorial optimization problems. In this paper we first give a mathematical description of the algorithm and discuss its behaviour from both a theoretical and a practical point of view. We illustrate the practical use of the algorithm by discussing the application to a number of combinatorial optimization problems. In addition, we cite applications in such diverse areas as design of integrated circuits, image processing, code design and neural network theory, and discuss computational experience with the algorithm.

Keywords: Combinatorial optimization, simulated annealing

1   Introduction

In this paper we are concerned with *combinatorial optimization* [12], [16], i.e. the search for optima of functions of discrete variables. Combinatorial optimization problems are nowadays ubiquitous in such diverse areas as computer science, engineering and operations research.

When dealing with a combinatorial optimization problem, one can try to construct either an *optimization algorithm*, i.e. an algorithm that returns a globally optimal solution for every instance of the problem, or an *approximation algorithm*, i.e. an algorithm that "merely" returns a preferably near-optimal solution for each instance [5].

Unfortunately, many combinatorial optimization problems are NP-hard [5], i.e. they belong to the class of problems for which it is commonly believed that no optimization algorithm can be constructed that solves each instance of such a problem to optimality in polynomial time. Consequently, solving large problem instances to optimality is im-

J. Manley et al. (eds.), Proceedings of the Third European Conference on Mathematics in Industry, 45–57.
© 1990 Kluwer Academic Publishers and B. G. Teubner Stuttgart.

practicable. For that reason many combinatorial optimization problems are tackled by constructing approximation rather than optimization algorithms. In that case the goal is to construct an approximation algorithm that runs in low-order polynomial time and has the property that final solutions are "close" to globally optimal ones, i.e. that is both *efficient* and *effective*. For many combinatorial optimization problems such algorithms are nowadays available, but usually they are tailored to the particular problem for which they are designed. As soon as a new problem arises, a new algorithm has to be constructed. Generally applicable approximation algorithms that are able to find near-optimal solutions for a wide variety of problems are rare. The *simulated annealing algorithm* is such a generally applicable approximation algorithm.

In this paper we first give a description of the algorithm in §2, where we also address the asymptotic convergence of the algorithm as well as its convergence in finite time. In §3 we illustrate the practical use of the algorithm by discussing its application to a number of combinatorial optimization problems. Computational experience with the algorithm is discussed in §4 and the paper is ended with some conclusions and remarks.

2 The Simulated Annealing Algorithm

2.1 Introduction

A combinatorial optimization problem can be formalized as a pair $(\mathcal{R}, C)$, where $\mathcal{R}$ is the set of *solutions* or *configurations* and C a *cost function*, $C : \mathcal{R} \to \mathbb{R}$. For convenience, we consider minimization problems only; the problem then is to find a configuration for which C takes its minimum value.

The simulated annealing algorithm can be viewed as a generalization of the well-known *iterative improvement* approach to combinatorial optimization problems. The application of an iterative improvement algorithm presupposes the definition of a mechanism for generating a transition from one configuration to another. Such a generation mechanism defines a *neighbourhood* for each configuration, consisting of all configurations that can be reached in a single transition.

The iterative improvement algorithm can now be formulated as follows. We start off at a given configuration and search for a configuration in its neighbourhood with lower cost. If such an improved configuration exists, we adopt it and repeat the neighbourhood search from the new configuration. The algorithm terminates when a configuration is obtained whose cost is no worse than any of its neighbours.

Iterative improvement is more or less generally applicable, since its main ingredients are usually easy to define. The main disadvantage of iterative improvement algorithms is that, by definition, they terminate in the first local minimum they encounter; generally, such a local minimum deviates substantially in cost from a global minimum. To overcome this disadvantage, one might think of an algorithm which also accepts, in some limited way, transitions corresponding to an increase in cost. Simulated annealing is an example of the latter approach: cost-increasing transitions are accepted with a non-zero probability, which gradually decreases during the course of the algorithm.

The precise way in which cost-increasing transitions are accepted is suggested by the analogy between the problems of finding the *ground state* of a solid and of finding a globally minimal configuration in a combinatorial optimization problem. In condensed matter physics, *annealing* denotes a physical process by which the ground state of a solid can be found. The simulated annealing algorithm takes its name from the fact that it is based on an algorithm to simulate (parts of) the physical annealing process.

Starting off at a given value of the temperature, the annealing process can be described as follows. At each temperature value, the solid is allowed to reach *thermal equilibrium*. At thermal equilibrium the probability of occurrence of a state i with energy E_i is given by the *Boltzmann distribution*, i.e. it is proportional to $\exp(-E_i/k_B T)$, where T is the *temperature* and k_B the *Boltzmann constant*. As the temperature decreases, the Boltzmann distribution concentrates on the low-energy states and finally, when the temperature approaches zero, only the minimum-energy states have a non-zero probability of occurrence.

To simulate the evolution of a solid to thermal equilibrium, Metropolis *et al.* [14] proposed a *Monte Carlo method*, which generates sequences of states of the solid in the following way. Given the current state of the solid, characterized by the positions of its particles, a small, randomly generated, perturbation is applied, i.e. a small displacement of a randomly chosen particle. If the perturbation results in a lower energy state of the solid ($\Delta E < 0$), then the process is continued with the new state. If $\Delta E \geq 0$, then the probability of accepting the perturbed state is given by $\exp(-\Delta E / k_B T)$. This rule for accepting new states is referred to as the *Metropolis criterion*. Guided by this criterion, the solid eventually evolves into thermal equilibrium.

The Metropolis algorithm can also be used to generate sequences of configurations of a combinatorial optimization problem. In that case, the configurations assume the role of the states of a solid, while the cost function C and the *control parameter* c assume the roles of energy and temperature, respectively. The simulated annealing algorithm can be viewed as a sequence of Metropolis algorithms evaluated at decreasing values of the control parameter. It can thus be described as follows. Initially, the control parameter is given a large value and a sequence of trials is generated using a generation mechanism as in the iterative improvement algorithm. Thus, in each trial, a configuration j is generated by choosing at random an element from the neighbourhood of the current configuration i. If $\Delta C_{ij} = C(j) - C(i)$, then the probability of configuration j being the next configuration in the sequence equals 1, if $\Delta C_{ij} \leq 0$, and $\exp(-\Delta C_{ij}/c)$, if $\Delta C_{ij} > 0$ (the Metropolis criterion). This sequence of trials is continued until equilibrium is approximately reached, i.e. until the probability distribution of the configurations approaches the Boltzmann distribution.

The control parameter is lowered in steps until it approaches 0, with the system being allowed to approach equilibrium for each step by generating a sequence of trials in the previously described way. After termination, the final 'frozen' configuration is taken as the solution of the problem at hand.

Thus, as with iterative improvement, we have a generally applicable approximation algorithm: configurations, a cost function and a neighbourhood structure are the only prerequisites for the application of simulated annealing.

2.2 Mathematical Model of the Algorithm and Asymptotic Convergence

Simulated annealing can be described mathematically by means of a *Markov chain*: a sequence of trials, where the outcome of each trial (the new configuration) only depends on the outcome of the previous trial (the current configuration). A Markov chain is determined by a set of conditional probabilities $P_{ij}(k)$, one for each pair of configurations

(i,j); $P_{ij}(k)$ is the probability that the k-th trial results in a transition from configuration i to configuration j. In the case of simulated annealing, $P_{ij}(k)$ is given by the product of two probabilities:

1. the generation probability, i.e. the probability to generate j from i (which is, of course, 0 if j is not a neighbour of i and the reciprocal of the number of neighbours of i otherwise);

2. the acceptance probability, i.e. the probability to accept j once it is generated from i, which is given by the Metropolis criterion.

For a fixed value of c, these probabilities do not depend on k; thus, the algorithm can be viewed as a procedure for generating a sequence of *homogeneous* Markov chains, one for each value of c.

It is fairly straightforward to show that, if the transition mechanism is such that for each pair of configurations (i,j) it is possible to construct a finite sequence of transitions leading from i to j (which is equivalent to the requirement that the matrix of generation probabilities be *irreducible*), then the stationary distributions of the homogeneous Markov chains exist [4], [10]. We recall that the stationary distribution of a Markov chain is defined as the probability distribution of the configurations after an infinite number of trials [4]. Moreover, the stationary distributions are such that they converge to a uniform distribution on the set of globally minimal configurations (in fact, the stationary distributions correspond to the Boltzmann distributions mentioned in §2.1).

Thus, from the theory of Markov chains we have the following asymptotic convergence result: if we have a sequence of values of the control parameter approaching 0 and if we allow an infinite number of trials for each value of c, then the probability to 'end' in a global minimum is 1, provided the matrix of generation probabilities is irreducible. For more details on asymptotic convergence results, the reader is referred to [1], [10].

2.3 Finite-time Behaviour of the Algorithm

From §2.2 we conclude that under a mild condition on the generation probabilities simulated annealing asymptotically behaves as an optimization algorithm, where asymptotically refers to an infinite number of trials. The latter cannot be realized in finite time; in that case a number of parameters should be specified in such a way that the asymptotic behaviour is closely imitated. These parameters are:

- the sequence of values of the control parameter, i.e. the initial value of the control parameter, a rule for changing the current value of the control parameter into the next one and a final value of the control parameter (a stop criterion);

- the number of trials at each value of the control parameter or, in other words, the length of each homogeneous Markov chain.

This set of parameters is usually referred to as a *cooling schedule*. The search for adequate cooling schedules has been addressed in many papers during the last few years. The literature provides a number of conceptually simple schedules that are similar to the original schedule proposed by Kirkpatrick *et al.* [8]: the algorithm starts off at an experimentally determined value for c_1 for which the *acceptance ratio* $\chi(c_1)$ is close to 1 ($\chi(c_k)$ is the ratio between the number of accepted transitions and the number of proposed transitions at a given value c_k). Next, a sequence of Markov chains is generated at descending values of c_k, where $c_{k+1} = \alpha \cdot c_k$, with α a constant smaller than but close to 1 (typically between 0.9 and 0.99). The length L_k of each Markov chain is usually determined by a minimum number of accepted transitions; a ceiling is put on L_k to avoid extremely long Markov chains for low values of c_k. Execution of the algorithm is terminated if the observed improvement in cost over a number of consecutive Markov chains is small.

More elaborate cooling schedules are given by a number of authors [10]; for some of these schedules it is possible to show that the algorithm based on such a schedule can be executed in polynomial time [2]. In those cases, we have a polynomial-time approximation algorithm. Such a result with respect to the *efficiency* of the algorithm is only worthwhile in combination with results on its *effectivity*, viz. on the difference in cost between solutions returned by the algorithm and globally minimal ones. These criteria are discussed in more detail in §5; in the following section we illustrate the practical use of the algorithm by discussing the application to a number of combinatorial optimization problems.

3   Simulated Annealing in Practice

Application of the simulated annealing algorithm to a combinatorial optimization problem requires the specification of three distinct items: (i) a concise representation of the
problem in terms of configurations and a cost function; (ii) a transition mechanism, and
(iii) a cooling schedule. The last item was already discussed in §2.3; in this section we
use the remaining items as a guideline in the discussion of some combinatorial optimization problems as examples of problems to which the simulated annealing algorithm can
be applied. These examples are chosen such that they illustrate a number of characteristic features of the application of simulated annealing such as the problem formulation,
incremental computation of the cost differences and the use of penalty functions.

(1) The Travelling Salesman Problem (TSP)

Given n cities and an $n \times n$-distance matrix $D = [d_{ij}]$, whose elements d_{ij} denote the distance between city i and city j, the TSP is to find the shortest tour visiting all cities just
once. Here, a solution can be represented by a cyclic permutation $\pi = (\pi(1), \dots, \pi(n))$,
where $\pi(i)$ denotes the successor of city i in the tour represented by π. The cost of
such a tour is obtained by summing the entries $d_{i,\pi(i)}$ over all i. A transition can be
generated by choosing two arbitrary cities i and j, and reversing the sequence in which
the cities in between i and j are traversed. This mechanism is known in the literature
as the 2-change generation mechanism and was introduced by Lin [13]. It is easy to see
that such a transition amounts to the replacement of two arcs in the current tour by two
other arcs. Thus, the difference in cost can be computed *incrementally* by subtracting
and adding the corresponding entries of the distance matrix from the cost of the current
tour. Since the cost difference has to be computed for each transition, it is one of the
most time-consuming parts of simulated annealing. Such computations are done preferably incrementally; for most applications this is by far the fastest way to carry out the
computation.

Finally, we mention that the 2-change mechanism is such that the matrix of generation
probabilities is irreducible; thus, asymptotic convergence to a global minimum is guaranteed.

(2) The Independent Set Problem (ISP)

Given a graph $G = (V, E)$, the ISP is to find the largest independent set, i.e. to find the largest set $V' \subset V$, such that for all $v, w \in V'$ the edge $\{v, w\}$ is not in E.

The ISP is an example of a problem, where we have a set of constraints that must be satisfied by each solution. For such problems it is often appropriate to consider *feasible* solutions, i.e. solutions satisfying the constraints, and *infeasible* solutions, i.e. solutions not satisfying the constraints. Infeasibility of solutions can often be used advantageously by the simulated annealing algorithm.

For the ISP, any solution can be represented by a partition of the set V into sets V' and $V \setminus V'$. The cost of such a solution is then given by $|V'| - \lambda |E'|$, where E' denotes the set of edges $\{v, w\} \in E$ with $v, w \in V'$, and λ a *weighting factor* (larger than 1; note that we are now dealing with a maximization problem). Feasible solutions will contribute only to the first term of the cost function. Infeasible solutions will also contribute to the second term and thus lower the cost (for partitions with equal cardinality). Hence the second term can be viewed as a *penalty function*: it penalizes the presence of edges between the subsets of the partition.

A transition can be generated by randomly choosing a vertex $v \in V$ and moving it from V' to $V \setminus V'$ if $v \in V'$ or *vice versa* if otherwise.

Clearly the solution space for the independent set problem could have been restricted to the set of feasible solutions. However, a mechanism generating a new feasible solution from a current one would be quite complicated and time consuming. Furthermore, the presence of the infeasible solutions leads to a smoothing of the cost function 'landscape' which enables the simulated annealing algorithm to escape more easily from local optima: there are more solutions with small cost differences in the vicinity of a local maximum, thus increasing the escape probability.

Finally, we mention that the transition mechanism is again such that the matrix of generation probabilities is irreducible.

(3) The Placement Problem (PP)

The placement problem is a well-known problem in the field of *VLSI layout* [18], [20]. It is typically one of those 'dirty' practical problems for which it is very hard to construct efficient and effective approximation algorithms.

Given a set of n rectangular blocks and a set of weights $w_{ij}, i, j = 1, \ldots, n$, the PP is to find a minimal *placement* of the blocks, i.e. positions of the blocks in a rectangular grid, such that (i) the blocks do not overlap, and (ii) a weighted sum of the area of the rectangle enveloping all blocks and a connectivity term is minimal. The connectivity term is obtained by summing the quantity $w_{ij}d_{ij}$ over all pairs of blocks, where d_{ij} denotes the distance between blocks i and j in a given placement (the weights w_{ij} correspond to connectivity properties in the VLSI layout problem).

Here, a solution can be represented by a set of positions, one for each block. The cost of a solution is usually defined as a weighted sum of (i) the area of the enveloping rectangle, (ii) the aforementioned connectivity term, and (iii) the amount of overlap in the given placement. To ascertain feasibility of a final solution, the overlap term is often divided by the control parameter: since the value of this parameter is monotonically decreased in the course of the algorithm, the penalty for overlap increases as c becomes small. Hence overlap is likely to be entirely removed as the algorithm proceeds. Moreover, in the same way as for the ISP, experience shows that allowing infeasibility (overlap) results in a better performance of the algorithm: fast convergence to high-quality solutions.

A transition can be generated by assigning a subset of the blocks to new positions. This can be done by local rearrangement of a single block, such as translations, rotations or inversions, or by swapping two blocks.

4   The Merits of Simulated Annealing

To judge an algorithm on its merits, one should consider such criteria as effectivity, efficiency, simplicity, ease of implementation and flexibility. Here, we make a few remarks on the last three criteria; the first two are discussed in the remainder of this section.

The simplicity of simulated annealing hardly calls for any further comment - it is part of its attraction. It is also an easy-to-implement algorithm - simulated annealing algorithms typically consist of a few hundred lines of computer code - though it is not always easy to formulate a problem in a way that lends itself to application of the algorithm [1], [10], [20]. Flexibility refers to the ability of an algorithm to handle problem variations and different problems. The claim that simulated annealing is a very flexible approximation technique is strongly supported by the rich variety of problems to which the algorithm has been applied in the past five years. We mention just a few examples (for an extensive review of applications, the reader is referred to [1], [10], [20]): classical problems in combinatorial optimization, such as graph colouring [15] and job shop scheduling [11], placement problems in VLSI design [19], design of codes [3], inverse problems in image processing [6] and seismology [17], and optimization in neural computing [1].

From a theoretical point of view, very little is known about the effectivity of simulated annealing in finite time; indeed, we consider finding analytical expressions for the worst-case or average-case difference in cost between the solution returned by the algorithm and a global minimum the major remaining problem in the theoretical analysis of simulated annealing. At the same time, there are many empirical results available, originating from a large amount of computational studies. For a number of reasons, however, it is difficult to draw conclusions from these data: instead of averaging over a number of runs of the algorithm, results are often obtained from one single run; the applied cooling schedules are often too simple, not getting the best out of the algorithm; results are often not pitted against those obtained with other (tailored) algorithms.

Nevertheless, we can make the following general observations:

- The algorithm has a potential for finding high-quality solutions, though the required amount of computation time to realize this potential is usually quite large.

- The probabilistic element of the algorithm (the acceptance of cost-increasing transitions with a non-zero probability) makes simulated annealing a significantly better technique than the iterative improvement algorithm on which it is based, even if the latter is allowed the same amount of computation time by applying it to a large number of (randomly generated) initial solutions.

- For classical problems in combinatorial optimization (the TSP being the best-known example) sophisticated tailored algorithms are competitive with and often superior to simulated annealing [7], [9]. Usually, this means that simulated annealing finds solutions of the same quality as those found by a tailored algorithm but in much longer running times.

- For many practical problems, such as those arising in VLSI design, no good approximation algorithms are available. Here, simulated annealing seems to be a panacea: the long running times are made up for by the consistent high quality of the returned solutions.

- Experience shows that the performance of simulated annealing algorithms depends almost as much on the skill and effort applied to the implementation as on the algorithm itself. Indeed, the choice of an appropriate neighbourhood structure and an efficient cooling schedule can substantially improve the effectivity and the efficiency of the algorithm.

5 Conclusion

Ever since its introduction in 1982, simulated annealing has attracted attention widely and literally hundreds of papers on the subject have appeared. As early as 1986 we concluded that 'the theoretical basis of the algorithm had reached a certain level of saturation and that major contributions were to be expected predominantly with respect to new applications' [10]. Since then we have had no reason to change our view. The theoretical analysis of the finite-time effectivity of simulated annealing is still an open problem, whereas the area of application is all but open-ended - new applications are reported almost monthly. It is for that reason that we conjecture that the answer to the question in the title is affirmative.

References

[1] Aarts, E.H.L.; Korst, J.H.M.: Simulated Annealing and Boltzmann Machines: A Stochastic Approach to Combinatorial Optimization and Neural Computing. Chichester: Wiley 1988.

[2] Aarts, E.H.L.; Laarhoven, P.J.M. van: Statistical Cooling: A General Approach to Combinatorial Optimization Problems. Philips J. of Research 40 (1985) 193-226.

[3] El Gamal, A.; Hemachandra, L.A.; Shperling, I.; Wei, V.K.: Using Simulated Annealing to Design Good Codes. IEEE Trans. Inf. Theory 33 (1987) 116-123.

[4] Feller, W.: An Introduction to Probability Theory and Applications, vol. 1. New York: Wiley 1950.

[5] Garey, M.R.; Johnson, D.S.: Computers and Intractability: A Guide to the Theory of NP-Completeness. San Francisco: W.H. Freeman and Co. 1979.

[6] Geman, S.; Geman, D.: Stochastic Relaxation, Gibbs Distributions, and the Bayesian Restoration of Images. IEEE Proc. Pattern Analysis and Machine Intelligence 6 (1984) 721-741.

[7] Johnson, D.S.; Aragon, C.R.; McGeoch, L.A.; Schevon, C.: Optimization by Simulated Annealing: an Experimental Evaluation (Part I). Submitted to Operations Research (1987).

[8] Kirkpatrick, S.; Gelatt Jr., C.D.; Vecchi, M.P.: Optimization by Simulated Annealing. Science 220 (1983) 671-680.

[9] Laarhoven, P.J.M. van: Theoretical and Computational Aspects of Simulated Annealing. Ph.D. Thesis, Erasmus University, Rotterdam, and Amsterdam: Centre for Mathematics and Computer Science 1988 (in press).

[10] Laarhoven, P.J.M. van; Aarts, E.H.L.: Simulated Annealing: Theory and Applications. Dordrecht: Kluwer 1987.

[11] Laarhoven, P.J.M. van; Aarts, E.H.L.; Lenstra, J.K.: Job Shop Scheduling by Simulated Annealing. Submitted to Operations Research (1988).

[12] Lawler, E.L.: Combinatorial Optimization: Networks and Matroids. New York: Holt, Rinehart and Winston 1976.

[13] Lin, S.: Computer Solutions of the Traveling Salesman Problem. Bell Systems Technical Journal. 44 (1965) 2245-2269.

[14] Metropolis, N.; Rosenbluth, A.; Rosenbluth, M.; Teller, A.; Teller, E.: Equation of State Calculations by Fast Computing Machines, J. of Chem. Physics 21 (1953) 1087-1092.

[15] Morgenstern, C.A.; Shapiro, H.D.: Chromatic Number Approximation Using Simulated Annealing. Department of Computer Science, The University of New Mexico, Albuquerque, Technical Report CS86-1 1986.

[16] Papadimitriou, C.H.; Steiglitz, K.: Combinatorial Optimization: Algorithms and Complexity. New York: Prentice-Hall 1982.

[17] Rothman, D.H.: Nonlinear Inversion, Statistical Mechanics, and Residual Statistics Estimation. Geophysics $\underline{50}$ (1985) 2784-2796.

[18] Soukup, J.: Circuit Layout. Proc. IEEE $\underline{69}$ (1981) 1281-1304.

[19] Sechen, C.; Sangiovanni-Vincentelli, A.: The TimberWolf Placement and Routing Package. IEEE J. on Solid State Circuits $\underline{30}$ (1985) 510-522.

[20] Wong, D.F.; Leong, H.; Liu, C.L.: Simulated Annealing for VLSI Design. Boston: Kluwer 1988.

Authors' addresses:

Peter J.M. van Laarhoven: Centre for Quantitative Methods
Nederlandse Philips Bedrijven BV
P.O. Box 218, 5600 MD Eindhoven,
The Netherlands.

Emile H.L. Aarts: Philips Research Laboratories,
P.O. Box 80.000, 5600 JA Eindhoven,
The Netherlands, or
Eindhoven University of Technology
P.O. Box 513, 5600 MB Eindhoven,
The Netherlands.

Domain decomposition for a generalized Stokes problem

Alfio Quarteroni - Alberto Valli

Introduction.

Domain decomposition for fluid dynamical problems is a very active research area nowadays. Besides their capability of exploiting parallel computing resources, numerical approaches of zonal type intrinsically allow a more precise treatment of regions having different flow patterns. Actually, a first attempt could be the using of numerical methods enjoying different features within the different flow-regions. An other (even more extreme) approach is to adopt different set of equations to model the flow regions, ensuring a proper interaction across the (fictitious) subregion interface boundary. A remarkable application of the latter approach is the one to the simulation of a compressible viscous flow in the presence of a body. As a matter of fact, if the viscous terms are neglected far from the boundary layer, one is left with the problem of coupling the full Navier-Stokes system nearby the body with its inviscid form in the "far field".

Domain decomposition methods have been formerly developed and analyzed for elliptic equations (see [7] and the references therein), and then applied to the fluid dynamics of incompressible viscous problems. However, so far only very few attempts have been done to provide them with a rigorous mathematical justification. Application of domain decomposition methods to compressible flows is still in its infancy; a weak trace can be found in [4]. Lately, a mathematical analysis has been carried out in [11] for hyperbolic systems of conservation laws (e.g., Euler equations for compressible inviscid flows).

In this report we consider a model linear problem (see (1.1) below) for the simulation of compressible viscous flows. This generalized Stokes problem stems from the use of a fractional step method that was proposed in [2] to advance in time the full Navier-Stokes equations for compressible viscous flows. According to this method, at each time-step the nonlinear terms are decoupled from the linear ones. This yields a nonlinear advection-diffusion like problem, and two generalized Stokes problems of the form (1.1).

First of all, we explore here which sets of inflow and outflow boundary conditions are allowed for the problem at hand. Then we devise a domain decomposition method consisting in partitioning the computational domain into disjoined subdomains, ensuring at each interface the continuity of the velocity field together with the one of both normal and shear stresses. Furthermore, we propose an iteration-by-subdomain procedure that allows a separate resolution of the Stokes problem within each subdomain. The convergence analysis for our iterative procedure is carried out in the Appendix.

J. Manley et al. (eds.), Proceedings of the Third European Conference on Mathematics in Industry, 59–74.
© 1990 *Kluwer Academic Publishers and B. G. Teubner Stuttgart.*

We also consider the inviscid counterpart of problem (1.1), and we endow it with a suitable set of boundary conditions. Then we propose a domain decomposition method for such an inviscid problem, and we apply the above iteration-by-subdomain algorithm with the obvious modifications. The interest of considering a generalized Stokes problem for inviscid flows is twofold. In most numerical applications, the far field boundary is dealt with as a characteristic (i.e., inviscid) boundary even by those codes devised for simulation of viscous flows. Hence, it is useful to make precise the correct treatment of computational boundary for inviscid flows. Moreover, as it was previously mentioned, a very challenging problem is the coupling of viscous and inviscid mathematical models to be used in different regions of the computational domain. For the problem at hand, our goal is to adopt the viscous Stokes problem in the vicinity of the body (in order to capture the boundary layer) and its inviscid counterpart in the complementary region. Our next step towards the achievement of this goal will be the analysis of the interface conditions which ensure a correct transition from the viscous and the inviscid regime. (A rigorous interface analysis for the coupling of one-dimensional hyperbolic and parabolic systems is given in [5]).

We conclude noticing that the domain decomposition techniques here discussed can be applied as well to the finite dimensional approximations to the problems considered in this report. We give a very short account in Section 5, and refer the interested reader to the forthcoming paper [12] for the details.

1.   The generalized Stokes problem.

Let us concentrate our attention on the following generalized Stokes problem on an open domain Ω of $\mathbf{R}^d$ (d = 2 or 3):

$$(1.1) \qquad \begin{cases} \alpha\sigma + \operatorname{div} u = g & \text{in } \Omega \ , \\ \alpha u - a\mu\Delta u + \beta\nabla\sigma = f & \text{in } \Omega \ , \end{cases}$$

where α, β, a and μ are positive constants and f and g are given functions. The meaning of the various constants and functions appearing in (1.1) depends on what physical problem the model system (1.1) stands for. In the case in which (1.1) represents an intermediate step of the linearization process for the full Navier-Stokes equations for compressible flows (as we mentioned in the Introduction), u denotes the velocity field, σ is the logarithm of the density, f, g, a and β depend on previously known time-levels, α is the inverse of the time-step, and μ is an average of the kinematic viscosity.

Several sets of boundary conditions render (1.1) a mathematically well posed boundary value problem. Among these, we consider those supplementing (1.1) in the simulation of external flows around an airfoil. In such a case, the computational domain Ω is the complement of the airfoil profile, truncated in the far field by the boundary Γ_∞ (see Fig. 1.1).

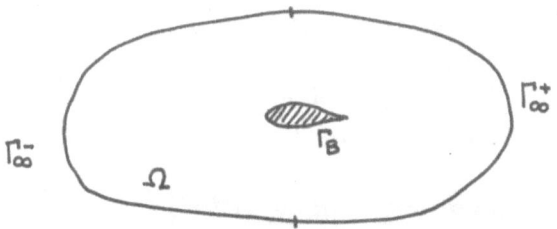

Fig. 1.1. The computational domain for the flow around an airfoil.

We have denoted by Γ_B the boundary of the solid body, and by

$$\Gamma_\infty^- \equiv \{x \in \mathbf{R}^d \mid x \in \partial\Omega \setminus \Gamma_B, \; u_\infty \cdot n < 0\}, \quad \Gamma_\infty^+ \equiv \partial\Omega \setminus (\Gamma_B \cup \Gamma_\infty^-),$$

the inflow and outflow external boundaries, respectively. Clearly, u_∞ and n denote the free stream velocity and the unit outward normal vector to $\partial\Omega$, respectively.

The boundary conditions we will consider take the following form

$$(1.2) \quad \begin{cases} u = 0 \quad \text{on } \Gamma_B, \\ u = u_\infty \quad \text{on } \Gamma_\infty^-, \\ S(u,\sigma) \equiv a\mu \dfrac{\partial u}{\partial n} - \beta \sigma n = 0 \quad \text{on } \Gamma_\infty^+, \end{cases}$$

where $(\frac{\partial u}{\partial n})_i \equiv \sum_j (\partial_j u_i) n_j$. Remark that we don't impose any condition on σ on Γ_∞^-, while m

conditions are enforced on Γ_∞^+. As a matter of fact, the last condition in (1.2) gives raise to the

vanishing of both the normal and shear stresses, which are given respectively by $a\mu \dfrac{\partial u}{\partial n} \cdot n - \beta\sigma$

and $a\mu \dfrac{\partial u}{\partial n} \cdot \tau^s$ ($s = 1, ..., d - 1$), τ^s being a base of the tangent space to $\partial\Omega$.

In order to show that problem (1.1)-(1.2) is well posed, we give its weak formulation. We begin by choosing $u_\infty = 0$, and later we will go back to the general situation. Define

$$(1.3) \qquad V_0 \equiv \{v \in H^1(\Omega) \mid v = 0 \text{ on } \Gamma_\infty^- \cup \Gamma_B\}, \qquad \Sigma_0 \equiv L^2(\Omega).$$

As usual, $H^s(\Omega)$ ($s \in \mathbf{N}$) denotes the Sobolev space of measurable functions whose distributional derivatives of order less or equal than s belong to the space $L^2(\Omega) \equiv H^0(\Omega)$. Moreover, it is possible to extend this definition constructing the Sobolev space $H^s(\Omega)$ for $s \in \mathbf{R}$ (see, e.g., Lions-Magenes [8]). The norm in this space will be denoted by $\| \cdot \|_s$. It is easy to see that (1.1), (1.2) can be formulated in a variational way as

$$(1.4) \quad \begin{cases} u \in V_0, \; \sigma \in \Sigma_0 : \\ \int (\alpha\sigma + \operatorname{div} u)\varphi = \int g\varphi \qquad \forall \; \varphi \in \Sigma_0, \\ \int (\alpha u \cdot v + a\mu \nabla u \cdot \nabla v - \beta\sigma \operatorname{div} v) = \int f \cdot v \qquad \forall \; v \in V_0. \end{cases}$$

If not otherwise specified, the integrals are extended to the whole domain Ω.

Consider the bilinear form on $V_0 \times \Sigma_0$

(1.5) $A[(u,\sigma),(v,\varphi)] \equiv \int (\alpha\beta\sigma\varphi + \beta \,\text{div}\, u\, \varphi + \alpha u \cdot v + a\mu \nabla u \cdot \nabla v - \beta\sigma \,\text{div}\, v)$,

which is associated to the variational problem (1.4). It is easy to see that A is continuous and coercive in $V_0 \times \Sigma_0$, since

(1.6) $A[(u,\sigma),(u,\sigma)] = \alpha\beta \|\sigma\|_0^2 + \alpha \|u\|_0^2 + a\mu \|\nabla u\|_0^2$.

Hence, owing to the Lax-Milgram lemma, for each $f \in H^{-1/2}(\Omega)$, $g \in L^2(\Omega)$ our problem admits a unique solution and we have the estimate

(1.7) $\alpha \|u\|_0^2 + 2a\mu \|\nabla u\|_0^2 + \alpha\beta \|\sigma\|_0^2 \le \alpha^{-1} (\|f\|_{-1/2}^2 + \beta \|g\|_0^2)$.

Let us consider now the case of a non-homogeneous inflow, namely we assume $u_\infty \neq 0$. We construct a function U_∞ defined in Ω and such that its value on Γ_∞^- is u_∞ and on Γ_B is 0. This can be done in several ways, of course. If (as we suppose) $u_\infty \in H^{1/2}(\Gamma_\infty^-)$, we obtain $U_\infty \in H^1(\Omega)$ and $\|U_\infty\|_1 \le c \|u_\infty\|_{1/2,\Gamma_\infty^-}$, the constant c depending only on Ω and Γ_∞^-. Here and in the sequel we will denote by $\|\cdot\|_{s,\Gamma_j}$ the norm in the Sobolev space $H^s(\Gamma_j)$ ($s \in \mathbf{R}$) for functions defined on the boundary Γ_j (see, e.g., Lions-Magenes [8]).

Now we search for

(1.8) $\begin{cases} u \in V_0, \ \sigma \in \Sigma_0 : \\ \int (\alpha\sigma + \text{div}\, u_0)\varphi = \int (g - \text{div}\, U_\infty)\varphi \quad \forall \ \varphi \in \Sigma_0, \\ \int (\alpha u_0 \cdot v + a\mu \nabla u_0 \cdot \nabla v - \beta\sigma \,\text{div}\, v) = \int (f \cdot v - \alpha U_\infty \cdot v - a\mu \nabla U_\infty \cdot \nabla v) \quad \forall \ v \in V_0. \end{cases}$

Since $U_\infty \in H^1(\Omega)$, we can repeat the same argument as before, and we find a unique $(u_0,\sigma) \in V_0 \times \Sigma_0$. The solution to the non-homogeneous problem is now given by $u = u_0 + U_\infty$. Furthermore, we have the estimate

(1.9) $\|u\|_0 + \|\nabla u\|_0 + \|\sigma\|_0 \le c (\|f\|_{-1/2} + \|g\|_0 + \|u_\infty\|_{1/2,\Gamma_\infty^-})$,

where c depends on Ω, Γ_∞^-, α, β, a and μ.

Let us remark that problem (1.1), (1.2) could be faced in a different way, i.e., solving first $(1.1)_1$ with respect to σ, then inserting the result into $(1.1)_2$. In this way we obtain (formally)

(1.10) $\begin{cases} \sigma = \alpha^{-1}(g - \text{div}\, u) , \\ \alpha u - a\mu \Delta u - \beta\alpha^{-1}\nabla \,\text{div}\, u = f - \beta\alpha^{-1}\nabla g . \end{cases}$

Equations $(1.10)_2$ define a strongly elliptic system in the sense of Agmon-Douglas-Nirenberg [1], and the boundary conditions

$$(1.11) \quad \begin{cases} u = 0 & \text{on } \Gamma_B \ , \\ u = u_\infty & \text{on } \Gamma_\infty^- \ , \\ a\mu\dfrac{\partial u}{\partial n} + \beta\alpha^{-1}\text{div } u \ n = \beta\alpha^{-1}g \ n & \text{on } \Gamma_\infty^+ \ , \end{cases}$$

are complementing. Finally, going back to the point of view we already utilized, one can easily verify that the weak formulation of $(1.10)_2$-(1.11) is associated to a coercive functional on V_0.

Remark 1.1. (Alternative sets of boundary conditions).

As we have proven, boundary conditions (1.2) make the generalized Stokes problem (1.1) well posed. However, these boundary conditions don't seem to be correct for supplementing the full nonlinear compressible Navier-Stokes equations. Actually, in this case one needs to add to (1.2) one more condition on the inflow boundary Γ_∞^-. Typically, the

density is assigned on Γ_∞^- (see, e.g., Serrin [13] and also Valli-Zajaczkowski [14]).

On the other hand, this set of admissible boundary conditions for the full nonlinear Navier-Stokes equations is unsuitable for the reduced problem (1.1). As a matter of fact, (1.1) and (1.2) provide spontaneously the value of σ on Γ_∞^-, which might be (and in general is) incompatible

with the a-priori prescribed density.

Concerning this matter, we would like to recall here that in [2] the following set of boundary conditions is proposed

$$\begin{cases} \sigma = 0 & \text{on } \Gamma_\infty^- \ , \\ \dfrac{\partial \sigma}{\partial n} = 0 & \text{on } \Gamma_\infty^+ \ , \\ u = 0 & \text{on } \Gamma_B \ , \\ u = u_\infty & \text{on } \Gamma_\infty^- \ , \\ \dfrac{\partial u}{\partial n} = 0 & \text{on } \Gamma_\infty^+ \ . \end{cases}$$

It seems that these conditions are not proper for either the generalized Stokes problem (1.1) and for the genuine nonlinear compressible Navier-Stokes equations. Actually, for the linear problem there are too many conditions on the inflow boundary (as pointed out in our previous discussion), while for the nonlinear problem the treatment of the outflow boundary looks incorrect. ∎

Remark 1.2.

Problem (1.1) with the boundary conditions (1.2) can also be used in geometrical situations different than the one displayed in Fig. 1.1.

64

An interesting example is provided by the case reported below:

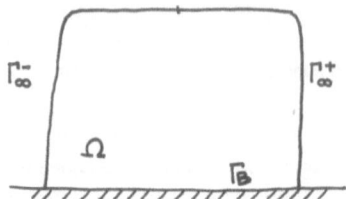

Fig. 1.2. The computational domain for a model boundary layer problem.

This rather simple situation is frequently considered for simulating boundary layer flows over a flat plate (see, e.g., Canuto-Hussaini-Quarteroni-Zang [3], ch. 7). All mathematical considerations carried out so far apply to the current situation as well.   ∎

2.   A domain decomposition method for the generalized Stokes problem.

Our aim is now to find a correct domain decomposition procedure for the solution of problem (1.4). Let us assume for simplicity that $u_\infty = 0$ (if this is not the case, we can argue as before), and that the computational domain is subdivided into two subdomains, as indicated in Fig. 2.1.

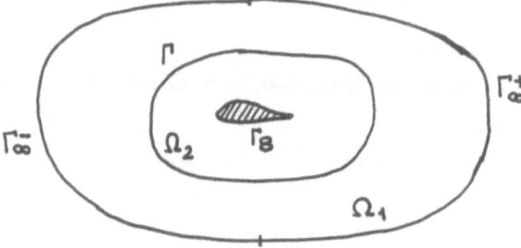

Fig. 2.1. The subdomain decomposition of the computational domain of Fig. 1.1.

We define

(2.1) $V_1 \equiv \{ v \in H^1(\Omega_1) \mid v = 0 \text{ on } \Gamma_\infty^- \}$,

(2.2) $V_2 \equiv \{ v \in H^1(\Omega_2) \mid v = 0 \text{ on } \Gamma_B \}$,

(2.3) $V_{0,k} \equiv \{ v \in V_k \mid v = 0 \text{ on } \Gamma \}$, $k = 1, 2,$

and consider the multidomain problem

$$u^1 \in V_1 , \ \sigma^1 \in L^2(\Omega_1), \ u^2 \in V_2 , \ \sigma^2 \in L^2(\Omega_2) :$$

$$\int_{\Omega_1} (\alpha \sigma^1 + \text{div } u^1) \varphi = \int_{\Omega_1} g \varphi \quad \forall \varphi \in L^2(\Omega_1) ,$$

$$\int_{\Omega_1} (\alpha u^1 \cdot v + a \mu \nabla u^1 \cdot \nabla v - \beta \sigma^1 \text{div } v) = \int_{\Omega_1} f \cdot v \quad \forall v \in V_{0,1} ,$$

$u^1 = u^2$ on Γ,

(2.4) $\qquad \int_{\Omega_2} (\alpha\sigma^2 + \text{div } u^2)\psi = \int_{\Omega_2} g\psi \quad \forall \psi \in L^2(\Omega_2)$,

$\qquad \int_{\Omega_2} (\alpha u^2 \cdot w + a\mu \nabla u^2 \cdot \nabla w - \beta\sigma^2 \text{div } w) = \int_{\Omega_2} f \cdot w \quad \forall w \in V_{0,2}$,

$\qquad \int_{\Omega_2} (\alpha u^2 \cdot R_2\chi + a\mu \nabla u^2 \cdot \nabla R_2\chi - \beta\sigma^2 \text{div } R_2\chi) = \int_{\Omega_2} f \cdot R_2\chi + \int_{\Omega_1} f \cdot R_1\chi -$

$\qquad\qquad - \int_{\Omega_1} (\alpha u^1 \cdot R_1\chi + a\mu \nabla u^1 \cdot \nabla R_1\chi - \beta\sigma^1 \text{div } R_1\chi) \quad \forall \chi \in H^{1/2}(\Gamma)$,

where R_1 and R_2 are any possible couple of (continuous) operators from $H^{1/2}(\Gamma)$ to V_1 and V_2, respectively, which satisfy $R_k\chi = \chi$ on Γ, $k = 1, 2$. Let us point out moreover that from (2.4)7 it follows (formally) that $S(u^1, \sigma^1) = S(u^2, \sigma^2)$ on Γ.

We want to show that (1.4) and (2.4) are equivalent. First of all, if we have a solution (u, σ) to (1.4) we set

$$u^k \equiv u_{|\Omega_k} \quad , \quad \sigma^k \equiv \sigma_{|\Omega_k} \quad , \quad k = 1, 2 \quad ,$$

and it is easy to verify that this is a solution to (2.4).

On the contrary, let u^k and σ^k ($k = 1, 2$) be a solution to (2.4). Since $u^1 = u^2$ on Γ, we can define

(2.5) $\qquad u \equiv \begin{cases} u^1 \text{ in } \Omega_1 \\ u^2 \text{ in } \Omega_2 \end{cases} \quad , \qquad \sigma \equiv \begin{cases} \sigma^1 \text{ in } \Omega_1 \\ \sigma^2 \text{ in } \Omega_2 \end{cases} \quad ,$

and we have $u \in V_0$, $\sigma \in \Sigma_0$. Moreover, (1.4)2 is satisfied, since

$$\int_\Omega (\alpha\sigma + \text{div } u)\varphi = \sum_k \int_{\Omega_k} (\alpha\sigma^k + \text{div } u^k)\varphi = 0 \quad \forall \varphi \in L^2(\Omega) \ .$$

Take now $v \in V_0$ and let us set $\chi \equiv v_{|\Gamma} \in H^{1/2}(\Gamma)$. Define $R\chi \in V_0$ such that $R\chi = R_k\chi$ in Ω_k, $v^k \in V_0$ (k=1,2) as: $v^k_{|\Omega_k} = v_{|\Omega_k} - R_k\chi$, and $v^k_{|\Omega/\Omega_k} = 0$.

Note that $v^k_{|\Omega_k} \in V_{0,k}$, $k = 1, 2$, and that $v = v^1 + v^2 + R\chi$. Thus we get

$$\int_\Omega (\alpha u \cdot v + a\mu \nabla u \cdot \nabla v - \beta\sigma \text{ div } v) = \sum_k \int_{\Omega_k} (\alpha u^k \cdot v^k + a\mu \nabla u^k \cdot \nabla v^k - \beta\sigma^k \text{div } v^k) +$$

$$+ \sum_k \int_{\Omega_k} (\alpha u^k \cdot R_k\chi + a\mu \nabla u^k \cdot \nabla R_k\chi - \beta\sigma^k \text{div } R_k\chi) = \sum_k \int_{\Omega_k} f \cdot (v^k + R_k\chi) = \int_\Omega f \cdot v \ ,$$

having used (2.4)3, (2.4)6 and (2.4)7.

Since problem (1.4) has a unique solution, it follows that uniqueness also holds for (2.4).

In order to make the above domain decomposition approach appealing in view of numerical computation, we propose an iterative procedure that allows the resolution of two subproblems, one within Ω_1, the other in Ω_2.

Assume that we are given an initial guess for u at the interface Γ. We go from the step m-1 to the next step m by solving the two following problems:

(A) within Ω_1 solve (1.1) with the boundary conditions

$$(2.6) \quad \begin{cases} u^1_m = u_\infty & \text{on } \Gamma^-_\infty \;, \\ S(u^1_m, \sigma^1_m) = 0 & \text{on } \Gamma^+_\infty \;, \\ u^1_m = \theta_m u^2_{m-1} + (1 - \theta_m) u^1_{m-1} & \text{on } \Gamma \;. \end{cases}$$

Here θ_m is a positive acceleration parameter.

(B) then in Ω_2 solve problem (1.1) with the boundary conditions

$$(2.7) \quad \begin{cases} S(u^2_m, \sigma^2_m) = S(u^1_m, \sigma^1_m) & \text{on } \Gamma \;, \\ u^2_m = 0 & \text{on } \Gamma_B \;. \end{cases}$$

(We omit here, for simplicity, to write the corresponding weak formulations, which however will be reported in the Appendix). We have:

Theorem 2.1. There exists a positive constant $\theta^* \in \,]0,1]$ such that, if $\sup \theta_m < \theta^*$ and $\inf \theta_m > 0$, then the sequences u^1_m, u^2_m, σ^1_m, σ^2_m converge to the solution u^1, u^2, σ^1, σ^2 of problem (2.4). Convergence is in the $H^1(\Omega)$-norm for velocity and $L^2(\Omega)$-norm for density.

The proof of this result is given in the Appendix.

We stress that the above multidomain approach, as well as the relative iterative procedure, can also be pursued in the framework of a finite dimensional approximation of problem (1.1)-(1.2) (see Section 5).

3. The inviscid generalized Stokes problem.

We consider in this Section the inviscid counterpart of problem (1.1), which reads as follows

$$(3.1) \quad \begin{cases} \alpha \sigma + \text{div } u = g \;, \\ \alpha u + \beta \nabla \sigma = f \;; \end{cases}$$

(it is formally obtained by setting $\mu = 0$ in (1.1)). In analogy with the viscous case, we can write alternatively

$$(3.2) \quad \begin{cases} \sigma = \alpha^{-1}(g - \text{div } u) \ , \\ \alpha u - \beta\alpha^{-1}\nabla \text{div } u = f - \beta\alpha^{-1}\nabla g \ , \end{cases}$$

or , even better,

$$(3.3) \quad \begin{cases} u = \alpha^{-1}(f - \beta\nabla\sigma) \ , \\ \alpha^2\sigma - \beta\Delta\sigma = \alpha g - \text{div } f \ ; \end{cases}$$

the latter formulation will be mainly used in the sequel. (Let us remark that this formulation cannot be utilized in the viscous case, since $(3.3)_1$ does not hold).

We are going to consider these boundary conditions for problem (3.1), namely

$$(3.4) \quad \begin{cases} u \cdot n = u_\infty \cdot n & \text{on } \Gamma_\infty^- \ , \\ u \cdot n = 0 & \text{on } \Gamma_B \ , \\ \sigma = 0 & \text{on } \Gamma_\infty^+ \ . \end{cases}$$

As we shall see, (3.4) renders problem (3.1) a well posed one. Let us remark that $(3.4)_3$ corresponds to condition $(1.2)_3$ in the degenerate case $\mu = 0$.

By using $(3.3)_1$, the above boundary conditions can be written in term of σ only. Precisely, (3.4) becomes

$$(3.5) \quad \begin{cases} \beta\dfrac{\partial\sigma}{\partial n} = f \cdot n - \alpha u_\infty \cdot n & \text{on } \Gamma_\infty^- \ , \\ \beta\dfrac{\partial\sigma}{\partial n} = f \cdot n & \text{on } \Gamma_B \ , \\ \sigma = 0 & \text{on } \Gamma_\infty^+ \ . \end{cases}$$

The existence theorem for problem $(3.3)_2$ with boundary conditions (3.5) is well known since it is a mixed Dirichlet-Neumann problem for an elliptic equation. Assuming $f \in L^2(\Omega)$, $g \in L^2(\Omega)$, $u_\infty \cdot n \in (H_{00}^{1/2}(\Gamma_\infty^-))'$ (see, e.g., Lions-Magenes [8] for the definition of this space), we find $\sigma \in H^1(\Omega)$, hence $u \in L^2(\Omega)$ with $\text{div } u \in L^2(\Omega)$ and

$$(3.6) \quad \|u\|_0 + \|\text{div } u\|_0 + \|\sigma\|_1 \le c \ (\|f\|_0 + \|g\|_0 + \|u_\infty \cdot n\|_{(H_{00}^{1/2}(\Gamma_\infty^-))'}).$$

One can also formulate this result in a variational form. Precisely, proceeding as before we have found (u,σ) such that $(u - \omega_\infty) \in W_0$ and $\sigma \in S_0$, where $\omega_\infty \in L^2(\Omega)$ with $\text{div } \omega_\infty \in L^2(\Omega)$ is such that $\omega_\infty \cdot n = u_\infty \cdot n$ on Γ_∞^- and $\omega_\infty \cdot n = 0$ on Γ_B,

$$(3.7) \quad W_0 \equiv \{v \in L^2(\Omega) \mid \text{div } v \in L^2(\Omega) , v \cdot n = 0 \text{ on } \Gamma_\infty^- \cup \Gamma_B\} \ ,$$

$$(3.8) \quad S_0 \equiv \{s \in H^1(\Omega) \mid s = 0 \text{ on } \Gamma_\infty^+\} \ ,$$

and (u,σ) satisfies

$$(3.9) \quad \begin{cases} \int (\alpha\sigma + \text{div } u)\varphi = \int g\varphi & \forall \ \varphi \in L^2(\Omega) \ , \\ \int (\alpha u \cdot v - \beta\sigma \text{ div } v) = \int f \cdot v & \forall \ v \in W_0 \ . \end{cases}$$

Remark 3.1.

Let us remark that, if we consider the boundary conditions

(3.10)
$$\begin{cases} \sigma = \sigma_0 & \text{on } \Gamma_\infty^- \cup \Gamma_\infty^+ , \\ u \cdot n = 0 & \text{on } \Gamma_B , \end{cases}$$

we obtain

(3.11)
$$\|u\|_0 + \|\text{div } u\|_0 + \|\sigma\|_1 \le c \ (\|f\|_0 + \|g\|_0 + \|\sigma_0\|_{1/2, \Gamma_\infty^- \cup \Gamma_\infty^+}) .$$

Alternatively, if we consider the boundary conditions

(3.12)
$$\begin{cases} u \cdot n = u_\infty \cdot n & \text{on } \Gamma_\infty^- , \\ u \cdot n = 0 & \text{on } \Gamma_\infty^+ \cup \Gamma_B , \end{cases}$$

we get

(3.13)
$$\|u\|_0 + \|\text{div } u\|_0 + \|\sigma\|_1 \le c \ (\|f\|_0 + \|g\|_0 + \|u_\infty \cdot n\|_{(H_{00}^{1/2}(\Gamma_\infty^-))'}) .$$

The corresponding weak formulations can be easily written, too. ∎

4. A domain decomposition method for the inviscid generalized Stokes problem.

Following what we have already done for the viscous case, we can formulate the inviscid generalized Stokes problem in the following way. Define

(4.1) $\quad W_1 \equiv \{v \in L^2(\Omega_1) \mid \text{div } v \in L^2(\Omega_1), \ v \cdot n = 0 \text{ on } \Gamma_\infty^-\} ,$

(4.2) $\quad W_2 \equiv \{v \in L^2(\Omega_2) \mid \text{div } v \in L^2(\Omega_2), \ v \cdot n = 0 \text{ on } \Gamma_B\} ,$

(4.3) $\quad W_{0,k} \equiv \{v \in W_k \mid v \cdot n = 0 \text{ on } \Gamma\}, \ k = 1, 2,$

(4.4) $\quad S_1 \equiv \{\sigma \in H^1(\Omega_1) \mid \sigma = 0 \text{ on } \Gamma_\infty^+\} ,$

(4.5) $\quad S_2 \equiv H^1(\Omega_2) .$

We consider the multidomain problem (assuming for simplicity that $u_\infty = 0$)

$$u^1 \in W_1, \ \sigma^1 \in S_1, \ u^2 \in W_2, \ \sigma^2 \in S_2 :$$

$$\int_{\Omega_1} (\alpha \sigma^1 + \text{div } u^1)\varphi = \int_{\Omega_1} g\varphi \quad \forall \ \varphi \in L^2(\Omega_1) ,$$

$$\int_{\Omega_1} (\alpha u^1 \cdot v - \beta \sigma^1 \text{div } v) = \int_{\Omega_1} f \cdot v \quad \forall \ v \in W_{0,1} ,$$

$$u^1 \cdot n = u^2 \cdot n \quad \text{on } \Gamma ,$$

(4.6)
$$\int_{\Omega_2} (\alpha \sigma^2 + \text{div } u^2)\psi = \int_{\Omega_2} g\psi \quad \forall \ \psi \in L^2(\Omega_2) ,$$

$$\int_{\Omega_2} (\alpha u^2 \cdot w - \beta \sigma^2 \mathrm{div}\ w) = \int_{\Omega_2} f \cdot w \qquad \forall\ w \in W_{0,2}\ ,$$

$$\int_{\Omega_2} (\alpha u^2 \cdot R_2\chi - \beta \sigma^2 \mathrm{div}\ R_2\chi) = \int_{\Omega_2} f \cdot R_2\chi + \int_{\Omega_1} f \cdot R_1\chi\ -$$

$$- \int_{\Omega_1} (\alpha u^1 \cdot R_1\chi - \beta \sigma^1 \mathrm{div}\ R_1\chi) \qquad \forall\ \chi \in H^{1/2}(\Gamma)\ ,$$

where now R_1 and R_2 are any possible couple of (continuous) operators from $H^{-1/2}(\Gamma)$ to W_1 and W_2, respectively, which satisfy $R_k\chi \cdot n = \chi$ on Γ, $k = 1, 2$ (here n is the unit normal vector to Γ, which is directed from Ω_1 to Ω_2).

Following the same procedure used in the viscous case, it is not difficult to show that (3.1) and (3.4) are equivalent to (4.6). The only point that needs to be specified here is about the matching conditions on Γ. Since $u^1 \cdot n = u^2 \cdot n$ on Γ (see (4.6)$_4$), it easily follows that the function u built up as in (2.5) belongs to the space W_0 defined in (3.7). On the other hand the last condition in (4.6) states that $\sigma^1 = \sigma^2$ on Γ, hence, if we construct σ as in (2.5), such a function belongs to the space S_0 defined in (3.8).

As in the viscous case, let us introduce now a suitable iterative procedure for the solution of the multidomain problem. At each step we proceed as follows:

(A) in Ω_1 solve (3.1) with these boundary conditions

(4.7)
$$\begin{cases} u^1_m \cdot n = u_\infty \cdot n & \text{on } \Gamma^-_\infty\ , \\[6pt] \sigma^1_m = 0 & \text{on } \Gamma^+_\infty\ , \\[6pt] u^1_m \cdot n = \theta_m u^2_{m-1} \cdot n + (1 - \theta_m) u^1_{m-1} \cdot n & \text{on } \Gamma\ . \end{cases}$$

(B) Then in Ω_2 solve problem (3.1) with the following boundary conditions:

(4.8)
$$\begin{cases} \sigma^2_m = \sigma^1_m & \text{on } \Gamma\ , \\[6pt] u^2_m \cdot n = 0 & \text{on } \Gamma_B\ , \end{cases}$$

(as before, we omit here to write the corresponding variational formulations; for that, see the Appendix).

The following convergence result will be proven in the Appendix:

Theorem 4.1. There exists a positive constant $\theta^*_1 \in\ ]0,1]$ such that, if $\sup \theta_m < \theta^*_1$ and $\inf \theta_m > 0$, then the sequences u^1_m, u^2_m, σ^1_m, σ^2_m converge to the solution u^1, u^2,

σ^1, σ^2 of problem (4.7). Namely, we have that u_m^k and div u_m^k converge in $L^2(\Omega_k)$ to u^k and div u^k, respectively, and that σ_m^k converges in $H^1(\Omega_k)$ to σ^k.

5. Numerical approximation.

For both viscous and inviscid Stokes problems considered so far, in the two-dimensional case we have investigated numerical approximations based on finite elements and spectral collocation methods. Then we have adapted to the finite dimensional problems the domain decomposition we have proposed here for the differential problems, as well as the associated iteration-by-subdomain procedures.

In both viscous and inviscid cases, we have proven that the single and multidomain finite dimensional solutions are stable and convergent. Moreover the iterative procedures converge with a rate independent of the finite elements mesh size (or the polynomial degree of the solution in the case of spectral approximation).

Due to the absolute lack of space, the description of the finite dimensional problems and the proofs of their stability and convergence properties will be reported in the forthcoming paper [12], together with several numerical experiments.

Appendix. The convergence of the iterative procedure.

(i) The viscous case.
First of all, let us define the bilinear forms

(A.1) $\quad a_k[(u^k, \sigma^k), (v, \varphi)] \equiv \int_{\Omega_k} (\beta \alpha \sigma^k \varphi + \beta \operatorname{div} u^k \varphi + \alpha u^k \cdot v + a\mu \nabla u^k \cdot \nabla v - \beta \sigma^k \operatorname{div} v)$,

which are continuous and coercive on $H^1(\Omega_k) \times L^2(\Omega_k)$, $k = 1, 2$. Let us remark however that these forms are not symmetric, hence they don't define a scalar product in $H^1(\Omega_k) \times L^2(\Omega_k)$.

Then define the extension operators $E_k : H^{1/2}(\Gamma) \to V_k \times L^2(\Omega_k)$ in the following way:

(A.2) $\quad \begin{cases} E_k \chi \in V_k \times L^2(\Omega_k), \\ a_k[E_k \chi, (v, \varphi)] = 0 \ \forall \ v \in V_{0,k}, \ \varphi \in L^2(\Omega_k), \\ (E_k \chi)_1|_\Gamma = \chi, \end{cases}$

The iterative scheme we have introduced in (2.6), (2.7) can be formulated in a variational form in the following way: solve for $m \geq 1$

(A.3) $\quad \begin{cases} u_m^1 \in V_1, \ \sigma_m^1 \in L^2(\Omega_1), \\ a_1[(u_m^1, \sigma_m^1), (v, \varphi)] = (f, v) + \beta(g, \varphi) \ \forall \ v \in V_{0,1}, \ \varphi \in L^2(\Omega_1), \\ u_m^1|_\Gamma = \theta_m u_{m-1}^2|_\Gamma + (1 - \theta_m) u_{m-1}^1|_\Gamma \equiv g_{m-1} \quad \text{on } \Gamma, \end{cases}$

$$\text{(A.4)} \quad \begin{cases} u_m^2 \in V_2 \ , \ \sigma_m^2 \in L^2(\Omega_2) \ , \\[2mm] a_2[(u_m^2,\sigma_m^2),(w,\psi)] = (f,w) + \beta(g,\psi) \quad \forall \ w \in V_{0,2} \ , \ \psi \in L^2(\Omega_2) \ , \\[2mm] a_2[(u_m^2,\sigma_m^2),E_2\chi] = (f,(E_2\chi)_1) + \beta(g,(E_2\chi)_2) - \\[2mm] \qquad - a_1[(u_m^1,\sigma_m^1),E_1\chi] + (f,(E_1\chi)_1) + \beta(g,(E_1\chi)_2) \quad \forall \ \chi \in H^{1/2}(\Gamma) \ , \end{cases}$$

where $(\, , \,)$ denotes the scalar product in $L^2(\Omega_1)$ or $L^2(\Omega_2)$, and the initial guess g_0 can be arbitrarily chosen in $H^{1/2}(\Gamma)$. Here θ_m is a positive relaxation parameter which will be determined in the sequel in such a way that the iterative scheme converges.

Let us remark that (A.4) is equivalent to the problem

$$\begin{cases} u_m^2 \in V_2 \ , \ \sigma_m^2 \in L^2(\Omega_2) \ , \\[2mm] a_2[(u_m^2,\sigma_m^2),(w,\psi)] = (f,w) + \beta(g,\psi) - a_1[(u_m^1,\sigma_m^1),E_1(w_{|\Gamma})] + \\[2mm] \qquad + (f,E_1(w_{|\Gamma})_1) + \beta(g,E_1(w_{|\Gamma})_2) \quad \forall \ w \in V_2 \ , \ \psi \in L^2(\Omega_2) \ . \end{cases}$$

Trivially, the right hand side is a continuous linear functional on $V_2 \times L^2(\Omega_2)$.

Let us define now the following norms:

(A.5) $\qquad [u^1,\sigma^1]^2 \equiv a_1[(u^1,\sigma^1),(u^1,\sigma^1)] \qquad$ for $(u^1,\sigma^1) \in H^1(\Omega_1) \times L^2(\Omega_1)$,

(A.6) $\qquad [[u^2,\sigma^2]]^2 \equiv a_2[(u^2,\sigma^2),(u^2,\sigma^2)] \qquad$ for $(u^2,\sigma^2) \in H^1(\Omega_2) \times L^2(\Omega_2)$,

(A.7) $\qquad |\!|\!| \chi |\!|\!|^2 \equiv a_1[E_1\chi,E_1\chi] = [E_1\chi]^2 \ , \ ((\chi,\lambda)) \equiv a_1[E_1\chi,E_1\lambda] \quad$ for $\chi, \lambda \in H^{1/2}(\Gamma)$;

(remark that $((\cdot,\cdot))$ is symmetric, though a_1 isn't). These norms are trivially shown to be equivalent to the usual norms in $H^1(\Omega_k) \times L^2(\Omega_k)$ and $H^{1/2}(\Gamma)$. Let us make precise this last assertion. By an estimate like (1.9) and the continuity of the trace operator from $H^1(\Omega_k)$ to $H^{1/2}(\Gamma)$ we have

$$|\!|\!| \chi |\!|\!| = [E_1\chi] \le c_1 \| \chi \|_{1/2,\Gamma} \le c_2 \, [[E_2\chi]] \ ,$$

$$[[E_2\chi]] \le c_3 \| \chi \|_{1/2,\Gamma} \le c_4 \, [E_1\chi] = c_4 \, |\!|\!| \chi |\!|\!| \ .$$

Hence we can define

(A.8) $\qquad \eta \equiv \sup\{ \ |\!|\!|\lambda|\!|\!|^2 / [[E_2\lambda]]^2 \mid \lambda \in H^{1/2}(\Gamma) \ , \ \lambda \neq 0 \ \} \ ,$

(A.9) $\qquad \omega \equiv \sup\{ \ [[E_2\lambda]]^2 / |\!|\!|\lambda|\!|\!|^2 \mid \lambda \in H^{1/2}(\Gamma) \ , \ \lambda \neq 0 \ \} \ .$

By following the same procedure employed by Marini-Quarteroni [9], [10] we can prove:

Lemma A.1. If $\inf \theta_m > 0$ and $u_{m|\Gamma}^1$ converges in $H^{1/2}(\Gamma)$, then (u_m^k,σ_m^k) and $u_{m|\Gamma}^2$ converge in $H^1(\Omega_k) \times L^2(\Omega_k)$ and $H^{1/2}(\Gamma)$, respectively, and the limit functions (u^k,σ^k) are the solution to (2.4).

Proof. Since $(u_m^1 - u_h^1, \sigma_m^1 - \sigma_h^1) = E_1(u_m^1 - u_h^1)_{|\Gamma}$, we have the following identity

$$[u_m^1 - u_h^1, \sigma_m^1 - \sigma_h^1]^2 = a_1[E_1(u_m^1 - u_h^1)_{|\Gamma}, E_1(u_m^1 - u_h^1)_{|\Gamma}] = |\!|\!| (u_m^1 - u_h^1)_{|\Gamma} |\!|\!|^2 \ ,$$

hence (u_m^1, σ_m^1) converges in $H^1(\Omega_1) \times L^2(\Omega_1)$.

From $(A.3)_3$ and inf $\theta_m > 0$ we have moreover that $(u_m^2|_\Gamma - u_m^1|_\Gamma)$ converges to zero in $H^{1/2}(\Gamma)$.

Finally, as $(u_m^2 - u_h^2, \sigma_m^2 - \sigma_h^2) = E_2(u_m^2 - u_h^2)|_\Gamma$, we have

$$[[u_m^2 - u_h^2, \sigma_m^2 - \sigma_h^2]]^2 = a_2[(u_m^2 - u_h^2, \sigma_m^2 - \sigma_h^2), E_2(u_m^2 - u_h^2)|_\Gamma] =$$

$$= - a_1[(u_m^1 - u_h^1, \sigma_m^1 - \sigma_h^1), E_1(u_m^2 - u_h^2)|_\Gamma] = - (((u_m^1 - u_h^1)|_\Gamma, (u_m^2 - u_h^2)|_\Gamma)) \leq$$

$$\leq |||(u_m^1 - u_h^1)|_\Gamma||| \; |||(u_m^2 - u_h^2)|_\Gamma||| \; ,$$

which conclude the proof. ∎

Introduce now the operator $T: H^{1/2}(\Gamma) \to H^{1/2}(\Gamma)$ defined in the following way

(A.10) $$T\lambda \equiv z^2|_\Gamma \; ,$$

where

(A.11) $$\begin{cases} z^2 \in V_2 \; , \; \eta^2 \in L^2(\Omega_2) \; , \\ a_2[(z^2, \eta^2), (w, \psi)] = 0 \; \forall \; w \in V_{0,2} \, , \psi \in L^2(\Omega_2) \; , \\ a_2[(z^2, \eta^2), E_2\chi] = - a_1[E_1\lambda, E_1\chi] \; \forall \; \chi \in H^{1/2}(\Gamma) \; . \end{cases}$$

We can write in particular

(A.12) $$(z^2, \eta^2) = E_2(z^2|_\Gamma) = E_2 T\lambda \; .$$

Then, for any positive θ we set

(A.13) $$T_\theta \lambda \equiv \theta T\lambda + (1 - \theta)\lambda \qquad \forall \; \lambda \in H^{1/2}(\Gamma) \; .$$

By following the method proposed by Marini-Quarteroni [9], [10], we can establish the following result:

Theorem A.2. There exists a positive constant $\theta^* \in \,]0,1]$ such that for each $\theta \in \,]0, \theta^*[$ the map T_θ is a contraction.

Before proving this Theorem, let us remark that the convergence of $u_m^1|_\Gamma$ follows from it. In fact it is easily checked that we can write $(u_m^1 - u^1)|_\Gamma = T_{\theta_m}(u_{m-1}^1 - u^1)|_\Gamma$.

Proof. We have

$$|||T_\theta \lambda|||^2 = \theta^2 \; |||T\lambda|||^2 + 2\theta(1 - \theta) \; (((T\lambda, \lambda))) + (1 - \theta)^2 \; |||\lambda|||^2 \; .$$

Moreover, by choosing $\chi = T\lambda$ in $(A.11)_3$

(A.14) $$(((T\lambda, \lambda))) = a_1[E_1 T\lambda, E_1\lambda] = - a_2[E_2 T\lambda, E_2 T\lambda] = - [[E_2 T\lambda]]^2 \; .$$

By using (A.7) and (A.8) we have, for $0 \leq \theta \leq 1$,

$$|||T_\theta \lambda|||^2 \leq \theta^2 \eta \; [[E_2 T\lambda]]^2 + (1 - \theta)^2 \; |||\lambda|||^2 - 2\theta(1 - \theta) \; [[E_2 T\lambda]]^2 \; .$$

Finally, from (A.14) and (A.8) we get

$$[[E_2 T\lambda]]^2 \leq |||T\lambda||| \; |||\lambda||| \leq \eta^{1/2} \; [[E_2 T\lambda]] \; |||\lambda||| \; ,$$

and, choosing $\chi = \lambda$ in $(A.11)_3$,

$$|||\lambda|||^2 = a_1[E_1\lambda, E_1\lambda] = -a_2[E_2T\lambda, E_2\lambda] \le [[E_2T\lambda]] \, [[E_2\lambda]] \le \omega^{1/2} |||\lambda||| \, [[E_2T\lambda]] \ .$$

Hence

(A.15)
$$|||T_\theta\lambda|||^2 \le [\theta^2\eta^2 + (1-\theta)^2 - 2\theta(1-\theta)\omega^{-1}] \, |||\lambda|||^2 \ .$$

Set

(A.16)
$$k(\theta) \equiv [\theta^2\eta^2 + (1-\theta)^2 - 2\theta(1-\theta)\omega^{-1}] \ ;$$

it is easily seen that

(A.17)
$$k(\theta) < 1 \ \text{iff} \ 0 < \theta < 2(\omega+1)(\eta^2\omega + \omega + 2)^{-1} \ . \qquad \blacksquare$$

(ii) The inviscid case.

We will just underline the differences between this case and the viscous case. Define at first the following bilinear forms

(A.18)
$$b_k[(u^k,\sigma^k),(v,\varphi)] \equiv \int_{\Omega_k} (\beta\alpha\sigma^k\varphi + \beta\text{div } u^k\varphi + \alpha u^k\cdot v - \beta\sigma^k\text{div } v) \ ,$$

$k = 1, 2$, which correspond to (A.1) in the degenerate case $\mu = 0$. These forms are continuous in $H(\text{div};\Omega_k) \equiv \{u \in L^2(\Omega_k) \mid \text{div } u \in L^2(\Omega_k)\} \times L^2(\Omega_k)$, but they are not coercive. As a consequence, we have already seen that the existence theorem for the variational problem associated to these forms cannot be obtained via the Lax-Milgram lemma, but requires a different approach (see Section 3).

Define now the (continuous) operators $F_k : H^{-1/2}(\Gamma) \to W_k \times S_k$, $k = 1, 2$, in the following way:

(A.19)
$$\begin{cases} F_k\chi \in W_k \times S_k \ , \\ b_k[F_k\chi,(v,\varphi)] = 0 \ \forall \ v \in W_{0,k} \, , \ \varphi \in L^2(\Omega_k) \ , \\ (F_k\chi)_1|_\Gamma \cdot n = \chi \ , \end{cases}$$

As in (A.5)-(A.7), define the following norms

(A.20) $]u^1,\sigma^1[^2 \equiv b_1[(u^1,\sigma^1),(u^1,\sigma^1)]$ for $(u^1,\sigma^1) \in H(\text{div};\Omega_1) \times L^2(\Omega_1)$,

(A.21) $]]u^2,\sigma^2[[^2 \equiv b_2[(u^2,\sigma^2),(u^2,\sigma^2)]$ for $(u^2,\sigma^2) \in H(\text{div};\Omega_2) \times L^2(\Omega_2)$,

(A.22) $|\chi|^2 \equiv b_1[F_1\chi, F_1\chi] =]F_1\chi[^2$, $\langle\chi,\lambda\rangle \equiv b_1[F_1\chi, F_1\lambda]$ for $\chi, \lambda \in H^{-1/2}(\Gamma)$.

The norms defined in (A.20)-(A.21) are easily seen to be equivalent to the norms in $L^2(\Omega_k) \times L^2(\Omega_k)$. The norm given by (A.22) is equivalent to the norm in $H^{-1/2}(\Gamma)$. In fact, by an elliptic estimate like (3.6) and the continuity of the operators : $u \to u\cdot n$ from $H(\text{div};\Omega_k)$ to $H^{-1/2}(\Gamma)$ (see, for instance, Girault-Raviart [6]), we have

$$|\chi| =]F_1\chi[\le c_5 \, (\|(F_1\chi)_1\|_0 + \|\text{div } (F_1\chi)_1\|_0) \le c_6 \| \chi \|_{-1/2,\Gamma} \le$$
$$\le c_7 \, (\|(F_2\chi)_1\|_0 + \|\text{div } (F_2\chi)_1\|_0) \le c_8 \,]]F_2\chi[[\ ,$$

and conversely

$$]]F_2\chi[[\le c_9 \| \chi \|_{-1/2,\Gamma} \le c_{10} | \chi | \ .$$

Hence we can define, as in (A.8)-(A.9),

(A.23) $\eta_1 \equiv \sup\{ |\lambda|^2 /]]F_2\lambda[[^2 \mid \lambda \in H^{-1/2}(\Gamma) , \lambda \ne 0 \}$,

(A.24) $\omega_1 \equiv \sup\{]]F_2\lambda[[^2 / |\lambda|^2 \mid \lambda \in H^{-1/2}(\Gamma) , \lambda \ne 0 \}$.

From now on, one can proceed in the same way than in the viscous case, making the obvious changes where it is necessary.

References.

[1]   Agmon, S.; Douglis, A.; Nirenberg, L.: *Estimates near the boundary for solutions of elliptic partial differential equations satisfying general boundary conditions. II.* Comm. Pure Appl. Math., **17** (1964), 35-92.

[2]   Bristeau, M. O.; Glowinski, R.; Mantel, B.; Periaux, J.: *Numerical methods for the Navier-Stokes equations. Applications to the simulation of compressible and incompressible viscous flows*. Preprint 1987 (partly issued from *Finite elements in Physics*. R. Grüber Ed., Computer Physics, Report 144. Amsterdam: North Holland 1987).

[3]   Canuto, C.; Hussaini, M. Y.; Quarteroni, A.; Zang, T. A.: *Spectral Methods in Fluid Dynamics*. New York: Springer-Verlag 1988.

[4]   Chan T. et Al., Eds.: *Domain Decomposition Methods for Partial Differential Equations, II*. Philadelphia: SIAM, in press.

[5]   Gastaldi, F.;Quarteroni, A.: *On the Coupling of Hyperbolic and Parabolic Systems: Analytical and Numerical Approach.*. I.A.N.-C.N.R. Publication n° 619, Pavia, 1988 (to appear in Appl. Numerical Math.).

[6]   Girault, V.; Raviart, P.-A.: *Finite element methods for Navier-Stokes equations. Theory and Algorithms*. Berlin Heidelberg: Springer-Verlag 1986.

[7]   Glowinski R. et Al., Eds.: *Domain Decomposition Methods for Partial Differential Equations, I*. Philadelphia: SIAM 1988.

[8]   Lions, J. L.; Magenes, E.:*Problèmes aux limites non homogènes et applications, 1*. Paris: Dunod 1968.

[9]   Marini, L. D.; Quarteroni, A.: *An iterative procedure for domain decomposition methods: a finite element approach*. In [7], 129-143.

[10]   Marini, L. D.; Quarteroni, A.:*A relaxation procedure for domain decomposition methods using finite elements*. I.A.N.-C.N.R. Publication n° 577, Pavia, 1987.

[11]   Quarteroni, A.: *Domain Decomposition Methods for Systems of Conservation Laws: Spectral Collocation Approximations*. ICASE Publication, NASA Langley, Hampton, VA, to appear.

[12]   Quarteroni, A.; Valli, A.; Zanolli, P.: in preparation.

[13]   Serrin, J.: *On the uniqueness of compressible fluid motions*, Arch. Rational Mech. Anal., **3** (1959), 271-288.

[14]   Valli, A.; Zajaczkowski, W. M.:*Navier-Stokes Equations for Compressible Fluids: Global Existence and Qualitative Properties of the Solution in the General Case*. Comm. Math. Phys., **103** (1986), 259-296.

Alfio Quarteroni, Dipartimento di Matematica, Università Cattolica, Via Trieste 17, 25121 Brescia, Italy & Istituto di Analisi Numerica del C.N.R., Corso Carlo Alberto 5, 27100 Pavia, Italy.

Alberto Valli, Dipartimento di Matematica, Università di Trento, 38050 Povo (Trento), Italy.

Mathematical Simulation and Optimization of Chemical Plants

F. Kokert, L. Peer, Hj. Wacker

1. Introduction

In recent years considerable effort has been devoted to the mathematical simulation of chemical plants using flowsheeting techniques. There are three stages to consider:

i) synthesizing a total flowsheet (see e.g. M.D.Lu, R.L.Motard [8])

ii) Mathematical Simulation (see e.g. H.P.Hutchinson, D.J.Jackson, W.Morton [6], L.M.Timár, F.Simon, Z.Csermely, J.Siklós, S.Bacskai, .'.Edes [12] and Fredenslund [5])

iii) Optimization

In this paper we only deal with ii) and iii) and in addition we consider only steady state processes. The aim of our work is to study and improve the mathematical algorithms especially with respect to numerical aspects. To sketch the idea what we want to discuss we present the highly simplified flowsheet of an Oxygen Plant.

Figure 1: Oxygen Plant

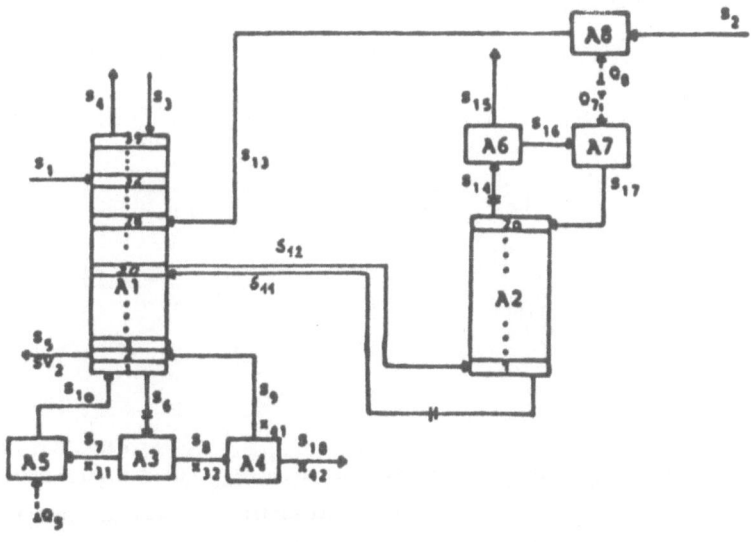

J. Manley et al. (eds.), Proceedings of the Third European Conference on Mathematics in Industry, 75–90.
© 1990 *Kluwer Academic Publishers and B. G. Teubner Stuttgart.*

The flowsheet consists of a number of units: plate absorption columns (A1, A2), heater (A5, A8), cooler (A7), splitter (A3, A4, A6) and connecting streams (18). Our model is based on mass balances, energy balances, liquid/vapor equilibrium conditions.

The specific form of the functions depends on the underlying thermodynamic model, e.g. UNIQUAC.

2. Mathematical Simulation of Chemical Plants – Flowsheeting

Even the simplified example above gives an idea why the Mathematics behind flowsheeting problems might be cumbersome. Beside simple units like mixers, splitters, etc. a flowsheet may contain heat exchangers, reactors and plate distillation columns. The mathematical model for, say, a plate distillation column with N plates and M components leads to a nonlinear equational system with $(2M + 1) \cdot N$ variables. Even for some few components some hundred variables have to be solved.

2.1 Flowsheeting

The flowsheet of a chemical plant can be seen as a directed graph where the knots are defined by the units and the edges are given by the (directed) streams. The graph is described by the incidence matrix $I(G)$:

$$I(G)_{ij} := \begin{cases} 1 & \text{if there is an edge from knot } i \text{ to knot } j; \\ 0 & \text{else} \end{cases}$$

Figure 2: Example of a Flowsheet

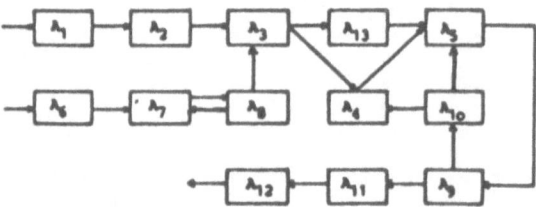

A chemical plant, resp. a flowsheet, can be determined sequentially if there exists a numbering of the units such that the resulting incidence matrix becomes right upper triangular. In general, however, most chemical plants involve feed back connections. Therefore, only certain subsets of the flowsheet can be determined sequentially. Technically this means that the incidence matrix is ordered in such a way that there results a right upper block triangular matrix. It is required that the diagonal blocks cannot be split up into smaller subblocks. This means that we have to search for irreducible subsystems (IRS) of the graph. To do this we need elementary cycles. An elementary cycle is a path where the starting knot coincides with the ending knot. All knots of the cycle occur only once except the first and the last one. For our example we have the following three elementary cycles: $\{A_7, A_8\}, \{A_5, A_9, A_{10}\}, \{A_5, A_9, A_{10}, A_4\}$.

Tiernan [11] gave an algorithm by which one can find all elementary circuits of a directed graph. It can be shown that an IRS is either a set consisting of one element only or it is a union of elementary cycles. Irreducible here means that the subsystem is not allowed to split up into jointly void union of elementary circuits. In our example we have the following IRS: $\{A_1\}, \{A_2\}, \{A_3\}, \{A_4\}, \{A_7, A_8\}, \{A_4, A_5, A_9, A_{10}\}, \{A_{11}\}, \{A_{12}\}$.

When including restrictions (e.g. a certain temperature is specified etc.) it might be necessary to unite certain subsystems in order to make the number of free parameters correspond to the number of restrictions. In most practical cases the resulting IRS s are still large because of feed back structures. There are two main strategies to cope with this situation.

a) *Equational Approach*

Each IRS is described and solved in total by one (i.g. large) nonlinear equational system. Numerical experiments clearly show that this approach is highly efficient. As a draw back one finds that available library programs for units (reactors, heat exchangers etc.) cannot be used directly but must be included in the system. An example of this strategy is given by the QUASILIN package proposed by Hutchinson et al [6].

b) *Sequential Modular Approach*

The IRS is split up into smaller parts by cutting certain connecting streams. Starting values are given to these torn streams such that the IRS can be determined sequentially. As an example we consider the IRS $\{A_4, A_5, A_7, A_{10}\}$ from Figure 2.

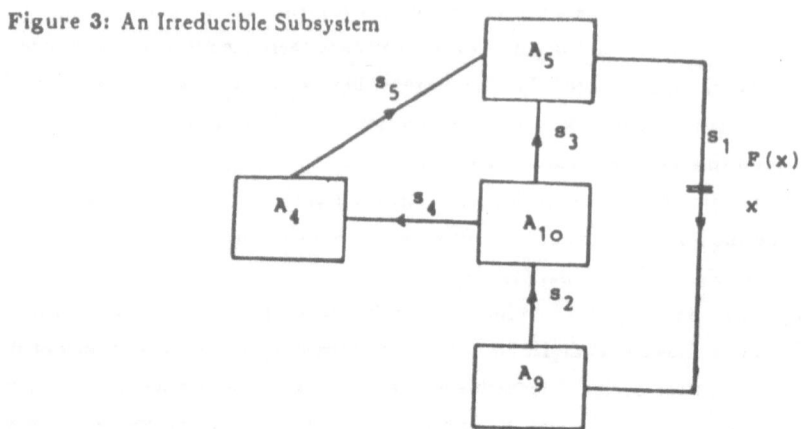

Figure 3: An Irreducible Subsystem

The stream s_1 (from A_5 to A_9) is split up. Starting with x as input. The IRS can then be computed sequentially: $x \rightarrow A_9(x) \rightarrow A_{10}^1(A_9(x)) \rightarrow A_4\left(A_{10}^1(A_9(x))\right)$, $A_5 = A_5(A_4, A_{10}^2)$. To preserve continuity we have to observe the fixed point equation $x = F(x)$. An IRS therefore is solved by two types of iterations:

  i) the outer cycle is determined by the cutting streams leading to a fixed point type nonlinear equation;

 ii) the inner cycle is defined by the units which themselves are usually solved iteratively.
The splitting can be done automatically, see e.g. Pho/Lapidus [9].

Of course it is up to one's own decision what is defined as a unit. In our program, e.g., both plate distillation columns with and without extensions like side strippers and/or pumparounds can be defined as units, for instance.

2.2 Newton's Method for the Solution of Irreducible Subsystems

In the following we want to propose a special solution technique based on the sequential modular approach. We want to solve the system

$$x = F_k(x) \qquad k = 1,\ldots,n \quad (n\text{: number of nontrivial IRSs})$$

In the case of restrictions there are additional parameters $P = (P_1, \ldots, P_r)$ to be determined. Examples for restrictions are given purity of the top product of a plate distillation column at unknown reflux or given output temperature of a mixer at an unknown heat input etc. One gets:

(1)
$$\begin{cases} z - F(z) & = 0 \\ R(z, P) & = 0 \end{cases}$$

We assume that we have at least directional derivatives both for the units and the restriction function R. We have to make sure that we have a 'legal' set of variables/equations, i.e. it is necessary that the 'input' variables together with the specified variables uniquely define the 'output' variables. Under our assumption we may use Newton's method. The main effort when using Newton's method is to be devoted to get the Jacobian. To do this analytically is out of reach: to determine a heat exchanger, for instance, a nonlinear twopoint boundary value problem has to be solved which can be done only numerically. Because of the numerical effort also divided differences are to costly. Under the assumption that the gradients of each single unit are available - this can be organized when computing the units - the Jacobian is easily determined by help of the chain rule. We demonstrate this for the following IRS:

Figure 4:

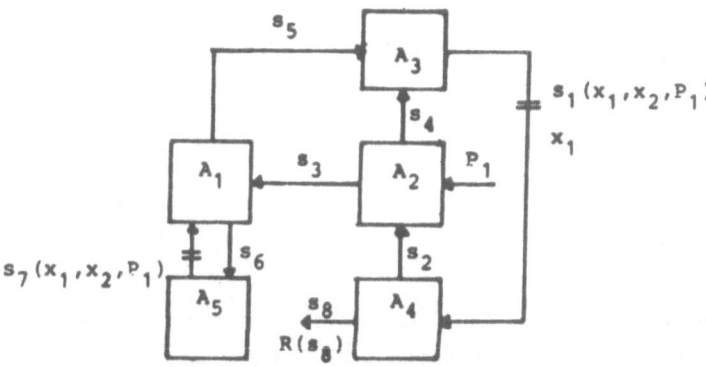

The system consists out of 5 units. In the unit A_2 we have the free parameter P_1 which corresponds to the restriction $R(s_8)$ in dependence on the output stream s_8. We have three elementary cycles: $\{A_1, A_5\}$, $\{A_1, A_3, A_4, A_2\}$, $\{A_3, A_4, A_2\}$. Using the tearing streams $\{s_1, s_7\}$ we get:

$$s_1(x_1, x_2, P_1) - x_1 = 0$$

(2)
$$s_7(x_1, x_2, P_1) - x_2 = 0$$

$$R(x_1, x_2, P_1) \quad = 0$$

To get the influence of x_i on $s_j : \frac{\partial s_j}{\partial x_i}$ we consider all paths starting at x_i and ending at the cutting stream s_j. The sensitivity factor $\frac{\partial s_j}{\partial x_i}$ is then given by summing up the products of all sensitivity factors of each path. To get $\frac{\partial s_1}{\partial x_1}$, for instance, we have the two paths: $x_1 \rightarrow s_2 \rightarrow s_4 \rightarrow s_1$ and $x_1 \rightarrow s_2 \rightarrow s_3 \rightarrow s_5 \rightarrow s_1$:

$$\frac{\partial s_1}{\partial x_1} = \frac{\partial s_1}{\partial s_4} \cdot \frac{\partial s_4}{\partial x_1} + \frac{\partial s_1}{\partial s_5} \cdot \frac{\partial s_5}{\partial x_1}$$

further

$$\frac{\partial s_4}{\partial x_1} = \frac{\partial s_4}{\partial s_2} \cdot \frac{\partial s_2}{\partial x_1} \quad \text{and} \quad \frac{\partial s_5}{\partial x_1} = \frac{\partial s_5}{\partial s_3} \cdot \frac{\partial s_3}{\partial s_2} \cdot \frac{\partial s_2}{\partial x_1}$$

In the case where restrictions are involved one has to use (2), for instance:

$$\frac{\partial R}{\partial x_1} = \frac{\partial R}{\partial s_8} \cdot \frac{\partial s_8}{\partial x_1} \quad \text{and} \quad \frac{\partial R}{\partial x_2} = 0$$

The numerical effort compared to the evaluation of the flowsheet (= function evaluation) is small. Newton's method then gives a correction $\Delta z_k = (\Delta x_k, \Delta P_k)$. We use a damped version:

$$z_{k+1} = z_k + \lambda_k \Delta z_k$$

In addition it might be that some variables must be projected to remain feasible, i.e. $T(^\circ K) \geq 0$ etc.

2.3 Gradients of units

Some simple units like mixers, splitters etc. allow for an analytical computation of gradients. For more complex units like plate distillation columns, flash drums, compressors, heat exchangers, reactors the chain rule again proves useful.

2.3.1 Numerical Determination of Gradients

For demonstration we discuss a flash unit

Figure 5: Flash Unit

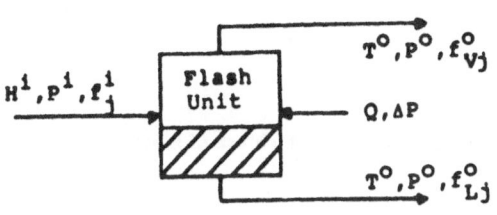

H: enthalpy

P: pressure

f: mass stream

T: Temperature

Q: heat stream

L: liquid

V: vapor

A given input stream (H^i, P^i, f_j^i) is split into vapour (V) and liquid (L) components. The thermodynamic equilibrium is controlled by the heat flux Q and the variation of pressure $\triangle P$. We get the following system:

$$P^0 = P^i - \triangle P$$

i : input

o : output

i) mass balances: $f_{Lj}^0 + f_{Vj}^0 - f_j^i = 0, j = 1, \ldots, N$

ii) energy balance: $H_L(f_{L1}^0, \ldots, f_{LN}^0, T^0, P^0) + H_V(f_{V1}, \ldots, f_{VN}, T^0, P^0) - H^i - Q = 0$

iii) equilibrium conditions: $k_j(f_{L1}^0, \ldots, f_{LN}^0, f_{V1}^0, \ldots, f_{VN}^0, T^0, P^0) \cdot \frac{f_{Lj}^0}{L} - \frac{f_{Vj}^0}{V} = 0, j = 1, \ldots, N$

where

$$L := \sum_{i=1}^{N} f_{Li}^0, \qquad V := \sum_{i=1}^{N} f_{Vi}^0$$

(i)—(iii) give a nonlinear system $F(x, p) = 0$ with the variables $x = (f_{L1}^0, \ldots, f_{VN}^0, T)$ and the parameter $p = (H^i, P^i, f_1^i, \ldots, f_N^i, Q, \triangle P)$. We use the identity $F(x(p), p) = 0$ where $x(p)$ solves our system in dependence on p. We have:

(*)
$$0 = \frac{dF(x(p), p)}{dp} = \frac{\partial F(x(p), p)}{\partial x} \cdot \frac{\partial x(p)}{\partial p} + \frac{\partial F}{\partial p}$$

82

As we use Newton's method to solve $F(x,p) = 0$ for x the Jacobian $\frac{\partial F}{\partial x}$ is already (even factorized) at our disposal. Therefore, one only has to compute $\frac{\partial F}{\partial p}$ and solve (*) for $\frac{\partial x(p)}{\partial p}$.

In the special case of a flash unit $\frac{\partial F}{\partial H^i}$, $\frac{\partial F}{\partial dQ}$ $(=\frac{\partial F}{\partial H^i})$, $\frac{\partial F}{\partial f_j}$ are determined analyticaly whereas $\frac{\partial F}{\partial p^i}$ $(=\frac{\partial F}{\partial p\sigma})$, and $\frac{\partial F}{\partial \Delta P}$ $(=\frac{-\partial F}{\partial p^i})$ may be approximated by forward differences.

2.4 Special Units

In the following we give some examples of units which are of a more complex nature.

2.4.1 Plate Distillation Columns

We consider configurations having one main column with a few pumparounds and a small number of side strippers which are linked with the main column only. The basic model for the main column (see Fig. 6) consists of three sets of equations.

Figure 6: Main Column

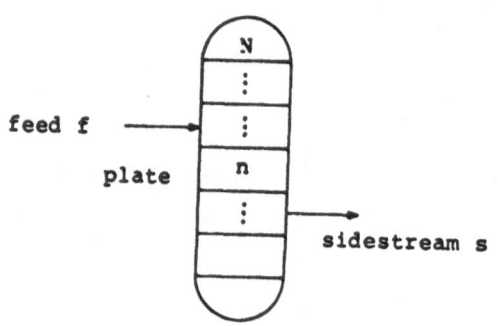

N: number of plates
M: number of components

feed f

plate n

sidestream s

Component mass balances (all plates, all components)

$$b_{nj} \equiv (1 + \frac{S_n^L}{L_n})l_{nj} + (1 + \frac{S_n^V}{V_n})v_{n,j} - l_{n+1,j} - v_{n-1,j} - f_{n,j} = 0$$

$$L_n = \sum_j l_{n,j} \qquad V_n = \sum_j v_{n,j}$$

Enthalpy balances: (all plates)

$$E_n \equiv (L_n + S_n^L)H_1^L + (V_1 + S_n^V)H_n^V - L_{n+1}H_{n+1}^L - V_{n-1}H_{n-1}^V - F_n H_n^L - \triangle E_n = 0$$

Vapor-liquid equilibrium relationship: (all plates, all components)

$$g_{n,j} \equiv k_{nj}\frac{l_{nj}}{L_n} - \frac{v_{nj}}{V_n} = 0$$

In addition H_n^L, H_n^V and k_{nj} are given functions of (known) pressure, temperature and the compositions of the liquid and vapor phase. Their particular form is based on the UNIQUAC model.

$$H_n^L = \sum_j (a_j + b_j T)\frac{l_{nj}}{L_n} + H_n^E(l_{n,1},\ldots,l_{n,m},T_n)$$

$$H_n^V = \sum_j (c_j + d_j T_n)\frac{v_{nj}}{V_n}$$

$$k_{nj} = \frac{par_j(T_n)}{P_n} \cdot \frac{\gamma_j(l_{n1},\ldots,l_{nM},T_n)}{\varphi_j(v_{n,1},\ldots,v_{n,M},T_n)}$$

It is easy to see that by ordering the unknowns resp. equations of plate n in the way $(l_{n,1},\ldots,l_{n,M},T_n,v_{n,1},\ldots v_{n,M})$ resp. $(b_{n1},\ldots,b_{nM},E_n,g_{n1},\ldots,g_{nM}) = 0$ and by arranging these blocks according to the natural ordering of the plates the Jacobian results to be N-by-N block tridiagonal with $(2M+1)$-by-$(2M+1)$ submatrices.

In the case of extension like pumparounds and/or side strippers there arise additional blocks outside the block tridiagonal part of the Jacobian. The numerical solution is described in [1]. An analysis of the numerical effort and numerical experiments show that the system should be linearized by Newton's method (no Quasi Newton technique!). The linearized system can be solved by L–U factorization with incomplete partial pivoting under explotion of the special structure of the Jacobian. As an alternative one may use iterative methods which proved superior especially in the case of extensions. Both these methods were favorably compared with existing solution techniques, see [1], [13].

2.4.2 Tubular Reactors

We discuss reactions of the following type:

$$\sum_{i=1}^{N} a_i^R A_i \rightarrow \sum_{i=1}^{N} a_i^P A_i$$

a_i^P, a_i^R: stoichiometric constants

A_i: reaction components $\qquad (N$ components$)$

We assume that the concentration of the reaction components velocity is proportional to the concentration of the reaction components (time law). Further we confine ourselves to simple reactions, i.e. we assume that the stoichiometric constants give directly the exponentials in the time law.

Figure 9: Tubular Reactor

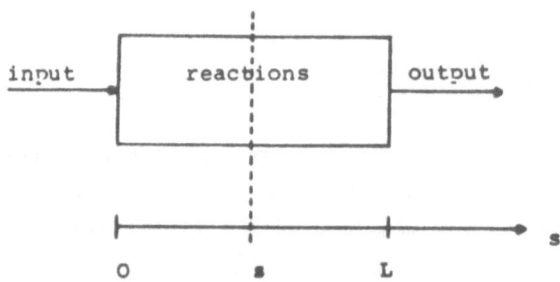

Though we only are interested in the stationary case we describe the process dynamically, i.e., in dependence on the position s in the reactor. We get:

$$\frac{d\xi_j}{ds} = A \cdot \left(k_j^R(T) \prod_{i=1}^{N} c_i^{a_{ij}^P} - k_j^P(T) \prod_{i=1}^{N} c_i^{a_{ij}^R} \right) \qquad j = 1, \dots, M$$

$$\frac{dT}{ds} = \frac{-\sum_{i=1}^{N} \frac{dH_g}{dm_{i_g}} \cdot \frac{dm_i}{ds} + \sum_{i=1}^{N} \frac{dH_s}{dm_{i_s}} \cdot \frac{dm_{i_s}}{ds} + \frac{dH_p}{dp} \cdot \frac{dp}{ds} + k_w \cdot U(T - T_w)}{\frac{dH_p}{dT}}$$

$$\xi_j(0) = 0 \qquad\qquad T(0) = T_0$$

From this we determine the mass streams m_i

ξ: reaction intensity $\qquad (M$ reactions$)$; $\quad A$: cross section of the tube

c: concentration; $\qquad\qquad\qquad\qquad U$: periphery of the tube

P, R: direction of reaction

k: velocity constant

H_g, H_s, H_F: Enthalpy of gases, resp. solids, resp. components

k_w, T_A: conductivity of the wall, outside temperature

This system of ODEs may be extremely stiff. For the numerical solution we use the semi implicit midpoint rule proposed by Deuflhard [3]. We confine ourselves to give an example, for details see [2].

$$CO + H_2O \leftrightarrow CO_2 + H_2$$

$$2COS \leftrightarrow CO_2 + CS_2$$

$$\frac{d\xi_1}{ds} = A \cdot (k_1^P c_{CO_2} c_{H_2} - k_1^R c_{CO} c_{H_2O})$$

$$\frac{d\xi_2}{ds} = A \cdot (k_2^P c_{CO_2} c_{S_2} - k_2^R c_{COS}^2)$$

[Kmol/h]	CO	CO_2	H_2O	COS	H_2	CS_2
Input	4.15	4.15	2.8	0.5	.	.
output	1.35	6.95	0.00	0.05	2.80	0.00

s	$0.5 \cdot 10^{-6}$	$0.4 \cdot 10^{-4}$	$0.4 \cdot 10^{-3}$	$0.3 \cdot 10^{-2}$	1.
ξ_1	0.01297	0.81981	2.47336	2.79914	2.79930
ξ_2	$0.5 \cdot 10^{-7}$	$0.4 \cdot 10^{-5}$	$0.4 \cdot 10^{-5}$	$0.3 \cdot 10^{-5}$	$0.3 \cdot 10^{-5}$
T	1316	1367	1465	1483	1483

The reaction is finished already at $s = 0.0005$ m. The number of integration steps (13) corresponds to the number of evaluations of the Jacobian. Computing time: 15.8 sec (BASF 7/78).

2.5   Example: Oxygen Plant

An oxygen plant splits up air into its components: nitrogen, oxygen and argon. The flowsheet is given in Figure 10.

Units: plate absorption column (AC) (3), mixer (M), cooler (C), heater (H) (17), splitter (S) (10), compressor (Co) (2); in addition dew point / bubble point computations are simulated by bubdew (BD) (5) (dotted).

86

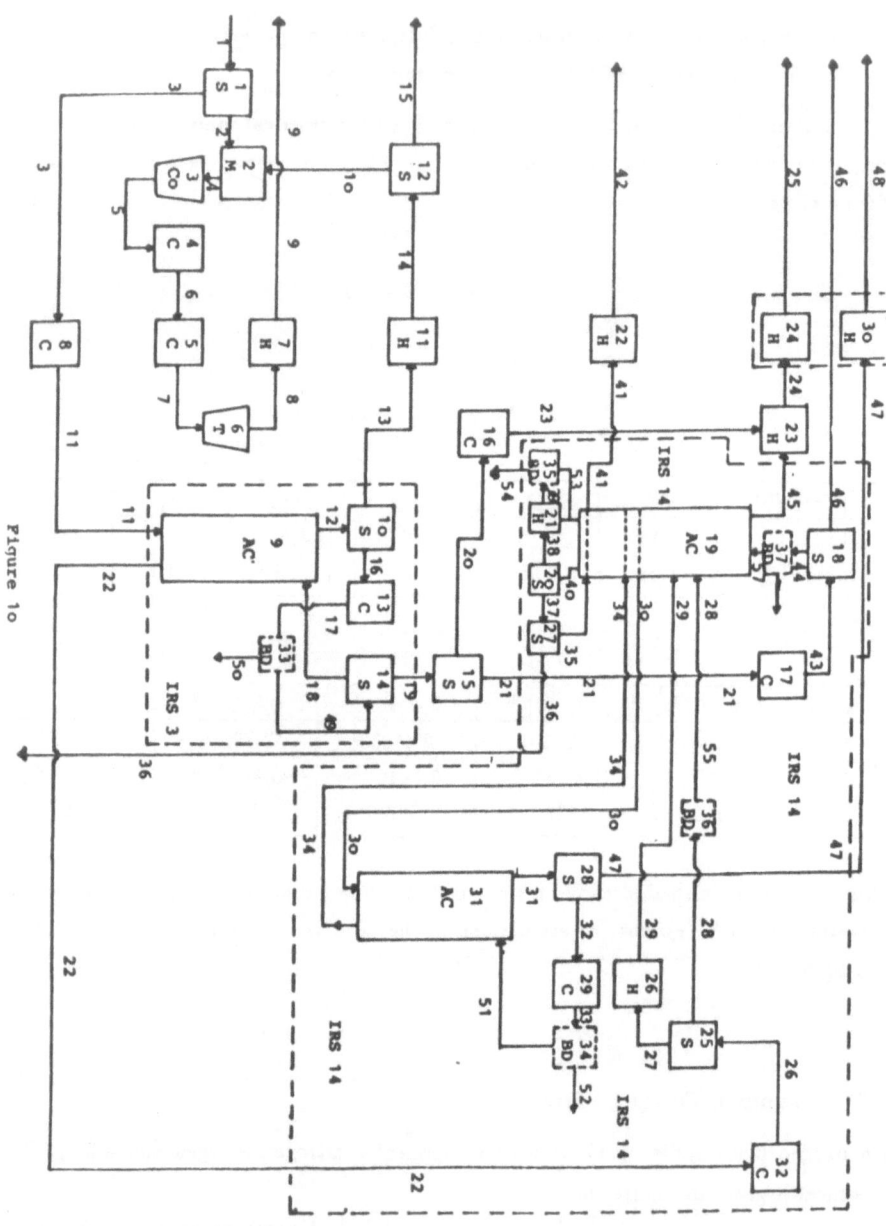

Figure 1o

Streams: feed (1), product streams (12), connecting streams (45)

Restrictions: 25

e.g. $T_{11} - 108 = 0$ (temperature)

$\qquad Q_5 + Q_7 + Q_8 + Q_{11} + Q_{22} + Q_{24} + Q_{30} = 0$ (heat)

$\qquad D_{29} - 0.9999 = 0$ (stream 29 is mostly vapor)

$\qquad F_{13} - 655.4 = 0$ (total flowrate of stream 13 is given)

$\qquad z_{12}^{16} + z_{12}^{13} = 1$ (stream 12 is split up into stream 16 and stream 13)

$\qquad \vdots$

Free Parameters: 25

a) Heating/Cooling duties at the units: 4,5,7,8,11,13,17,21,22,23,24,26,29,30,32

b) Distribution factors at the units: 10,14,20,25,28

The flowsheet can be split up into 17 IRSs:

IRS	Units	Tear Streams	Restrictions
1	1	–	–
2	8	–	1
3	33,14,9,10,13	17	17,18,21,22,9
		$\vdots$	
16	23	–	16
17	24,30	–	7,8

In the flowsheet the nontrivial IRSs 3 and 14 are given by dotted lines.

We confine ourselves to present some few results: IRS 14:

Tear Stream: 30 (Ar, N_2, O_2):

$\qquad$ starting point 12.81, 16.22, 257.80 [mol/h]

$\qquad$ result (6 It.) 14.85, 0.12, 290.83 [mol/h]

Product stream 42:

$\qquad$ Ar: 37.08, N_2: 0.09, O_2: 945.32 [mol/h]

$\qquad$ Enthalpy: $-0.18544.10^6$ [J/h], Temp.: 292 [°K];

Computing time: 1'14" (BASF 7/78)

3. Optimization

There are two aspects of optimization with respect to flowsheeting

i) One might be interested in the structure of a flowsheet (e.g. inserting a heat exchanger, determining the number of plates for a distillation column etc.) and

ii) one might like to know how to choose certain process parameters (e.g. reflux of a distillation column, input of heat etc.) such that certain goals (e.g. purity of products, costs for heating etc.) are achieved as good as possible. Here we only deal with the second aspect.

In Chapter 2 we described how a function evaluation for the steady state of a chemical plant is to be performed by Newton's method. The Jacobian is determined effectively by the chain rule. If we want to make full use of this solution technique we use Sequential Quadratic Optimization techniques, compare f.e. Fletcher [4]. $G^{(k)}$ is approximated by Quasi Newton technique making full use of the gradients.

$$f(x+s)^{(k)} \approx q^{(k)}(s) := f^{(k)} + s^T g^{(k)} + \frac{1}{2} s^T G^{(k)} s$$

$$q^{(k)}(s) = \text{Min!}$$

$$a_i^T s = b_i - a_i^T x^{(k)} \qquad i \in E \qquad \text{(equality constraints)}$$

$$a_i^T s \geq b_i - a_i^T x^{(k)} \qquad i \in J \qquad \text{(inequality constraints)}$$

In our case we only consider box-constraints for the parameters. To achieve convergence we use trust region technique.

We confine ourselves to give an example, for details see [10].

Example: Optimization of a methanol-ethanol-water plate distillation column.

Figure 11:

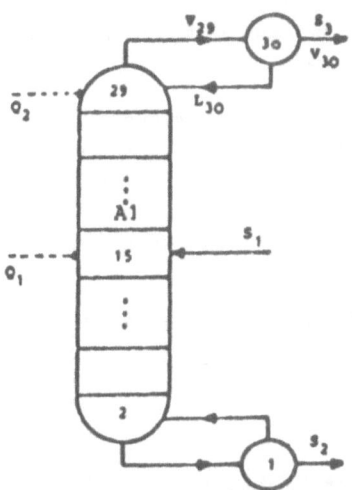

Objective: We want to minimize the percentage of water in the distillate:

$$f(x) \equiv x_{H_2O}(DEST, RFLX, Q_1, Q_2) = \text{ Min!}$$

distillate: DEST: $= V_{30}$
Reflux relation: RFLX: $= L_{30}/V_{30}$
heat influx: Q_1 (plate 15), Q_2 (plate 29)

Constraints:

$$1 \leq Dest < 20 \quad [mol/s] \qquad 0.2 \leq RFLX \leq 30$$

$$-1 \leq Q \leq 1 \quad [J/s] \qquad 0 \leq Q_2 \leq 300000 \quad [J/s]$$

Result:

$$x_0 = (19, 10, 0, 0), \quad f_0 = 7.16969$$

$$x_* = (8.4215, 30, 1.84 \cdot 10^{-4}, 0)^T, \quad f_* = 0.00211$$

Acknowledgement: Research was supported by the Austrian Science Foundation Fonds (projects P5729 and P6674P), the Austrian Ministry of Research (Sect.Research) and the VOEST ALPINE AG (Abt. FAT).

References:

[1] D.Auzinger, L.Peer, Hj.Wacker, W.Zulehner: Numerical Calculation of Separation Processes, in: Case Studies in Industrial Mathematics, ECMI II, (ed. Engl / Wacker / Zulehner), 1988, pp.131-154

[2] G.Bachleitner: Simulation of a Tubular Reactor, Diploma Thesis, Univ.Linz, Math. Dept., Febr. 1988, pp.1-105 (in cooperation with VOEST-ALPINE AG, Abt. FAT) (in German)

[3] P.Deuflhard: Recent Progress in Extrapolation Methods for ODEs., Univ.Heidelberg, Techn.Report Nr. 224 (1983)

[4] R.Fletcher: Practical Methods of Optimization, Vol.2: Constrained Optimization, John Wiley & Sons, New York, pp. 1-224

[5] A.Fredenslund, J.Gmehling, P.Rasmussen: Vapor-Liquid Equilibria using UNIFAC, Elsevier, Amsterdam, 1977

[6] H.P.Hutchinson, D.J.Jackson, W.Morton: The Development of an Equation Oriented Flowsheet Simulation and Optimization Package I, II, Comp. Chem.Eng. Vol 10, Nr. 1, (1986) pp 19-29, 31-47

[7] M.Leitner: Mathematical Simulation of a Matrix Heat Exchanger, Diploma Thesis, Univ.Linz, Math.Dept., August 1988 (in German)

[8] M.D.Lu, R.L.Motard: Computer-Aided Total Flowsheet Synthesis. Comp. Chem. Eng., Vol.9, No. 5, (1985), pp.431-445

[9] T.K.Pho, L.Lapidus: An Optimum Tearing Algorithm for Recycle Systems, AICHE Journal, Vol. 19, Nr.6 (1973), pp. 1170-1180

[10] J.Synka: Optimization of Chemical Plants with Box Constraints (Plate Distillation Columns), Diploma Thesis, Univ.Linz, Math.Dept., June 1988, pp.1-15o (in German)

[11] J.C.Tiernan: An efficient search algorithm to find elementary ?? of a Graph, Comm. of the ACM, Vol. 13, Nr. 12, 1970, pp. 722-726

[12] L.Timár, F.Simon, Z.Csermely, J.Siklós, S.Bácskai, J.Édes, Comp. Chem. Eng., Vol. 8, No. 3/4, (1984), pp.185-194

[13] Hj.Wacker: Mathematical Research and Industrial Consulting in Linz: Power Plants - Chemical Columns - Heating of Slabs, in: 1st Europ.Symp.on Math. in Industry, ECMI I (eds. Hazewinkel/Mattheij/ van Groessen), 1988, pp.163-185

MINI-SYMPOSIUM 1
(organiser: P. W. Gaffney)

Bergen Scientific Centre

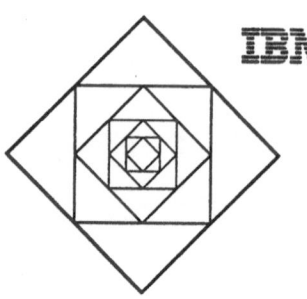

Numerically Intensive Computing
A Mini–symposium of

The European Conference on Mathematics in Industry (ECMI)
Glasgow, 28–31 August 1988
arranged by

P. W. Gaffney
Bergen Scientific Centre

Abstract

This mini-symposium focuses on four industrial applications that require both sophisticated *numerical* techniques and *intensive* computing power for their solution. Hence, the title of the symposium *numerically intensive computing*. The presentations are:

Elastic Modeling on the IBM 3090 Vector Multiprocessor	**Solving a Model Interface Problem for the Laplace Operator by Boundary Collocation and Applications**
Aladin H. Kamel	*Lothar Reichel*
Electroplating Simulation	**Formulation of a Sea Model with Continuous Density Stratification in the Vertical**
L. J. Gray et al.	*G. K. Furnes*

The four application areas covered by these presentations are representative of the industrial research conducted at the IBM Bergen Scientific Centre in Norway. They reflect three of the *strategic* areas of research highlighted by the Norwegian Government namely, *Information Technology*, *Offshore Technology*, and *Biotechnology and Fish Farming*.

Keywords

industrial applications, numerically intensive computing, supercomputing, vector and parallel processing

J. Manley et al. (eds.), Proceedings of the Third European Conference on Mathematics in Industry, 92.
© 1990 *Kluwer Academic Publishers and B. G. Teubner Stuttgart.*

Bergen Scientific Centre

FORMULATION OF A SEA MODEL WITH CONTINUOUS
DENSITY STRATIFICATION IN THE VERTICAL

G. K. Furnes

Bergen Scientific Centre, Allégaten 36, 5007 Bergen, Norway

Abstract

The linear hydrodynamical equations are formulated for a two-layered model designed to represent the motion of a density stratified sea, driven by atmospheric forces and tides. The model can deal with any distribution of vertical basic density profile. An eddy viscosity formulation has been adopted: the eddy viscosity is made proportional to the reciprocal of the static stability. By expanding the dependent variables in terms of continuous functions over the stratified part of the depth, a continuous representation of the current field and vertical displacement is obtained. In the deeper constant density part of the water column, only a two dimensional depth mean structure is recovered from the model. Examples of applications of the model will be given in the presentation.

The solution of the equations requires a numerical approximation procedure. The calculations are performed to second order accuracy in space and time on a finite difference, staggered grid. In order to resolve internal motions caused by the presence of density gradients (baroclinic modes), the horizontal grid spacing has to be less than the internal Rossby deformation radius, typically 1–5 km.

Practical calculations over a larger sea region call for a nesting technique in which the stratified model covers a limited (internal) area where density effects are considered to be important. The open boundary conditions may be obtained from a constant density (barotropic, external) model which can be integrated on a much coarser horizontal grid. A flux relaxation method has been tested that allows the internal solutions to be gradually adjusted towards the external solutions over a transition zone. Using this method it is possible to obtain a one way transfer of information, namely from the external to the internal model area.

Experiments that have been performed have shown encouraging results for the transferring of a barotropic external (e.g. tidal) solution into the baroclinic model area.

J. Manley et al. (eds.), Proceedings of the Third European Conference on Mathematics in Industry, 93.
© 1990 *Kluwer Academic Publishers and B. G. Teubner Stuttgart.*

ELECTROPLATING SIMULATION

L. J. Gray, Bergen

G. E. Giles, J. S. Bullock, P. W. McKenzie, Oak Ridge

Summary: A simulation of a three dimensional electroplating process which includes a model of the heat flow generated during the plating is presented. The steady state temperature distribution in the cell is coupled to the electric field through the nonlinear polarization (overvoltage) boundary conditions and the temperature dependence of the electrolyte conductivity. As the conductivity directly influences the deposition rate, variations due to temperature can have an impact on the shape of the plated part. The thermal and electrostatic problems are simultaneously solved using the boundary element method, and a simple iterative scheme is employed to satisfy the boundary conditions. A newly developed boundary element procedure for crack geometries is used to cope with the thin shielding present in the cell. Utilizing this technique, a computationally expensive and cumbersome multidomain decomposition of the region is avoided.

1 Introduction

Electrochemical deposition and machining have many advantages in the manufacture of precision parts, but they are generally employed only in situations where other techniques fail (e.g., when the part geometry or the plating material is particularly difficult). The basic problem with electrochemical methods is that, in most situations, it is exceedingly difficult to determine either a plating cell or a machining tool configuration which will produce an acceptable part geometry and surface morphology. Electrochemical methods have been very much an 'art', relying primarily on past experience to design the production equipment by a cumbersome trial and error procedure. Not only is this scheme very slow (involving setup, experimental run, and inspection time), but also very costly in terms of labor and materials. These factors necessarily limit the number of trial iterations; conformation with the prescribed part specifications is therefore difficult to achieve, and the part must be subjected to additional processing steps. The high cost and uncertain success of this design process has limited the application and development of electrochemical methods as industrial manufacturing processes.

Numerical simulation of an electrochemical process offers the possibility of an efficient design tool: compared to performing the corresponding laboratory work, a computer analysis of a proposed configuration is fast and inexpensive. Moreover, simulation provides an experimental device that can aid in understanding the intricacies of the process itself. Numerical experiments can provide information about plating under new circumstances (e.g., high speed plating or new materials) where past experience is lacking, and can indicate which of the many parameters have an important impact

J. Manley et al. (eds.), Proceedings of the Third European Conference on Mathematics in Industry, 95–105.
© 1990 *Kluwer Academic Publishers and B. G. Teubner Stuttgart.*

on the final result. An improved understanding of the process would almost certainly lead to wider applications in the industrial environment. In this paper, a mathematical algorithm modeling a three dimensional electroplating process will be described, and computational results for realistic plating configurations will be presented. This model takes into account the nonlinear boundary conditions required by the electrochemistry at the electrode surfaces, but at present neglects the effects of electrolyte flow. In addition, an incomplete model of the heat generated in the cell is incorporated. The temperature in the cell is potentially an important variable because the conductivity of the electrolyte, which factors directly into the rate of material deposition, is temperature dependent. The heat produced by the electrochemical reactions at the electrode surfaces is included in the calculations presented below, but resistance heating in the electrolyte due to current flow is, at present, ignored.

The Laplace equation controls both the (electric) potential and temperature distributions in the cell, and the boundary element method is used to obtain the solutions. As has been previously discussed [2], this technique is particularly advantageous for plating applications. The additional complication considered herein, that of determining the temperature distribution, is also easily and economically incorporated into the algorithm. Moreover, it should also be noted that, as there are far fewer unknowns with boundary elements (as compared to finite elements or finite differences), it is easier to converge to the solution of the nonlinear boundary conditions.

Another significant issue in the application of boundary elements to plating simulation concerns insulated shielding. Shielding is commonly employed as a convenient device for controlling the plating rate: the cell designer can adjust the geometry and location of a shield so as to achieve a desired current density distribution. The shield is generally a very thin piece of plastic, and thus appears in the geometry as a slit or crack. The 'standard' boundary element approach to cracks is to divide the domain into several subdomains without cracks by using fictitious internal boundaries [3]. However, this method is highly inefficient for three-dimensional problems, especially for a shield that is 'porous' (i.e., has many holes). This electrochemistry application has motivated the development of a new boundary element technique for treating crack geometries [5-8]. This method does not utilize extra internal surfaces, and is therefore highly efficient, and convenient for use in a production code. With this new method, the treatment of shields is considerably simplified, and complicated shield geometries, including porous ones, are easily handled. A very brief discussion of this approach to cracks will be given;

a shield is included in the plating calculations reported in this paper.

2   Equations

As the time scale of the plating process is very long, it is reasonable to consider the steady state temperature profile in the cell. As a first step, resistance heating in the electrolyte will be ignored, and thus the potential $V(X)$ and temperature $T(X)$ distributions, $X = (x, y, z)$, are both governed by the Laplace equation

$$\nabla^2 V = 0,$$

(2.1)

$$\nabla^2 T = 0.$$

For boundary elements, this differential expression is replaced by an equivalent integral equation[7] (written out for V):

(2.2)
$$\psi(X_0)V(X_0) + \int_{\partial R} V(X)\nabla G(X, X_0) \bullet \vec{n} \, dS = \int_{\partial R} G(X, X_0)\nabla V(X) \bullet \vec{n} \, dS.$$

Here ∂R is the boundary of the region, X is the surface integration variable, $X_0 \in \partial R$ and $G(X, X_0)$ is a Green's function or fundamental solution of the Laplace equation. The point source potential

(2.3)
$$G(X, X_0) = \frac{1}{4\pi\|X - X_0\|},$$

$\|X - X_0\|$ the distance between X and X_0, is the standard choice for G, but is not the only one possible.

The insulated shielding in the cell is sufficiently thin that it can be treated mathematically as a crack. As described in detail elsewhere [5], incorporating a crack directly into Eq. (2.2) leads to a singular system of equations. To remedy this situation without resorting to subdividing the domain, Eq. (2.2) is augmented by an additional relationship for nodes X_0 on the crack surface. Differentiating Eq. (2.2) with respect to X_0 in the direction $\vec{N}$, the normal to the crack at X_0, yields

$$\psi(X_0)\nabla_{X_0} V(X_0) \bullet \vec{N} + \int_{\partial R} V(X)\nabla_{X_0} \{\nabla_X G(X, X_0) \bullet \vec{n}\} \bullet \vec{N} \, dS$$

(2.4)
$$= \int_{\partial R} \left[\nabla_{X_0} G(X, X_0) \bullet \vec{N}\right]\left[\nabla_X V(X) \bullet \vec{n}\right] dS.$$

This second integral equation is somewhat of a problem due to the presence of two derivatives of the Green's function on the left hand side. This term is more singular than usual boundary element integrals, and requires special treatment. The details can be found elsewhere [5,9].

By decomposing the boundary surface into small elements and approximating the potential and flux ($\nabla V \cdot \vec{n}$) over each element, Eqs. (2.2) and (2.4) can be reduced to a matrix equation relating the values of the potential and flux at a finite number of boundary points $\{X_k\}$,

(2.5a)
$$H\Phi = Z\Phi^n.$$

The $M \times M$ matrices H and Z are determined by the geometry and choice of Green's function, and Φ and Φ^n are the column matrices of potential and flux values,

(2.5b)
$$\Phi = \begin{pmatrix} \phi_1 \\ \vdots \\ \phi_K \end{pmatrix} \quad \Phi^n = \begin{pmatrix} \phi_1^n \\ \vdots \\ \phi_K^n \end{pmatrix},$$

where $\phi_k = V(X_k)$ and $\phi_k^n = \frac{\partial V}{\partial n}(X_k)$. For each X_k, the boundary conditions supply either the value of ϕ_k or ϕ_k^n; the remaining unknown values on the boundary can then be obtained from Eq. (2.5a).

The most time consuming part of the calculation (for a moderately sized problem) is the construction of the matrices H and Z. However, as H and Z are independent of the boundary conditions, these matrices are the same for both problems posed in Eq. (2.1), and the additional work involved in obtaining T is therefore rather minimal.

Due to various physical processes occurring at the electrode (cathode and anode) surfaces, the boundary conditions for V are quite complicated. These reactions, lumped together under the term *overvoltage*, reduce the amount of the applied potential difference which can go into actual plating. If V_C and V_A denote the potentials on the cathode and anode, then

(2.6)
$$V_C = V_C^0 - \eta_C$$
$$V_A = V_A^0 - \eta_A,$$

where V^0 and η are the respective applied potentials and overvoltage. The overvoltage is a function of, among other things, the local current density I,

(2.7)
$$I = \kappa \frac{\partial V}{\partial n},$$

where κ the conductivity of the electrolyte. Thus, although the applied potentials are generally constant, the overvoltage is spatially dependent. Moreover, as η is a function of the flux, the problem is nonlinear. For the calculations presented below, the overvoltage is composed of two terms, activation and concentration,

(2.8)
$$\eta = \eta_{act} + \eta_{con}.$$

The concentration overvoltage can be evaluated directly:

(2.9)
$$\eta_{con} = \frac{RT}{n_v F} \log \left(1 - \frac{|I|}{I_1} \right)$$

In this expression, R and F are the gas constant and Faraday's constant, n_v is the valence of the ion being plated, and I_1 is the limiting current. The activation overvoltage however is determined through the Butler-Volmer equation

(2.10)
$$I = I_0 \left[e^{\frac{\alpha_A F}{RT} \eta_{act}} - e^{-\frac{\alpha_C F}{RT} \eta_{act}} \right],$$

and must be solved for numerically.

The temperature and potential distributions are linked through several mechanisms. First, the temperature appears explicitly in the expressions (2.9) and (2.10) for the overvoltage terms. It also contributes indirectly via the current density, as the conductivity in Eq. (2.6) given by

(2.11)
$$\kappa = \kappa(T) = e^{[\beta_0 + \beta_1/T]}$$

where β_0 and β_1 are constants. Finally, the heat generated at the electrode surfaces by the plating is related to the activation overvoltage, yielding a flux boundary condition of the form

(2.12)
$$\frac{\partial T}{\partial n} = \gamma \eta_{act} I$$

(the constant γ depending upon the particular electrode/electrolyte system).

The algorithm for solving this collection of equations first determines the potential field assuming zero overvoltage and a constant initial temperature T^0. This solution is then used to compute the overvoltage, which in turn supplies the boundary conditions for the heat flow problem and the new boundary conditions for V. A new potential field is then obtained, and updated overvoltages, now using the calculated temperature distribution, are calculated. This cycle is then repeated until convergence. From past experience, it is known that, without including the temperature variations, this iterative procedure converges. It is of course hoped that these new modifications will not alter this result.

3   Calculation

The electroplating cell to be modeled consists of a square tank, four cylindrical anodes running from the top to the bottom of the tank in each corner, the cathode, and the shield. The cathode is partially surrounded by the shield and suspended from

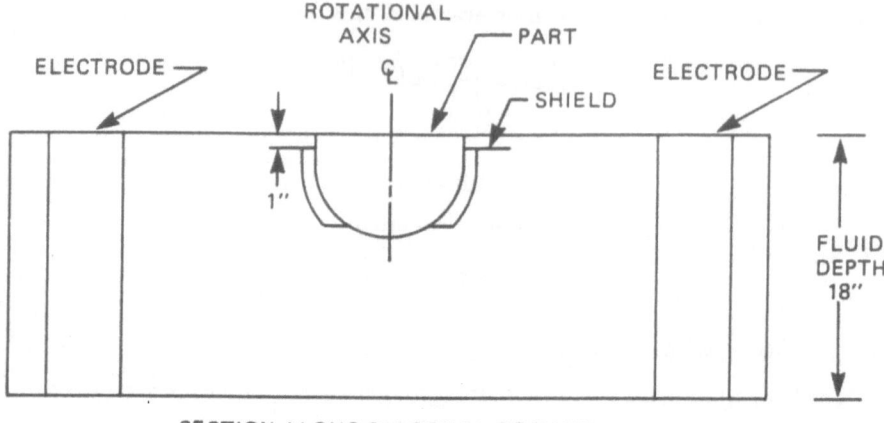

ROTATIONAL
AXIS
— PART
ELECTRODE —
₵
ELECTRODE —
— SHIELD
1"
FLUID
DEPTH
18"

SECTION ALONG DIAGONAL OF TANK

Figure 1: A view of the plating cell configuration along the diagonal of the square tank. The cathode and shield are at the center, and the anodes are at the corners of the tank.

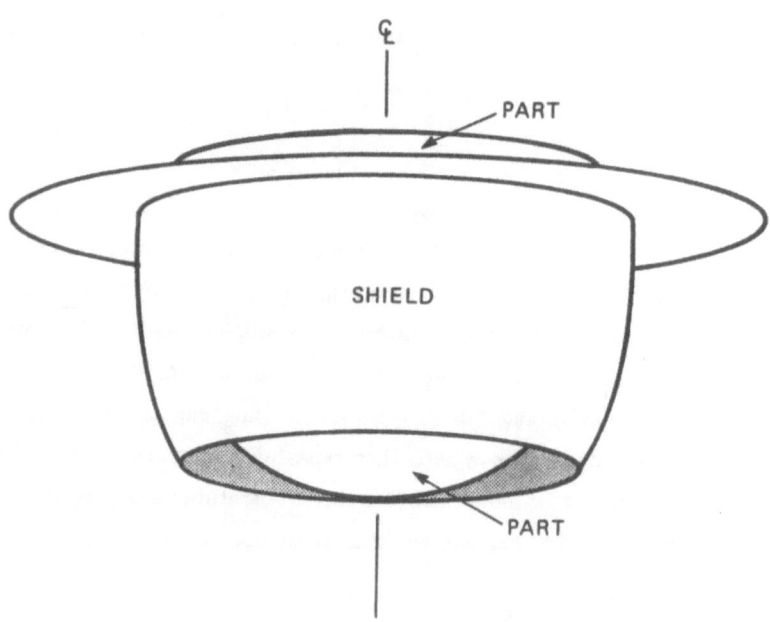

₵
— PART
SHIELD
PART

Figure 2: The cathode/shield configuration.

the top center of the cell. A cross-sectional view along a diagonal of the tank is shown in figure 1. Figure 2 shows a view of the cathode/shield configuration. The shield is suspended in the tank by means of the annular flange at the top. Further details about the electroplating cell can be found in the paper of Bullock et. al [2].

A discretization of a one-eighth section of the cell into triangular elements is shown in figure 3. Note that the symmetry planes $y = 0$ and $x = y$ are absent from the model, as is the top of the tank (electrolyte surface), $z = 0$. A Green's function satisfying the zero flux boundary conditions on these planes will be used; consequently, these planes are not required in the calculation. For the potential problem, the boundary conditions on the remaining tank wall and floor is zero flux. For the heat flow problem, these surfaces are constant temperature, namely T^0. A list of all the parameters used in the calculation is given in Table 1.

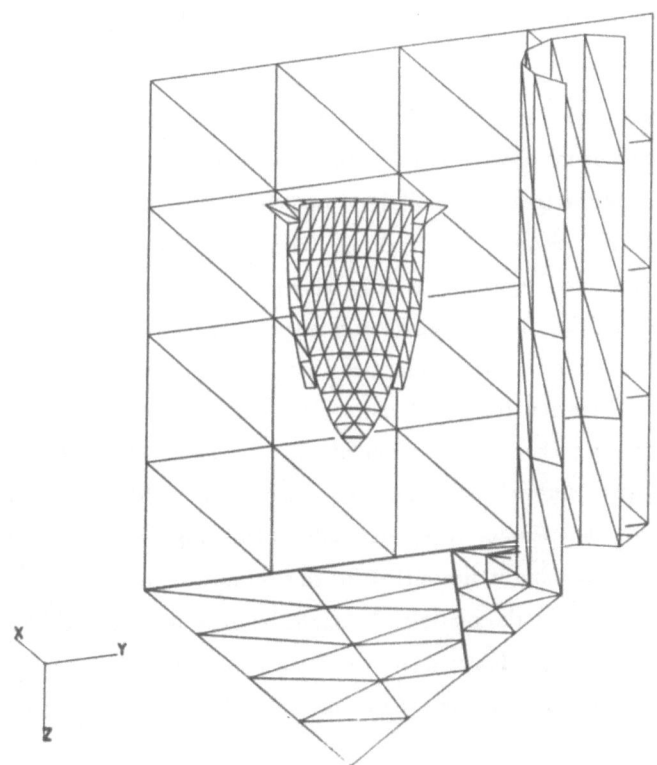

Figure 3: Discretization of a one-eighth section of the plating cell; in the center is the cathode, and off to the right is the cylindrical anode. The shield is behind the cathode.

Tab. 1   **Electroplating Parameters**

Name	Symbol	Value
Faraday's Constant	F	96487.0
Gas Constant	R	8.314
Ion Valence	n_v	2.0
Initial Temperature	T^0	306.0
Applied Voltage	V_C^0	-1.54
Applied Voltage	V_A^0	0.0
Transfer Coefficient	α_C	0.5
Transfer Coefficient	α_A	0.6
Exchange Current	I_0	0.0168
Heat Flux Constant	γ	0.5
Conductivity Constant	β_0	2.86832
Conductivity Constant	β_1	-1225.86

In order to get a good representation of the part (cathode) surface and an accurate calculation of the current density on the cathode, a highly refined mesh was employed. The current density changes quite rapidly as one emerges from the 'shadowing' of the shield, and thus small elements are required. There were 605 elements in the model, and the size of the matrix system was 461. This relatively large model was solved on the IBM 3090/200. Using the parallel fortran language, it was quite straightforward to adapt the serial boundary element code to run efficiently in parallel.

The nonlinear iteration converged quite nicely. Figure 4 displays the calculated current density on the cathode as a function of z, at the 'center' of the 45° section shown in figure 3. The three curves represent the current obtained without overvoltage, with overvoltage but constant temperature, and finally with temperature variation included. These results are typical and indicate that for this setup, the temperature distribution does not strongly influence the plating. As might be expected, the shield reduces the current flow, and thus η_{act} is fairly small over most of the cathode. Consequently, the heat flow into the cell is minimal and the temperature distribution is almost constant.

4   Discussion

With very little extra effort, it is possible to include heat generation within electroplating simulations. The task of computing $T(X)$ is easily merged with the boundary element procedure for the potential problem, and no difficulties were encountered in converging to the solution of the nonlinear boundary conditions. For this particular plating configuration, it appears that the temperature plays a relatively minor role. It must be emphasized however, that only one part of the heat generation has been treated so far. Including the resistance heating in electrolyte requires adding a source term to the right hand side of the Eq. (2.1) for $T(X)$. However, provided the source term can be expressed in terms of the boundary, this Poisson equation can be treated in much the same way as the Laplace equation and both the temperature and potential can still be obtained in an efficient manner by boundary elements.

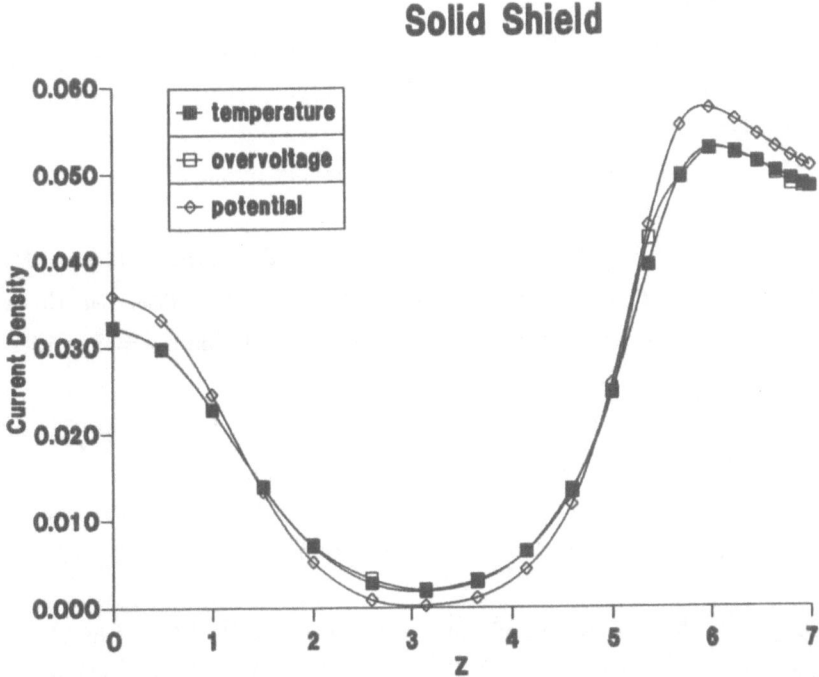

Figure 4: Current density values on the cathode employing constant applied potential (*potential*), overvoltage with constant temperature (*overvoltage*), and overvoltage with temperature distribution (*temperature*).

Accurate modeling of the plating process, especially in the presence of one or more shields, will require an extensive amount of computing. The boundary element method is ideally suited to parallel computation, and this should be exploited to obtain reasonable run times. The calculations reported here using a shared memory machine, together with previous work on a local memory hypercube [4], indicate that boundary elements is likely to function well in most architectures.

The calculation described herein is one time step of a complete simulation. The amount of material plated onto the cathode is determined by the current density I. Using this solution, a new cathode geometry is obtained and the entire process repeated. Unfortunately, a plating run using the configuration shown in figures 1-3 has not yet been completed. After the experiment is performed and the part inspected, the results will be compared to the simulation predictions.

5 References

[1]. Brebbia, C.A.; Telles, J.C.F.; Wrobel, L.C.: *Boundary Element Techniques*. Berlin: Springer Verlag 1984.

[2]. Bullock, J.S.; Giles, G.E.; Gray, L.J.: Simulation of an electrochemical plating process. Topics in Boundary Element Research Vol. 5, (Ed. Brebbia C.A.), Berlin: Springer Verlag (in press)

[3]. Cruse, T.A.; Wilson, R.B.: Advanced applications of boundary-integral equation methods. *Nucl. Engr. Design* 46 (1978) 223-234.

[4]. Drake, J.B.; Gray, L.J.: Parallel Implementation of the Boundary Element Method. *Proceedings of the 2nd International Conference on Parallel and Vector Computing.* (in press)

[5] Gray, L.J.: Boundary element method for regions with thin internal cavities. *Engineering Analysis* (in press).

[6]. Gray, L.J.; Giles, G.E.: Boundary element method for regions with thin internal cavities II. *Engineering Analysis* (submitted)

[7]. Gray, L.J.; Askew, A.L.; Giles, G.E.: Contact heat transfer by the boundary element method. *Proceedings of the European Boundary Element Conference*, Brussels 1988 (in press).

[8] Gray, L.J.; Giles, G.E.: Application of the Thin Cavity Method in Electroplating. *Proceedings of the Boundary Element 10 Conference* Southampton: Computational Mechanics 1988.

[9]. Gray, L.J.: Evaluation of Hypersingular Integrals in the Boundary Element Method. *Computers and Mathematics with Applications* (submitted).

6 Acknowledgments

The work of the first author was sponsored by the IBM Bergen Scientific Centre. Additional support was received from the Y-12 Development Division, Martin Marietta Energy Systems, Inc., under contract DE-AC05-84OR21400 between the U.S. Department of Energy and Martin Marietta Energy Systems, Inc. The authors would like to thank Aladin Kamel and Johnny Petersen of BSC for their assistance in getting the program running on the IBM 3090/200.

L. J. Gray
IBM Bergen Scientific Centre
Allégaten 36
5007 Bergen Norway

G. E. Giles
Computing and Telecommunications Division
Martin Marietta Energy Systems
Oak Ridge, TN 38731

J. S. Bullock and P. W. McKenzie
Y-12 Development Division
Martin Marietta Energy Systems
Oak Ridge, TN 38731

ELASTIC MODELING ON THE IBM 3090 VECTOR MULTIPROCESSOR

Aladin H. Kamel, *IBM Bergen Scientific Centre, Allegaten 36, N-5007, Bergen, Norway*

Abstract. Computerized seismic prospecting is an echo-ranging technique usually targeted at mapping accurately oil and gas reservoirs. In seismic surveys an impulsive source, often an explosive charge, located at the earth's surface, generates elastic waves which propagate in the subsurface and are scattered by the earth's geological discontinuities back to the surface where an array of receivers register the reflected signals. The subsurface imaging of the geological structures is obtained by means of complex mathematical inversion and modeling techniques amongst which the process of forward elastic modeling of the wave field plays a major role. The aim of this paper is to present formulations of pseudo-spectral as well as finite differences elastic models. Vector multiprocessor implementations of both formulations exploiting the IBM 3090 architecture together with its associated software are then described. Performance results are also reported. The results summarized in this paper show the suitability of this class of geophysical computations to efficient implementation on vector multiprocessors: combined vector/parallel speedups around 20-30 are in fact observed.

Key words: forward modeling vector multiprocessing pseudo-spectral finite differences

J. Manley et al. (eds.), Proceedings of the Third European Conference on Mathematics in Industry, 107–117.
© 1990 *Kluwer Academic Publishers and B. G. Teubner Stuttgart.*

1. Introduction

Seismic prospecting for hydrocarbon detection aims at determining the geologic structure of the earth from indirect measurements obtained at the earth's surface. In seismic prospecting elastic wave fields (acoustic in first approximation) are generated in a controlled fashion at the surface, penetrate the earth, and are backscattered by the earth's inhomogeneities to an array of receivers where they are recorded. The seismic interpreters correlate what is seen in the seismic data to the earth's structure.

A complementary tool in relating the seismic data to the earth's inhomogeneities is provided by seismic numerical models which assume a subsurface structure and compute the seismic data which would be collected in a field survey, by solving the direct problem of exploration geophysics.

The aim of this paper is to present formulations of pseudo-spectral as well as finite differences elastic models together with vector multiprocessor implementations of both formulations exploiting the IBM 3090 architecture and its associated software. Performance results are also reported.

The paper is organized as follows: section 2. deals with formulation of the modeling problem. Section 3. describes vector multiprocessing implementations and performance analysis. Concluding remarks are given in Section 4.

2. Formulation

2.1. Describing Equations

The equations describing elastic wave motion are given by [1, 11] :

$$\ddot{\sigma} = A\sigma + Bf \tag{1}$$

where

$$\sigma = \begin{vmatrix} \sigma_{xx} \\ \sigma_{yy} \\ \sigma_{zz} \\ \sigma_{xy} \\ \sigma_{xz} \\ \sigma_{yz} \end{vmatrix} \quad , A = LD^T\frac{1}{\rho}D, \quad B = LD^T, \tag{2}$$

$$L = \begin{vmatrix} \lambda + 2\mu & \lambda & \lambda & 0 & 0 & 0 \\ \lambda & \lambda + 2\mu & \lambda & 0 & 0 & 0 \\ \lambda & \lambda & \lambda + 2\mu & 0 & 0 & 0 \\ 0 & 0 & 0 & \mu & 0 & 0 \\ 0 & 0 & 0 & 0 & \mu & 0 \\ 0 & 0 & 0 & 0 & 0 & \mu \end{vmatrix} \quad , \tag{3}$$

$$D = \begin{vmatrix} \partial_x & 0 & 0 & \partial_y & \partial_z & 0 \\ 0 & \partial_y & 0 & \partial_x & 0 & \partial_z \\ 0 & 0 & \partial_z & 0 & \partial_x & \partial_y \end{vmatrix} \quad , \quad \text{and } f = \begin{vmatrix} f_x \\ f_y \\ f_z \end{vmatrix} \tag{4}$$

where x, y and z are Cartesian coordinates, respectively horizontal (x, y) and vertical (z), σ_{xx}, σ_{yy}, σ_{zz}, σ_{xy}, σ_{xz} and σ_{yz} are the six stress components, f_x, f_y and f_z represent the body forces, $\rho(x, y, z)$ represents the density, and $\lambda = \lambda(x, y, z)$ and $\mu = \mu(x, y, z)$ represent Lame's elastic parameters. Here a dot above a variable represents a time derivative and T denotes matrix transposition. The stresses serve as the main unknowns whereas the rock parameters ρ, λ, and μ, as well as the body forces $f_x(x, y, z, t)$, $f_y(x, y, z, t)$ and $f_z(x, y, z, t)$ are known quantities.

2.2. Numerical Scheme

2.2.1. Time marching

In order to solve numerically Equation (1), a discretization in time is performed, using a leapfrog technique:

$$\sigma(t + \Delta t) - 2\sigma(t) + \sigma(t - \Delta t) = = (\Delta t)^2 [A\,\sigma(t) + B\,f(t)] \tag{5}$$

2.2.2. Spatial differentiation

- Pseudo-spectral approach [9, 11]

The spatial derivatives are accurately computed in the Fourier (spectral) domain, substituting the differential operators $\partial / \partial x$, $\partial / \partial y$ and $\partial / \partial z$ with D_x, D_y and D_z where

$$D_\gamma \equiv F_\gamma^{-1}\, i\, K_\gamma\, F_\gamma \text{ (with } \gamma = x, y, z) \tag{6}$$

F_γ and F_γ^{-1} represent the direct and inverse Fourier transform operators; K_γ is the operator multiplying each Fourier coefficient by its wave number, and $i \equiv \sqrt{-1}$. The stability criterion of this scheme is [11]:

$$\Delta t \leq \frac{1}{\sqrt{\dim}\,\pi} \frac{h_{\min}}{V_{\max}}$$

where $h_{\min} = \min(h_x, h_y, h_z)$, $V_{\max}$ is the maximum compressional velocity in the model and dim is the spatial dimension ($\dim = 1, 2, 3$).

- Finite differences approach [10].

The spatial derivatives are computed by approximating the differential operators ∂_x, ∂_y and ∂_z with d_x, d_y and d_z where

$$d_\gamma = \frac{1}{2h_\gamma}\left(\frac{1}{6}E_\gamma^{-2} - \frac{4}{3}E_\gamma^{-1} + \frac{4}{3}E_\gamma^{1} - \frac{1}{6}E_\gamma^{2}\right) \tag{7}$$

with E_γ representing the translation operator and h_γ the grid spacing in the direction $\gamma = x, y, z$. The stability criterion of the above scheme is [7]:

$$\Delta t \leq \frac{0.729}{\sqrt{\dim}} \frac{h_{\min}}{V_{\max}}$$

2.2.3. Absorbing boundaries

An absorbing boundary condition based on the gradual reduction of wave amplitudes in the vicinity of the boundary [4, 5] has been used. The values of stresses and their time derivatives are slightly reduced after each time step in a region adjacent to the external boundary of the domain. This reduction is gradually tapered from a zero value in the interior region, by the use of a reduction factor of the form $\exp(-ad^2)$ where a is the absorption coefficient and d is the distance to the interior layer of the boundary region. An absorption coefficient of 0.015 and a boundary region width of 7% of the mesh suffices to reduce wraparounds to unnoticeable values.

2.2.4. Complexity analysis

- Pseudo-spectral approach

  The computational cost per time step of the pseudo-spectral modeling algorithm is

  $$\frac{C}{N_t} = 20N_xN_z\log_2(N_xN_z) + 40N_xN_z + 12(N_xNS_z + N_zNS_x) \tag{8}$$

- Finite differences approach

  The computational cost per time step of the finite difference modeling algorithm for the 2-D case is

  $$\frac{C}{N_t} = 80N_xN_z + 12(N_xNS_z + N_zNS_x) \tag{9}$$

  In the 3-D case the cost is

  $$\frac{C}{N_t} = 188N_xN_yN_z + 24(N_xN_yNS_z + N_xN_zNS_y + N_yN_zNS_x) \tag{10}$$

  where N_γ is the grid size and NS_γ is the thickness of the absorption layer in the direction γ.

3. Vector Multiprocessing

3.1. Implementations

In order to make an efficient use of the IBM 3090 Vector Facility, we have used routines from the ESSL library [6] to compute the direct and inverse Fourier transforms. In addition, we have used the VS FORTRAN Vectorizing Compiler to vectorize the remaining computations [13 – 14].

With respect to the parallel decomposition of the algorithms, we have exploited a simplified domain decomposition technique. This strategy makes use of the fact that with computational arrays oriented along the cartesian coordinates, space derivatives operate on individual columns/rows/sections independently of one another. Thus the problem domain can be decomposed into strips, and each processor handles the operation for its strip. A switching of the parallel strips orientation is needed when switching the direction of the partial derivative. Equivalently, the data arrays could be transposed and the parallel strips kept always oriented column-wise as recommended for faster execution in FORTRAN. This constitutes an important difference between pseudo-spectral and finite difference schemes. The pseudo-spectral schemes are non-local, in the sense that in order to compute derivatives for a direction at a given point, one must access all the data corresponding to that direction at the point. On the contrary, finite difference schemes are local, and, for each point, only data in a relatively small neighborhood have to be considered. This difference has an important implication when trying to parallelize the algorithm. For finite difference schemes, the parallel processors can be permanently assigned a data subdomain, and only require to exchange or share the data at the boundaries of the subdomains. In Fourier methods, on the contrary, the parallel processors need to access (by example in the x differentiation), data previously assigned to other processors (in the z sweep), thus leading to a greater volume of communication.

With respect to the software support for parallelization, we have used Parallel FORTRAN (PF [12]) constructs which allow a fine level of parallelism, down to the grain of a DO-loop.

As an example of the numerical results obtained with the elastic models, Figures 1-3 show the response of a geological model corresponding to a corner structure.

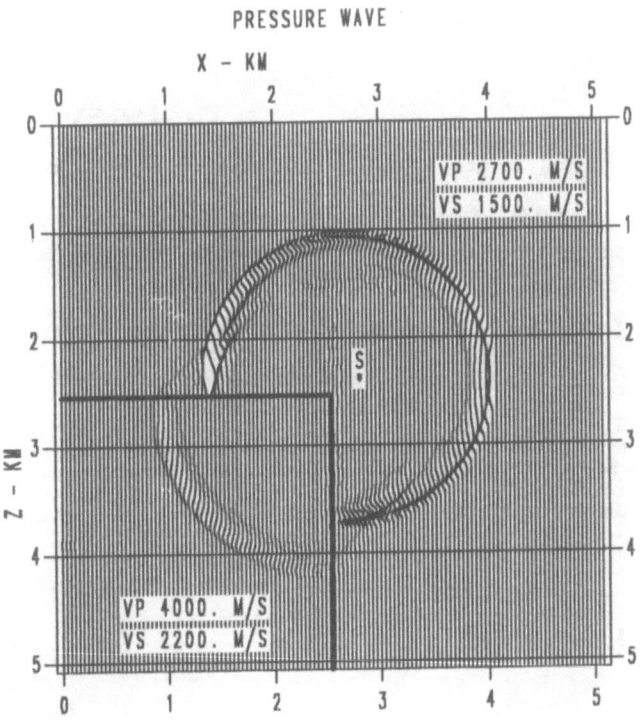

Figure 1. Pressure wave snapshot for the corner problem (high velocity inside the corner) as computed by the 2-D elastic pseudo-spectral model. VP and VS refer to pressure and shear velocities respectively and S indicates the source location. Calculations were made using a pressure line source oriented along the y direction; with a high-cut frequency of 40 Hz for a numerical mesh of 256 × 256 nodes with grid spacing of 20 meters in both x and z directions and 0.001 second time step.

PRESSURE WAVE

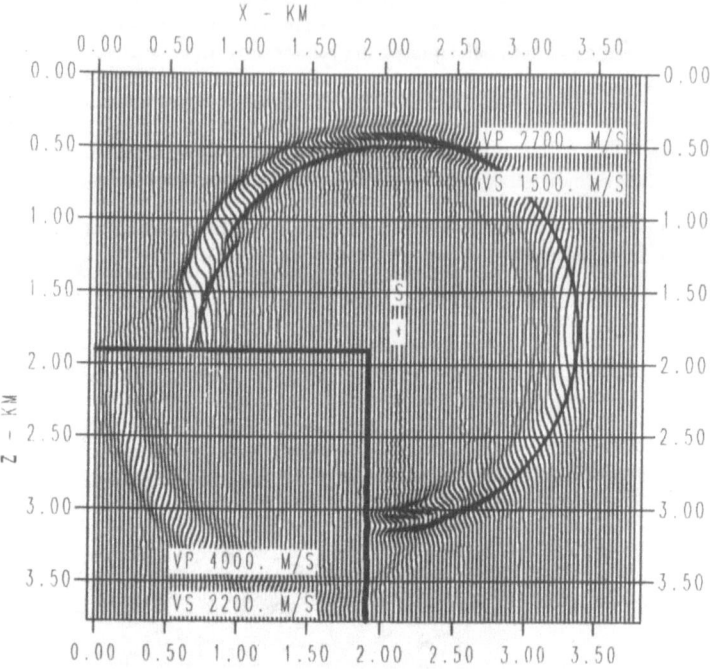

Figure 2. Pressure wave snapshot for the corner problem (high velocity inside the corner) as computed by the 2-D elastic finite differences model. VP and VS refer to pressure and shear velocities respectively and S indicates the source location. Calculations were made using a pressure line source oriented along the y direction; with a high-cut frequency of 10 Hz for a numerical mesh of 255 × 255 nodes with grid spacing of 15 meters in both x and z directions and 0.001 second time step.

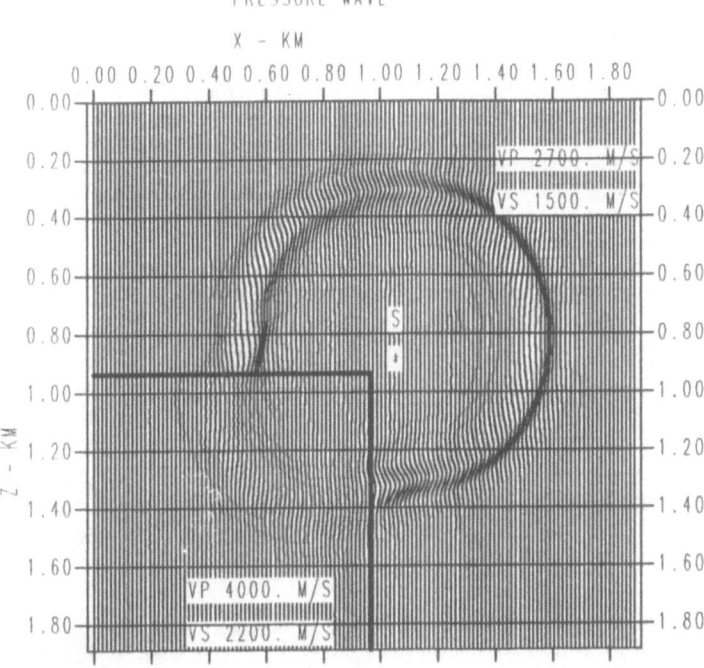

PRESSURE WAVE

Figure 3. Pressure wave snapshot for the corner problem (high velocity inside the corner) as computed by the elastic finite differences 3-D model. VP and VS refer to pressure and shear velocities respectively and S indicates the source location. Calculations were made using a pressure point source with a high-cut frequency of 10 Hz for a numerical mesh of 127 × 127 × 127 nodes with grid spacing of 15 meters in the three spatial directions and 0.001 second time step.

3.2. Performance analysis

The scalar and vector performance of the pseudo-spectral algorithm is reported in Table I, which includes the sustained performance in millions of floating point operations per second (Mflop/s) for different resolutions. The Mflop/s rates are obtained using Equation (8) to calculate the number of floating point operations per time step. Equation (8) considers only floating point additions and multiplications, and therefore neglects integer arithmetic, transpositions, and loads/stores, although it is well known that they play a non-negligible role in the balance.

Speed *(Mflop/s)*	*Grid Size*			
Configuration	64^2	256^2	512^2	1024^2
3090 scalar uniprocessor	4.6	5.30	5.52	4.54
3090 VF uniprocessor	21.5	31.1	34.3	26.9

Table I. Performance of the pseudo-spectral elastic code for different grid sizes on the IBM 3090-200 VF [2] (main storage = 64 Mbytes; expanded storage = 128 Mbytes).

As it is shown in Table I, the results obtained by running the vectorized algorithm on the 3090 VF, indicate very good performance results, with a 3090 vector/scalar ratio of 5.21. It is worth noting that for relatively short vector lengths (the 64^2 case) the vector performance is degraded. Moreover, for the cases in which the problem size does not fit in real storage (last column of Table I), the data are allocated to the expanded storage of the 3090 [2] (the specific configuration used in this experiment had 64 Mbytes of real storage and 128 Mbytes of expanded storage). The operating system then pages the data in and out of real storage at a rate of approximately 75 μs per page fault. This leads to a degradation in performance as can be observed from the results shown in Table I for the 1024 × 1024 problem size. Nevertheless, this performance is still very satisfactory: moreover, since the operating system takes care of paging, it is possible to run very large models without explicit memory management.

Configuration	Processors	Time (s)	Mflop/s	Speedup
3090-400E, PF	1 (1 VF)	2.71	38.6	1.00
3090-400E, PF	2 (2 VF)	1.38	75.8	1.96
3090-400E, PF	3 (3 VF)	0.93	112.8	2.89
3090-400E, PF	4 (4 VF)	0.73	144.2	3.73

Table II. Performance on the IBM 3090-400E VF of the pseudo-spectral elastic modeling code for a resolution of 512 × 512, as a function of the number of processors, using PF for parallel execution.

Table II shows performance data on the four-way 3090-400E for a grid size of 512 × 512, in the vectorized and parallelized case of the pseudo-spectral code. An almost linear speedup is observed using PF.

	Grid Size || Memory(MByte)		
	255^2 || 5.12	511^2 || 20.48	1023^2 || 81.52
Configuration	Time (s) | Speed (Mflop/s)		
3090-400E (1VF)	0.164 | 32.2	0.682 | 31.0	2.767 | 30.6
3090-400E (2VF)	0.0847 | 62.4	0.351 | 60.3	1.454 | 58.2
3090-400E (3VF)	0.5964 | 88.53	0.2369 | 89.24	0.970 | 87.25
3090-400E (4VF)	0.466 | 113.27	0.182 | 116.17	0.746 | 113.49

Table III. Performance of the finite difference 2-D elastic code for different grid sizes on the IBM 3090-400E VF (main storage = 128 Mbytes; expanded storage = 256 Mbytes) using PF for parallel execution.

The sustained performance of the different versions of the 2-D finite difference code for different grid sizes is given in Table III. Vector/scalar speedups of the order of 3 are observed together with an almost linear speedup when the code is parallelized with PF. It should be noted that the vector/scalar speedup of the finite differences code (~ 3) is less than the corresponding speedup for the pseudo-spectral code (~ 5.2). This is attributed to that the number of flops per memory reference is less than the pseudo-spectral case; beside using the highly optimized ESSL library with the pseudo-spectral code. Detailed pseudo-spectral - finite differences comparisons could be found in [7, 8].

Configuration	Time (sec) | *Speed (Mflop/s)*
3090-400E scalar uniprocessor	13.578|*11.71*
3090-400E 1VF	4.908|*32.4*
3090-400E 2VF	2.535|*62.72*
3090-400E 3VF	1.714|*92.78*
3090-400E 4VF	1.327|*119.8*

Table IV. Performance of the finite differences 3-D elastic code for the grid size 127 × 124 × 60 on the IBM 3090-400E VF (main storage = 128 Mbytes; expanded storage = 256 Mbytes) using PF for parallel execution.

Configuration	Time (sec) | *Speed (Mflop/s)*
3090-400E scalar uniprocessor	31.345|*10.55*
(2 scalar proc.)	17.05|*19.39*
(3 scalar proc.)	11.82|*27.98*
(4 scalar proc.)	9.303|*35.54*
3090-400E (1 VF)	12.71|*26.01*
2 VF	6.842|*48.33*
3 VF	4.892|*67.59*
4 VF	4.195|*78.82*

Table V. Performance of the finite differences 3-D elastic code for the grid size 127 × 124 × 120 on the IBM 3090-400E VF (main storage = 128 Mbytes; expanded storage = 256 Mbytes) using PF for parallel execution and user routines to control paging.

Tables IV and V report the sustained performance of the 3-D finite differences elastic model. The case in Table IV fits in real storage (~ 100 *Mbytes*) and its performance is similar to that reported in Table III. On the contrary, the case of Table V does not fit in real storage (~ 200 *Mbytes*) and the results reported are those obtained by using user controlled paging routines [3] to overcome the synchronous nature of expanded storage paging. Figure 4 shows, for a part of the 3-D finite differences code, the synergism between parallel and vector execution.

```
                    C*DIREC  PREFER MAXCHUNK(40) MINCHUNK(40)
0487 PARA               parallel loop 1975 k = 3,nz-2
                    C*DIREC  PREFER SERIAL
0488 ELIG  +-----        do 1975 j = 3,nx-2
           |    C*DIREC  PREFER VECTOR
0489 VECT  |+ ----       do 1975 i = 3,ny-2
0490     ||         tfx(i,j,k) = (c1y*(sxy(i-2,j,k)-sxy(i + 2,j,k))
0490     ||         & + c2y*(sxy(i + 1,j,k)-sxy(i-1,j,k))
0490     ||         & + c1x*(sxx(i,j-2,k)-sxx(i,j + 2,k))
0490     ||         & + c2x*(sxx(i,j + 1,k)-sxx(i,j-1,k))
0490     ||         & + c1z*(sxz(i,j,k-2)-sxz(i,j,k + 2))
0490     ||         & + c2z*(sxz(i,j,k + 1)-sxz(i,j,k-1)))*fiden(i,j,k)
0491     ||         tfy(i,j,k) = (c1x*(sxy(i,j-2,k)-sxy(i,j + 2,k))
0491     ||         & + c2x*(sxy(i,j + 1,k)-sxy(i,j-1,k))
0491     ||         & + c1y*(syy(i-2,j,k)-syy(i + 2,j,k))
0491     ||         & + c2y*(syy(i + 1,j,k)-syy(i-1,j,k))
0491     ||         & + c1z*(syz(i,j,k-2)-syz(i,j,k + 2))
0491     ||         & + c2z*(syz(i,j,k + 1)-syz(i,j,k-1)))*fiden(i,j,k)
0492     ||         tfz(i,j,k) = (c1x*(sxz(i,j-2,k)-sxz(i,j + 2,k))
0492     ||         & + c2x*(sxz(i,j + 1,k)-sxz(i,j-1,k))
0492     ||         & + c1y*(syz(i-2,j,k)-syz(i + 2,j,k))
0492     ||         & + c2y*(syz(i + 1,j,k)-syz(i-1,j,k))
0492     ||         & + c1z*(szz(i,j,k-2)-szz(i,j,k + 2))
0492     ||___     & + c2z*(szz(i,j,k + 1)-szz(i,j,k-1)))*fiden(i,j,k)
         |___
```

Figure.4 Using the PF constructs and directives to parallelize the execution of the 3-D finite differences code. Synergism between parallel and vector execution in the code compiled with PF.

4. Conclusions

The seismic computations presented in this paper share some common characteristics: large amounts of data have to be handled, high floating point content and heavy use of arrays. Furthermore each problem can be naturally decomposed into a set of independent tasks which need minimum synchronization.

The cases addressed in this paper indicate that in seismic computations there is a very large potential for coarse-grain parallelism at the algorithmic level, which can be enhanced by one or more suitable integral transformations (for example Fourier transforms). Parallelism appears at multiple levels which can be simultaneously exploited, rendering possible and useful the application of parallel decomposition techniques like domain decomposition and (at the lowest level of parallelism) vectorization.

It is concluded that the algorithms presented are very well suited to vector multiprocessors, on which important performance improvements are obtained by simultaneously vectorizing the innermost loops and parallelizing the outer loops. The results summarized in this paper show their suitability to efficient implementation on the IBM 3090 vector multiprocessor: combined vector/parallel speedups around 20-30 are in fact observed.

References

1. Achenbach, J.D., *Wave Propagation in Elastic Solids* (North Holland Publishing Company, The Netherlands, 1975)

2. Buchholz, W., The IBM System/370 vector architecture, *IBM Systems Journal 25* (1986) 1, 51-62.

3. Carnevali, P., Kindelan, M., Sguazzero, P., Vitaletti, M. and Kamel, A., Numerically intensive computing with large storage sizes on the IBM 3090: user-controlled paging under MVS/XA, IBM Technical Report, (IBM European Center for Scientific and Engineering Computing, 1988).

4. Cerjan, C., Kosloff, D., Kosloff, R. and Reshef, M., A non reflecting boundary condition for discrete acoustic and elastic wave equations, *Geophysics 50* (1985) 705-708.

5. Clayton, R. and Enquist, B., Absorbing boundary conditions for acoustic and elastic wave equations, *Bull. Seism. Soc. Am. 67* (1977) 1529 - 1540.

6. Engineering and Scientific Subroutine Library, General Information, Order No. GC23-0182, IBM.

7. Fornberg, B., The Pseudospectral method: Comparisons with finite differences for the elastic wave equation, *Geophysics*, 52, 483-501 (1987).

8. Hestholm, S. and Kamel, A., Numerical differentiation in large scale simulation of wave phenomena, Workshop meeting on seismic waves in laterally inhomogeneous media III. Prague, Czechoslovakia, June 13-18 1988.

9. M. Kindelan, P. Sguazzero and A. Kamel, "Elastic modeling with Fourier methods on the IBM 3090 vector multiprocessor", in *Scientific Computing on IBM Vector Multiprocessors*, R. Benzi and P. Sguazzero Eds., IBM European Center for Scientific and Engineering Computing, (1987), 635-674

10. Kamel, A., Kindelan, M. and Sguazzero, P., Seismic Computations on the IBM 3090 vector multiprocessor, *IBM Systems Journal 27* (1988) 4, 510-527.

11. Kosloff, D., Reshef, M. and Loewenhal, D., Elastic wave calculations by the Fourier method, *Bull. Seism. Soc. Am. 74* (1984) 875 - 899.

12. Parallel FORTRAN Language and Library Reference, Order No. SC23-0431, IBM.

13. Scarborough, R. G. and Kolsky, H.G., A vectorizing FORTRAN compiler, *IBM J. Res. Develop. 30* (1986) 2, 163-171.

14. VS FORTRAN Version 2, General Description, Order No. GC26-4219, IBM.

Solving a Model Interface Problem for the Laplace Operator by Boundary Collocation and Applications *

L. Reichel
Bergen Scientific Centre IBM
Allégaten 36   and
N–5007 Bergen, Norway

University of Kentucky
Department of Mathematics
Lexington, KY40506, USA

Abstract. Boundary collocation is a method for computing solutions to boundary problems for linear partial differential equations, for which complete families of particular solutions are explicitly known. The method is simple to implement on a computer and has been found to be competitive for many problems of electrical engineering.

An application of boundary collocation requires several choices, such as the choices of subspace and basis of particular solutions, and the choice of collocation points. The numerical aspects of these choices have so far only received little attention. This paper investigates how subspace, basis and collocation points should be chosen when solving a model interface problem. The purpose of this investigation is to obtain guidelines for these choices applicable also to other boundary problems. These guidelines have been used to solve two problems that arose in Swedish industry: an investigation of microwave heating of food with the aim of designing equipment to achieve uniform heating, and a study of the heat conduction of a rock important for the design of a hot water storage near Uppsala for water heated by solar collectors.

Key words: boundary collocation, partial differential equation, the Helmholtz equation, microwave heating.

1. Introduction

Boundary collocation (BC) is a method for computing approximate solutions to linear partial differential equations, where one

- chooses a finite dimensional subspace of particular solutions to the partial differential equation, or if the domain is divided into subdomains, then one chooses one subspace for each subdomain,

- chooses a basis for each subspace,

- determines a linear combination of these basis functions that minimizes the error in the given boundary conditions, or in the given continuity conditions across an interface, at selected points, so–called *collocation points*. The error is most conveniently measured by a least squares norm.

The BC method may be viewed as a generalization of the separation of variables technique. The wider applicability of BC stems from that the boundary or interface(s) may be of fairly general shape. Moreover, BC allows the use of subspaces other than those that can be obtained by separation of variables technique. Applications of BC can be found, e.g., in the electrical engineering literature, where the method often is called *point matching*, see [2], [3], [7] and references there. In both [2] and [3] the BC method is found to be competitive with other numerical schemes.

The possibility of choosing subspace, basis and collocation points leads to the desire to make choices that are suitable from a numerical point of view. We would like to achieve rapid convergence with increasing number of basis functions, and we want the linear system of algebraic equations to be well–conditioned. Moreover, the basis functions should be well–conditioned and enable rapid evaluation. Finally, the the collocation points should be easy to determine.

---

* Research supported in part by NSF under Grant DMS-8704196.

J. Manley et al. (eds.), Proceedings of the Third European Conference on Mathematics in Industry, 119–130.
© 1990 *Kluwer Academic Publishers and B. G. Teubner Stuttgart.*

In order to study the choices of subspace, basis and collocation points, we introduce the following *model interface problem*. Let $\Omega \subset \mathbf{R}^2$ be an open bounded simply connected region with an analytic boundary $\partial\Omega$. Let $\Omega_c := \mathbf{R}^2 \backslash (\Omega \cup \partial\Omega \cup \{\infty\})$. Determine a function $u = u(x, y)$ such that

$$(1.1) \begin{cases} u \text{ is twice continuously differentiable in } \Omega \text{ and } \Omega_c, \\ \Delta u = 1 \text{ in } \Omega, \\ \Delta u = 0 \text{ in } \Omega_c, \\ u, \frac{\partial u}{\partial n} \text{ are continuous across } \partial\Omega, \text{ where } \frac{\partial}{\partial n} \text{ denotes the normal derivative,} \\ u(x, y) = \alpha \log(x^2 + y^2) + o(1) \text{ as } x^2 + y^2 \to \infty, \text{ for some constant } \alpha \in \mathbf{R}. \end{cases}$$

It is easy to verify that (1.1) has the unique solution

$$u^*(x, y) = \frac{1}{4\pi} \iint_\Omega \log((x - \xi)^2 + (y - \eta)^2) \, d\xi d\eta.$$

Example 1.1. Let $\Omega := \{(x, y) : x^2 + y^2 < 1\}$. Then the solution of (1.1) can be written

$$\begin{cases} u^*(x, y) = \frac{1}{4} r^2 - \frac{1}{4}, & 0 \le r \le 1, \\ u^*(x, y) = \frac{1}{2} \log r, & r > 1, \end{cases}$$

where $r := (x^2 + y^2)^{1/2}$. □

By Example 1.1 it follows that the solution u^* of (1.1) can be written $u^*(x, y) = \frac{1}{4} r^2 + u_0(x, y)$ in Ω, where u_0 is a (unique) harmonic function. When solving (1.1) by BC, we approximate u_0 in Ω and u^* in Ω_c, i.e. we choose a finite–dimensional subspace of functions harmonic in Ω, and a finite–dimensional subspace of functions harmonic in Ω_c. The functions u of the latter subspace are also required to satisfy $| u(x, y) | / \log(x^2 + y^2) \le \beta$ as $x^2 + y^2 \to \infty$ for some constant β.

The model problem (1.1) has previously been introduced in [10], where we studied the selection of collocation points for given subspaces. In the present paper we consider the choices of subspaces and of their bases. We also present a more concise treatment than in [10] of the choice of collocation points. The purpose of our investigation is to obtain guidelines for how these choices should be made for boundary and interface problems which are more complicated than the model problem (1.1). The computed examples of Section 5 shows that experience gained from solving (1.1) by BC, indeed carries over to other problems. In fact, the guidelines obtained in this paper and in [10] have been applied to the solution of the following two problems that arose in Swedish industry.

Problem 1: Microwave Heating of Food. A manufacturer of microwave ovens wanted to investigate how food is heated by microwaves. Generally, food is heated very nonuniformly in a microwave oven. The aim of the investigation was to find out how the shape of the oven, or the frequency of the microwaves, could be altered in order to yield more uniform heating. The problem requires the solution of an interface problem for the Maxwell equations and the determination of the power density of the electric field inside the object being heated. A detailed description of this problem can be found in [8], where also computations based on the separation of variables technique are presented. This technique enables studying microwave heating of meat balls and (infinite) sausages (whose cross section is a disk). Computations based on BC are presented in [11] and enable studying microwave heating of food of more general shape. □

Problem 2: Heat Conduction of a Rock. A torus shaped hot water storage in a rock was planned near the city of Uppsala. The water was to be heated by solar energy, and its temperature therefore would vary periodically with time. The temperature variation in the rock surrounding the storage was computed in order to assess the structural stability of the storage. Temperature changes in the rock pillar surrounded by the water was of particular interest. The problem required the solution of a Helmholtz equation in the exterior of the torus. Rotationally symmetric basis functions were used. This made it sufficient to allocate collocation points on a plane curve, and reduced the original rotationally symmetric three–dimensional problem to a one–dimensional approximation problem. Details and computational results can be found in [9]. □

This paper is organized as follows. Section 2 reformulates problem (1.1) into an equivalent complex interface problem. This enables us to use complex variable methods in our analysis. Section 3 discusses the selection of subspace, basis and collocation points. In Section 4 we present some computed examples for the complex interface problem. Finally, Section 5 shows some computed examples for an exterior Dirichlet problem for the Helmholtz equation. These examples illustrate that guidelines obtained for the selection of subspace, basis and collocation points from the model problem (1.1) are applicable to a larger class of problems.

2. A Complex Interface Problem

In this section we derive a complex interface problem that is equivalent with problem (1.1). Identify the complex plane $\mathbb{C}$ with $\mathbb{R}^2$ by letting $z = x + iy \in \mathbb{C}$ for $(x, y) \in \mathbb{R}^2$ and $i := \sqrt{-1}$. Let $\Omega, \partial\Omega$ and Ω_c be the same as in problem (1.1). The *Schwarz function* $S(z)$ for $\partial\Omega$ is defined as the complex valued function such that $S(z) = \bar{z}$ for $z \in \partial\Omega$, where the bar denotes complex conjugation. We will express the solution of the complex interface problem using $S(z)$. Several applications of the Schwarz function are described in [5].

Let $\psi(z)$ be the analytic function which maps $\Omega_c \cup \{\infty\}$ conformally and univalently onto $\{w : |w| > 1\}$, so that $\psi(\infty) = \infty$ and $\psi'(\infty) > 0$. By the analyticity of $\partial\Omega$ we can continue ψ to be analytic in an open set containing $\partial\Omega$ in such a way that the extension remains univalent. We denote this extension also by ψ, and let ψ^{-1} be its inverse. The Schwarz function can be expressed as

$$S(z) = \overline{\psi}^{-1}(1/\psi(z)), \quad z \in \partial\Omega,$$

and by the analyticity of ψ in a neighborhood of $\partial\Omega$, it follows that $S(z)$ can be extended to an analytic function in a neighborhood of $\partial\Omega$. This extension is also denoted by $S(z)$. Introduce

$$(2.1) \qquad f(z) := \frac{1}{4\pi i} \int_{\partial\Omega} \frac{S(\zeta)}{\zeta - z} d\zeta, \quad z \in \Omega,$$

$$(2.2) \qquad g(z) := -\frac{1}{4\pi i} \int_{\partial\Omega} \frac{S(\zeta)}{\zeta - z} d\zeta, \quad z \in \Omega_c,$$

where the integration in (2.1) is carried out in the positive direction with respect to Ω, and the integration in (2.2) is carried out in the positive direction with respect to Ω_c.

Lemma 2.1. Let $\partial\Omega$ be an analytic Jordan curve, and let f and g be defined by (2.1) - (2.2). Then there are extensions of f and g, which we also denote by f and g, that satisfy

$$(2.3a) \qquad f \text{ is analytic in an open set containing } \Omega \cup \partial\Omega,$$

$$(2.3b) \qquad g \text{ is analytic in an open set containing } \Omega_c \cup \partial\Omega \cup \{\infty\}, \text{ and } g(\infty) = 0,$$

$$(2.3c) \qquad f(z) - g(z) = \frac{1}{2}\bar{z}, \quad z \in \partial\Omega.$$

Proof. The statements (2.3) follow from the analyticity of S in a neighborhood of $\partial\Omega$. □

We call the problem of determining functions f and g that satisfy (2.3) the *complex interface problem*.

Lemma 2.2. The complex interface problem (2.3) is equivalent with the model interface problem (1.1).

Proof. Let f and g satisfy (2.3). Then integration of f and g yields the solution u^* of (1.1). Conversely, let u^* satisfy (1.1). Then the functions

$$(2.4) \qquad f(z) := -\left(\frac{\partial}{\partial x} - i\frac{\partial}{\partial y}\right)\left(u^*(x, y) - \frac{1}{4}(x^2 + y^2)\right), \quad (x, y) \in \Omega,$$

$$(2.5) \qquad g(z) := -\left(\frac{\partial}{\partial x} - i\frac{\partial}{\partial y}\right)u^*(x, y), \quad (x, y) \in \Omega_c,$$

where $z = x + iy$, satisfy (2.3). This shows the lemma. □

The equivalence between the model interface problem and the complex interface problem implies, in particular, that the latter problem also has a unique solution. In the following two sections we discuss how to compute approximate solutions to the complex interface problem.

3. Choice of Subspaces, Bases and Collocation Points

Let Π_n denote the linear space of polynomials of degree of at most n. Let $\{\zeta_j\}_{j=1}^m$ be a point set in Ω, and introduce a subspace of rational functions, that vanish at infinity and have fixed poles at the points ζ_j,

$$(3.1) \qquad Q_m := \operatorname{span}\{(z-\zeta_1)^{-1}, (z-\zeta_1)^{-1}(z-\zeta_2)^{-1}, \ldots, \prod_{j=1}^m (z-\zeta_j)^{-1}\}.$$

We wish to find a pair $\{f_n, g_m\} \in \Pi_n \times Q_m$ that is a good approximate solution of the complex interface problem (2.3), and seek to determine such a pair by requiring that that $\{f_n, g_m\} \in \Pi_n \times Q_m$ satisfies

$$(3.2) \qquad f_n(z_j) - g_m(z_j) = \frac{1}{2}\bar{z}_j, \qquad 0 \le j \le m+n,$$

where $\{z_j\}_{j=0}^{m+n}$ is a set of preselected distinct points on $\partial\Omega$. The points z_j are called *collocation points*.

We note that while we require f_n to be singular only at infinity, we are free to select the singular points of g_m in Ω. The possibility of making the location of the singular points ζ_j of g_m depend on the shape of Ω is essential in order to achieve rapid convergence. For Ω convex it generally suffices to let f_n be a polynomial. This can be seen by studying level curves (3.14) of certain Green's functions. In the present paper we restrict ourselves to develop computational schemes for problems with convex regions Ω.

We first consider the *allocation of collocation points* z_j and assume that a set of poles $\{\zeta_j\}_{j=1}^n$ defining Q_m is given. Consider the approximation errors

$$(3.3) \qquad \varepsilon(z) := f(z) - g(z) - (f_n(z) - g_m(z)),$$
$$(3.4) \qquad \varepsilon_f(z) := f(z) - f_n(z),$$
$$(3.5) \qquad \varepsilon_g(z) := g(z) - g_m(z).$$

Assume that $\partial\Omega$ is an analytic Jordan curve. Then f and g are analytic in a neighborhood of $\partial\Omega$. Therefore there are Jordan curves Γ in Ω and Γ_c in Ω_c, such that both f and g are analytic in the closure of the annular point set between Γ and Γ_c. In the computations we explicitly choose a set of poles $\{\zeta_j\}_{j=1}^l \subset \Omega$ for some fixed fairly small integer $l \ge 1$, and define the remaining poles $\zeta_j, l < j \le m$, by

$$(3.6) \qquad \zeta_{j+k\ell} := \zeta_j, \qquad 1 \le j \le \ell, \quad k = 1, 2, 3, \ldots \quad .$$

We therefore may assume that Γ is chosen so that all poles ζ_j lie in the open interior of Γ. Denote the *open* annular point set between Γ and Γ_c by $\delta\Gamma$. We are now in a position to bound the approximation errors (3.3) - (3.5). By Walsh [15, Theorem 2, p. 186] we have

$$(3.7) \qquad \varepsilon(z) = \frac{1}{2\pi i} \int_{\Gamma \cup \Gamma_c} \frac{\omega(z)}{\omega(\zeta)} \frac{f(\zeta) - g(\zeta)}{\zeta - z} d\zeta, \quad z \in \delta\Gamma,$$

where

$$(3.8) \qquad \omega(z) := \frac{\displaystyle\prod_{j=0}^{m+n} (z - z_j)}{\displaystyle\prod_{j=0}^m (z - \zeta_j)}.$$

Theorem 3.1. Let $\partial\Omega$ be analytic, and assume that the poles $\zeta_j \in \Omega$ satisfy (3.6) for some integer $\ell \ge 1$ independent of m and n. Let $G_j(z)$ denote the Green's function for the Laplace operator for Ω with a logarithmic singularity at $z = \zeta_j$. Then $G_j(z)$ satisfies i) $\Delta G_j(z) = 0$ in $\Omega \setminus \{\zeta_j\}$, where $z = x + iy$, ii) $G_j(z) = 0$ on $\partial\Omega$, and iii) $\frac{1}{2\pi} \int_{\partial\Omega} \frac{\partial G_j}{\partial n}(z) \mid dz \mid = -1$, where integration is carried out in the counterclockwise direction.

Let $G(z)$ denote the Green's function for Ω with a logarithmic singularity of infinity. Then i) $\Delta G(z) = 0$ in Ω_c, where $z = x + iy$, ii) $G(z) = 0$ on $\partial\Omega$, and iii) $\frac{1}{2\pi} \int_{\partial\Omega} \frac{\partial G}{\partial n}(z) \mid dz \mid = 1$.

Assume that the limit

$$(3.9) \qquad \kappa := \lim_{\substack{m+n\to\infty \\ m,n\geq 0}} \frac{n+1}{m+n+1}$$

exists, and define the *density function* σ for the collocation points by

$$(3.10) \qquad \sigma(z) := \frac{\kappa}{2\pi} \frac{\partial G}{\partial n}(z) - \frac{1-\kappa}{2\pi\ell} \sum_{j=1}^{\ell} \frac{\partial G_j}{\partial n}(z), \quad z \in \partial\Omega.$$

Let the collocation points $z_k, 0 \leq k \leq n+m$, be equidistributed with respect to σ on $\partial\Omega$, i.e. let $z_0 \in \partial\Omega$ be arbitrary but fixed, and determine the remaining points $z_k \in \partial\Omega$ from

$$(3.11) \qquad \int_{z_0}^{z_k} \sigma(z) \mid dz \mid = \frac{k}{m+n+1}, \quad 1 \leq k \leq m+n.$$

Integration in (3.11) is carried out along $\partial\Omega$ in the counter–clockwise direction. Let $\omega(z)$ be defined by (3.8). Then

$$(3.12) \qquad \lim_{\substack{m+n\to\infty \\ m,n\geq 0}} \mid \omega(z) \mid^{\frac{1}{m+n+1}} = \begin{cases} (\text{cap } \Omega)^{\kappa} \exp\left(\frac{1-\kappa}{\ell} \sum_{j=1}^{\ell} G_j(z)\right), & z \in \Omega, \\ (\text{cap } \Omega)^{\kappa} \exp(\kappa G(z)), & z \in \Omega_c. \end{cases}$$

The expression cap Ω denotes the *capacity* of Ω and equals $\lim_{|z|\to\infty} \mid z \mid \exp(-G(z))$. The capacity depends on the scaling of Ω.

Proof. The bound (3.12) follows from Walsh [15, Theorem 10, p. 212], which Walsh applies to the case $\ell = 1$ and $\zeta_1 = \infty$, see [15, pp. 214– 216]. The formulas (3.10) and (3.12) are slight modifications of the formulas in [15, pp. 214-216]. The collocation points z_k in (3.11) are uniquely determined by z_0 and σ, since the analyticity of $\partial\Omega$ and the Hopf maximum principle yield that $\frac{\partial G}{\partial n}(z) > 0$ for $z \in \partial\Omega$, and $\frac{\partial G_j}{\partial n}(z) < 0$ for $z \in \partial\Omega$ and $1 \leq j \leq \ell$. Hence, $\sigma > 0$ on $\partial\Omega$, and therefore the z_k, $k \geq 1$, are uniquely determined by (3.11). $\square$

In applications we choose a constant $\kappa \in [0,1]$, and then generate sequences of increasing m-values and n-values that satisfy (3.9). For instance, we may choose $m = m(j)$ and $n = n(j)$ defined by

$$(3.13) \qquad n(j) := \lfloor \kappa j \rfloor - 1, \quad m(j) := j - \lfloor \kappa j \rfloor,$$

for increasing integer values of $j \geq 1/\kappa$. Here $\lfloor \alpha \rfloor$ denotes the integer part of $\alpha \geq 0$.

Corollary 3.1. Let $\partial\Omega$ be analytic, and assume that the poles ζ_j satisfy (3.6) for some integer $\ell \geq 1$ independent of m and n. Assume that (3.9) holds, and let the collocation points be defined by (3.10) – (3.11). Then there is an open annular point set $\delta\Gamma_0$ that contains $\partial\Omega$, and is such that $\varepsilon(z), \varepsilon_f(z)$ and $\varepsilon_g(z)$ converge to 0 on $\delta\Gamma_0$ with a geometric rate of convergence, uniformly for $z \in \delta\Gamma_0$, as $m + n \to \infty, n \geq 0, m \geq 0$.

Proof. Let $\delta\Gamma$ be as in (3.7), and let the constant $\alpha > 0$ be sufficiently small so that the level curve

$$(3.14a) \qquad \gamma(\alpha) := \{z : \frac{1-\kappa}{\ell} \sum_{j=1}^{\ell} G_j(z) = \alpha\}$$

lies in $\Omega \cap \delta\Gamma$, and the level curve

$$(3.14b) \qquad \gamma_c(\alpha) := \{z : \kappa G(z) = \alpha\}$$

lies in $\Omega_c \cap \delta\Gamma$. Let $\delta\Gamma_0$ be the closed bounded annular point set with boundary curves $\gamma(\alpha/2)$ and $\gamma_c(\alpha/$ By (3.7) we have

$$(3.7') \qquad \varepsilon(z) = \frac{1}{2\pi i} \int\limits_{\gamma(\alpha) \cup \gamma_c(\alpha)} \frac{\omega(z)}{\omega(\zeta)} \frac{f(\zeta) - g(\zeta)}{\zeta - z} d\zeta, \quad z \in \delta\Gamma_0.$$

The geometric convergence of $\varepsilon(z)$ towards zero as $m+n \to \infty, m \geq 0, n \geq 0$, now follows by substituting (3.12) into (3.7'). Convergence of $\varepsilon_f(z)$ to zero then follows from the Cauchy integral formula

$$\varepsilon_f(z) = \frac{1}{2\pi i} \int\limits_{\gamma_c(\alpha)} \frac{\varepsilon(\zeta)}{\zeta - z} d\zeta, \quad z \in \delta\Gamma_0.$$

Convergence of $\varepsilon_g(z)$ in $\delta\Gamma_0$ can also be shown using a Cauchy integral formula. □

The next lemma shows how the density function (3.10) can be computed.

Lemma 3.2. The integral equation

$$(3.15) \qquad \begin{cases} q + \int\limits_{\partial\Omega} \ln | z - \zeta | \, \sigma(\zeta) | \, d\zeta | = \dfrac{1 - \kappa}{\ell} \sum\limits_{j=1}^{\ell} \ln | z - \zeta_j |, \quad z \in \partial\Omega, \\[2mm] \int\limits_{\partial\Omega} \sigma(\zeta) | \, d\zeta | = 1, \end{cases}$$

has a unique solution $\{\hat\sigma, \hat q\} \in L^2(\partial\Omega) \times \mathbb{R}$, where $\hat\sigma$ is given by (3.10).

Proof. The Green's functions G and G_j satisfy

$$\ln | z - \zeta_j | = - \int\limits_{\partial\Omega} \ln | z - \zeta | \frac{1}{2\pi} \frac{\partial G_j}{\partial n}(\zeta) | \, d\zeta |, z \in \partial\Omega, \quad 1 \leq j \leq \ell,$$

$$\tilde q = \int\limits_{\partial\Omega} \ln | z - \zeta | \frac{1}{2\pi} \frac{\partial G}{\partial n}(\zeta) | \, d\zeta |, z \in \partial\Omega,$$

for some constant $\hat q$. This and property iii) of Theorem 3.1 shows that σ defined by (3.10) satisfies (3.15). The unicity of the solution of (3.15) has been shown, e.g., in [10]. □

We turn to the *selection of subspace* Q_m. Assume that $\kappa < 1$ and that the poles ζ_j satisfy (3.6). Then we only have determine the set $\{\zeta_j\}_{j=0}^{\ell} \subset \Omega$. Introduce the norm for $\varphi \in \mathbb{C}(\partial\Omega)$,

$$(3.16) \qquad \| \varphi \|_{\partial\Omega} := \max_{z \in \partial\Omega} | \varphi(z) | .$$

By the proof of Corollary 3.1, the rate of convergence of $\| \varepsilon \|_{\partial\Omega}$ to zero as $m+n$ increases ($m \geq 0, n \geq 0$) depends on the shape of the level curves $\gamma(\alpha)$, see (3.14). We would like to allocate the poles $\zeta_j \in \Omega, 1 \leq j \leq \ell$, so that

$$(3.17a) \qquad \min_{z \in \partial\Omega} -\frac{1}{\ell} \sum_{j=1}^{\ell} \frac{\partial G_j}{\partial n}(z) \text{ is not tiny},$$

$$(3.17b) \qquad \text{the poles } \zeta_j \text{ are not close to } \Omega_c,$$

$$(3.17c) \qquad \ell \geq 1 \text{ is fairly small}.$$

Requirement (3.17a) stems from that we do not know where in Ω the function g is singular. If g is singular at a point at which $\frac{1-\kappa}{\ell} \sum\limits_{j=1}^{\ell} G_j$ is "tiny", then, by construction of the level curves (3.14) in the proof at Corollary 3.1, the convergence of $\| \varepsilon \|_{\partial\Omega}$ to zero as $m+n$ increases ($m \geq 0, n \geq 0$) could be very slow.

The purpose of requirement (3.17a) is to avoid that $z \to \frac{1}{\ell} \sum_{j=1}^{\ell} G_j(z)$ grows very slowly when moving z from $\partial\Omega$ into Ω along a normal to $\partial\Omega$. Requirement (3.17b) is to avoid that a slow rate of convergence is obtained because the ζ_j are close to $\partial\Omega$. Requirement (3.17c), finally, is included because for certain problems for the Helmholtz equation a small value of ℓ enables rapid evaluation of the basis functions, see Section 5 for an illustration.

Assume that the region Ω is convex with an analytic boundary curve. For many such regions the normal derivative $\frac{\partial G}{\partial n} > 0$ does not become "tiny" anywhere on $\partial\Omega$. A lower bound for $\min_{z \in \partial\Omega} \frac{\partial G}{\partial n}(z)$, that increases with the distance that G can be continued harmonically into Ω, can be derived by applying a distortion theorem for conformed mappings [4, Theorem, p. 186]. We omit the details.

Assume now that $\min_{z \in \partial\Omega} \frac{\partial G}{\partial n}(z)$ is not very small. We may then replace requirement (3.17a) by the requirement that

(3.17a) $$\left\| \frac{\partial G}{\partial n} + \frac{1}{\ell} \sum_{j=1}^{\ell} \frac{\partial G_j}{\partial n} \right\|_{\partial\Omega} \text{ is small .}$$

Such an allocation of the poles ζ_j can be determined fairly simply for many regions Ω.

Example 3.1. Let $\partial\Omega$ be a lemniscate generated by the $\zeta_j, 1 \leq j \leq \ell$, i.e.

(3.18) $$\partial\Omega := \{z : \prod_{j=1}^{\ell} | z - \zeta_j | = \alpha\}$$

for some constant $\alpha > 0$. Then the Green's function G is given by

$$G(z) = \frac{1}{\ell}\Big(\sum_{j=1}^{\ell} \ln | z - \zeta_j | - \ln(\alpha)\Big).$$

By properties i) – iii) of the Green's functions G_j stated in Theorem 3.1, it follows that the function

$$H(z) := \frac{1}{\ell} \sum_{j=1}^{\ell} G_j(z) + \frac{1}{\ell}\Big(\sum_{j=1}^{\ell} \ln | z - \zeta_j | - \ln(\alpha)\Big)$$

has removable singularities in Ω, and therefore can be extended to be harmonic in Ω. Since, moreover, H vanishes on $\partial\Omega$, we obtain

(3.19) $$G(z) = -\frac{1}{\ell} \sum_{j=1}^{\ell} G_j(z).$$

Hence, if $\partial\Omega$ is a lemniscate (3.18) generated by the points $\zeta_j, 1 \leq j \leq \ell$, then we choose these points to be poles, and define the remaining poles by (3.6). For this choice of poles (3.19) is valid, and therefore the norm in (3.17a') vanishes. □

If the boundary $\partial\Omega$ is an analytic Jordan curve, then we first approximate $\partial\Omega$ by a lemniscate

(3.20) $$L(\beta) := \{z : \prod_{j=1}^{q} | z - \xi_j | = \beta\}.$$

We then choose $\ell := q$ and the poles $\zeta_j := \xi_j, 1 \leq j \leq \ell$. If $L(\beta)$ is a good approximation of $\partial\Omega$, then by the discussion of Example 3.1, the norm in (3.17a') is small. The possibility of constructing a lemniscate $L(\beta)$ that approximates an analytic curve well is shown by [15, Theorem 2, p. 68]. A numerical scheme for computing such a lemniscate is presented in [12].

In case we dot not have any a priori knowledge about the location of the singular points of f and g, we choose $\kappa = 1/2$. This yields $m = n + 1$, i.e. the dimensions of Π_n and Q_m are the same. Assuming that

$\kappa = 1/2$ and that (3.19) holds, the growth of $|\,\omega(z)\,|^{\frac{1}{m+n+1}}$ is nearly the same on each side of $\partial\Omega$ as z moves away from $\partial\Omega$ along a normal. We formulate this as a lemma.

Lemma 3.3. Assume that $\partial\Omega$ is an analytic Jordan curve. Let $\kappa := 1/2$ and assume that (3.19) is valid. Let $\hat{z} \in \partial\Omega$, let h be a unit normal to $\partial\Omega$ into Ω_c at $\hat{z}$, and let $\varepsilon > 0$ be sufficiently small. Then

$$\lim_{\substack{m+n\to\infty \\ m,n\geq 0}} |\,\omega(\hat{z}+\varepsilon h)\,|^{\frac{1}{m+n+1}} = \lim_{\substack{m+n\to\infty \\ m,n\geq 0}} |\,\omega(\hat{z}-\varepsilon h)\,|^{\frac{1}{m+n+1}} + O(\varepsilon^2).$$

Proof. The analyticity of $\partial\Omega$ implies that G and the G_j are analytic in a neighborhood of $\partial\Omega$. Therefore we obtain by (3.19) that for $\varepsilon > 0$ sufficiently small

$$(3.21) \qquad\qquad G(z+\varepsilon h) = \frac{1}{\ell}\sum_{j=1}^{\ell} G_j(z-\varepsilon h) + O(\varepsilon^2).$$

The lemma now follows by substituting (3.21) into (3.12) and using $\kappa = 1/2$. $\quad\square$

We conclude this section by introducing a fairly well-conditioned basis of $\Pi_n \times Q_m$. Let

$$(3.22) \qquad \begin{cases} p_0(z) := \beta_0, \\[2mm] p_k(z) := \beta_k \displaystyle\prod_{j=1}^{k}(z-\zeta_j), \quad 1 \leq k \leq n, \\[4mm] p_{-k}(z) := \beta_{-k} \displaystyle\prod_{j=1}^{k}(z-\zeta_j)^{-1}, \quad 1 \leq k \leq m, \end{cases}$$

where the $\beta_j > 0$ are scaling factors such that $\|\,p_j\,\|_{\partial\Omega} = 1$ for $-m \leq j \leq n$. Then $\{p_k\}_{k=-m}^n$ is a basis of $\Pi_n \times Q_m$. We introduce definitions analogous to those used by Gautschi [6] in his investigation of polynomial bases on intervals on the real axis. Let $\mathbf{a} := [a_k]_{k=-m}^n$, and define the mapping $F : \mathbb{C}^{m+n+1} \to \Pi_n \times Q_m$ by

$$(3.23) \qquad\qquad (F\mathbf{a})(z) := \sum_{k=-m}^{n} a_k p_k(z).$$

Equip the domain of F with the norm

$$\|\,\mathbf{a}\,\|_\infty := \max_{-m\leq k\leq n} |\,a_k\,|,$$

and the range of F with the norm (3.16). Let $\|F\|$ and $\|F^{-1}\|$ denote the induced operator norms of F and F^{-1}, respectively. The basis $\{p_k\}_{k=-m}^n$ is said to be well–conditioned if the condition number of F, given by

$$\mathrm{cond}(F) := \|F\|\,\|F^{-1}\|,$$

grows slowly with m and n.

Lemma 3.4. Let F be defined by (3.22) – (3.23). Assume that $\partial\Omega$ is a lemniscate, i.e. $\partial\Omega$ has the representation (3.15). Let the points ζ_j in (3.18) and (3.22) agree and let (3.6) be valid. Then there is a constant M depending on $\{\zeta_j\}_{j=1}^\ell$, but independent of m and n, such that

$$(3.24) \qquad\qquad \mathrm{cond}(F) \leq M(m+n+1).$$

Proof. In [13] a bound for $\mathrm{cond}(F)$ is presented for the case $n = 0$. Inequality (3.24) can be shown analogously. $\quad\square$

If $\partial\Omega$ is not a lemniscate, but is approximated well by one, then $\mathrm{cond}(F)$ might still grow slowly with m and n. Bounds for $\mathrm{cond}(F)$ are presented in [13] for the case $n = 0$. These bounds can easily be generalized to cover $n > 0$, as well.

4. Computed Examples for the Complex Interface Problem

The computations of the examples of this section involve the following steps:

(1) Solve integral equation (3.15) for $\kappa = 1$, e.g., by the method described in [10], [14]. Denote the solution by $\{\sigma^*, q^*\}$.

(2) Determine a lemniscate (3.20) that approximates $\partial\Omega$ well, e.g., by the scheme in [11]. Choose ℓ and the set of poles $\{\zeta_j\}_{j=1}^{\ell}$ as suggested in the discussion following (3.20). The poles ζ_j for $j > \ell$ are defined by (3.6).

(3) First assume that we have chosen $\kappa \in [0, 1[$. If the function $z \to \frac{1-\kappa}{\ell} \sum_{j=1}^{\ell} \ln |z - \zeta_j|$ is not nearly constant on $\partial\Omega$, then solve the integral equation (3.15). Denote the solution by $\{\tilde{\sigma}, \tilde{q}\}$, and determine a set of collocation points $\{z_k\}_{k=0}^{m+n}$ by (3.11), with $\sigma := \tilde{\sigma}$ in (3.11).

If the ζ_j are determined as described in step (2), then the function $z \to \frac{1-\kappa}{\ell} \sum_{j=1}^{\ell} \ln |z - \zeta_j|$ is, generally, nearly constant on $\partial\Omega$, and we can determine the collocation points by (3.11), with $\sigma := \sigma^*$ in (3.11), where σ^* is computed in step (1). Hence, in general, the integral equation (3.15) does not have to be solved for $\{\tilde{\sigma}, \tilde{q}\}$. Finally, if $\kappa = 1$, then $\tilde{\sigma} = \sigma^*$.

(4) Solve the linear system of equations (3.2) using basis (3.22) for $\Pi_n \times Q_m$.

The examples were computed on a DEC–10 computer in single precision arithmetic, i.e. with 8 significant digits.

Example 4.1. Let $\Omega := \{(x, y) : \frac{1}{2}x^2 + 2y^2 < 1\}$. We do not have to determine a set $\{\zeta_j\}_{j=1}^{\ell}$ computationally, but choose $\ell = 3$, $\zeta_1 = 0$, $\zeta_2 = 3 \cdot 2^{-3/2}$ and $\zeta_3 = -\zeta_2$. These ζ_j are the zeros of $T_3(z)$, the Chebyshev polynomial of the first kind of degree three for the interval $\left[-\sqrt{\frac{3}{2}}, \sqrt{\frac{3}{2}}\right]$ between the foci of Ω. The boundary $\partial\Omega$ can be approximated well by a level curve of $|T_3(z)|$. This is illustrated by [13, Figure 3.7]. We therefore can allocate the collocation points z_j by using the function σ^* in (3.11), where σ^* is obtained from step (1). We let $\kappa := 1/2$, $m := 15$ and $n := 16$. Figure 4.1 shows the boundary curve $\partial\Omega$, the poles $\zeta_j, 1 \leq j \leq 3$ (marked with crosses), and 32 collocation points (marked with dashes). We obtain

$$(4.1) \qquad \| \varepsilon \|_{\partial\Omega} := 3 \cdot 10^{-3}, \quad \| \varepsilon_f \|_{\partial\Omega} := 2 \cdot 10^{-3}, \quad \| \varepsilon_g \|_{\partial\Omega} := 2 \cdot 10^{-3}.$$

The solution of the complex interface problem is

$$f(z) := \frac{1}{6}z, \quad g(z) := -(z + (z^2 - 3/2)^{1/2})^{-1},$$

where we choose a branch of the square root that is analytic exterior to the interval $\left[-\sqrt{\frac{3}{2}}, \sqrt{\frac{3}{2}}\right]$.

For comparison, we change the subspace Q_m and now let $\ell := 1$ and $\zeta_1 := 0$. We generate collocation points for $\kappa = 1/2$ using the function $\tilde{\sigma}$ as described in step (3). The rate of convergence of ε with increasing m and n is now much slower than above, and the basis (3.22) and the linear system of equations (3.2) are severely ill–conditioned for $m = 15$ and $n = 16$. The combination of slow convergence and an ill–conditioned linear system of equations made it impossible to obtain an error smaller than $\| \varepsilon \|_{\partial\Omega} = 0.1$. This illustrates the importance of the choice of subspace and basis. □

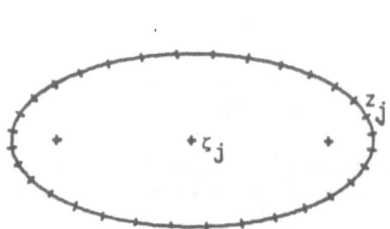

Figure 4.1: Collocation points and poles for Example 4.1.

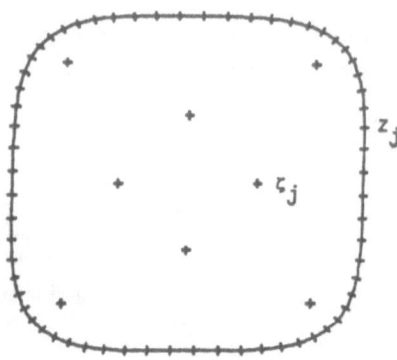

Figure 4.2: Collocation points and poles for Example 4.2.

Example 4.2. Let $\Omega := \{(x,y) : x^4 + y^4 < 1\}$. We choose $\ell = 8$ and the eight poles ζ_j given by $\pm 04i$, $\pm 0.4i$ and $(\pm 1 \pm i)/\sqrt{2}$. These poles are marked with crosses in Figure 4.2. They have been determined as described in step (2). The product $\prod_{j=1}^{8} |z - \zeta_j|$ is nearly constant on $\partial\Omega$, and we therefore use the function σ^* obtained from step (1) to allocate the 64 collocation points z_j shown on Figure 4.2. We obtain Table 4.1.

m	n	$\|\varepsilon\|_{\partial\Omega}$
15	16	$5 \cdot 10^{-3}$
31	32	$7 \cdot 10^{-5}$

Table 4.1: Error $\|\varepsilon\|_{\partial\Omega}$. □

5. Examples for the Helmholtz Equation

In this section we present computed examples for an exterior Dirichlet problem for the Helmholtz equation. Such boundary value problems arise, e.g., from scattering problems for electromagnetic or acoustic waves. The computed examples suggest that an allocation of singular points ζ_j and collocation points z_j that is suitable for the complex interface problem is also appropriate for Dirichlet problems for the Helmholtz equation.

Introduce the Hankel functions of the first kind

(5.1a) $$H_0(\lambda, r) := -i\frac{\pi}{2}\Big(J_0(\lambda^{1/2}r) + iY_0(\lambda^{1/2}r)\Big),$$

(5.1b) $$H_k(\lambda, r) := -i\Big(\frac{\lambda^{1/2}}{2}\Big)^k \frac{\pi}{(k-1)!}\Big(J_k(\lambda^{1/2}r) + iY_k(\lambda^{1/2}r)\Big), \quad k \neq 0,$$

which are scaled so that for $z = re^{i\theta}$

(5.2) $$\begin{cases} H_k(\lambda, r)e^{-ik\theta}/z^k \to 1 & \text{as } r \to 0, \quad k \neq 0, \\ H_0(\lambda, r)/\ln(z) \to 1 & \text{as } r \to 0, \end{cases}$$

see [1, Chapter 9] for details. Let $\Omega := \{(x,y) : \big(\frac{x}{2}\big)^2 + y^2 < 1\}$ and introduce

$$g(z) := H_0(\lambda, |z - \sqrt{3}|) + H_0(\lambda, |z + \sqrt{3}|).$$

Then g is singular at the foci of Ω. We would like to solve the exterior Dirichlet problem

(5.3a) $$\Delta u + \lambda u = 0 \text{ in } \Omega_e,$$

(5.3b) $$u = g \text{ on } \partial\Omega,$$

(5.3c) $$\lim_{r \to \infty} r^{1/2}\left(\frac{\partial u}{\partial r} - i\lambda^{1/2}u\right) = 0,$$

where $z = x + iy = re^{i\theta}$. The radiation condition (5.3c) is assumed to hold uniformly in θ as r tends to infinity.

Example 5.1. We seek a solution of (5.3) of the form

(5.4) $$u_n(z) := \sum_{k=-\lfloor n/2\rfloor}^{\lfloor n/2\rfloor} a_k \, H_k(\lambda, r)e^{-ik\theta}, \quad z = re^{i\theta}.$$

Then u_n satisfies (5.3a) and (5.3c) for any coefficients $a_k \in \mathbb{C}$. The only singular point in the finite plane of u_n is $z = 0$. By (5.2) the asymptotic behaviour of u_n is given by

$$u_n(z) \sim a_0 \ln(z) + \sum_{\substack{k=-\lfloor n/2\rfloor \\ k \neq 0}}^{\lfloor n/2\rfloor} a_k \, z^{-|k|} \quad \text{as } z \to 0.$$

This suggests the use of the same collocation points as when we approximate a function analytic in $\Omega_c \cup \{\infty\}$ by elements from $Q_{\lfloor n/2\rfloor} := \{z^{-1}, z^{-2}, \ldots, z^{-\lfloor n/2\rfloor}\}$. We therefore let $\kappa := 0$ in the integral equation in step (3), and obtain $\tilde{\sigma}$, which we use to determine m collocation points z_k by (3.11). This yields a linear system of equations

(5.5) $$u_n(z_j) = g(z_j), \quad 1 \leq j \leq m,$$

for the coefficients $\{a_k\}_{k=-\lfloor n/2\rfloor}^{\lfloor n/2\rfloor}$ of (5.4). If $m > 2\lfloor n/2\rfloor + 1$, then the linear system (5.5) is solved in the least squares sense.

n	5	9	17	33
m				
n	$3.3 \cdot 10^{-1}$	$1.9 \cdot 10^{-1}$	$6.2 \cdot 10^{-2}$	$7.1 \cdot 10^{-3}$
$2(n-1)+1$	$2.6 \cdot 10^{-1}$	$1.2 \cdot 10^{-1}$	$3.4 \cdot 10^{-2}$	$3.7 \cdot 10^{-3}$
$4(n-1)+1$	$2.4 \cdot 10^{-1}$	$1.2 \cdot 10^{-1}$	$3.4 \cdot 10^{-2}$	$3.7 \cdot 10^{-3}$

Table 5.1: $\| u_n - g \|_{\partial\Omega}$

The errors in Table 5.1 are computed by evaluating $u_n - g$ at many more than $4n$ points on $\partial\Omega$. Table 5.1 indicates that convergence takes place. Moreover, the small decrease in the error when m is chosen larger than n suggests that the collocation points are well allocated. The table indicates that $m \approx 2n$ is suitable. □

Example 5.2. In this example we solve the Dirichlet problem (5.3) using a different subspace than in Example 5.1. We let the approximate solution be singular at points analogous to the ζ_k of Figure 4.1, i.e. we let $\zeta_1 := 0$, $\zeta_2 := \frac{2}{3}$ and $\zeta_3 := -\zeta_2$. These ζ_j are the zeros of the Chebyshev polynomial of the first kind of degree three for the interval between the foci of Ω. Let

(5.6) $$u_n(z) := \sum_{j=1}^{3}\left(\sum_{k=-\lfloor (n-1)/3\rfloor}^{\lfloor (n-1)/3\rfloor} a_{jk} \, H_k(\lambda, r_j)e^{-ik\theta_j}\right),$$

where $z - \zeta_j =: r_j e^{i\theta_j}$. With this choice of the ζ_j, the product $\prod_{j=1}^{3} | z - \zeta_j |$ is nearly constant on $\partial\Omega$. We use σ^* obtained from step (1) to compute the collocation points z_k, distributed as shown in Figure 5.1.

We note that we would like a subspace with only few distinct singularities ζ_j, since this enables rapid evaluation of the Hankel functions by a three term recurrence relation. We solve the linear system of equations (5.5) with u_n now defined by (5.6) in the least squares sense.

n m	9	15	27
$2n - 3$	$1.9 \cdot 10^{-2}$	$2.5 \cdot 10^{-3}$	$1.7 \cdot 10^{-4}$
$4n - 9$	$1.9 \cdot 10^{-2}$	$2.4 \cdot 10^{-3}$	$1.7 \cdot 10^{-4}$

Table 5.2: $\| u_n - g \|_{\partial\Omega}$

The errors shown in Table 5.2 are computed by evaluating $u_n - g$ at many more than $4n$ points on $\partial\Omega$. These errors appear to converge to zero more rapidly than the errors of Table 5.1. In the computation for Table 5.2 the linear system of equations is numerically singular for $m = n$. □

6. Conclusion

An investigation of a model problem has shed light on how to select subspace, basis and collocation points in the boundary collocation method. This has enabled us to solve some engineering problems in industry. However, much work remains in order to fully understand the boundary collocation method.

References

[1] Abramowitz, M; Stegun, I.E., eds.: Handbook of Mathematical Functions. National Bureau of Standards 1972.

[2] Bates, R.H.T.: Analytic constraints on electromagnetic field computations. IEEE Trans. Microwave Theory Tech. MTT-23 (1975) 608–623.

[3] Bates, R.H.T.; Ng, F.L.: Point matching computation of transverse resonances. Int'l J. Numer. Meth. Engng. 6 (1973) 155–168.

[4] Bieberbach, L: Conformal Mapping. New York: Chelsea 1964.

[5] Davis, P.J.: The Schwarz Function and its Applications. Math. Assoc. Amer. 1974.

[6] Gautschi, W.: Conditions of polynomials in power form. Math. Comput. 33 (1979) 343–352.

[7] Jones, D.S.: Methods in Electromagnetic Wave Propagation, vol. 1. Oxford: Oxford University Press 1987.

[8] Ohlsson, T.; Risman, P.O.: Temperature distribution of microwave heating – spheres and cylinders. J. Microwave Power 13 (1978) 303–310.

[9] Rehbinder, G.; Reichel, L.: Heat conduction in a rock mass with an annular hot water storage. Int'l J. Heat and Fluid Flow 5 (1984) 131–137.

[10] Reichel, L.: On the determination of boundary collocation points for solving some problems for the Laplace operator. J. Comput. Appl. Math. 11 (1984) 175–196.

[11] Reichel, L.: On the numerical solution of some 2–d electromagnetic interface problems by the boundary collocation method. Comput. Math. Appl. Mech. Engng. 53 (1985) 1–11.

[12] Reichel, L.: Numerical methods for analytic continuation and mesh generation. Constr. Approx. 2 (1986) 23–39.

[13] Reichel, L.: Some computational aspects of a method for rational approximation. SIAM J. Sci. Stat. Comput. 7 (1986) 1041–1057.

[14] Reichel, L.: A fast method for solving certain integral equations with application to conformal mapping. J. Comput. Appl. Math. 14 (1986) 125–142.

[15] Walsh, J.L.: Interpolation and Approximation by Rational Functions in the Complex Domain, 5th ed. Providence: Amer. Math. Society 1969.

MINI-SYMPOSIUM 2
(organiser: R. Mattheij)

Minisymposium

On flow-induced vibrations of overhead
transmission lines.

Organiser: R. Mattheij (Eindhoven)
Speakers : S. Rienstra (Eindhoven), P. Hagedorn (Darmstadt),
A. van der Burgh (Delft)

The history of flow-induced vibrations dates back to the ancient Greek and their aeolian harp with wind driven strings. Indeed, one technically important type of vibration of overhead transmission lines is exactly due to this very effect of a Von Kármán street of shed vortices coupled to elastic vibrations of the cable. Other aeroelastic phenomena associated with overhead transmission lines are galloping, wake-induced flutter, and buffeting.

All these vibrations may in some way damage the line or cause complete failure of the line (short circuit, fatigue), so there is a need to know more about these vibrations and to prevent or reduce their occurence by improved design rules or damping devices.

Although discussions on the deterministic or stochastic nature of aeolian vibrations of quasi-cylindrical conductors are still going on, rather simpel mathematical models have been developed on the basis of which for instance the acting of Stockbridge dampers is understood fairly well. The galloping phenomenon is still far from understood. The study of this low frequency vibration requires consideration of transverse and torsional vibrations as well as conductor stretching.

Still little is known on instability, resonance and limiting mechanisms Nevertheless, as a growing interest for this fascinating subject may be observed, of which the most sophisticated models concern systems of nonlinear stochastic partial differential equations, there is hope that model studies and experiments will contribute to the development of a useful theory which may solve the many practical problems of vibrating transmission lines.

In the papers presented at the symposium some aspects of aeolian and galloping type of vibration are studied. As the aeolian vibration is rather well understood the theory may include source, elastic response, and damping mechanism. As a result this vibration can be controlled. On the other hand, the galloping phenomenon is far from understood. It is therefore wise to reduce the problem to its essential mechanism, or to consider only the dynamics of the associated free vibration.

J. Manley et al. (eds.), Proceedings of the Third European Conference on Mathematics in Industry, 132.
© 1990 *Kluwer Academic Publishers and B. G. Teubner Stuttgart.*

NON-LINEAR FREE VIBRATIONS
OF COUPLED SPANS OF SUSPENDED CABLES

S.W. Rienstra

1. Summary

The problem of free in-plane nonlinear nearly harmonic vibrations of elastic suspended cables is investigated, with particular emphasis on the configuration of multiple spans, coupled via suspension strings, which is relevant in the context of overhead transmission lines. A systematic asymptotic theory is developed, for a suitable set of small parameters based on a shallow geometry and the presence of only transversal waves. The finally obtained reduced set of equations is solved by a variant of the Lindstedt-Poincaré technique. The (non-trivial) solutions for multiple spans appear to be gravity waves, considerably different from the elasto-gravity waves in the symmetric single span configuration (which is included for reference). An internal resonance is discovered giving a new explanation to the practically observed asymmetry of the vertical displacement. Application of the theory to describe the reaction force induced to a suspension string is indicated.

2. Introduction

In the present paper we will present a study of non-linear free vibrations of suspended elastic cables.

The problem was motivated by research on an aero-elastic instability of overhead transmission lines, called galloping. This galloping is generated by a combination of wind and ice rain and results in a slow almost vertical periodic cable motion. For high enough amplitudes neighbouring conductors may touch each other, causing a short circuit and structural damage to the cables [10]. Although recognised and studied for more than fifty years, the problem is far from being solved. For example, it appears to be still not possible to design the system of towers and suspended electricity cables, possibly equipped with dampers, to be free of galloping.

An important observation is that galloping is a motion of the cable very close to a free vibration, since the forces (wind) are only small. Various aspects of galloping are therefore inherent to the free motion the cable is close to, and the study of the dynamics of free vibrating suspended cables is essential for understanding galloping. For example, the coupling between harmonics, the relation between tension and displacement, and the existence of internal resonances can be found by studying the equivalent free vibration.

J. Manley et al. (eds.), Proceedings of the Third European Conference on Mathematics in Industry, 133–144.
© 1990 Kluwer Academic Publishers and B. G. Teubner Stuttgart.

Although the related theory of tensed strings is a classic and well established part of theoretical mechanics [5], the theory of a vibrating heavy elastic suspended cable is relatively new. A number of investigations on the problem have been published for the geometry of a single span, i.e., with fixed ends [1,2,3,4,7,9]. However, the geometry we will mainly consider, which is the most relevant in overhead transmission line practice, consists of a series of coupled spans, and to our knowledge this has not yet been treated in the literature. Nevertheless, our theory is equally well applicable to the single span configuration, and this will therefore be included for reference.

The analysis will consist of three parts. First, we establish the model adopted, with differential equations and boundary conditions. This is relatively standard. Then we derive an asymptotically approximate problem by introducing a small parameter based on assumptions on sag, instationary amplitude, and transversal and longitudinal wave length, which are essentially the same assumptions necessary for the well-known parabola approximation of the stationary solution. The resulting (still nonlinear) problem is similar or nearly similar to that of other studies. We believe, however, that our approach is more systematic and consistent. Finally, we solve the equations by a variant of the Lindstedt-Poincaré technique [6]. For this we assume the existence of a periodic nearly harmonic solution, and expand the dependent variables in a perturbation amplitude power series (on practical grounds restricted here to three terms). The full solution includes a variety of standing and propagating waves, and some additional conditions of symmetry and regularity are applied to define the solution further.

In addition, the reaction force in the suspension string, relevant to measurements of galloping, is briefly discussed.

3 Model

3.1 Differential equations and boundary conditions

Consider a cable, fixed at the outer ends, and divided into N equal spans by $N-1$ supports. These supports are inextensible suspension strings of length a and negligible weight, suspended from fixed pivots separated by a distance S, the span size. The suspension string allows the span end to describe a circle of radius a, and thus provides a coupling between adjacent spans. Through all spans the cable properties are the same. The cable is linearly elastic, with negligible bending stiffness and friction effects, of uniform undeformed cross-sectional area A, mass per unit length m, and Young's modulus E, and with a length per span L when the cable is free of tension. We parametrise the position along the cable (per span) by the variable $\ell \in [0, L]$, such that this is just the arc length when the cable is unstretched. The time variable is $\bar{t}$. The cable moves in a vertical plane provided with a Cartesian coordinate system orientated such that the gravity vector points into the negative "y"-direction. The cable's position is

$(\overline{x}(\ell,\overline{t}),\overline{y}(\ell,\overline{t}))$ with a corresponding tension $\overline{\tau}(\ell,\overline{t})$.

The equations determining $\overline{x},\overline{y}$, and $\overline{\tau}$ are Hooke's law relating stress and strain, and Newton's law applied to the condition of equilibrium of internal and external forces on a cable element. The finally resulting equations are

(3.1)
$$\frac{\partial}{\partial \ell}\left(\frac{\overline{\tau}}{1+\overline{\tau}/EA}\frac{\partial \overline{x}}{\partial \ell}\right)=m\frac{\partial^2 \overline{x}}{\partial \overline{t}^2}$$

(3.2)
$$\frac{\partial}{\partial \ell}\left(\frac{\overline{\tau}}{1+\overline{\tau}/EA}\frac{\partial \overline{y}}{\partial \ell}\right)=mg+m\frac{\partial^2 \overline{y}}{\partial \overline{t}^2}$$

(3.3)
$$\left(\frac{\partial \overline{x}}{\partial \ell}\right)^2+\left(\frac{\partial \overline{y}}{\partial \ell}\right)^2=\left(1+\frac{\overline{\tau}}{EA}\right)^2$$

while it may be noted that

(3.4)
$$\frac{\partial \overline{x}}{\partial \ell}=(1+\overline{\tau}/EA)\cos\psi,\quad \frac{\partial \overline{y}}{\partial \ell}=(1+\overline{\tau}/EA)\sin\psi.$$

g denotes the gravity acceleration, and ψ the angle between cable tangent and the horizontal. The boundary and coupling conditions are at

(i) rigid supports

(3.5)
$$\overline{x}=0,\ \overline{y}=0\qquad(\ell=0)$$
$$\overline{x}=S,\ \overline{y}=0\qquad(\ell=L)$$

(ii) suspension strings

(3.6)
$$\overline{x}^2+(\overline{y}-a)^2=a^2\qquad(\ell=0)$$
$$(\overline{x}-S)^2+(\overline{y}-a)^2=a^2\qquad(\ell=L)$$

(3.7)
$$[\overline{\tau}\cos(\phi-\psi)]_{\pm}=0$$

The brackets $[.]_{\pm}$ denote the difference between the value of the quantity at the right and the left side of the supports; ϕ is the angle of the suspension string with the vertical. The force necessary to maintain the string end at its position is usefully split up into two orthogonal components. One, tangential to the circle described by the string end, cannot be sustained by the string and has to vanish, resulting into equation (3.7). The other, however, is directed along the string, and induces the reaction force

(3.8)
$$F_{ss}=[\overline{\tau}\sin(\phi-\psi)]_{\pm}.$$

3.2 Small parameters and modes of vibration

The type of motion we are interested in is further specified by
- The ratio of sag D and cable length L is small (typically 1/30), so

$$\varepsilon = D/L \to 0.$$

- The vertical displacement is of the order of the sag, so

$$\bar{y}/L = O(\varepsilon).$$

- The transversal wave length λ_T is of the order of L, so

$$\lambda_T/L = O(1).$$

- The longitudinal wave length ("sound") λ_L is large compared to L. Consistent with practice is the estimate

$$\lambda_L/L = O(1/\varepsilon).$$

- The string length a is of the order of the sag, so

$$a/L = O(\varepsilon).$$

Furthermore, we assume a nearly harmonic vibration, with a single dominating frequency $\bar{\omega}$. Higher harmonics not generated by the first ($\sim$ eigensolutions) will be excluded. We will only investigate the effect of non-trivial coupling. For example, a sequence of single span solutions in phase, although a valid solution, will not be considered for the coupled configuration.

4 Asymptotic analysis

4.1 Reduced problem

The basic small parameter ε will be utilised to reduce the above general problem to the asymptotically leading order problem, i.e., in terms of dimensionless variables of $O(1)$, independent of ε.

Since the longitudinal wave speed is $c_L = (EA/m)^{\frac{1}{2}}$, we have $\lambda_L/L = c_L/\bar{\omega}L = O(1/\varepsilon)$, so the scaled frequency and corresponding time variable are given by

$$\bar{\omega} = \omega\varepsilon(EA/m)^{\frac{1}{2}}/L, \qquad \bar{t} = tL(m/EA)^{\frac{1}{2}}/\varepsilon.$$

The transversal wave velocity is $c_T = (\bar{\tau}/m)^{\frac{1}{2}}$, so $\lambda_T/L = c_T/\bar{\omega}L = O(1)$. This yields for the spatial coordinate and the tension

$$\ell = sL,$$

$$\bar{\tau} = \varepsilon^2 EAT.$$

Since $\bar{y}/L = O(\varepsilon)$, we introduce

$$\bar{y} = \varepsilon LY.$$

With the above estimates substituted in (3.3) we obtain $\bar{x}_\ell = 1 + O(\varepsilon^2)$, and so

$$\bar{x} = L(s + \varepsilon^2 X).$$

Finally, we investigate the role of gravity. Substitute the present results into equation (3.2) to find the term $mgL/EA\varepsilon^3$ next to terms of $O(1)$. So it has to be $O(1)$ or smaller. Suppose it is small, then the stationary solution would be to leading order $Y \equiv 0$, so $D = 0$, which is contradictory to our assumptions. So the term is $O(1)$ and we introduce

$$\mu = mgL/8EA\varepsilon^3 = O(1).$$

Since we will only consider the ε-approximation to leading order, we may as well assume $\omega, T, Y,$ and X to be ε-independent with an (a priori) relative error of $O(\varepsilon^2)$.

The basic equations (3.1-3) are then reduced to

$$(4.1) \qquad \frac{\partial}{\partial s}T = 0, \quad \frac{\partial}{\partial s}\left(T\frac{\partial}{\partial s}Y\right) = 8\mu + \frac{\partial^2}{\partial t^2}Y, \quad \frac{\partial}{\partial s}X + \frac{1}{2}\left(\frac{\partial}{\partial s}Y\right)^2 = T$$

It is convenient to split up the solution into a stationary and instationary part:

$$X = X_0 + x, \ Y = Y_0 + y, \ T = T_0 + \tau,$$

with boundary conditions for the stationary part $Y_0(0) = Y_0(1) = 0$, $Y_0(\frac{1}{2}) = -1$, $X_0(0) = 0$, $X_0(1) = S_0$, where S_0 is given by $S = L(1 + \varepsilon^2 S_0 + O(\varepsilon^4))$. The stationary solution is the well-known parabola shape

$$(4.2) \qquad Y_0 = -4(s - s^2), \quad T_0 = \mu, \quad X_0 = \mu - \frac{4}{3}(1 + (2s - 1)^3), \quad S_0 = \mu - \frac{8}{3}.$$

If we substitute (4.2) into (4.1) we obtain our *fundamental instationary problem*

$$(4.3) \qquad \frac{\partial \tau}{\partial s} = 0$$

$$(4.4) \qquad (\mu + \tau)\frac{\partial^2 y}{\partial s^2} + 8\tau = \frac{\partial^2 y}{\partial t^2}$$

$$(4.5) \qquad \frac{\partial x}{\partial s} + 4(2s - 1)\frac{\partial y}{\partial s} + \frac{1}{2}\left(\frac{\partial y}{\partial s}\right)^2 = \tau$$

The boundary conditions at rigid supports follow readily:

$$(4.6) \qquad\qquad x(0) = x(1) = y(0) = y(1) = 0$$

For the conditions at the suspension string we observe that $a\phi/L = O(\varepsilon^2)$ as it is of the order of the x-variations, so $\phi = O(\varepsilon)$, and we obtain the conditions

$$(4.7) \qquad\qquad y(0) = y(1) = 0, \quad [x]_\pm = 0, \quad [\tau]_\pm = 0.$$

An immediate consequence of (4.3) and (4.7) is that τ is constant in s and the same in all spans. This is of course only true if $N = O(1)$, since otherwise a small deflection from the constant would accumulate on a larger scale. Another way to portray this is by considering, in a matched expansion terminology, the present $O(1)$ domain as the inner region of the longitudinal wave regime of $O(1/\varepsilon)$.

Finally, the reaction force (3.8) is (to a relative error of $O(\varepsilon^2)$)

$$(4.8) \qquad\qquad F_{ss} = -EA\varepsilon^3 T \left[\frac{\partial Y}{\partial s}\right]_\pm = mgL + EA\varepsilon^3 f_{ss}$$

with $f_{ss} = 8\tau - (\mu + \tau)[y]_\pm$. The constant mgL is just the weight of a span.

4.2 Amplitude power series expansion

The Lindstedt-Poincaré technique [6] involves the assumption of a periodic solution with fundamental frequency ω, the introduction of the amplitude of the linearised solution as a small parameter δ, and then expanding the full solution into a δ-power series, starting with $O(\delta)$. Of course, we could have included the foregoing stationary solution (4.2) as the first, $O(1)$, term in this series, but this stationary part is so important in itself that we have taken it apart.

Since ω will also depend on δ it is convenient to introduce $t' = \omega t$. Now we assume

$$(4.9) \qquad\qquad y = \delta y_1 + \delta^2 y_2 + \delta^3 y_3 + \cdots$$

and similarly for x and τ, and

$$(4.10) \qquad\qquad \omega = \omega_0 + \delta^2 \omega_2 + \cdots$$

Note that $\omega_1 = 0$ since ω should not depend on the sign of δ. Introduce the notation $' \equiv \partial/\partial s$, $\dot{} \equiv \partial/\partial t'$. The equations that result from substitution of (4.9,10) into (4.3-5) and collecting like powers of δ are then

$$(4.11) \qquad\qquad \begin{cases} \mu y_1'' + 8\tau_1 - \omega_0^2 \ddot{y}_1 = 0 \\ x_1' + 4(2s - 1)y_1' = \tau_1 \end{cases}$$

$$(4.12) \quad \begin{cases} \mu y_2'' + 8\tau_2 + \tau_1 y_1'' - \omega_0^2 \ddot{y}_2 = 0 \\ x_2' + 4(2s-1)y_2' + \frac{1}{2}(y_1')^2 = \tau_2 \end{cases}$$

$$(4.13) \quad \begin{cases} \mu y_3'' + 8\tau_3 + \tau_2 y_1'' + \tau_1 y_2'' - \omega_0^2 \ddot{y}_3 - 2\omega_0\omega_2 \ddot{y}_1 = 0 \\ x_3' + 4(2s-1)y_3' + y_1'y_2' = \tau_3 \end{cases}$$

with $\tau_1, \tau_2, \tau_3, \cdots$ constant in s. The interesting solutions here are based on the harmonic solutions of the linearised problem, which generate via the nonlinear coupling sub- and super harmonics in the higher order terms. So we put

$$(4.14) \quad y_1 = y_{11}\sin(t'), \quad y_2 = y_{20} + y_{22}\cos(2t'), \quad y_3 = y_{31}\sin(t') + y_{33}\sin(3t'),$$

and similarly for x and τ. Substitution of (4.14) into (4.11-13) and collecting the harmonics yields

$$(4.15) \quad \begin{cases} \mu y_{11}'' + 8\tau_{11} + \omega_0^2 y_{11} = 0 \\ x_{11}' + 4(2s-1)y_{11}' = \tau_{11} \end{cases}$$

$$(4.16) \quad \begin{cases} \mu y_{20}'' + 8\tau_{20} + \frac{1}{2}\tau_{11}y_{11}'' = 0 \\ x_{20}' + 4(2s-1)y_{20}' + \frac{1}{4}(y_{11}')^2 = \tau_{20} \end{cases}$$

$$(4.17) \quad \begin{cases} \mu y_{22}'' + 8\tau_{22} - \frac{1}{2}\tau_{11}y_{11}'' + 4\omega_0^2 y_{22} = 0 \\ x_{22}' + 4(2s-1)y_{22}' - \frac{1}{4}(y_{11}')^2 = \tau_{22} \end{cases}$$

$$(4.18) \quad \begin{cases} \mu y_{31}'' + 8\tau_{31} + \omega_0^2 y_{31} + (\tau_{20} - \frac{1}{2}\tau_{22})y_{11}'' + \tau_{11}(y_{20}'' - \frac{1}{2}y_{22}'') + 2\omega_0\omega_2 y_{11} = 0 \\ x_{31}' + 4(2s-1)y_{31}' + y_{11}'(y_{20}' - \frac{1}{2}y_{22}') = \tau_{31} \end{cases}$$

$$(4.19) \quad \begin{cases} \mu y_{33}'' + 8\tau_{33} + 9\omega_0^2 y_{33} + \frac{1}{2}\tau_{22}y_{11}'' + \frac{1}{2}\tau_{11}y_{22}'' = 0 \\ x_{33}' + 4(2s-1)y_{33}' + \frac{1}{2}y_{11}'y_{22}' = \tau_{33} \end{cases}$$

It may be noted that in the usual formulation of the Lindstedt-Poincaré technique ω_2 is determined by the condition of vanishing secular terms. Here we have simplified this to the condition that there exists a solution of the present type.

5 Solution

5.1 Coupled spans

A most important property of vibrating coupled spans is the fact that to leading order the tension vanishes and contains no first harmonic. This makes the motion of a coupled span considerably different from the symmetric modes of a single span. In a coupled span the restoring force is basically gravity, without tension variation to leading order ($\sim$ first harmonic), while in a single span symmetric mode both gravity and elasticity are equally important. The respective motion may therefore be called: gravity and elasto-gravity waves.

By standard techniques we find for the i-th span ($1 \leq i \leq N$) the general solution

$$(5.1) \qquad \tau_{11} = 0, \quad y_{11} = A_i \sin(ks), \quad \text{where} \quad k = \omega_0/\sqrt{\mu} = (2n+1)\pi$$

$$x_{11} = B_i - 4(2s-1)A_i \sin(ks) - 8A_i \cos(ks)/k$$

$$B_i = 8\big(A_i + 2(A_1 + A_2 + \cdots + A_{i-1})\big)/k$$

$$(5.2) \qquad \tau_{20} = \tfrac{3}{8}\omega_0^2 \overline{A^2}/(16+3\mu), \qquad y_{20} = 4\tau_{20}(s-s^2)/\mu$$

$$x_{20} = C_i + \tau_{20}s + \tfrac{8}{3}\tau_{20}(1+(2s-1)^3)/\mu - \tfrac{1}{8}k^2 A_i^2(s+\sin(2ks)/2k)$$

$$C_i = (i-1)\tau_{20}(1+16/3\mu) - \tfrac{1}{8}k^2(A_1^2 + \cdots + A_{i-1}^2)$$

$$(5.3) \qquad \tau_{22} = \tfrac{1}{8}k^2\omega_0^2 \overline{A^2}/(16-\omega_0^2), \qquad y_{22} = 2\tau_{22}(\cos(2ks)-1)/\omega_0^2$$

$$x_{22} = D_i + \tau_{22}s - 8\tau_{22}\big(1+(2s-1)\cos(2ks)-\sin(2ks)/k\big)/\omega_0^2 + \tfrac{1}{8}k^2 A_i^2(s+\sin(2ks)/2k)$$

$$D_i = -(i-1)\tau_{22}(16/\omega_0^2 - 1) + \tfrac{1}{8}k^2(A_1^2 + \cdots + A_{i-1}^2)$$

$$(5.4) \qquad \tau_{31} = 0, \qquad y_{31} = 0$$

$$x_{31} = E_i + A_i\big(4k\tau_{20}(2s-1)\sin(ks) + (8\tau_{20}+\tau_{22})\cos(ks) + \tfrac{1}{3}\tau_{22}\cos(3ks)\big)/\mu k$$

$$E_i = -B_i(\tau_{20} + \tfrac{1}{6}\tau_{22})/\mu$$

$$(5.5) \qquad \tau_{33} = 0, \qquad y_{33} = \tfrac{1}{16}\tau_{22}A_i\sin(ks)/\mu$$

$$x_{33} = F_i - \tau_{22}A_i\big(\tfrac{1}{4}k(2s-1)\sin(ks) + \tfrac{3}{2}\cos(ks) + \tfrac{1}{3}\cos(3ks)\big)/\mu k$$

$$F_i = \tfrac{11}{48}\tau_{22}B_i/\mu$$

where $\overline{A^2} = (A_1^2 + A_2^2 + \cdots + A_N^2)/N$. The frequency shift necessary for a solution y_{31} is

$$(5.6) \qquad \omega_2 = \tfrac{1}{4}\omega_0(2\tau_{20} - \tau_{22})/\mu,$$

and the amplitudes A_i are further restricted by the condition

(5.7) $$A_1 + A_2 + \cdots + A_N = 0.$$

Note the breakdown of the solution if $\omega_0 = 4$ by the vanishing denominator of τ_{22}. Although further research is necessary to reveal the character of this singularity, it is to be interpreted as a resonance between first and second harmonic.

Equation (5.7) does not determine a unique solution (except for $N = 2$). Some special cases, however, are useful to be considered in detail.

A most regular solution is found if N is *even* with

(5.8) $$A_i = (-1)^i,$$

since this is independent of N. As every even suspension string is now motionless, it is in fact just a sequence of (N=2)-solutions. Furthermore, the suspension string reaction force (4.8) in δ-expanded form only depends on the tension and simplifies to $f_{ss} = 8\delta^2\tau_{22}\cos(2t') + O(\delta^4)$, since by symmetry $[y'_{11}]_\pm = [y'_{33}]_\pm = 0$, and inertial effects are absent.

If N is *odd*, we cannot have all A_i equal in magnitude, but an almost uniform middle region $2 \leq i \leq N - 1$ is obtained with $A_1 = A_N = -\frac{1}{2}$, $A_i = (-1)^i$ $(2 \leq i \leq N - 1)$.

5.2 Single span

The asymmetric modes of a single span are very similar to the periodic solution (5.8) for an even number of coupled spans. The *only* difference is the wave number k. Instead of an odd it is now an even multiple of π : $k = 2n\pi$.

The symmetric modes are considerably different, and not reported to occur often with galloping. However, it is the configuration considered most in the literature, so we will, for reference, briefly present it here in the context of our analysis.

It is convenient to introduce $\hat{\tau} = \tau/\mu$, and $z = s - \frac{1}{2}$ where τ is any $\tau_{11}, \cdots$. The solution is then

(5.9) $$\hat{\tau}_{11} = \tfrac{1}{8}k^2\cos(\tfrac{1}{2}k), \qquad y_{11} = \cos(kz) - \cos(\tfrac{1}{2}k)$$

where $k = \omega_0/\sqrt{\mu}$ is a solution of the equation $\tan(\tfrac{1}{2}k) = \tfrac{1}{2}k(1 - \tfrac{1}{64}\omega_0^2)$

$$x_{11} = \hat{\tau}_{11}\mu z + 8\left(\sin(kz) - kz\cos(kz)\right)/k$$

(5.10) $$\hat{\tau}_{20} = \tfrac{3}{8}k^2\left(3\sin(k)/k - 2\cos(k) - 1\right)/(16 + 3\mu)$$

$$y_{20} = -\tfrac{1}{2}\hat{\tau}_{11}(\cos(kz) - \cos(\tfrac{1}{2}k)) - \hat{\tau}_{20}(4z^2 - 1)$$

$$x_{20} = \hat{\tau}_{20}\mu z + \tfrac{64}{3}\hat{\tau}_{20}z^3 - \tfrac{1}{2}k\cos(\tfrac{1}{2}k)(\sin(kz) - kz\cos(kz)) + \tfrac{1}{16}k(\sin(2kz) - 2kz)$$

$$(5.11) \qquad \hat{\tau}_{22} = \tfrac{1}{24}k^3(5\sin(k) - 2\tan(k) - 3k)/(\omega_0^2 - 16 + 16\tan(k)/k)$$

$$y_{22} = 2(\hat{\tau}_{22} + \tfrac{2}{3}\hat{\tau}_{11}^2)(\cos(2kz) - \cos(k))/k^2\cos(k) - \tfrac{1}{6}\hat{\tau}_{11}(\cos(kz) - \cos(\tfrac{1}{2}k))$$

$$x_{22} = \hat{\tau}_{22}\mu z - \tfrac{4}{3}\hat{\tau}_{11}(\sin(kz) - kz\cos(kz))/k - \tfrac{1}{16}k(\sin(2kz) - 2kz)$$
$$+ 8(\hat{\tau}_{22} + \tfrac{2}{3}\hat{\tau}_{11}^2)(\sin(2kz) - 2kz\cos(2kz))/k^3\cos(k)$$

and a frequency shift of

$$(5.12) \qquad \omega_2 = \tfrac{2}{3}\omega_0\hat{\tau}_{11}\cos(\tfrac{1}{2}k)\Big[(5\hat{\tau}_{22} + 18\hat{\tau}_{20} + \tfrac{19}{3}\hat{\tau}_{11}^2)(2\tan(\tfrac{1}{2}k) - k)$$

$$- (3\hat{\tau}_{22} - 6\hat{\tau}_{20} + 5\hat{\tau}_{11}^2)k\tan^2(\tfrac{1}{2}k) + 4(\hat{\tau}_{22} + \tfrac{2}{3}\hat{\tau}_{11}^2)(\tan(\tfrac{1}{2}k) - \tan(k) + \tfrac{1}{2}k)\Big]$$

$$/k^3\left(2 + \cos(k) - 3\sin(k)/k\right)$$

The other third order terms become increasingly lengthy and complex, and can be found in [8]. The linear solution (5.9) is similar to what is presented in for example [2,3,9]. The higher order corrections are new. Nonlinear extensions to the linear theory have been described in [2,4,7], but for an equation in spatially averaged variables using an assumed shape function.

6 Discussion and examples

A most interesting result of the present analysis for coupled spans is the resonance frequency $\omega_0 = 4$, where the higher order terms become comparable to the first, and the solution breaks down. When ω_0 is not too far from 4, the singular denominator easily amplifies the second harmonic enough to render the displacement y asymmetrically upward or downward, depending on the sign of $\omega_0 - 4$ (i.e., the driven second harmonic is in or out of phase with its source, the first harmonic). The common situation for coupled spans of transmission lines seems to be a relative elasticity μ smaller than $16/\pi^2 = 1.62$, and so a frequency ω_0 smaller than 4, resulting into an (indeed reported) asymmetry upwards. However, by only changing the parameters a little we might as well have an asymmetry downwards. This is completely different from the always upward asymmetry of a single span symmetric mode, which is usually a smaller effect and has a displacement maximum always accompanied by a tension minimum.

We have plotted examples of a coupled multiple and a single span configuration. The vertical displacement dY at $s = \tfrac{1}{2}$ and tension variation dT are shown for a cable, given by $S = 300$ m, $mg = 10$ N/m, $EA = 15.10^6$ N, $D = 9$ m with $\varepsilon = 0.03$ and $\mu = 0.93$. In figure 1 we have a 2-span (eq.(5.8)), with $\omega_0 = 3.03$, and in figure 2 a single span, with $\omega_0 = 7.22$. Absence of the first harmonic in the 2-span tension, presence of other harmonics, asymmetry in dY, etc., are clear.

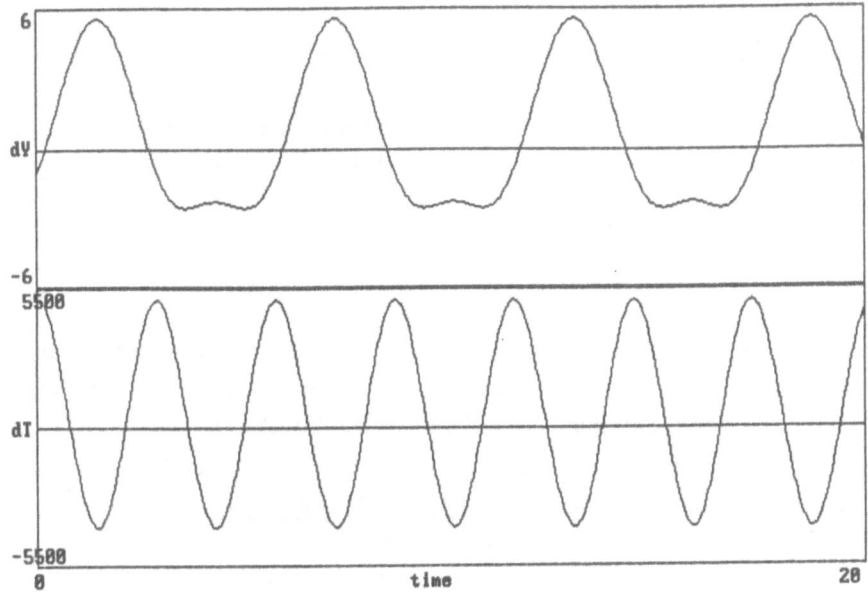

Figure 1. 2 span in 1st mode at $s = \frac{1}{2}$ ($\delta = .45$, frequency$=.1731$)

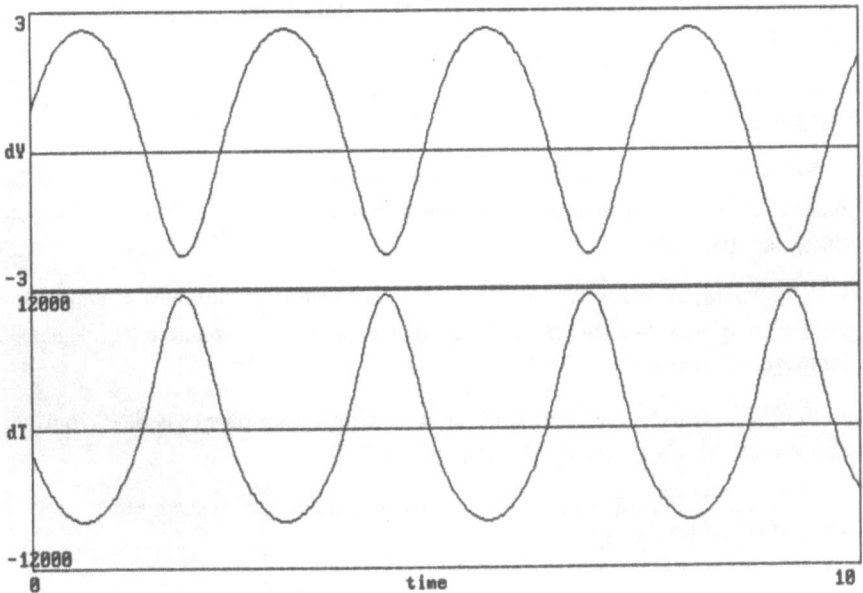

Figure 2. 1 span in 1st mode at $s = \frac{1}{2}$ ($\delta = .13$, frequency$=.4081$)

144

Acknowledgement

This paper is based on a study, prepared at the Mathematical Consulting Department, University of Nijmegen, under contract for the NV Provinciale Limburgse Elektriciteits Maatschappij PLEM, Maastricht. The support, interest, and stimulating discussions of (in alphabetical order) A.H.P. van der Burgh, P.H.Leppers, and J.Molenaar are gratefully acknowledged.

References

[1] A.H.P. van der Burgh, An asymptotic theory for the free vibrations of an iced two-conductor bundled transmission line, Asymptotic Analysis II, Lecture Notes in Mathematics 985, Springer-Verlag, Berlin, 413-430, 1983.

[2] P. Hagedorn and B. Schäfer, On non-linear free vibrations of an elastic cable, International Journal of Non-linear Mechanics, 15, 333-340, 1980.

[3] H.M. Irvine and T.K. Caughey, The linear theory of free vibrations of a suspended cable, Proceedings of the Royal Society of London A341, 299-315, 1974.

[4] A. Luongo, G. Rega, and F. Vestroni, Planar non-linear free vibrations of an elastic cable, International Journal of Non-linear Mechanics, 19(1), 39-52, 1984.

[5] P.M. Morse and H. Feshbach, Methods of Theoretical Physics, Mc Graw-Hill, New York, 1953.

[6] A.H. Nayfeh and D.T. Mook, Nonlinear Oscillations, John Wiley & Sons, New York, 1979

[7] G. Rega, F. Vestroni, and F. Benedettini, Parametric analysis of large amplitude free vibrations of a suspended cable, International Journal of Solids and Structures 20(2), 95-105, 1984.

[8] S.W. Rienstra, A nonlinear theory of free vibrations of single and coupled suspended elastic cables, Report WD 88-06, Katholieke Universiteit Nijmegen, The Netherlands, 1988

[9] A. Simpson, On the oscillatory motions of translating elastic cables, Journal of Sound and Vibration, 20(2), 177-189, 1972.

[10] A. Simpson, Wind-induced vibration of overhead power transmission lines, Sci. Prog. Oxford 68, 285-308, 1983.

S.W. Rienstra, Department of Mathematics and Computing Science,
University of Technology, P.O.box 513, 5600 MB Eindhoven, The Netherlands

Vibrations of Overhead Transmission Lines:
Computations and Experiments

P. HAGEDORN & M. KRAUS
Institut für Mechanik, TH Darmstadt, W.-Germany

1.   Introduction

While overhead transmission lines to the superficial spectator seem to be mainly static structures, different interesting dynamical phenomena reveal themselves to the more thorough observer.

These phenomena can for example be classified with respect to the time constants or with respect to dominant frequencies, as follows. Fast transient oscillations occur in long-rod insulator chains, when one of the porcelain insulators fails. This leads to a dynamic redistribution of the load in the remaining insulators, which may be essentialy finished after 40 milliseconds, for example. The high accellerations involved may cause failure of additional insulators or even of the complete chain. Mathematically the motion of the insulator chain can be modelled by a system of strongly nonlinear ordinary differential equations; the transport of mechanical energy (vibrations) in the cable – away from the chain – is essentially a damping mechanism for the chain. The computer simulations carried out during recent years have been checked against experiments, and excellent agreement has been observed (see [1]). Of course mechanical insulator failure is an exceptional phenomenon which almost never takes place, except under very special circumstances (accident during the construction of the line, people shooting with firearms at the insulator, etc.).

Somewhat slower but also transient phenomena are the short-circuit oscillations. Under normal working conditions electromagnetic forces are completely negligible in overhead transmission lines. For very high currents, such as those occuring during short circuits – often caused by lightning, for example – these forces may, however, be quite large. Particularly in conductor bundles the individual conductors oscillate violently during short-circuits; they sometimes hit each other destroying their surface by impact. Also the spacers may fail due to short circuit oscillations. This transient phenomenon has a typical duration of a few seconds (the short-circuit itself is usually

145

J. Manley et al. (eds.), Proceedings of the Third European Conference on Mathematics in Industry, 145–157.
© 1990 *Kluwer Academic Publishers and B. G. Teubner Stuttgart.*

much shorter). Mathematically the problem can be described by a system of coupled integro-partial differential equations. The numerical solution of these equations presents great difficulties due to the different time scales involved, since a relatively slow but large-amplitude transverse motion of the cables is superimposed to high frequency longitudinal oscillations and to localized fast transverse vibrations. The problem has been treated via finite elements and by other means, but so far no satisfatory solutions have been found for the general case (see [2]).

While both of the transient phenomena mentioned above are present only in rare, exceptional situations, this is completely different for the vortex excited oscillations in the frequency range of 10 to 100 Hz, which will be discussed in some details in the next sections.

Finally, at the other end of the frequency scale, there are the galloping oscillations with frequencies of the order of 0.1 Hz. These low frequency oscillations are due to an aeroelastic instability occuring almost exclusively under very special meteorological conditions and with ice deposits formed on the cables. They are extremely difficult to damp out and lead to isolated but catastrophic breakdowns of transmission lines in Europe and in North America, almost every winter. They can be mathematically modelled by nonlinear differential equations in different ways, but all the models in use seem to be very sensitive to changes in the parameters. In the authors' opinion this implies that the only hope for a reasonable solution of the actual technical problem consists in finding ways to introduce additional damping in the low frequency modes of the cables, by developing appropriate dampers (see [3]).

2. Vortex excited oscillations of cables

In a single cable exposed to a stationary transverse flow, vibrations are excited by the von Karman vortex shedding (see [4]), which may cause failure due to material fatigue. Since approximately 1930, different versions of the Stockbridge damper have been used to supress these oscillations. The vibrations in a first approximation occur in a vertical plane and can be described by the partial differential equation

$$EI\, w''''(x,t) - T\, w''(x,t) + \rho \ddot{w}(x,t) = q(x,t) + \bar{d}(w,\dot{w},t), \qquad (1)$$

where EI is the cable's bending stiffness, T its tension force, ρ is the mass per unit length, $w(x,t)$ is the transverse displacement of the cable, $q(x,t)$ is the wind force due to vortex shedding and $\bar{d}$ represents the cable's structural damping (Fig.1). The primes stand for partial differentiation with respect to x and the dots indicate differentiation with respect to the time t. If the cable carries a damper at the location $x = l_1$, equation (1) holds for values of x satisfying $0 \leq x < l_1$ and $l_1 < x \leq 1$; the ends at $x = 0,1$ corresponding to the suspension clamps, which are assumed to be fixed in this simplified analysis.

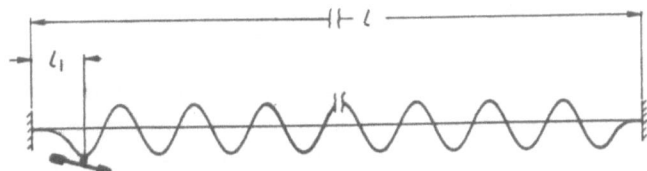

Fig.1: Oscillating cable with Stockbridge damper

The cable vibrations are sometimes studied with the Ritz–Galerkin method, where the eigenfunctions of the homogeneous problem corresponding to (1) can be used in the expansion. In this case, the frequency range of 10 to 50 Hz lies typically between the 100th and the 500th eigenfrequency, so that a very large number of modes has to be considered in the computations in order to obtain good results.

Since the spacing of the eigenfrequencies is very close, of the order of 0.1 Hz, a different approach is usually taken, in which the discrete spectrum, being relatively dense, is approximated by the continuous spectrum of an infinite or semi-infinite cable. Comparative studies showed that the error due to the substitution of the discrete spectrum by a continuous one is negligible for practical purposes. Fig.2 shows the excellent agreement between a solution for the bending strains in the cable obtained in this manner vs. an "exact solution". Only the envelope of the resonance curve shown in Fig.2 is of practical importance, since the maxima and minima lie extremely close together and their exact position is irrelevant. This envelope is however obtained with good accuracy by means of the energy balance method, which is shortly outlined in what follows.

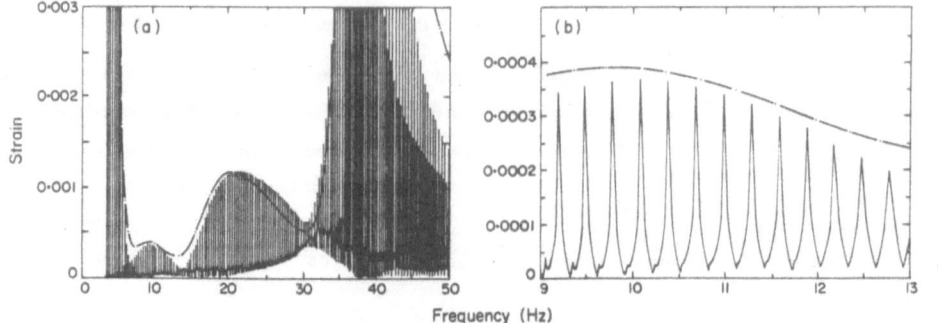

Fig.2: Comparison between the "exact solution" and the solution obtained by energy balance

While in equation (1) the parameters on the left-hand side are easily determined (only the small cable's bending stiffness EI is somewhat difficult to measure), the right-hand side is not well known: experimental data are available on the aerodynamic forces acting on a rigid cylinder oscillating in a planar stationary flow, and to some extent also on structural damping; they are, however, far from complete. Moreover, the wind velocity as a function of space and time is not known and the flow is obviously in general not stationary, laminar and orthogonal to the cable's vertical plane. In (1) the aerodynamic forces q(x,t) should of course actually not only be a function of x and t only, but they are at least also functionals of the functions w(x,t) and $\dot{w}(x,t)$.

Due to these facts, and also due to a number of other reasons, certain simplifications are made, in order to gain a simplified model, which should nevertheless retain the essential features of the original problem. First, the wind speed is assumed to change slowly with respect to the cable's oscillations in the frequency range of 10 to 100 Hz, so that stationary oscillations can be assumed. For a given constant speed of the transverse wind a vertical force then acts on the cable as a harmonic time function. Its frequency is proportional to the transverse wind speed, as given by the Strouhal relation

$$f_s = \frac{S\,v}{D}\,,\tag{2}$$

v being the wind velocity, D the cable's diameter and S ≈ 0,20 the Strouhal number. The cable is assumed to oscillate with this frequency.

Obviously for these steady state vibrations an energy balance of the type

$$P_W = P_D + P_C \tag{3}$$

holds, P_W being the power of the aerodynamic forces, P_D the power corresponding to the dissipation of mechanical energy in the damper and P_C the power of the structural damping in the cable. Of course the power is a function of time and the expressions in (3) are assumed to be the averages of these time functions. Each of these terms is then a function of both the frequency f and of the amplitude A of the cable oscillations. Since for a given wind speed the frequency f is known, (3) is a nonlinear algebraic equation in the only remaining unknown, namely in the amplitude A.

The numerical solution of equation (3) is a relatively simple task and the approximate results shown in Fig.2 were obtained in this manner. Once the vibration amplitude in the far field, i.e. far away from the damper clamp and from the suspension clamps is known, the bending strains in the cable can easily be computed as shown in [4, 5]. Moreover, if statistical information on the wind profiles is available for the site of a given transmission line, this information can be used to compute the time intervals during which the different strain levels are present in the cable and in this manner estimates for the cable's useful life can be obtained. The reliability of these results does of course strongly depend on how well the actual power transfer is described by the functional expressions used in (3).

For a cable without a Stockbridge damper the term P_D in (3) vanishes and the amplitude A is then simply obtained from $P_W = P_C$. If however a damper is present, it usually dissipates much more power than the power P_C due to the structural damping of the cable, i.e. $P_D \gg P_C$. In this case the amplitude can often be obtained simply from $P_W = P_D$.

The power P_C dissipated by structural damping in the cable is sometimes measured by cable manufacturers and also in other vibration labs. It is a function not only of the amplitude and of the frequency (or wavelength) but also of the tension in the cable. For higher tension the structural damping usually decreases. Different laboratories usually approximate P_C by different analytical expressions (see e.g. [5]). We will not examine the term P_C further in the present paper.

The damper power P_D is a function of the motion of the damper clamp which can easily be measured in the laboratory. If the dynamic behavior of the Stockbridge damper can be approximated by linear elements, the power follows immediately from its complex impedance. But also in the case of a "nonlinear damper" the experimental determination of the dependence of P_D on the clamp's motion in a first approximation poses no great problem, at least as long as the clamp motion is harmonic. The measurement of the complex impedance of a Stockbridge damper today is a standard technique used in vibration laboratories. In [4, 5] it is shown how the function $P_D(\cdot)$ can be calculated if the damper impedance is known; of course it contains also the damper location l_1 as a parameter. The function $P_D(\cdot)$ can therefore be assumed to be well known and it will not be examined further in this paper.

The situation is different for the power P_W. It is not very difficult to measure the transversally pulsating force acting on a fixed rigid cylinder immersed in a wind tunnel under stationary laminar flow, and these experiments have been carried out by several authors. The situation is more complicated for a rigid cylinder in harmonic transversal motion: here the forces will depend on the vibration amplitudes and also on the ratio between the oscillation frequency and the Strouhal frequency given by (2). They become particularly large if the two frequencies are approximately equal. Moreover, for a forced harmonic motion of the cylinder with a frequency in the neighborhood of f_s, the vortices will be generated with the driving frequency instead of the Strouhal frequency. The force is no longer harmonic, particularly for very small vibration amplitudes, where both frequencies, the driving frequency and the Strouhal frequency, are present in the oscillating force. The dependence of the force on the amplitude A is very important for large amplitudes. It is not surprising that for very large amplitudes the power P_W becomes even negative: in this case the oscillating cylinder gives up energy to the fluid and we have the opposite phenomenon as in the wind-excited vibration of a cable. For "reasonably large" vibration amplitudes (in transmission lines the amplitude can be of the order of the diameter of the cylindric cable) the aerodynamic force can be assumed to be harmonic with the same frequency as the oscillation, and the "wind-power" can be written as

$$P_W(f,A) = 1f^3D^4F(A/D). \tag{4}$$

In (4) the diameter D of the cylinder is used to form the nondimensional amplitude A/D.

Even for rigid cylinders the expressions obtained by different experimentalists vary considerably, as shown in Fig.3 (see [6]). The reasons for this disagreement are several. The experiments are usually designed in such a way as to approximate the behavior of an infinitely long cylinder in planar flow; this goal is attained to a higher or lesser degree in the different laboratories. Also, the degree of turbulence, present in all the experiments, varies from one wind-tunnel to the other, etc.

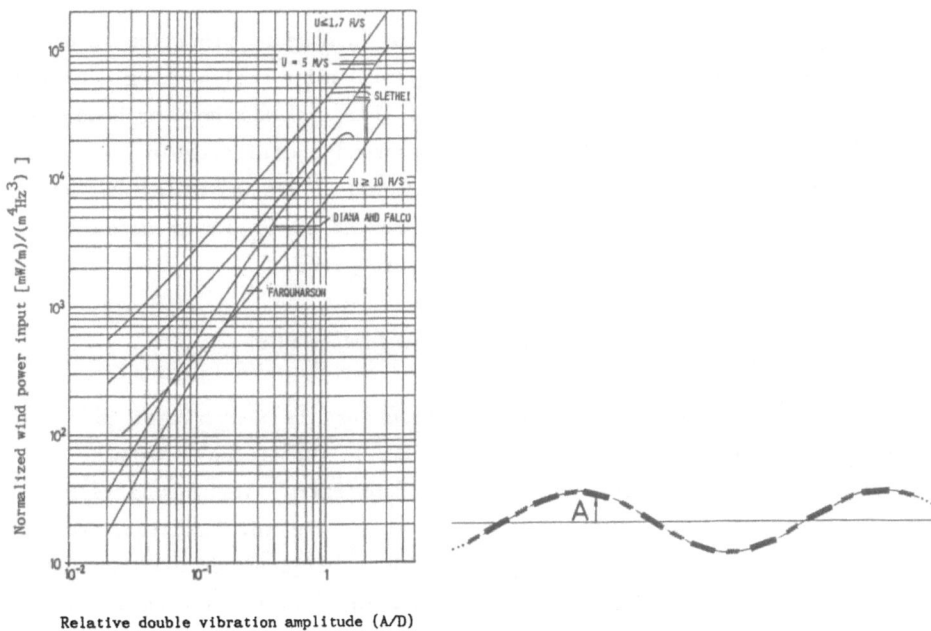

Fig.3: Normalized wind power as obtained by different authors

Fig.4: On the approximate calculation of the wind power for an oscillating cable

In estimating the amplitudes of the cable vibrations via equation (3) additional approximations have to be made in order to use (4). In the oscillating cable the transverse flow pattern can of course never be perfectly two-dimensional, as was the case with the rigid cylinder, since there are nodes and antinodes in the cable. If the wave-length of the cable oscillations is very large with respect to D, each element of the cable is approximated by a small rigid cylinder with inclined axis, as shown in Fig.4. The function $P_W(\cdot)$ in (3) is then obtained by computing the local power (4) and averaging with respect to the local amplitude.

These few remarks already show that the function $P_W(\cdot)$ to be used in (3) is insufficiently known. It should of course also depend on the degree of turbulence of the wind, which is known to be an important factor. On the other hand, not only life time estimates of the cables are based on the vibration levels computed via (2), but also the decision whether to use damping devices, and the number of dampers to be used. Considerable costs are therefore involved and a good approximation in estimating the vibration amplitudes is highly desirable.

Very few attempts have so far been made to measure vibration levels and wind data in actual cables of transmission lines under real working conditions and comparing them to computational results. This task has however become more manageable in recent years due to the widespread and affordable use of computers in the automatic recording and evaluation of measurements. In the next section, a new test facility for the measurement of P_W in an actual transmission line is described and first results are given.

3. The test facility at Büren, description and first results

A computerized, fully automated test station for the measurement of vortex-excited oscillations has been set up in Büren (Westfalen), Germany, by a consortium of utilities (Bayernwerk, PreussenElektra, Vereinigte Elektrizitätswerke Westfalen VEW) and by RIBE, Schwabach. In this station, data are collected on the conductor oscillations, meteorological conditions and the power P_W transmitted by the wind to the cable in an actual transmission line. The Institut für Mechanik of the Darmstadt Technical University is giving scientific support to the project and is analyzing the results of the measurements.

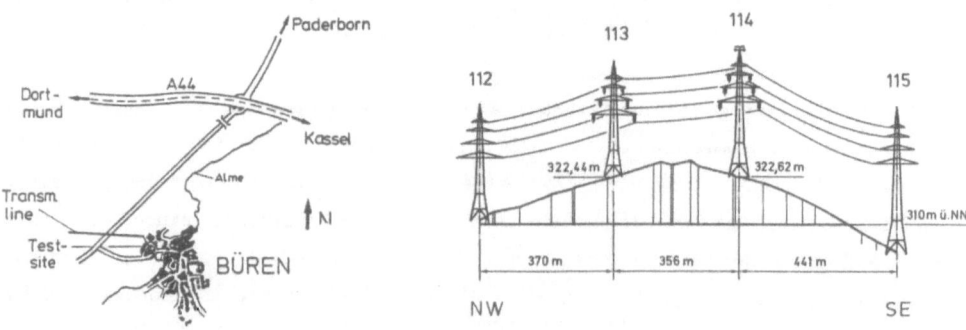

Fig.5: Location and elevation of the test site

The earth wire (cable 240/40, diameter 21,9 mm) of the 220 kV transmission line Lippborg-Büren is used for the measurements between the masts 113
and 114; Fig.5 shows location and the elevation of the test site. The following variables are measured:

-   wind temperature,
-   wind speed perpendicular and parallel to the transmission line,
-   tension in the earth wire,
-   vertical acceleration at four discrete points of the wire
    (distance from the clamps ca. 2-4 m).

The wind speed is measured by means of an ultrasonic anemometer on mast 114,
the tension in the cable is registered with a force transducer at the clamp on
mast 113 and the acceleration is determined by means of accelerometers. The
analog signals obtained by these transducers are amplified and transmitted to
the test station, where they are filtered and digitized in a 12-bit A/D converter. These pre-processed data are then further processed by with an HP
9000/310 computer system. This system consists of the computer itself, with a
Motorola 68010 processor and 1 MB RAM, a 20 MB hard disk and a 788 kB floppy
disk drive, an ink jet-printer and of a DMA controller card, necessary for the
fast data transfer between the A/D converter and the computer. The computer
works with the HP BASIC 4.0 programming language and operating system.
The recording of the data sets and their evaluation is controlled by the
software package MOVICON, especially designed for that purpose.

After a set-up procedure concerning the choice, description and calibration of the data channels, FFT parameters, A/D converter adjustments,
trigger parameters, etc., the variables described above are measured with
prespecified time intervals (5-10 minutes) between two measurements. As a
result of the chosen parameters (number of channels, FFT block size, sampling
frequency, number of averages (10)) the measurement of these data frames takes
about 20 seconds. During this phase the digitized data are recorded without
time delay on a hard disk. These data are then processed further in a second
phase taking ca. 300 seconds. The mean value, minimal and maximal values are
computed together with the standard deviation for each channel from the
recorded time history, averaged over the choosen number of frames. Since the
measured vibrations are of course not monofrequent, a nominal "dominating"
frequency is calculated using the centroid of the acceleration spectrum (see
Fig.6) for all four acceleration channels. The data resulting from these com-

putations are then stored on a disk – the real time data are not recorded, but
deleted – and later analyzed in Darmstadt.

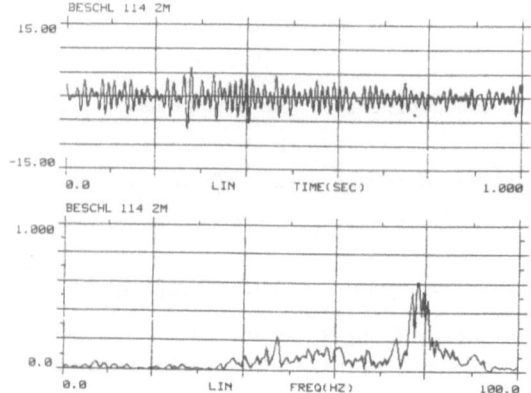

Fig.6: Conductor acceleration (time history and corresponding spectrum)

The first measurements, taken in October and November 1987, were used
mainly for checking the test facility. Due to a lack of wind only two out of
eight data disks could be analyzed.

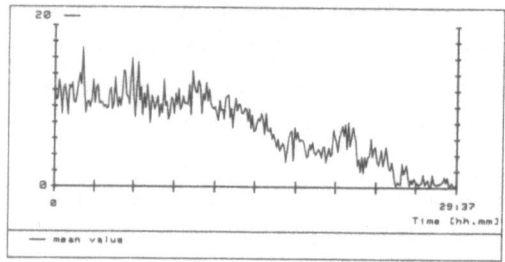

Fig.7: Wind speed component perpendicular to the cable

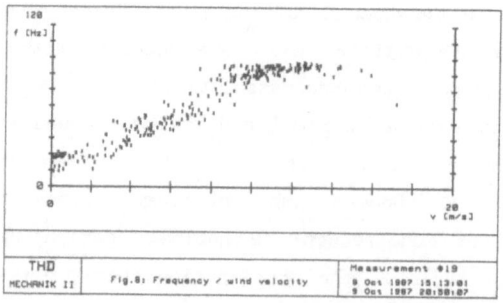

Fig.8: Dominating frequency as a function of the wind velocity

Fig.7 shows a typical wind speed component perpendicular to the cable as a function of time during the first measurement. The oscillation frequency as a function of the wind velocity is shown in Fig.8; the slope comes to S/D (At 100 Hz the low-pass filter cuts the spectrum). It can be clearly seen, that for very low wind speeds (2) is no longer valid. Checking the relationship between the Strouhal number and the Reynolds number

$$Re = \frac{v\, D}{v} ,$$
(5)

v being the wind velocity, D the wire's diameter and v the kinematic viscosity of air, it can be observed (Fig.9) that for Re $< \approx$ 3000 the Strouhal number is much greater than 0.20 and there is no reliable vortex-shedding (The curve has been computed from [7]. An other interesting variable is the turbulence level of the wind speed, here defined as'

$$\xi = \frac{\sigma_v}{m_v} .$$
(6)

σ_v being the standard deviation and m_v the mean value of the wind velocity computed during the process phase described above, since it contains information about the laminarity of the wind flow. During our measurement the turbulence level was of the order of 10-20 % (Fig.10), a value much larger than in most wind tunnel experiments (1-2%), but realistic for real transmission lines.

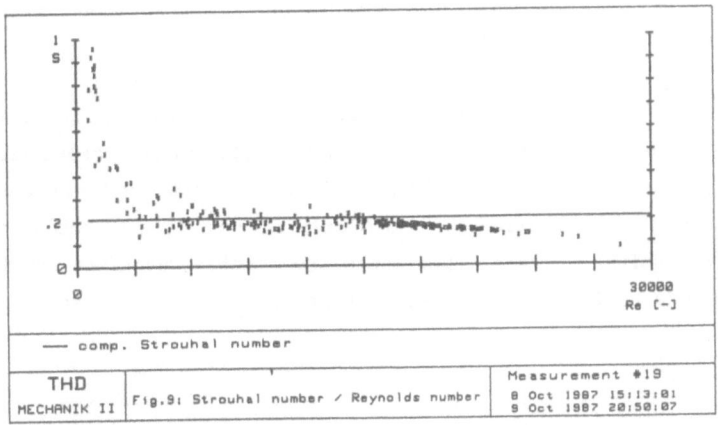

Fig.9: Strouhal number as a function of the Reynolds number

156

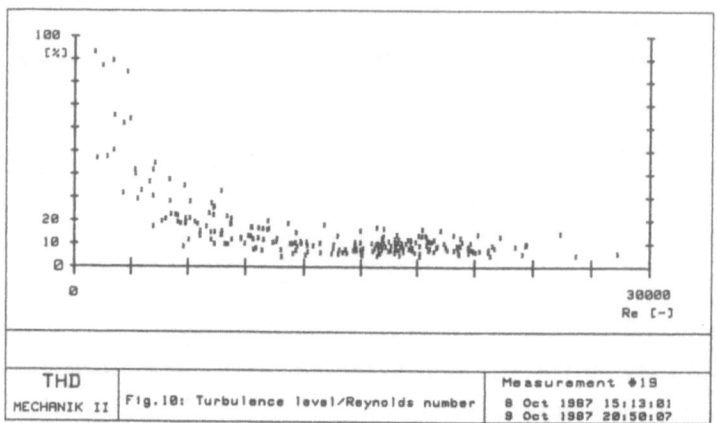

THD
MECHANIK II | Fig.18: Turbulence level/Reynolds number | Measurement #19
8 Oct 1987 15:13:81
9 Oct 1987 20:50:07

Fig.10: Turbulence level as a function of the Reynolds number

During the current year a large volume of data is being accumated, which will then permit a more detailed and representative statistical evaluation.

4.   Final remarks

In this paper some of the vibration problems related to overhead transmission lines were briefly discussed together with their mathematical models. Of these phenomena the most common one corresponds to the oscillations due to von Karman vortex shedding. The mathematical model for these oscillations was described in this paper and it was shown that insufficient data are available on the wind-power input into the vibrating cable. The data from wind tunnel experiments with rigid cylinders give only very uncertain information for an actual transmission line.

Vibration levels and wind data are being measured at a new test facility in Germany, in an actual transmission line. With this additional information it should be possible to improve the mathematical models actually being used in the computation of these vibrations, particularly with respect to the wind-power input. First results of these experiments are briefly described in the paper. A more complete evaluation of these current tests should be available at the end of 1988.

References

[1] Hagedorn, P. & Idelberger, H.; Dynamische Vorgänge bei Lastumlagerung in Abspannketten von Freileitungen, etz-Archiv, 2 (1980), 1-11

[2] Neumann, U.; Zum Problem der Kurzschlußschwingungen von Leiterseilen, Doctoral Dissertation, TH Darmstadt (to appear)

[3] Hagedorn, P. & Meier-Dörnberg, K.-E.; Tanzschwingungen in Freileitungen und ihre Unterdrückung: Beschreibung eines laufenden Versuches, etz-Report 21, VDE-Verlag, Berlin 1986

[4] Hagedorn, P.; Ein einfaches Rechenmodell zur Berechnung winderregter Schwingungen an Hochspannungsleitungen mit Dämpfern, Ing.-Arch 49 (1980), 161-177

[5] Hagedorn, P.; On the Computation of Damped Wind-Excited Vibrations in Overhead Transmission Lines., J. of Sound & Vibrations, 83 (1982), 253-271

[6] Estimating aeolian vibration based on the energy balance principle,CIGRE, Study committee 22, WG 1 (1977/78)

[7] Sarpkaya, T.; Vortex-Induced Oscillations., J. of Applied Mechanics, Vol 46 (1979), 241-258

ON THE MODELING OF A CONTINUOUS OSCILLATOR

BY OSCILLATORS WITH A FINITE NUMBER

OF DEGREES OF FREEDOM

A.H.P. van der Burgh, Delft

Abstract: In an engineering approach continuous systems are often modeled by systems with a finite number of degrees of freedom. This way of modeling is usually based on arguments and principles from e.g. mechanics or physics. In this paper a continuous system and a system with a finite number of degrees of freedom are considered. Both systems are oscillators in a uniform wind field. An attempt is made to answer the question if on the basis of the behaviour of the solutions of the equations of motion of the oscillators, the engineering way of modeling may be understand or justified.

1   Introduction

The mathematical description of problems in flow-induced vibrations e.g. the interaction of a flowing medium with an elastic structure has gained a growing interest recent years.

In general this mathematical description concerns coupled equations from fluid dynamics and the theory of elasticity. However, for obtaining models which may be analysed analytically one has to consider simple elastic structures and simple equations and (empirical) laws from fluid dynamics. In this paper it will be shown that the rather simple model equations obtained in this way throw up interesting questions and problems. This paper partly reviews the recent work [1] and [2] and partly aims to contribute to the question of modeling continuous systems by systems with a finite number of degrees of freedom. In [1] an oscillator with two degrees of freedom in a uniform flowing medium is studied. The oscillator consists of a cylinder, with certain aerodynamic properties, hung from a number of linear springs. Special attention is paid to the interaction of the oscillations perpendicular to and in the direction of the flowing medium.

In [2] a stretched string in a uniform flow is considered. The analogous

159

J. Manley et al. (eds.), Proceedings of the Third European Conference on Mathematics in Industry, 159–170.
© 1990 *Kluwer Academic Publishers and B. G. Teubner Stuttgart.*

question of interaction of oscillations in two directions is studied in this work. The results obtained in [1] and [2] with respect to the qualitative and quantitative behaviour of special solutions of the equations of motion may be used for the question if the engineering way of modeling continuous systems by systems with a finite number of degrees of freedom may be understand or justified.

The paper is organized as follows. In section 2 the vertical motion of both an oscillator with one degree of freedom as a stretched string in a flowing medium is considered. The approach is based on ideas and principles published in [3]. The section is meant as an introduction to section 3 where the same approach is used for the study of oscillations in two directions. The derivations of the equations of motion in section 3 are omitted; the derivations are straightforward but laborious extensions of the ones presented in section 2. Qualitative results on the interaction of the oscillations in both directions or occurrence of resonance and quantitative results on the magnitudes of limit amplitudes may be used to evaluate the engineering way of modeling.

The paper ends with some conclusions on this modeling problem in section 4.

2   The vertical motion

In this section we consider the vertical wind induced motion of an oscillator with one degree of freedom as well as the vertical wind induced motion of a continuous oscillatory system. The one degree of freedom system consists of a rigid cylinder hung from springs in such a way that only vertical oscillations are possible, whereas the continuous system is a stretched finite string with fixed endpoints. Both systems are supposed to oscillate in a vertical plane perpendicular to the direction of a horizontal uniform airflow. In an engineering approach the one degree of freedom system is usually considered as a realistic physical model for the study of the continuous system.

In what follows model equations for both systems will be presented. Although both systems have a number of physical aspects in common, on the basis of which the continuous system could be modeled by the one degree of freedom system, the qualitative behaviour of the solutions of both the model equations is quite different. However, from recent quantitative results it follows that there are some similarities in frequencies and limit amplitudes. An attempt will be made to compare qualitative and quantitative aspects of the oscillatory behaviour of both systems.

Let the rigid cylinder and the taut string have the same cross-section as sketched in figure 1. The cross-section is only able to move along the z-axis that is in the vertical direction, the position and velocity of (the centre of) the circular cylinder is denoted respectively by $w(t)$ and w_t.

In order to have a driving force e.g. lift force in the z-direction a small symmetric ridge indicated with dots in figure 1 is fixed to the cylinder or string.

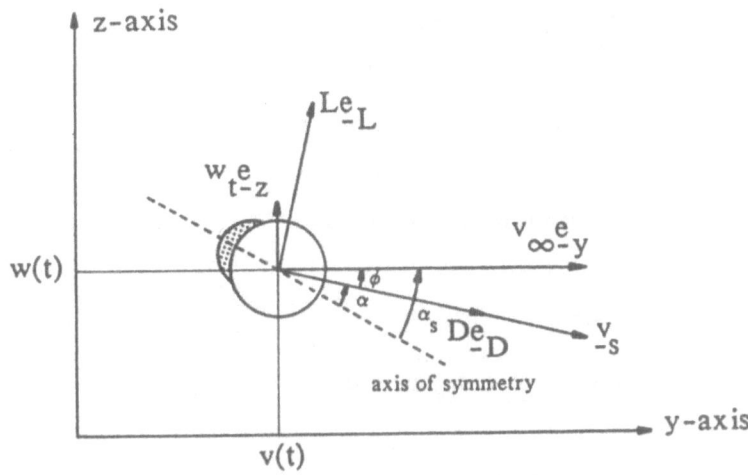

Figure 1. Aerodynamic forces acting on the cross-section of the cylinder and the string.

The position of this ridge is given by α_s; the angle between the axis of symmetry and the positive y-axis.

The airflow induces a drag and a lift force acting on the cross-section; the drag force $D\underline{e}_D$ has the direction of the resultant wind velocity $\underline{v}_s$ ($\underline{v}_s = v_\infty \underline{e}_y - w_t \underline{e}_z$) and the lift force $L\underline{e}_L$ is perpendicular to $\underline{e}_D$ in anti-clockwise direction.

$\underline{e}_y$, $\underline{e}_z$, $\underline{e}_D$ and $\underline{e}_L$ are unit vectors respectively along the y-axis, the z-axis, along $\underline{v}_s$ and perpendicular to $\underline{v}_s$.

ϕ is the angle between $\underline{e}_y$ and $\underline{e}_D$, positive in anti-clockwise direction (so in figure 1 ϕ is negative) and α, the angle of attack, is the angle between the axis of symmetry and $\underline{v}_s$ ($= v_s \underline{e}_D$). The magnitudes D and L of respectively the drag and lift force may be given by the empirical laws:

$$(2.1) \qquad D = \frac{1}{2} \rho_a A c_D(\alpha) v_s^{\,2}$$

$$(2.2) \qquad L = \frac{1}{2} \rho_a A c_L(\alpha) v_s^{\,2}$$

where ρ_a is the density of the air, A the area of the cross-section and where the aerodynamic coefficients $c_D(\alpha)$ and $c_L(\alpha)$ depend on the angle of attack. A typical variation of c_D and c_L with α, obtained from wind-tunnel measurements in a steady flow is depicted in figure 2.

The curves may be approximated around $\alpha = \alpha_s$ by:

$$(2.3) \qquad c_D = \text{constant}$$

$$(2.4) \qquad c_L = c_{L1}(\alpha - \alpha_s) + c_{L3} (\alpha - \alpha_s)^3.$$

For simplicity α_s, defining the position of the ridge and being fixed during the motion, is choosen to be the angle of attack where the lift force vanishes. From figure 1 one may derive that the driving aerodynamic force F_z in the positive z-direction equals:

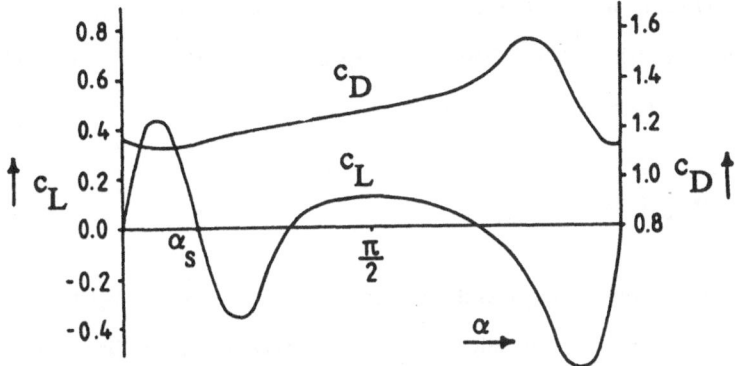

Figure 2. Prototype of variation of c_D and c_L with the angle of attack for a symmetric cross-section with small ridge.

(2.5) $F_z = D \sin \phi + L \cos \phi.$

By using the relations $\sin \phi = -w_t/v_s$, $\cos \phi = v_\infty/v_s$, $\alpha_s = \alpha - \phi$,
$v_s^2 = v_\infty^2 + w_t^2$, (2.1) and (2.2), F_z may be written as:

(2.6) $F_z = \frac{1}{2} \rho_a A \{v_\infty^2 + w_t^2\}^{1/2} \{-c_D w_t - v_\infty c_{L1} \ \text{arctg} \ (w_t/v_\infty) +$
$\qquad\qquad - v_\infty c_{L3} \ \text{arctg}^3 \ (w_t/v_\infty)\}$

Apparently F_z depends on w_t and a number of parameters.
Let $\overline{F}_z$ be the Taylor expansion of F_z with respect to w_t around $w_t = 0$, up to terms of the third degree:

(2.7) $\overline{F}_z = \frac{1}{2} \rho_a A v_\infty^2 \ \{-(c_D + c_{L1}) \ w_t/v_\infty +$
$\qquad\qquad - (\frac{1}{2} c_D + \frac{1}{6} c_{L1} + c_{L3})(w_t/v_\infty)^3\}.$

When $|w_t|$ is small $\overline{F}_z$ may be used as an approximation for F_z.
In what follows $\overline{F}_z$ will be used as a driving force for a cylinder-spring system respectively a stretched string.

164

A rigid cylinder, of which the cross-section is sketched in figure 1, with mass m per unit length is hung from linear springs with stiffness s per unit length. Application of Newton's law yields:

$$(2.8) \qquad mw_{tt} + sw = \overline{F}_z(w_t).$$

A stretched string with length ℓ and tension T, has a position along the x-axis. The x-axis is perpendicular to the picture plane of figure 1. The mass of the string is ρ per unit length. It is well-known that the equation of motion of this perfect string may be given by:

$$(2.9) \qquad \rho A W_{tt} - T A W_{xx} = \overline{F}_z(W_t)$$

where $W = W(x,t)$ is now a function of two independent variables and $c^2 = T/\rho$.

Suppose that the following conditions are satiesfied:

$$(2.10) \qquad\qquad c_D + c_{L1} < 0$$

$$(2.11) \qquad \tfrac{1}{2} c_D + \tfrac{1}{6} c_{L1} + c_{L3} > 0.$$

The first condition implies instability of the trivial solution in the linear approximation whereas the second condition implies non-linear positive damping. It is not difficult to show that when these conditions hold equations (2.8) and (2.9) may be transformed to respectively the Rayleigh equation and the Rayleigh wave equation:

$$(2.12) \qquad w_{tt} + w = \epsilon \left(w_t - \tfrac{1}{3} w_t^3 \right)$$

$$(2.13) \qquad W_{tt} - W_{xx} = \overline{\epsilon} \left(W_t - \tfrac{1}{3} W_t^3 \right)$$

where for the sake of convenience the same symbols for the variables have been used.

With respect to both equations the interesting question of existence of periodic solutions may be posed. For equation (2.12) it is well-known that for all $\epsilon > 0$ a unique periodic solution (limit cycle) exists, and that all solutions, except the trivial one which is unstable, approach this limit cycle for $t \to \infty$. However, little seems to be known on existence of time-periodic solutions for a boundary-value problem for the Rayleigh wave equation (2.13). Recently some progress has been made in the study of initial-boundary value problems for this equation.

A prototype of results concerns the evolution for $t \to \infty$ of monochromatic initial values with simple boundary conditions when $0 < \bar{\epsilon} < 1$, e.g.:

$W(0,t) = W(\ell,t) = 0$ and $W(x,0) = a \sin \frac{\pi x}{\ell}$, $W_t(x,0) = 0$.

Although up to now only well-posedness and validity of approximations has been established on a $O(\frac{1}{|\epsilon|})$ time-scale one may conjecture, also on the basis of numerical results, that the limit behaviour of the solution for $t \to \infty$ of this initial-boundary value problem is a standing triangular wave motion with (angular) frequency close to $\frac{c\pi}{\ell}$ which is the frequency of the unperturbed, that is $\bar{\epsilon} = 0$, initial-boundary value problem. A sketch of this standing motion is given in figure 3.

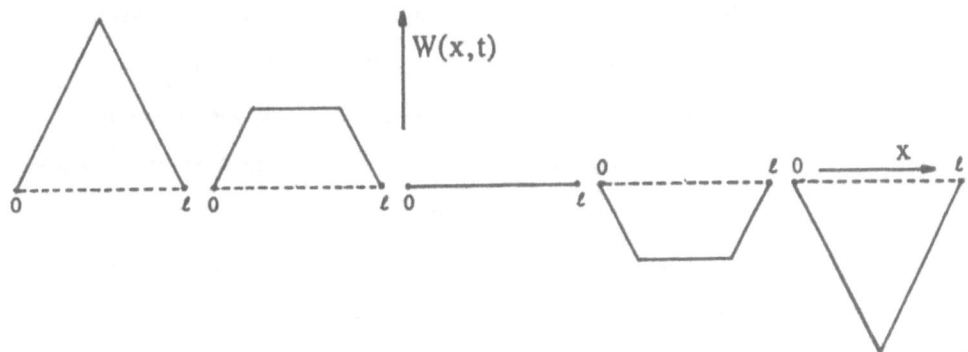

Figure 3. The standing triangular oscillation of the string.

An interesting quantitative result obtained by using perturbation methods concerns the ratio of the limit amplitudes of both oscillatory systems with identical cross-section and aerodynamical properties and the same angular frequencies in vacuum i.e. $\frac{c\pi}{\ell} = (\frac{s}{m})^{1/2}$.

Let $a_{\ell c}$ be the amplitude of the limit-cycle solution of (2.12) when $0 < \epsilon \ll 1$ and a_∞ be the limit amplitude of the standing triangular wave solution of the initial boundary value problem when $0 < \bar{\epsilon} \ll 1$. It can be shown that with good approximation $a_\infty = \frac{\pi}{4} \sqrt{3}\ a_{\ell c}$ which means that the limit-amplitude of the continuous system exceeds surprisingly the limit amplitude of the one degree of freedom system with about 36%. Apparently one small non-linear (cubic) term in the equations of motions gives rise to large differences in limit amplitudes. Especially in practical applications this result may be of interest.

3 The vertical and horizontal motion

In this section an oscillator with two degrees of freedom and a continuous oscillatory system which may oscillate in two directions are considered. The two degrees of freedom system is a simple extension of the one degree of freedom system introduced in section 2. The rigid cylinder is now hung from springs in such a way that oscillations in both vertical and horizontal direction are possible. The springs are supposed to be massless whereas the ratio of the spring stiffnesses s_y and s_z, s_y/s_z, is used as a parameter.

As a continuous system a stretched string analogous to the one in section 2 is considered. However, the string in vacuum is now allowed to have two families of modes; the vertical and the horizontal modes. The speed of the transversal wave motion of the string is supposed to be independent of the direction.

Let for both systems the displacements in horizontal (y-axis) and vertical (z-axis) direction be respectively v(t) and w(t) for the oscillator with two degrees of freedom and V(x,t) and W(x,t) for the stretched string. Then, by extending figure 1 with the new variables v(t) respectively V(x,t) and using the same modeling principles as in section 2 including truncation of terms of degree four and higher and including an appropriate transformation of all variables one arrives at the following sets of model equations for respectively the oscillator with two degrees of freedom:

$$(3.1) \qquad v_{tt} + \Omega^2 v = \epsilon \{a_{10} v_t + a_{20} v_t^2 + a_{02} w_t^2\}$$

$$(3.2) \qquad w_{tt} + w = \epsilon \{b_{01} w_t + b_{11} v_t w_t + b_{03} w_t^3\}$$

and the stretched string:

$$(3.3) \qquad V_{tt} - V_{xx} = \bar{\epsilon} \{a_{10} V_t + a_{20} V_t^2 + a_{02} W_t^2\}$$

$$(3.4) \qquad W_{tt} - W_{xx} = \bar{\epsilon} \{b_{01} W_t + b_{11} V_t W_t + b_{03} W_t^3\}$$

where $\Omega^2 = s_y/s_z$ and

$$a_{10} = -2c_D < 0, \ a_{20} = c_D > 0, \ a_{02} = \frac{1}{2} c_D - c_{L1}$$

$$b_{01} = -(c_D + c_{L1}) > 0, \ b_{11} = c_D + c_{L1} < 0, \ b_{03} = -(\frac{1}{2} c_D + \frac{1}{6} c_{L1} + c_{L3}) < 0.$$

The qualitative theory for systems of perturbed wave equations of which system (3.3), (3.4) is an example has not been developed fairly well. Nevertheless on the basis of some results regarding initial-value problems for this system it is aimed to contribute to the question whether the system of ordinary differential equations (3.1), (3.2) i.e. the oscillator with two degrees of freedom may be used for the study of the system of non-linearly coupled wave equations (3.3), (3.4) i.e. the taut string.

A prototype of results obtained for initial-boundary value problems for system (3.3), (3.4) is the following.

Consider a taut string along the x-axis of length ℓ and with fixed endpoints i.e. $V(0,t) = V(\ell,t) = W(0,t) = W(\ell,t) = 0$ and initial conditions

$$V(x,0) = \sum_{n=0}^{\infty} b_n \sin \frac{n\pi x}{\ell}, \quad V_t(x,0) = 0; \quad W(x,0) = a_1 \sin \frac{\pi x}{\ell}, \quad W_t(x,0) = 0. \text{ Then,}$$

on the basis of rigorous results obtained on a time-scale of $O(\frac{1}{\epsilon})$ one may

conjecture that for $0 < \bar{\epsilon} \ll 1$ $V(x,t) \to 0$ for $t \to \infty$ whereas $W(x,t)$ for

$t \to \infty$ evolves to a standing triangular wave as sketched in figure 3. It is

remarkable that all modes which are initially in horizontal direction present

will vanish. So the initial-boundary value problem defined above will result in

a vertical oscillation (that is an oscillation perpendicular to the flow direction)

only.

With respect to system (3.1), (3.2) the following results are of interest. For

the case that $\Omega = 2 + O(\epsilon)$ the non-linear interaction between the two

degrees of freedom results in a periodic or almost periodic motion of which a

prototype is sketched in figure 4. Apparently the unstable equilibrium position

evolves in a motion which involves $O(1)$ amplitudes in both directions. For the

case that $\Omega \neq 2 + O(\epsilon)$ only a vertical periodic oscillation, similar to the

periodic motion of the oscillator with one degree of freedom, exists (ampli-

tudes which are initially present in the horizontal direction will vanish).

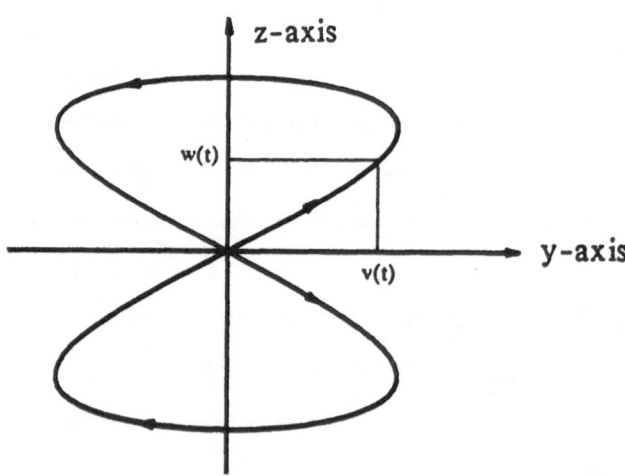

Figure 4. The 2 : 1 resonance for the two degrees of freedom oscillator.

On the basis of these results one may conclude that for the study of initial-boundary value problems for system (3.3), (3.4) with initial values $v(x,0) = b_n \sin \frac{n\pi x}{\ell}$ $n \neq 2$, $v_t(x,0) = 0$; $w(x,0) = a_1 \sin \frac{\pi x}{\ell}$, $w_t(x,0) = 0$ and fixed endpoints one may use system (3.1), (3.2). The order of magnitude of the amplitudes of both oscillators may differ considerably that is 36% as mentioned in section 2.

When $n = 2$ one may not use system (3.1), (3.2), the string will not oscillate in a way as sketched in figure 4.

4   Some conclusions

For the study of vertical oscillations of a stretched string in a uniform horizontal flow one may use a single degree of freedom system. The study should concern initial-boundary value problems with monochromatic initial values. The magnitudes of limit amplitudes obtained from the single degree of freedom system will be smaller (i.e. approximately 0.73) then those of the continuous system when both systems have the same natural frequencies and the same aerodynamical properties.

For the study of vertical and horizontal oscillations of the stretched string involving initial-boundary value problems with monochromatic initial values vertically and C^2 initial values horizontally one may also use the one degree of freedom system for the vertical oscillations as all horizontal initial values will be damped out.

As the two degrees of freedom oscillator may have oscillations which may be identified as a 2 : 1 (horizontal : vertical) resonance, and which have not been found in the continuous system, this system is less appropriate for the study of the continuous system.

170

References

[1]   van der Beek, C.G.A.; van der Burgh, A.H.P.: On the periodic wind-
      induced vibrations of an oscillator with two degrees of freedom. Nieuw
      Archief voor Wiskunde 2 (1987) 207-225.

[2]   van Horssen, W.T.: An asymptotic analysis of a class of nonlinear hyper-
      bolic equations. Ph.D. Thesis (1988), Delft University of Technology.
      Faculty of Technical Mathematics and Informatics.

[3]   van der Burgh, A.H.P.: On the galloping response of a simple aerolastic
      oscillator. Symp. on flow-induced vibrations, Am. Soc. of Mech. Eng. 2
      (1984) 37-53.

Delft University of Technology,

Faculty of Technical Mathematics and Informatics,

P.O. Box 356,

2600 AJ   Delft,

The Netherlands.

MINI-SYMPOSIUM 3
(organiser: J. Ockendon)

Mini Symposium

Organised by

J.R. Ockendon
Mathematical Institute
Oxford University

Some Novel Partial Differential Equations Problems Arising in Industry

These papers have been grouped together to illustrate how the application of differential equations to industrial problems can lead to new problems which both have substantial mathematical interest and can shed new light on physical processes.

The first paper (CJB,AW) is perhaps the least conventional mathematically; a relatively simple model for space charge effects in power lines, electrostatic precipitators etc. leads to a very little-studied third order scalar partial differential equation with one real and two complex characteristics. However, this equation is so much simpler than, say, for the incompressible Euler equations that several statements can be made about the existence and structure of the solutions. Hopefully this may lead eventually to a better general theory for such equations.

The second paper (SDH,ACF,) concerns the coupled electric and thermal response of a thermistor. This leads to a nonlinear elliptic/parabolic system which is almost of "high activation-energy" type because of the switch-like behaviour of the electrical conductivity as a function of temperature. The asymptotic solution can shed considerable light on the thermal field which is of great importance in deciding the stress and lifetime of the device.

Finally the paper by (AF) highlights an area which is undergoing intensive theoretical study at present, namely nonlinear partial differential equation systems which have the possibility of changing type within the domain of interest. Here a crucial argument for deciding the value of the model is whether or not such a change of type might violate causality. The two-phase flow models exhibited describe one of the simplest physical realisations of this kind of situation.

J. Manley et al. (eds.), Proceedings of the Third European Conference on Mathematics in Industry, 172.
© 1990 *Kluwer Academic Publishers and B. G. Teubner Stuttgart.*

MODELING CORONAS AND SPACE CHARGE PHENOMENA

Chris Budd
Oxford University
Computing Laboratory

Adam Wheeler
School of Mathematics
University of Bristol.

1. INTRODUCTION.

Many modern industrial processes rely for their operation upon the charged ions which are created when a high electric field is applied to a gas. Examples of such devices are the electrostatic precipitators used to remove dust from power station flue gases , cyclone separators , electrostatic scrubbers and granular bed filters. Other systems , for example high voltage DC power cables , create ions as an unwanted byproduct of their operation. These ions then migrate to earth and the resulting current causes significant loss of energy during transmission. (A summary of further examples of the industrial applications of electrostatic phenomena is given in Cross (1987).)

To understand the operation of these devices we must model the creation and the transport of charged ions in an electric field. Typically such a field is created by placing a conductor with a large (local) curvature into the gas and then applying a high potential to it. The resulting electric field takes its maximum value close to the conductor and in this region the molecules of the gas separate into electrons and positive ions. The electrons then accelerate toward the conductor if it is at a positive potential or away from it if it is at a negative potential. During this motion they collide with other molecules causing further ionisation and releasing photons. These photons maintain the ionisation process releasing further electrons by photoionisation and causing the gas to glow. The glowing region of the gas will be termed the corona for the remainder of

173

J. Manley et al. (eds.), Proceedings of the Third European Conference on Mathematics in Industry, 173–195.
© 1990 Kluwer Academic Publishers and B. G. Teubner Stuttgart.

this paper. If the conductor is at a positive potential the positive ions drift away from it and form a positive charge density - termed the space-charge in the region external to the corona. If the conductor is at a negative potential then the electrons reattach themselves to the air molecules to form negative ions. These then drift away from the conductor to form a negative space-charge density.

To model this behaviour we must separately consider two processes : the detailed calculation of the ionisation process inside the corona and then the calculation of the motion of the ions in the region exterior to it. To solve the latter problem we need only consider ions with one sign for their charge and the exterior of the corona will then form a (free) boundary for a somewhat simpler problem than the original. We shall term the problem of calculating the filed arising from one species of ion only as the space-charge problem. (Throughout this paper we shall only consider the electric field set up by a conductor or a set of conductors with a single polarity. It is also possible to study the much more subtle problem of the field set up by two conductors of opposite polarity. An example of this problem is given in Abdel-Salam et. al. (1982)

The interaction between the corona and the space-charge can be very delicate. It is possible to obtain steady solutions in which the corona boundary is fixed and others where the boundary varies periodically in time - with the space-charge density varying periodically in time and space. The steady state is the easiest case to analyse and we may deduce from a boundary layer calculation some appropriate boundary conditions for the space-charge problem. Using these we may then calculate the solution of the space-charge problem for a variety of two-dimensional geometries using a numerical technique described in Budd & Wheeler (1988a). In this paper we shall present some recent calculations for the electric field set up inside an electrostatic precipitator. These will be compared with the results of some experiments made by R. Corbin and J. Horrocks using a test electrostatic precipitator at the Central Electricity Research Laboratories , Leatherhead , Surrey.

The research described in this paper arose directly from the Mathematical Study Group with Industry held at Oxford University in 1985 . The problem was originally presented by S. Smith (ECRC) and it was studied by several of the participants of the Study Group. Further details of this work are given in the 1985 Study Group report .

2. A MODEL OF AN IONISED GAS.

For simplicity we shall describe a system in which a conductor at a positive potential is placed inside a gas. We shall refer to this conductor as Ω_c . This problem involves the study of two species of charged particles. (If the conductor is at a negative potential then the problem is more complex and involves three species of charged particles. Further details of this system are given in Morrow (1985) and a more general summary of corona behaviour is given in Loeb (1965) .) In such a system there is a region close to the conductor in which both electrons and positive ions are present. Further away from the conductor only positive ions are present. We shall denote the absolute value of the (negative) electron charge density by n the positive ion charge density by p and the electric field by $\underline{E}$. The ions and the electrons move under the action of the field , the effects of diffusion , electromagnetic forces and any external fluid motion of the gas. If we ignore the latter three effects then the resulting currents are given by

$$j^+ = \mu_+ \, p \, \underline{E}$$

and (2.1)

$$j^- = \mu_- \, n \, \underline{E}$$

Where $\mu_{+/-}$ are the respective mobilities of the ions and of the electrons. We shall assume that these are constant and that they have the following values

$$\mu_+ \approx 2 \times 10^{-4} \, m^2V^{-1}s^{-1} \; , \quad \mu_- \approx 4 \times 10^{-2} \, m^2V^{-1}s^{-1}$$

It follows from the above estimates that

$$\mu_- / \mu_+ \approx 200$$

which expresses the considerable difference in the speeds of motion of the ions and of the electrons. (The mobility of a negative ion is similar to that of a positive ion .)

Inside the corona electrons and ions are created by ionisation and it may be shown (see for example Morrow (1985)) that the charge densities of the positive ions and electrons satisfy the following conservation laws.

$$p_t + \nabla \cdot j^+ = [\, \alpha(|\underline{E}|) - \eta(|\underline{E}|) \,] \, |j^-| \qquad (2.2a)$$

and

$$n_t - \nabla \cdot j^- = [\, \alpha(|\underline{E}|) - \eta(|\underline{E}|) \,] \, |j^-| \qquad (2.2b).$$

Where $\alpha(|\underline{E}|)$ is the first Townsend coefficient of ionisation and $\eta(|\underline{E}|)$ the coefficient of recombination. These functions are tabulated for air in Sarma & Janischewskyj (1969). There is a critical value of $|\underline{E}| \equiv E_I$ below which the value of $\alpha(|\underline{E}|) - \eta(|\underline{E}|)$ is negligible and no ionisation occurs. This value is about $3MVm^{-1}$ in dry air at standard atmospheric temperatures and pressures. If we knew the distribution of the field $\underline{E}$ then the determination of the values of p and n from (2.2a,b) involves solving a coupled hyperbolic system. If we denote the conductor surface by

Γ_c then the characteristics for problem (2.2a) point away from Γ_c - the ions are repelled away from the conductor - whereas the characteristics for problem (2.2b) point towards Γ_c . An appropriate boundary conditions for problem (2.2a) is to set

$$p \mid \Gamma_c = 0 \qquad\qquad (2.3)$$

implying that no ions are emitted from the conductor surface itself . To impose appropriate boundary conditions for the problem (2.2b) we must consider the distribution of electrons on the boundary of the corona. We define this boundary to be the surface Γ_{cor} so that

$$\Gamma_{cor} = \{ \underset{\sim}{x} \; : \; |\underset{\sim}{E}(\underset{\sim}{x})| = E_I \}.$$

For the present we shall presume that Γ_{cor} does not intersect the conductor surface Γ_c . An electron freed by ionisation at Γ_{cor} accelerates toward Γ_c causing further avalanches of electrons as it collides with the air molecules in its path. It is thus appropriate to define a value for n on Γ_{cor} which is denoted at time t by $n(\Gamma_{cor},t)$. We shall presume that there are two contributions to this charge density namely

$$n(\Gamma_{cor},t) \;=\; e_r \;+\; n_p(\Gamma_{cor},t) \;. \qquad (2.4)$$

Here e_r is a residual electron density caused by the action of cosmic rays or by electron diffusion from previous avalanches. The quantity $n_p(\Gamma_{cor},t)$ represents the number of electrons released by photoionisation ie. by photons created in previous electron avalanches. If we consider a simple geometry in which Γ_C and Γ_{cor} are concentric circles of radii r_0 and r_1 then it is shown in Budd & Wheeler (1988c) that $n_p(\Gamma_{cor},t)$

may be approximately calculated from the following expression.

$$r_1 \, n_p(\Gamma_{cor},t) \approx \gamma \, [r_0 n(\Gamma_c,t) \, |\underline{E}(\Gamma_c,t)|/E_I - r_1 \, n(\Gamma_{cor},t-t_d)] \quad (2.5)$$

where t_d is the time taken for an electron created at Γ_{cor} to drift toward Γ_c , typically t_d is about 5ns . The coefficient γ is a measure of the efficiency of the photoionisation process and takes a value of about 1×10^{-3} in a transparent gas (Morrow (1985)) .

We close the above system by presuming that any electromagnetic effects may be ignored and hence

$$\nabla \cdot \underline{E} = (p - n) / \epsilon_0 \quad (2.6)$$

where ϵ_0 is the permitivity of free space and has the value $\epsilon_0 = 8.85 \times 10^{-12}$ Fm^{-1} . In the next section we shall describe some appropriate boundary conditions for the calculation of $\underline{E}$.

A scaling for this set of partial differential equations is to set $\hat{E} = \underline{E}/E_I$, to express all lengths in units of millimeters and to set $\hat{t} = t/t_m$ where t_m is the time taken for an electron moving in a field of magnitude E_I to travel 1mm so that $t_m = 1 \times 10^{-3}/ (\mu_- \, E_I)$. Rescaling the equations and dropping overbars for clarity we then find that

$$p_t + \nabla \cdot (p\underline{E}) /200 = n \, |\underline{E}| \, (\alpha - \eta)(|\underline{E}|) , \quad (2.7a)$$
$$n_t - \nabla \cdot (n\underline{E}) = n \, |\underline{E}| \, (\alpha - \eta)(|\underline{E}|) , \quad (2.7b)$$
$$\nabla \cdot \underline{E} = 37.7 \, (p - n) . \quad (2.7c)$$

The function $(\alpha - \eta)(|\underline{E}|)$ takes the following very approximate form

$$(\alpha - \eta)(|\underline{E}|) \approx \alpha \, (\, |\underline{E}| - 1 \,)^+ \quad (2.8)$$

where $\qquad \alpha \approx 20$.

The system (2.7,8) together with the boundary conditions (2.3,4,5) may be solved numerically for a simple two dimensional geometry which comprises two concentric cylinders. In this system Ω_c will be taken to be the smaller cylinder upon which we shall impose a positive potential V. The calculation of the corona boundary Γ_{cor} is then part of the solution process. Further details of this numerical procedure are given in Budd & Wheeler (1988c). In Figures 2.1a,b and 2.2a,b we present the resulting profiles of rp and $|\underline{E}|$ for such a calculation where $r = |\underline{x}|$. Here we consider two cylinders of radii 1mm and 10mm with a potential of 15kV applied between them . It is discovered from these calculations that if we start with a zero distribution of the quantity p then initially the system exhibits a transient burst phase in which a large burst pulse of charge is produced close to the coronating electrode. However as $t \to \infty$ we observe that two different forms of behaviour are possible depending upon the value we take for the residual electron density e_r in the formula (2.4). If $e_r > 1\times10^{-10}$ then the system tends toward a steady state as $t \to \infty$. In this case there is a fairly uniform positive space-charge density between the cylinders in which p varies smoothly between the values of 0.6 and 1 mC m^{-3} . This situation illustrated by Figures 2.1a,b models the phenomenon of a steadily glowing discharge. When e_r is reduced this steady state loses stability and the system tends toward a stable oscillating solution as $t \to \infty$. In this case the value of p on Γ_{cor} behaves like the output of a relaxation oscillator rising rapidly to a high peak value and then decaying to a near zero value. This situation illustrated by Figures 2.2a,b models the Trichel pulse phenomenon described (for example) in Sigmond (1978). In both of these asymptotic states the value of n increases exponentially as we approach Γ_c and it is negligible for points other than those very close to the conductor. This exponential increase in n models the avalanche

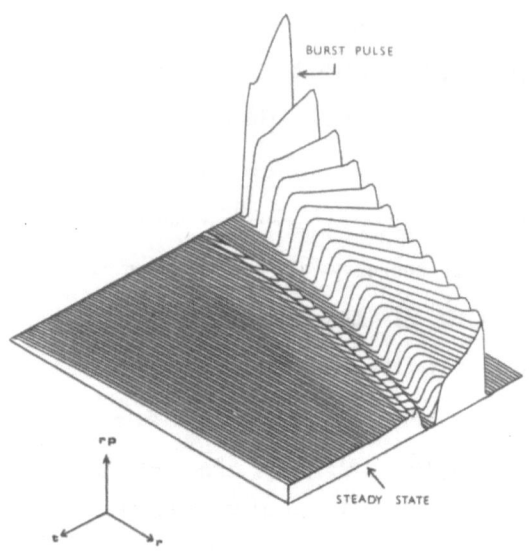

2.1a   The evolution of the charge density rp with time for a residual electron density of $e_r = 1 \times 10^{-9}$.

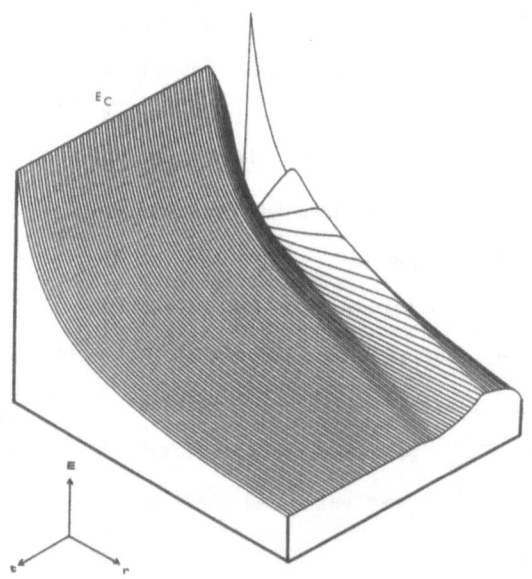

2.1b The evolution of the field E with time .

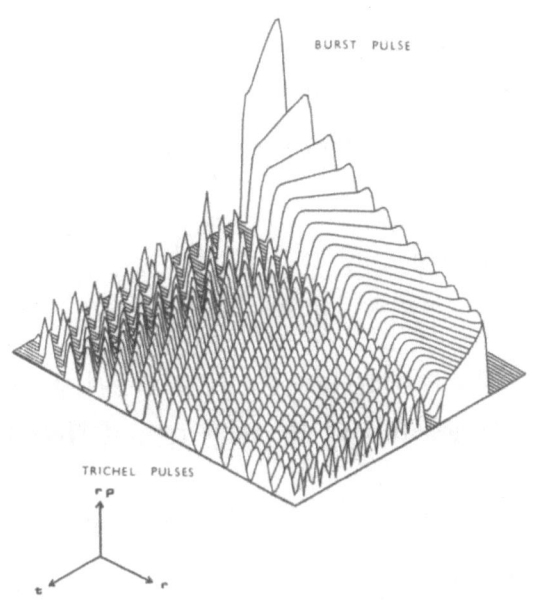

2.2a The evolution of the charge density rp with time for a residual electron density of $e_r = 1 \times 10^{-15}$.

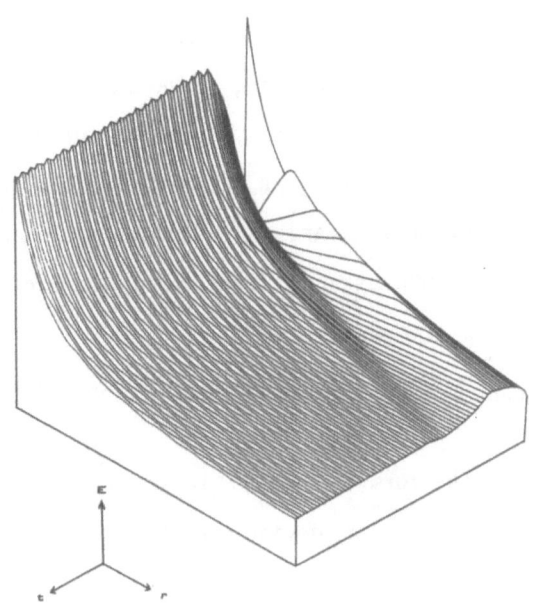

2.2b The evolution of the field E with time.

effect in the ionisation process and can be deduced from a direct integration of the equations (2.7b,8) using the method of characteristics. In particular for the steady solution of the previous geometry in which Γ_c and Γ_{cor} are concentric circles we find that

$$|\underline{E}(r)| \; r \; n(r) \; \approx \; r_1 \; n_{cor} \; \exp \int_r^{r_1} \alpha \; (\; |\underline{E}| - 1 \;)^+ \, dr \quad (2.9)$$

Details of the above calculations are given in Budd & Wheeler (1988c) where we also discuss the possibility of "blow up" solutions which correspond to sparks.

THE BOUNDARY CONDITIONS FOR THE STEADY STATE.

For the remainder of this paper we shall assume that the system (2.4,..,8) tends to the (stable) steady state illustrated in Figures 2.1a,b. We shall presume that the electron charge density at Γ_{cor} has the stable value of n_{cor} . Following Townsend (1914) we may deduce that the steady state is achieved when the number of electrons created by photoionisation on Γ_{cor} plus e_r precisely balances the number of electrons being accelerated toward Γ_c . For the geometry of two concentric circles we may then deduce that a steady state is achieved when

$$n_{cor} = e_r + \gamma \; [\; |\underline{E}(\Gamma_c)| \; n(\Gamma_c) \; r_0/r_1 - n_{cor} \;] . \quad (2.10)$$

By calculating the form of $n(\Gamma_c)$ it is then possible to deduce from (2.10) the asymptotic location of the corona boundary Γ_{cor} . In particular , if $r_1 = r_0 + \delta$
and if δ/r_0 and e_r/n_{cor} are both small, then the following asymptotic

estimates are derived in Budd & Wheeler (1988c).

$$\delta \approx [2 |\log \gamma | r_0 / \bar{a}]^{1/2} , \qquad (2.11a)$$

$$|\underline{E}(\Gamma_c)| \approx E_c \equiv 1 + [2 |\log \gamma | / (r_0 \bar{a})]^{1/2} , \qquad (2.11b)$$

$$r_0 n(\Gamma_c) \approx (r_0 + \delta) n_{cor} \exp (\alpha \delta^2 / 2 r_0) / E_c . \qquad (2.11c)$$

and

$$(r_0 + \delta) p(\Gamma_{cor}) \approx 200 [E_c r_0 n(\Gamma_c) - (r_0 + \delta) n_{cor}] \qquad (2.11d)$$

In particular , taking the values of $\gamma = 10^{-3}$ and $\bar{a} = 20$ then $\delta \approx [0.691 r_0]^{1/2}$. This value agrees closely with the empirical value for δ quoted by Peek (1929).

A photon sustained corona will be created if the field on the surface Γ_c formed in the absence of space-charge is greater than E_c . If this is not the case then even if $|\underline{E}| > E_I$ close to Γ_c a glowing steady state will not form as insufficient photons are released to maintain it. The rapid exponential rise of n as $\underset{\sim}{x}$ approaches Γ_c implies that the value of n may be neglected apart from a thin boundary layer of width $0(r_0^{1/2} / \bar{a}$ mm) close to the conductor although the corona itself occupies a region somewhat thicker than this boundary layer. We illustrate this situation in Figure 2.3.

We are now in a position to define a set of boundary conditions for the space-charge problem . Let Γ_c now be an irregularly shaped curve with local radius of curvature $r_0(\underset{\sim}{x})$ from which we may calculate a value of $E_c(\underset{\sim}{x})$ from the asymptotic formula (2.11b). Away from the thin boundary layer given above we may ignore the electron charge density n . If we consider the steady state situation we may then deduce that p and $\underline{E}$ satisfy the following system of equations

$$\nabla . (p \underline{E}) = 0 \qquad \text{and} \qquad (2.12a)$$

$$\nabla . \underline{E} \quad = 37.7 \text{ p} . \quad\quad\quad (2.12b),$$

where these equations apply in a region Ω_S in which the effect of the value of n may be neglected.

If we let $\underline{E} = - \underline{\nabla} \phi$ so that $\Delta \phi = - 37.7$ p then

$$\nabla . (\underline{\nabla} \phi \; \Delta \phi) = 0 . \quad\quad\quad (2.13)$$

We shall refer to the third order nonlinear partial differential equation (2.13) as the space-charge equation.

On Γ_c we shall take $\phi = V$ and we shall presume that the conductor Ω_c is placed inside an earthed conductor which has a boundary Γ_E upon which $\phi = 0$.

The region Ω_S will then be a subset of the space enclosed between the surfaces Γ_E and Γ_c bounded on the exterior by Γ_E and on the interior by a surface close to Γ_c . If we presume that the boundary layer close to Ω_c in which the value of n is significant has a thickness $0(r_0^{1/2}/ \bar{a} \text{ mm})$ which is small compared to the overall geometry of the system then we may approximate the interior boundary Γ_S of Ω_S to be coincident with Γ_c . For part of this boundary : Γ_{sc} the electric field will be sufficiently high to initiate a photon sustained corona and the resulting field strength will then take the value $E_c(\underline{x})$. If there is no corona then we may deduce from (2.3) that p = 0. Thus we shall model the corona by the following complementarity boundary condition

$$(|\underline{\nabla} \phi| - E_c(\underline{x})) \; p \quad = 0 \quad\quad\quad (2.14)$$

where $\quad |\underline{E}| = |\underline{\nabla}\phi| = E_c \quad$ and $\quad p > 0$ if $\underline{x} \in \Gamma_{sc}$

and $\quad\quad |\underline{E}| = |\underline{\nabla}\phi| < E_c \quad$ and $\quad p = 0$ if $\underline{x} \in \Gamma_c - \Gamma_{sc}$.

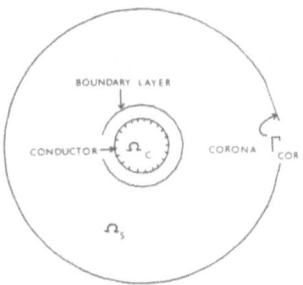

2.3   The configuration of the corona and the electron boundary layer about a circular conductor Ω_c .

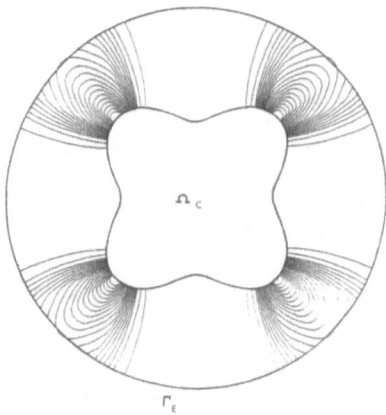

2.4   The contours of the space charge density formed in the region external to a conductor of varying curvature.

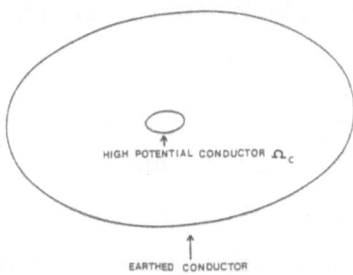

3.1   A doubly-connected geometry relevant to an overhead power cable.

We shall also insist that $\underline{E}$ is everywhere continuous. The calculation of the set Γ_{sc} is part of the solution procedure. We note that if δ/r_0 is reasonably small we may approximate the value of $E_c(\underline{x})$ by 1 (E_I) in our calculations.

A more complete discussion of this boundary condition is given in Budd & Wheeler (1988c) .

To illustrate this reasoning we present a figure showing the numerical solution of problem (2.13,14) for an irregularly shaped conductor Ω_c placed inside a circular container. Here we define Γ_c in polar coordinates to be the curve $r = 1. - 0.2 \cos (4 \theta)$, Γ_E the curve r = 2. and we apply a potential of 2kV between these two surfaces . In Figure 2.4 we show level curves of the resulting space-charge distribution. (An outline of the numerical method used to produce this figure will be given in section 4.) For such a system in the absence of space-charge a high field is developed at the points of highest curvature and hence the corona will be located close to these. The surface Γ_{sc} will not however necessarily extend to cover the whole of Γ_c . This behaviour is evident from an inspection of Figure 2.4 in which we see that Γ_{sc} is restricted to a neighbourhood of the points of highest curvature.

3. THE SPACE CHARGE PROBLEM AND
CALCULATIONS FOR AN ELECTROSTATIC PRECIPITATOR.

The model developed in section two and summarised by the formulae (2.13,14) allows us to calculate the distribution of space-charge external to a coronating electrode Ω_c . In figures 3.1,2 we illustrate two geometries relevant to some physical systems. Figure 3.1 is a cross section through a wire surrounded by an earthed conductor and it models an overhead

DC transmission line. In Figure 3.2 we consider an array of wires Ω_i at potential V placed between two earthed parallel plates. Some of the wires will be assumed to be coronating and thus sources of ions whereas others (of smaller curvature) will not act as sources of ions. This system models the geometry of an electrostatic precipitator. We shall refer to the systems with the geometries indicated in Figures 3.1,2 as problems I and II respectively. It is shown in Budd,Friedman,McLeod & Wheeler (1988) that if we replace the boundary condition (2.14) by the much simpler condition

$$p \mid \partial\Omega_i = g_i(\underset{\sim}{x}) \qquad (3.1)$$

then problem (2.13) has a weak solution $\phi(\underset{\sim}{x})$ for both problems I and II such that

$$\phi(\underset{\sim}{x}) \in C^{1,\alpha} (\Omega_S) ,$$

where the region Ω_S is defined as before. For problem I it may further be shown that $\phi(\underset{\sim}{x}) \in C^2(\Omega_S)$ and that the solution of problem (2.13,3.1) is unique. It is unclear at present whether the more realistic boundary condition (2.14) leads to a well defined problem although extensive numerical experiments imply this to be the case.

For certain geometries we can construct exact solutions of problem (2.13) . Indeed a solution has been known for some time (Townsend (1914)) for a system comprising two concentric cylinders. More recently some exact solutions have been constructed for other geometries notably by Eric Varley during the 1985 Mathematical Study Group by Smith (1987) and by Budd & Wheeler (1988b). To construct the latter solutions we consider functions $\phi(\underset{\sim}{x})$ which satisfy the partial differential equation (2.13) but which are functions of the single variable Φ where Φ is the solution of the much simpler problem

$$\Delta \Phi = 0 , \quad \Phi \mid \Gamma_C = V , \Phi \mid \Gamma_E = 0. \qquad (3.2)$$

188

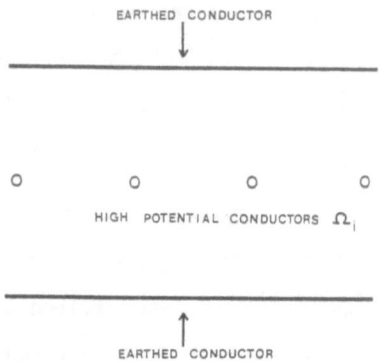

3.2 A multiply-connected geometry relevant to an electrostatic precipitator.

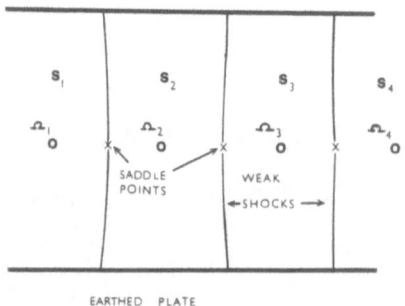

3.3 The division of the region interior to an electrostatic precipitator into subregions S_i which are bordered by weak shock lines.

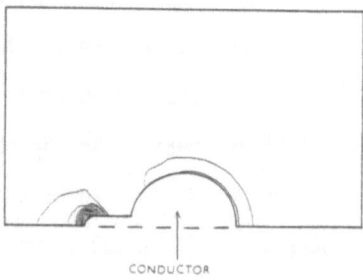

3.4 The distribution of the space-charge density in the region external to the spiked electrode used in a "cold" electrostatic precipitator.

All of these solutions are only valid in simply connected domains. In contrast the geometry for problem II is multiply connected and the corresponding solutions have a particularly rich mathematical structure. Let us consider such a problem comprising a series of conductors Ω_i placed between two parallel plates so that the conductors are alternatively charged (due to the effect of the corona) and uncharged. (In an electrostatic precipitator the latter conductors form part of the mechanical structure of the device and are not considered to be sources of ions.) It may then be shown that the equipotentials for $\phi(\underset{\sim}{x})$ have a series of saddle points lying between the conductors. There are field lines passing through these points which intercept the surface Γ_E of the two earthed parallel plates. These field lines divide the region Ω_S into a number of subregions S_i such that $\Omega_i \subset S_i$ and S_i is bounded by part of Γ_E and the two field lines through the saddle points on either side of Ω_i. Moreover , if $p \mid \partial\Omega_j = 0$ then it is shown in Budd,Friedman,McLeod & Wheeler (1988) that $p \equiv 0$ throughout S_j. Indeed the boundary of the region S_j is a weak shock line across which $\underline{\nabla}p$ is discontinuous .

A particular example of a system with this geometry is the experimental electrostatic precipitator used at CERL , Leatherhead . In this device electrodes with dimensions of around 3mm are raised to a potential of -45kV and are placed between two parallel plates separated by 150mm . We consider two problems , one where the electrodes are all cylinders with coronating spikes on their surface and another where the electrodes are alternatively coronating cylinders.

In Figure 3.4 we show a cross section through half of the electrode considered in the first problem and a numerical calculation of the resulting space-charge density. It is evident from this figure that the majority of the charge is produced from a region close to the tip of the spike. In

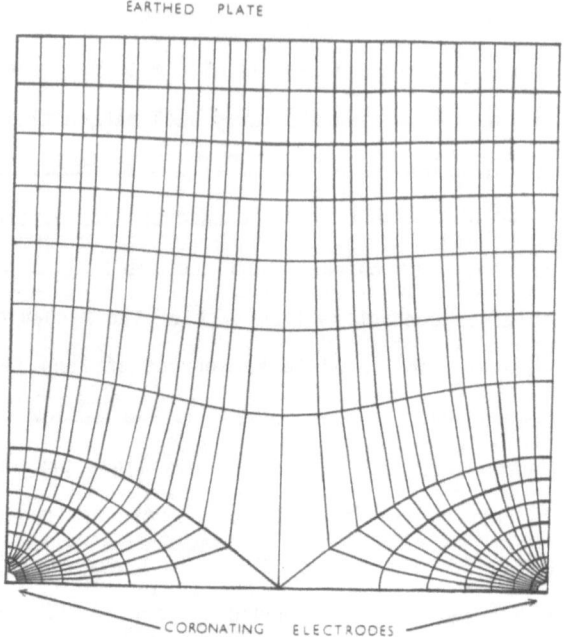

3.5 The field lines and equipotentials for a "cold" electrostatic precipitator geometry in which both electrodes are coronating spikes.

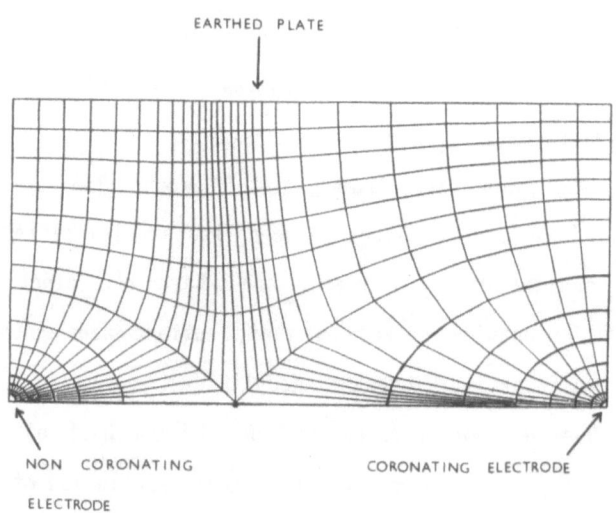

3.6 The field lines and equipotentials for a "hot" electrostatic precipitator geometry in which only the electrode on the right of the figure is coronating.

Figure 3.5 we present a picture of the field lines and equipotentials produced by an array of these electrodes placed between two parallel plates. In this figure the saddle point lies midway between the two conductors and the weak shock line divides the figure in half .

In Figure 3.6 we see in contrast the field line and equipotential distribution for the series of alternatively coronating electrodes. In this figure the coronating electrode is on the right hand side of the system and the shock line is displaced away from it and toward the non coronating electrode. This displacement follows because of the mutual repulsion of the space-charge in the region surrounding the coronating electrode.

We may further calculate the variation in the current densities j along the earthed plates and can compare these calculations with the values obtained experimentally by R.Corbin and J. Horrocks and presented in Corbin (1987) . In Figures 3.7,8 we present two graphs of these values corresponding to the systems illustrated in Figures 3.5,6. The agreement is reasonably good when we allow for the approximations made in the model.

It is also of interest when considering an electrostatic precipitator to calculate a solution of the problem (2.13) in a "parallel plate" geometry such as that described above but in a region distant from any electrode. Numerical and experimental evidence would suggest that if the two earthed plates are situated along the lines $y = +/- L$ and a coronating electrode is placed in a neighbourhood of the point $x = 0$ then , as $x \to \infty$, the solution of problem (2.13) is asymptotic to the function $\phi_S(x,y)$ where

$$\phi_S(x,y) = A \ e^{-\lambda x} \ y^{3/4} \ J_{-3/4}(\lambda y) \ .$$

Here A is a constant and $\lambda = j_{-3/4} / L$ where $j_{-3/4}$ is the first positive zero of the Bessel function $J_{-3/4}(z)$. In fact the function $\phi_S(x,y)$ is an exact similarity solution of the problem (2.13) . Further

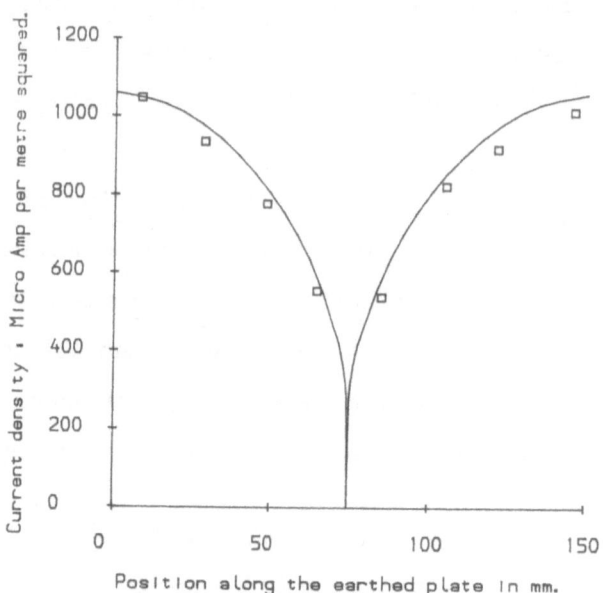

3.7   The current distribution along the earthed plate for the "cold" electrostatic precipitator geometry illustrated in Figure 3.5.

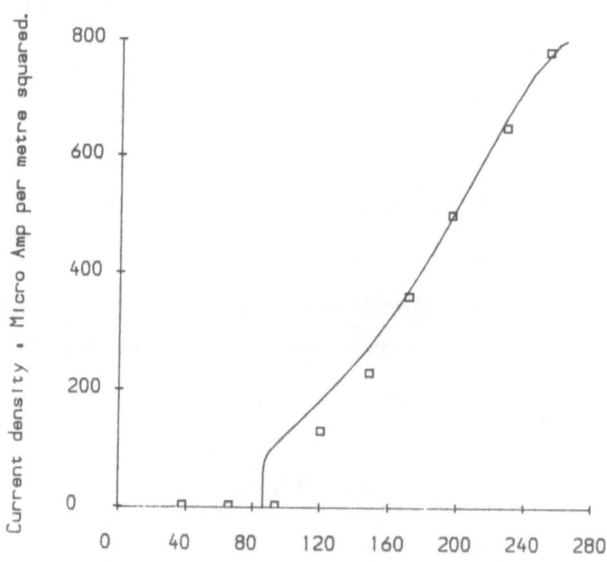

3.8   The current distribution along the earthed plate for the "hot" electrostatic precipitator geometry illustrated in Figure 3.6.

details of this calculation will be given in Budd & Norbury (1988) .

4. A NUMERICAL METHOD.

The numerical method used to produce the figures in this paper is an adaptation of the method described in Budd & Wheeler (1988a) . We base our numerical calculations upon a transformation of the problem (2.13) in which the independent variables x and y are expressed in terms of the potential ϕ and another variable ψ which is chosen to be constant along the field lines of the solution. The transformation maps the original (x,y) domain into a rectangle in the (ϕ,ψ) domain . Further the equation (2.13) is transformed into a simple non-linear elliptic system which is coupled to a simple ordinary differential equation. The transformed equations can then be solved using a simple algorithm based upon a finite difference discretisation. By using this approach we are essentially solving the partial differential equation (2.13) on an orthogonal body fitted mesh which is chosen to be coincident with the field lines and equipotentials of the solution. (The form of this mesh is evident from an inspection of Figures 3.5,6.) Indeed there is a very close relationship between the solution of the problem (2.13) and the techniques for orthogonal mesh generation described in Thompson et. al. (1974). It is in fact possible to use our algorithm for solving the space-charge equation as an automatic generator of an orthogonal body fitted mesh for a variety of different geometries.

A further advantage of a numerical method based upon the transformation described is the relative ease with which it can treat the complementarity boundary condition (2.14) and thus calculate the location of the corona boundary. Details of this calculation are given in Budd & Wheeler (1988a) . As an example the charge distribution illustrated in

Figure 2.4 was calculated using a 20 x 20 mesh and the run took about 1 minute on a VAX 11/780. We note at this point , however , that the precise location of the corona boundary depends upon many conditions not discussed in our model , notably the effects of diffusion , three dimensionality and the surface roughness. Thus the numerical calculations described above are likely to be somewhat inaccurate in their exact predictions of the corona boundary .

5. CONCLUSIONS

We have presented a simple (two dimensional) model for a coronating gas and have looked at a steady state solution showing how this gives us a simpler problem to solve , namely the space-charge problem (2.13) together with the complementarity boundary condition (2.14). This system can be efficiently solved numerically for complex geometries and the resulting predictions compare favorably with some experimental results.

The behaviour of a gas ionised by means of a coronating electrode can be very subtle and can take many different forms .The analysis presented here only begins to model some of this variety of structure. In particular we have concentrated upon steady two dimensional problems whereas in reality three dimensional effects are significant. Studies of this behaviour and also a closer investigation of the oscillating modes and of the effects of diffusion are now in progress.

ACKNOWLEDGEMENTS

We are grateful to R.Corbin and J.Horrocks for permission to use their experimental data. C.J.B. is also grateful to the Central Electricity Generating Board for their financial support during the course of this research.

6. REFERENCES .

[1]   Abdel-Salam,M.,Farghaly,M. & Abdel-Sattar,S.,
        "Monopolar corona on bundle conductors", 1982,
        IEEE Trans. PAS-101 ,pp 4079-4087.

[2]   Budd,C.,Friedman.A.,McLeod,J B. & Wheeler,A.
        "The space charge problem", 1988, Submitted.

[3]   Budd,C. & Norbury,J. " Similarity solutions of
        the space charge problem", 1988, in preparation.

[4]   Budd,C. & Wheeler,A. " A new approach to the
        space charge problem ", 1988a ,Proc. Roy. Soc
        A 417 , pp 389-415.

[5]   Budd,C. & Wheeler,A. " Exact solutions of the
        space charge problem using the hodograph
        transformation", 1988b, IMA J. Appl. Math.
        40, pp 1-14.

[6]   Budd,C. & Wheeler,A. " Coronas and the space
        charge problem", 1988c, in preparation.

[7]   Corbin,R.," Electrical performance of serrated
        strip electrodes in electrostatic precipitators",
        1987 , CEGB report TPRD/L/ES 0680/M87 .

[8]   Cross,J. " Electrostatics ", 1987 Adam - Hilger.

[9]   Loeb,L. "Electrical coronas ", 1965 , California.

[10]   Mathematical Study Group report , 1985, Oxford

[11]   Morrow,R. " Theory of negative corona in Oxygen"
        1985 , Phys. Rev. A 32 , pp 1799-1809.

[12]   Peek,F., " Dielectric phenomena in high voltage
        engineering ", 1929, McGraw - Hill .

[13]   Sarma,M. & Janischewskyj,W. " DC corona on smooth
        conductors in air", 1969, Proc IEE 116.

[14]   Smith,S.," Congruent harmonic and space-charge.
        electrostatic fields", 1987, IMA J. Appl. Math.
        39.

[15]   Sigmond,R., " Corona discharges" , 1978 ,
        Electrical breakdown in gases ed. J. Meek and
        J. Craggs , Wiley.

[16]   Thompson,J.,Thames,F. & Mastin,C. " Automatic
        numerical generation of body fitted curvilinear
        coordinate systems for fields containing any
        number of arbitrary two-dimensional bodies"
        1974 , J. Comp. Phys. 15 , pp 299-320.

[17]   Townsend,J. " The potentials required to maintain
        currents between coaxial cylinders", 1914 ,
        Phil Mag. 28, pp 83-90 .

Temperature Surges in Thermistors

AC Fowler and SD Howison

Mathematical Institute
24/29 St Giles
Oxford OX1 3LB

Abstract:

A thermistor is a nonlinear resistor whose resistivity increases with
temperature. We analyse a simple circuit containing such a device and
show that under certain circumstances rapid temperature surges can occur.

J. Manley et al. (eds.), Proceedings of the Third European Conference on Mathematics in Industry, 197–204.
© 1990 Kluwer Academic Publishers and B. G. Teubner Stuttgart.

1. Introduction

Thermistors are circuit components made from a ceramic material whose electrical resistivity $\rho(T)$ varies significantly with temperature T. In this paper we discuss the interaction between the heat generated in the device and the current flow through it, and the subsequent change in the current itself. In particular, we consider the behaviour of a positive-temperature-coefficient thermistor, i.e. one whose resistivity increases with temperature (as opposed to a negative-temperature-coefficient device whose resistivity decreases with increasing temperature). Such devices are frequently used to protect circuits, since any current surge leads to a temperature increase which in turn reduces the current by increasing the thermistor's resistance.

A simple circuit is shown in Fig. 1a. A short circuit is represented by closing the switch S, and we require a model for the subsequent evolution of the current. Two questions are of particular interest: (a) what is the dependence of the steady current I_0 on the external voltage V_0, and (b) can large temperature gradients occur inside the device. The second question is prompted by the experimental observation [1] that for large values of V_0 the device can crack; it is suspected that this may be caused by thermal stresses associated with high temperature gradients.

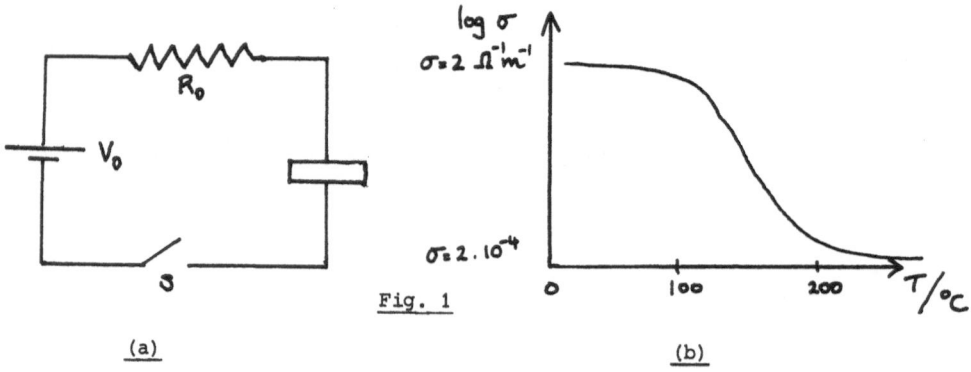

Fig. 1

(a)   (b)

It is convenient to work with the electrical conductivity $\sigma(T) = 1/\rho(T)$; the variation of $\log\sigma$ with T is sketched in Fig. 1b. The main result of this paper is to show that the temperature variation of σ and the external resistance R_0 can combine to produce large temperature gradients within the device. This fact was noted in [2]; previous work (see [2] for a review and references) had always ignored the effect of the external circuit.

In form, a typical thermistor is a cylinder whose thickness 2H is about 2mm, and whose radius is about 5mm. The two end surfaces are covered in thin metal contacts and onto these are soldered connecting wires. The net effect of this arrangement is that the heat loss from the thermistor is mostly through the top and bottom, and thus for simplicity we consider a one-dimensional model. With distance x measured from the centre plane, the temperature $T(x,t)$ and electric potential $\Phi(x,t)$ satisfy, for $t > 0$ and $|x| < H$, conservation of charge:

$$\frac{\partial}{\partial x}(\sigma(T)\frac{\partial\Phi}{\partial x}) = 0 , \tag{1.1}$$

and conservation of heat:

$$\bar{\rho}c\frac{\partial T}{\partial t} = k\frac{\partial^2 T}{\partial x^2} + \sigma(T)(\frac{\partial\Phi}{\partial x})^2 ; \tag{1.2}$$

the last term in (1.2) represents the Joule heating.
On the conducting surfaces $x = \pm H$, the potential satisfies

$$\Phi(\pm H, t) = \pm \Phi_0(t) \tag{1.3}$$

where Φ_0 satisfies the circuit equation

$$V_0 = I(t)R_0 + 2\Phi_0(t) , \tag{1.4}$$

in which the current $I(t)$ is given by

$$I(t) = \pi r^2 (\sigma(T)\frac{\partial\Phi}{\partial x})\Big|_{x = H} . \tag{1.5}$$

At t = 0, the temperature satisfies

$$T(x,0) \equiv T_a \qquad (1.6)$$

where $T_a \sim 20^\circ C$ is the ambient temperature; on $x = \pm H$, we model the heat transfer to the surroundings by a heat transfer law,

$$\pm k \frac{\partial T}{\partial x} + h(T-T_a) = 0 \qquad (1.7)$$

where h is a heat transfer coefficient. Lastly, we assume that the variation of σ with T is given by an exponential law:

$$\sigma = \sigma_0 \exp[-F(T)] \qquad (1.8)$$

where $\sigma_0 = \sigma(T_a)$ is the 'cold' value of σ and

$$F(t) = 0 \qquad T_a < T < T_a + \Delta T$$

$$= \frac{T - (T_a + \Delta T)}{\varepsilon \Delta T} \qquad T_a + \Delta T < T < T_a + 2\Delta T$$

$$= -1/\varepsilon \qquad T_a + 2\Delta T < T < \infty .$$

where $\Delta T \sim 100^\circ C$ is the increase in temperature needed before σ starts to decrease, and $\varepsilon \sim 10^{-1}$ is dimensionless.

Typical values for the parameters in these equations are as follows[1]:

$$\bar{\rho}c = 3 \times 10^6 \text{ J m}^{-3} \text{ K}^{-1} \qquad\qquad \sigma_0 = 2 \text{ m}^{-1} \Omega^{-1}$$

$$k = 2 \text{ W m}^{-1} \text{ K}^{-1} \qquad\qquad R_0 = 50 \Omega$$

$$H = 10^{-3} \text{ m} \qquad\qquad V_0 = 250V$$

$$r = 5 \times 10^{-3} \text{ m} \qquad\qquad \Delta T = 100K .$$

$$h = 10^2 \text{ W m}^{-2} \text{ K}^{-1} .$$

Using these parameters, and with some hindsight, we scale the variables as follows:

$$x = L \bar{x} \qquad\qquad t = (H^2 \rho c / k)\bar{t} \ , \ T - T_a = (\Delta T) u$$

$$\Phi = \tfrac{1}{2} V_0 \phi \qquad\qquad \sigma = \sigma_0 \bar{\sigma} \ .$$

Dropping the overbars, we have the following dimensionless system:-

For $\qquad -1 < x < 1, \qquad t > 0,$

$$(\sigma(u)\phi_x)_x = 0, \tag{1.9}$$

$$u_t = u_{xx} + \gamma \sigma(u) \phi_x^2 \ ; \tag{1.10}$$

on $\qquad x = \pm 1, \ \pm u_x + \beta u = 0; \tag{1.11}$

and $\qquad \phi = \pm (1 - \lambda \phi_x) \ ; \tag{1.12}$

and $\ u(x, 0) = 0 \ . \tag{1.13}$

The variation of $\sigma(u)$ with u is given by

$$\sigma(u) = \exp(-f(u)/\varepsilon) \tag{1.14}$$

where

$$f(u) = \begin{cases} 0 & 0 < u < 1 \\ u-1 & 1 < u < 2 \\ 1 & 2 < u < \infty \end{cases} \tag{1.15}$$

The dimensionless parameters here are

$$\gamma = \sigma_0 V^2 / 4k\Delta T \ \tilde{} \ 150 \ , \qquad\qquad \beta = hH/k \ \tilde{} \ 10^{-1} \ ,$$

$$\varepsilon \ \tilde{} \ 10^{-1} \ \text{(taken from data for } \sigma\text{)}, \ \lambda = \tfrac{1}{2} \, r^2 R_0 \sigma_0 / H \ \tilde{} \ 40 \ .$$

We analyse (1.9) - (1.15) in the following section.

2. Asymptotic analysis

All the dimensionless parameters in (1.9)-(1.15) are either large or small. In particular, the fact that ε is small suggests using the techniques of high-activation-energy asymptotics. We give a summary here; more details can be found in [2].

We begin by integrating (1.9) and substituting for ϕ_x into (1.10). The result is the non-local equation

$$u_t = u_{xx} + \frac{(\gamma/\sigma(u))}{[\lambda + \int_0^1 \frac{dx}{\sigma(u(x,t))}]^2}$$

$$= u_{xx} + \frac{\gamma \exp(f(u)/\varepsilon)}{[\lambda + \int_0^1 \exp(f(u)/\varepsilon)dx]^2} . \qquad (2.1)$$

with $\partial u/\partial x + \beta u = 0$ at $x = 1$

     $\partial u/\partial x = 0$ at $x = 0$ (by symmetry).

There are three stages in the evolution of u.

1. __Almost uniform increase in u__ until time t*, when $u(0,t*) = 1$. While $0 < u < 1$, $f(u) = 0$ and so u satisfies $u_t = u_{xx} + \gamma/(\lambda+1)^2$ with $u_x + \beta u = 0$ at $x = 1$. Because β is small, the solution is $u = \gamma t/(\lambda+1)^2 (1+0(\beta))$. Thus the dimensionless t* is approximately λ^2/γ.

2. __Acceleration in u near x = 0__ As soon as u reaches 1 at $x = 0$, the term $\exp(f(u)/\varepsilon)$ in the numerator of the heating term in (2.1) 'switches on' and produces locally large heating in a thin region near $x = 0$. This 'surge', reminiscent of thermal runaway problems, lasts until the integral in the denominator of (2.1) becomes larger than λ and reduces the heating term. The details of the unsteady development of this transition are complicated [2] and will not be considered here; we merely note that the timescale is thought to be $0 (\varepsilon\lambda^2/\gamma)$. This is the phase that may cause cracking via thermal stresses.

3. Equilibrium by conduction to a steady state

If the maximum temperature is u*, the heating term in (2.1) is only effective when $u* - u \sim 0(\varepsilon)$. This suggests that the thickness of the layer where the heating is effective is $x \sim 0(\beta/\varepsilon) \sim 0(1)$ for our parameter values, and then a balance of terms in (2.1) gives

$$u* \,\tilde{}\, \varepsilon \ln(\gamma/\varepsilon) + 1 ,$$

which for our parameters corresponds to a realistic maximum temperature of about $190°C$.

Our final point concerns the 'current-voltage characteristic', i.e. the dependence of the steady current on V_0. Clearly until u reaches 1 at $x = 0$ the thermistor resistance is unchanged, and so the steady current $I_\infty = V_0/(R_0+R_T)$, where R_T is the thermistor resistance $\pi r^2/\sigma_0 L$. However, when $u(0,\infty) > 1$, which occurs at voltages V_0 such that $\gamma/(1+\lambda)^2 > 2\beta/(2+\beta)$, the current starts to fall. In the steady state described above, the (dimensional) current is easily shown to be approximately

$$I_\infty \sim \frac{V_0}{R_0} \cdot \frac{\lambda}{\lambda+m\gamma/\varepsilon}$$

where m is an $\sigma(\varepsilon/\beta)$ constant depending on the precise details of the temperature profile. Substituting for λ,γ, one finds that since $\lambda << m\gamma/\varepsilon$,

$$I_\infty \sim \frac{4\pi\varepsilon H k \Delta T}{m} \cdot \frac{1}{V_0} ,$$

so that I_∞ changes from being proportional to V_0 to being inversely proportional as V_0 increases.

Acknowledgements

This problem was brought to Oxford in 1986 by Dr Miles Drake, then of STC Components. We would like to thank Colin Please, John Hinch and Andrew Lacey for helpful discussions, also Drs R Casselton and H Macartney of STC who took the problem over.

References

1. M. Drake, private communication.

2. A.C. Fowler, S.D. Howison and E.J. Hinch.
   'Temperature surges in current-limiting circuit devices', preprint, 1989.

MIXED HYPERBOLIC-ELLIPTIC SYSTEMS IN INDUSTRIAL PROBLEMS

A. D. Fitt, Mathematics Group, Royal Military College of Science,

Shrivenham, Swindon SN6 8LA England.

1.Introduction

In recent years there has been a large increase in the volume of literature pertaining to 'mixed' systems of partial differential equations. In order to fix the terminology which we shall use, consider a system of conservation laws

$$w_t + Aw_x = S \tag{1}$$

where A is an $N \times N$ matrix, w is a N-vector, S is an N-vector of source terms and subscripts denote differentiation. When the eigenvalues of A are non-zero, real and distinct, the system 1 is hyperbolic and many (though by no means all) of its mathematical properties are known (see, for example SMOLLER (1983)). In the strictest sense, the system 1 is said to be of 'mixed type' if any of the hyperbolicity conditions on the eigenvalues of A fail, so that for example two eigenvalues become equal, or an eigenvalue has a non-trivial zero. We shall be concerned mainly however with the much more serious case when some or all of the eigenvalues of A become complex, so that the problem is 'elliptic in time'. In this case it is tempting to discard the problem completely, arguing that the innate ill-posedness of the system and need for boundary conditions at $t = \infty$ precludes any meaningful analytical or numerical results. The fact remains however that such systems occur with surprising regularity. Because of this, a number of questions can be posed :

(a) Can such complex sound speeds ever occur in a *correct* mathematical model ?

(b) Even if complex eigenvalues *are* present, do they occur in regions of phase space which the model ever enters ?

(c) If complex eigenvalues can never occur in 'correct' models, is there ever any point in studying mixed problems for 'physical' systems ?

(d) If we do study such problems and ultimately attempt to find a numerical solution, what effects can we expect to arise from the ellipticity ?

As far as (a) above is concerned, we can be fairly sure that equations which turn out to be mixed in some regions have a 'mistake' somewhere in them. Although we distinguish between mixed systems where in some regions of phase space *all* of the eigenvalues are complex, so that

205

J. Manley et al. (eds.), Proceedings of the Third European Conference on Mathematics in Industry, 205–214.
© 1990 *Kluwer Academic Publishers and B. G. Teubner Stuttgart.*

the region is truly elliptic, and 'semi-elliptic' systems where only some of the eigenvalues become complex, leaving at least some information propagators, both must be considered to provide condemnation of our model. What is more difficult however in many cases is to see how the model should be changed. Often in a mixed system the physical assumption which has led to ellipticity is a constitutive one. In the equations of gas/particulate flow which we study in detail below, there is no doubt that the 'wrong' assumption is that a single pressure characterizes both phases - the pressure on the surface of a sphere moving through an inviscid fluid is simply not the same as the pressure far away from the sphere. Although modelling work is continuing with the aim of representing the interfacial pressure terms correctly, the fact is that at present the only equations available to us are 'incorrect' ones. As far as (b) above is concerned, it is certainly true that in some mixed systems the elliptic regions are unlikely ever to be entered, though it should be remembered that numerical solutions might nevertheless stray into them due to discretization errors. As we shall shortly see however for some equations the 'forbidden' regions cannot be avoided. It clearly makes sense therefore to study both physical mixed problems and also prototype systems of such equations. The points that we have made so far provide a rapid answer to (c); although we realize that a mixed system of conservation laws has serious faults, it may represent the best physical model which is available. To consider any worthwhile problems, we have to use the raw materials, however poor in quality they may be, until our modelling skill improves to a level where the equations can be trusted.

What effects do complex regions of phase space have on numerical schemes ? It is no surprise that the news here is nearly all bad. A quick consideration of the system

$$w_t + A(x_0, t_0)w_x = 0$$

where the matrix A has been 'frozen' at some time t_0 and position x_0 shows that if in the usual way we let P be the matrix of eigenvectors of A and D be the diagonal matrix composed of the eigenvalues of the 'frozen' matrix A, so that $A = PDP^{-1}$, then defining $w = P^{-1}v$ allows us to diagonalize the system. Taking the complex Fourier transform shows that the differential equation for V_i, the transform of v_i is

$$V_{it} + \lambda_i ik V_i = 0.$$

Under normal circumstances the solutions to this are purely oscillatory, but if one of the eigenvalues is complex then either it or its conjugate will inevitably lead to an exponentially growing mode. Thus any numerical inaccuracies will grow and eventually swamp the correct solution, their rate of growth being determined by the size of the imaginary part of the complex eigenvalue. Needless to say, it may also be shown that schemes which are TVD for hyperbolic systems become non-monotone. The only good piece of news is that, somewhat surprisingly, there are circumstances where strong source terms may help the stability when the eigenvalues are complex. This is in contrast to the situation generally encountered for strictly hyperbolic systems.

2. Mixed Systems in Industrial Problems

The discussion above has provided the motivation to study mixed problems, and this has been undertaken for both physical and prototype systems of conservation laws. One of the first systems to be recognized as mixed was the traffic flow problem described in BICK & NEWELL (1961). Whilst realizing the significance of the complex eigenvalues, they only considered the hyperbolic region, remarking that 'As yet we have found no satisfactory explanation of why the equations should be elliptic, nor any satisfactory suggestion as to what one should do about it'. JAMES (1980) considered a mixed system arising from waves in elastic bars, whilst HATTORI (1986) considered the flow of a Van Der Waal's fluid which led to a mixed system, showing that the concept of an 'admissible' solution could be physically motivated. Perhaps the most important contribution to the subject was that of BELL, TRANGENSTEIN & SCHUBIN (1986) who considered a model for the three phase flow of oil, gas and water in a porous medium. Their numerical calculations showed that when the initial states of a Riemann problem were chosen to be outside the elliptic region, the solution appeared to remain in the hyperbolic region for all time. They also found stable shocks connecting states inside the non-hyperbolic region with states outside.

As well as these physical problems, many authors have considered prototype mixed problems, most using the classical solution of the Riemann problem for 2×2 systems of conservation laws described in LIU (1974,1975) as a starting point. KEYFITZ & KRANZER (1983) studied Riemann problems for nonstrictly hyperbolic 2×2 systems where although the eigenvalues were real, they coalesced at an umbilic point. They found that in contrast to the strictly hyperbolic case where two waves are always sufficient, as many as four were necessary to construct a solution. A peculiar feature of some solutions was that for a given left state, the Hugoniot locus of the state was not necessarily connected to the state itself. This was also noted by SHEARER (1982) who solved a class of mixed conservation laws, producing a solution which involved stationary shocks not possessing viscous profiles, which left some doubts about their admissibility. SHEARER,

208

SCHAEFFER, MARCHESIN & PAS-LEME (1986) identified 'undercompressive' shocks which did not fully satisfy the Lax entropy condition, but for which viscous profiles could be found. More recently HOLDEN (1987) studied a mixed problem which was elliptic in a closed region of phase space. She found that although the system always had a weak solution which could be constructed in the normal way, it was not unique. Finally, mention should be made of the work of ISAACSON & TEMPLE (1985) and SCHAEFFER & SHEARER (1987) who have made a concerted attempt to solve and classify the general 2 × 2 system of non-strictly hyperbolic conservation laws with quadratic flux functions.

Interest in mixed systems, both of elliptic and non-strictly hyperbolic type, is therefore great. The theoretical work shows us that some novel numerical and analytical phenomena may be expected. To show that these are actually encountered in the study of physical systems of conservation laws, we now consider a specific example.

3. Two-Phase Gas/Particulate Flow

Let us now turn our attention to a specific mixed system, namely the equations of quasi one-dimensional two phase gas/particulate flow. These have been studied at length with the particular defence application of internal ballistics in mind. In this scenario, a highly energetic solid burns, becoming part of a gas phase. The classical equations which have been used by many authors are

$$(\rho_1 A_1)_t + (\rho_1 A_1 u_1)_x = m_p + \dot{m}$$

$$(\rho_2 A_2)_t + (\rho_2 A_2 u_2)_x = -\dot{m}$$

$$(\rho_1 A_1 u_1)_t + (\rho_1 A_1 u_1^2)_x = -A_1 p_x - D + \dot{m} u_2$$

$$(\rho_2 A_2 u_2)_t + (\rho_2 A_2 u_2^2)_x = -A_2 p_x + D - \dot{m} u_2 - S_x$$

$$(\rho_1 A_1 E_1)_t + (\rho_1 A_1 u_1 \left(E_1 + \frac{p}{\rho_1}\right))_x = -p(A_2 u_2)_x + m_p e_p - u_2 D + \dot{m} E_2$$
$$- q - Q$$

$$(N_2)_t + (N_2 u_2)_x = 0.$$

Here the subscript 1 refers to the gas phase, 2 to a solid phase, which we take to be a granular material and p to the primer gas which initiates combustion. We also use S = intergranular stress, D = interphase drag, q = interphase heat transfer, Q = heat loss to surrounding solid boundaries and $\dot{m}$ = gas mass addition due to burning, all of these terms being specified by various constitutive laws. Also we have $E = \frac{u^2}{2} + e$ where e is the internal energy and N_2 = number density of solid particles per unit length.

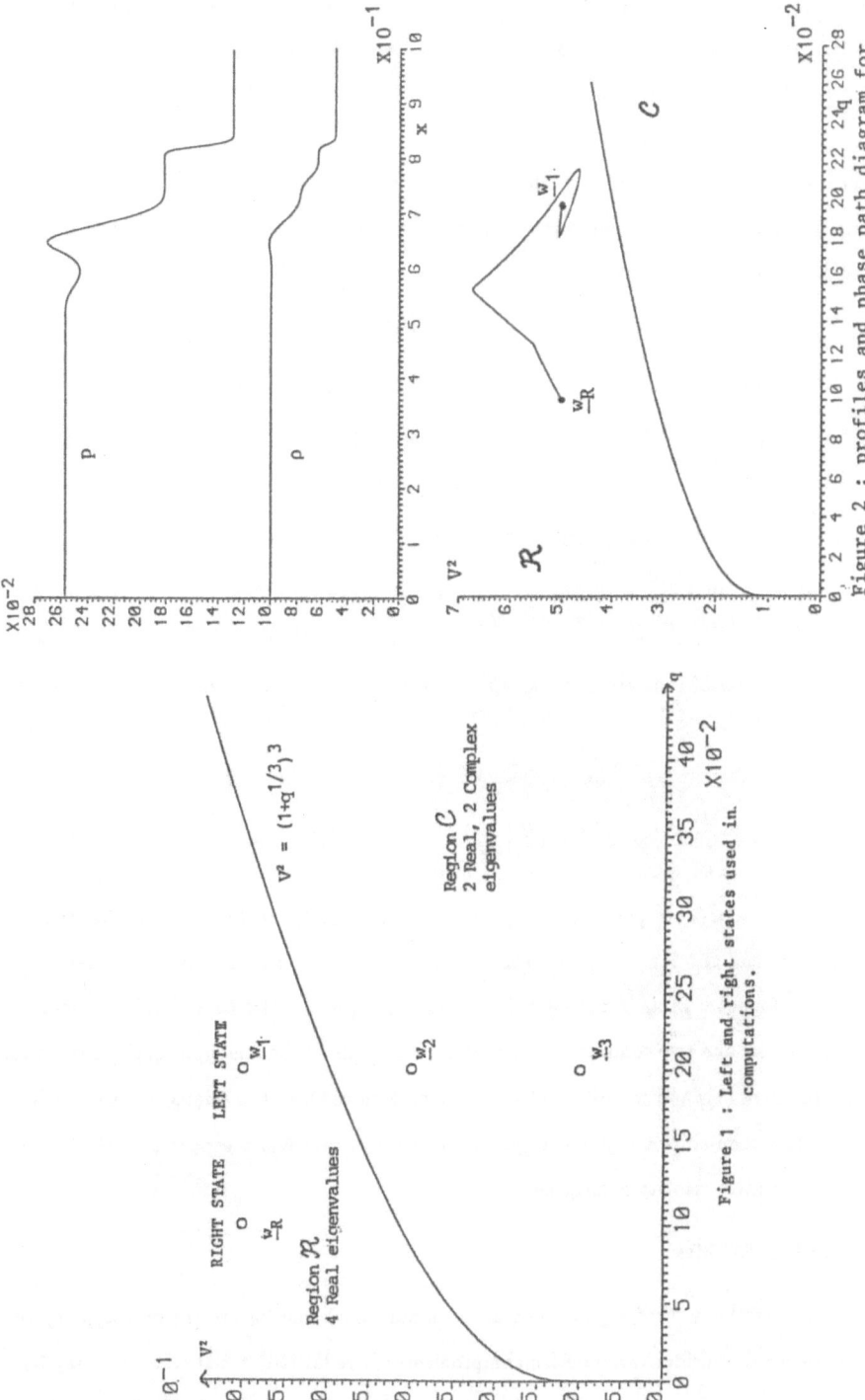

Figure 2 : profiles and phase path diagram for left state w_1

Figure 1 : Left and right states used in computations.

To this model we must also add an equation of state, which we take to be the co-volume gas law relating pressure to temperature and density, and suitable boundary conditions.

For the forthcoming discussion, it is convenient to produce a simpler (though identical in type) prototype system of equations by setting all the source terms to zero, assuming an ideal gas and an incompressible solid and allowing motion to take place only within fixed boundaries. (Normally one boundary will represent the base of an accelerating projectile so that the mesh expands). The equations become

$$(\rho_1 A_1)_t + (u_1 \rho_1 A_1)_x = 0$$
$$(A_2)_t + (u_2 A_2)_x = 0$$
$$(u_1)_t + u_1 u_{1x} + p_x/\rho_1 = 0 \qquad (2)$$
$$(u_2)_t + (u_2^2/2 + p/\rho_2)_x = 0$$
$$(p)_t + u_1 p_x + (p\gamma/A_1)(u_1 A_1 + u_2 A_2)_x = 0$$

It should be noted that the equations are not in conservation form. This is a consequence of the averaging which has been used and is unavoidable. Proceeding in the standard manner, we find that the eigenvalues of the system are given by $\lambda = yc + u_1$ where $c^2 = \gamma p/\rho$, and y satisfies the equation,

$$y(y^4 - 2Vy^3 + y^2(V^2 - q - 1) + 2Vy - V^2) = 0$$

where $V = (u_2 - u_1)/c$ and $q = (A_2\rho_1)/(A_1\rho_2)$. Clearly the zero sound speed corresponds to the incompressibility of the solid phase, but the behaviour of the roots of the remaining quartic is less clear. For $V = 0$ (both phases moving with the same velocity) we find trivially that there are two more zero and two real roots, but some elementary analysis shows that for non-zero V the equation has four real roots if and only if $V^2 > (1 + q^{\frac{1}{3}})^3$, and if this condition is not met then there are two real and two complex sound speeds. A diagram of the region in (q, V^2) phase space is shown in figure (1), and we note that for any q as soon as there is a relative velocity we must pass through a semi-elliptic region C of phase space before reaching a strictly hyperbolic region $\mathcal{R}$ when the relative velocity is large enough.

4. Numerical Results

To study the effects of the complex eigenvalues on numerical solutions to the prototype system, it is convenient to consider various Riemann problems where initially a 'left state' $w = w_L$ and a 'right state' $w = w_R$ are separated by a 'membrane' which bursts at time $t = 0$. Instead of using

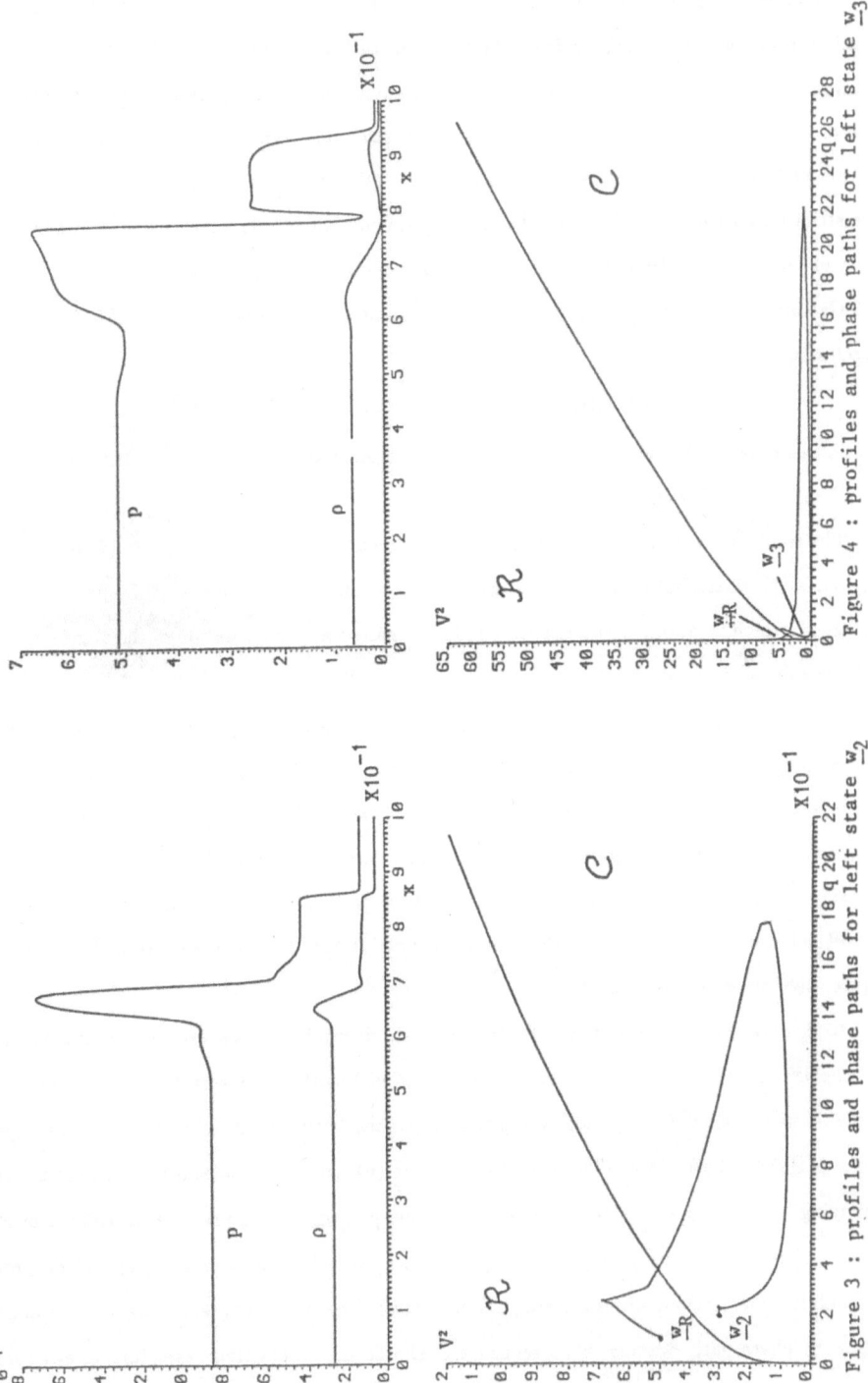

Figure 4 : profiles and phase paths for left state $\underline{w}_3$

Figure 3 : profiles and phase paths for left state $\underline{w}_2$

TVD methods or other more sophisticated schemes, the very simple Lax - Friedrichs scheme was used so that phenomena arising from the existence of the complex eigenvalues could be clearly distinguished. The conclusions apply however to any scheme which may be used. Of course the Lax-Friedrichs scheme is first order, and so for accuracy many (typically of the order of 1000) mesh points must be used. The scheme is also monotone (that is, free from any spurious oscillations) for strictly hyperbolic problems, so that any non-monotonic effects which we see must either be part of the actual solution, or numerical consequences of the complex eigenvalues.

Calculations were performed with 1000 space steps and a Courant number of 0.9, for a fixed right state

$$w_R = (\rho_1 A_1, A_2, u_1, u_2, p)^T = (0.05, 0.5, 5.0, 2.0, 0.12857)^T$$

and three different left states (see figure (1)) which started outside, moved a little way into C, and finally lay well within C.

The results of these computations at time $t = 0.05$ are shown in figures (2) to (4) respectively. In figure (2) where both initial states are in $\mathcal{R}$, it is clear that a smooth, physical solution is obtained, which is, as one would expect, similar in many ways to a solution of the standard shock tube gas dynamics problem. In figure (3) the left state lies just inside C, and whilst the solution still looks reasonably physical, the phase diagram shows that the complex region has been entered, and some puzzling non - monotonicities are present in the pressure and density profiles. These occur near to $x = 0.7$ and are magnified in figure (4) as the left state penetrates deeper into C. The velocity and phase diagrams for these two final cases are also most revealing - a potentially disastrous amplification is taking place and the phase path is extending further and further away from the left and right states into the complex region. It need hardly be said that the solution of figure (4) looks most unphysical.

To display the problems caused by the complex eigenvalue region even more clearly, figure (5) shows the phase diagram for a calculation made with both states very near to the right state $w_R \in \mathcal{R}$ used above. As expected, the problem exhibits 'well-posedness' in that the phase path remains close to both states. The situation in figure (6), when the states are close to each other but lying near to $w_3 \in C$ is very different. As time progresses, the pressure develops a bizarre shock system, which later computations (not shown) reveal to be a stable propagating profile (though in some circumstances growing unboundedly). The phase diagram of figure (6) confirms this - the phase paths wander, via a sequence of shocks and rarefactions, ever further away from the originally adjacent left and right states. It is worth pointing out the contrast between these

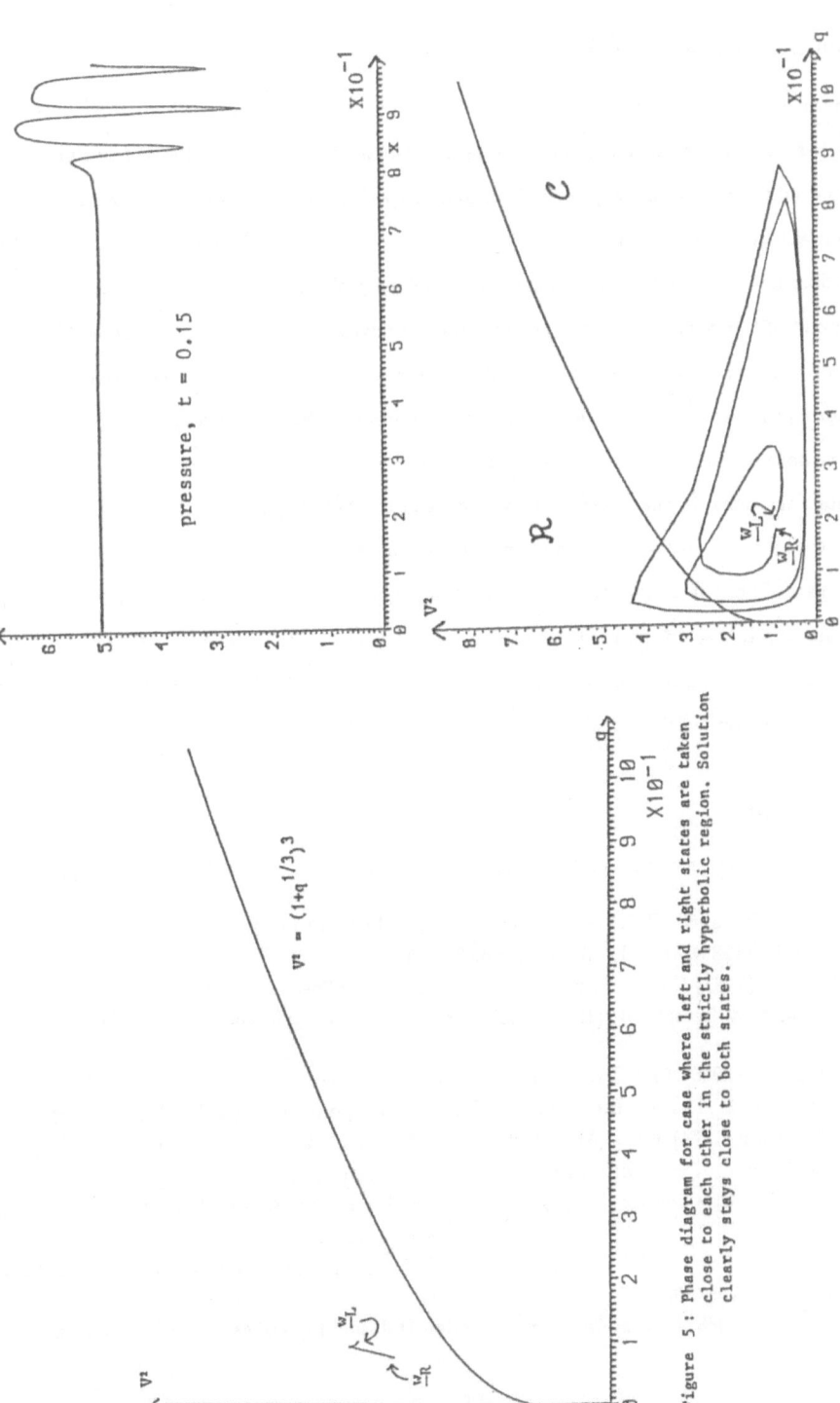

pressure, t = 0.15

Figure 6 : pressure profile and phase diagram for case
where states are adjacent in complex region

Figure 5 : Phase diagram for case where left and right states are taken
close to each other in the strictly hyperbolic region. Solution
clearly stays close to both states.

214

results and those of Bell, Trangenstein & Schubin, whose region of ellipticity was closed.

4. Conclusions and Discussion

We have seen that the equations of quasi one-dimensional two-phase flow which are frequently used as the basis for numerical calculations of internal ballistics flows have complex eigenvalues under certain conditions which are physically realisable. Experience has shown that numerical problems in internal ballistics codes have often been accompanied by profiles similar to those in figure (6). Space does not permit a full discussion of all the mixed problems which we have studied, but the main point of this paper is to encourage the study of mixed systems in their own right. Certainly models which lead to mixed systems have their faults, and in some cases the very existence of the complex eigenvalues may point the way to better modelling. For complicated problems however, often we have to make the best of the equations which we have. An alternative to this is is to add equations piecemeal to the system until we arrive at a totally hyperbolic system. This has been attempted by many authors and usually results in a model which is unphysical in some regions of phase space. Surely it is preferable to use equations whose underlying assumptions and derivation are clear, even if they do contain some complex eigenvalues. All that we must ensure is that their limitations, faults and possible associated problems are well understood. This last caveat, after all, is the litany on which all mathematical models are based.

References

[1] Bell, J.B., Trangenstein, J.A. & Schubin, G.R. (1986) SIAM J. Appl. Maths **46** no. pp. 1000-1017
[2] Bick, J.H. & Newell, G.F. (1961) Quart. J. Appl. Maths **18** pp. 191-204
[3] Hattori, H. (1986) Arch. R. Mech. **92** pp. 247-263
[4] Holden, H. (1987) Comm. Pure Appl. Math. **40** pp. 229-264
[5] Isaacson, E. & Temple, B. (1985) MRC Technical Summary Report no. 2892, University of Wisconsin
[6] James, R.D. (1980) Arch Rat. Mech. Anal. **13** pp. 125-128
[7] Keyfitz, B.L. & Kranzer, H.C. (1983) J. Differential Equations **47** pp. 35-65
[8] Liu, Tai-Ping (1974) Trans. Am. Math. Soc. **199** pp. 89-112
[9] Liu, Tai-Ping (1975) J. Differential Equations **18** pp. 218-234
[10] Schaeffer, D.G.& Shearer, M. (1987) Trans. Am. Math. Soc. **304** pp. 267-306
[11] Shearer, M. (1982) J. Differential Equations **46** pp. 426-443
[12] Shearer, M., Schaeffer, D.G., Marchesin, D. & Pas-Leme, P.L. (1986) Arch. Rational Mech. Anal. **97** pp. 299-320.
[13] Smoller, J. (1983) 'Shock Waves and Reaction-Diffusion Equations', Springer - Verlag

CONTRIBUTED PAPERS

Linear and Nonlinear Approximation of Power Density Spectra with Linear Dynamical Filter Systems

Dieter Ammon

Institute for Technical Mechanics, University of Karlsruhe
and
Daimler-Benz AG, Stuttgart, FRG

Abstract

Investigations on the behaviour of technical systems by means of computer simulations or laboratory tests require methods for the 'construction' of stochastic excition processes in order to imitate the natural effects as well as possible. In the stationary and gaussian case a process is completely described by its power density spectrum. W. Wedig suggested in 1985/6 the use of linear dynamical filter systems driven by White Noise for the artificial generation of the desired process. A linear approximation procedure is presented and discussed considering practical applications.

The direct approximation of the power density spectrum leads to a nonlinear and - in general - nonconvex optimization problem. It is shown that the direct adaptation can be performed in the time domain using the linear solutions as initial values. If the accuracy of these approximation solutions is not sufficiently high, evolution procedures may be applied to solve the resulting nonlinear optimization problem numerically. Some heuristic control concepts for evolution procedures based on statistical considerations are presented.

For piecewise constant excitation an equivalent discrete time system can be evaluated and transformed into an ARMA system. The generation of the process can be performed with a minimal amount of computation time using the ARMA system which is driven by a simple sequence of random numbers.

Key words:

Approximation of power density spectra, ARMA systems, artificial generation of stochastic processes, dynamical filter systems, evolution procedures, nonlinear optimization, road excitation of vehicles, step size control.

J. Manley et al. (eds.), Proceedings of the Third European Conference on Mathematics in Industry, 217–223.
© *1990 Kluwer Academic Publishers and B. G. Teubner Stuttgart.*

The classical approach to the artificial generation of stochastic processes is based on a system of harmonic functions with random phase angles (Shinozuka [4] et. al.). The amplitudes or the frequency distribution can be adapted to the given power density spectrum. This method yields periodical or almost periodical processes respectively. Because of the special properties of the above system of functions, it may not be suitable in some cases of On-line applications.

In 1986 an alternative approach has been suggested by Wedig [5], the approximation of the power density spectrum by complete filter systems driven by White Noise. In a first step the parameters of this linear dynamical model are adapted to the given spectrum. Then an equivalent ARMA system, corresponding to the continuous time system, is used to generate trajectories of the desired process with a minimal amount of computation time. Both methods have been extended to approximate multi-dimensional processes [4,1].

In the following we describe such a procedure to approximate one-dimensional processes. The linear method mentioned above is based on the adaptation of the corresponding differential operator to the given spectrum (target). Hence the solutions are sub-optimal. If the approximation of the spectrum is performed directly, a special non-linear optimization problem arises.

1 Target Process and System Model

A stationary Gaussian distributed process Y_t^v is completely characterized by its power density spectrum $S^v(\omega)$. The target spectrum can be defined as a smooth, piecewise linear curve in a double logarithmic scale.

The system model is a linear dynamical system of the order n (complete filter system). It is excited by a normalized stationary White Noise process ξ_t with vanishing mean. The properties of the output process Y_t of the system are depending on $2n + 1$ parameters a_i and b_i.

$$(1) \qquad L_N\{X_t\} = \sum_{i=0}^{n} a_i X_t^{(i)} = \xi_t; \quad Y_t = L_Z\{X_t\} = \sum_{i=0}^{n-1} b_i X_t^{(i)}; \quad X_t^{(i)} := \frac{d^i}{dt^i} X_t$$

Applying the Fourier-Transform to the differential operators L_N and L_Z in (1), we obtain the corresponding admittances (frequency responses) $N(j\omega)$ and $Z(j\omega)$ in the frequency domain.

$$(2) \qquad N(j\omega) = \sum_{i=0}^{n} a_i(j\omega)^i; \quad Z(j\omega) = \sum_{i=0}^{n-1} b_i(j\omega)^i; \quad j = \sqrt{-1}$$

With respect to the excitation properties the power density spectrum $S_y(\omega)$ of the process Y_t can be expressed in terms of ω and the system parameters. The parameters a_i and b_i have to be determined in such a way, that $S_y(\omega)$ approximates the target spectrum $S^v(\omega)$ as well as possible. This is decribed in the following sections.

In the case of a piecewise constant excitation process $\xi_t := Z_k = Z(k\,\Delta t)$, $k\,\Delta t \leq t < (k+1)\,\Delta t$, an ARMA system can be evaluated using the analytical solution of the continuous time system (1).

$$(3) \qquad Y(k+n) = \frac{1}{d_n}\left\{-\sum_{i=0}^{n-1} d_i Y(k+i) + \sum_{i=0}^{n-1} e_i Z(k+i)\right\}; \qquad k = 0,1,2,\ldots$$

2 Approximation

2.1 Direct Approximation in the Frequency Domain

Actually we have to perform an adaptation of the spectrum of the system $S_y(\omega)$ to the given spectrum $S^v(\omega)$. The application of a global L_2-norm results in the objective function $J^\omega(a_i, b_i)$, which is to be minimized with respect to the parameters a_i and b_i.

$$(4) \qquad J^\omega(a_i, b_i) = \int_0^\infty [S_y(\omega) - S^v(\omega)]^2\, d\omega \equiv minimum$$

The spectrum $S_y(\omega)$ of the system is a fractional rational function of a_i and b_i (2). Hence, for higher order systems the above formulation (4) yields a nonlinear and nonconvex optimization problem. In general, neither the existence nor the uniqueness of a solution can be proved.

2.2 The Linear Approximation Method

The problems above mentioned do not appear in the linear formulation suggested by W. Wedig [5]. A L_2-approximation applied to the correlation differential equation $L_N\{R_y(\tau)\}$ yields the system parameters a_i in a linear way.

$$(5) \qquad R_y(\tau) = E\{Y_t Y_{t+\tau}\}; \qquad L_N^j\{R_y(\tau)\} = \sum_{i=0}^n a_i R_y^{(i+j)}(\tau) = 0; \qquad \tau > 0$$

In order to shift the 'weighting' of the approximation to the range of lower frequencies we can also use integrated forms $(j = -1, -2, \ldots)$ of the operator $L_N\{R_y(\tau)\}$.

Since the correlation operator is homogenuous for $\tau > 0$, the corresponding L_2 objective function J^{lin} only contains the correlation $R^v(\tau)$ of the target.

$$(6) \qquad J^{lin}(a_i; j) = \int_{+0}^\infty \left[-\sum_{i=0}^n a_i R^{v\,(i+j)}(\tau)\right]^2 d\tau \equiv minimum; \qquad j = 0, -1, -2, \ldots$$

A necessary condition of (6) is that the partial derivatives of $J^{lin}(a_i; j)$ are vanishing. We obtain a linear system of equations to calculate the parameters a_i. The improper integrals in (6) can be solved analytically.

An exact adaptation of the system correlation $R_y(\tau)$ to the initial conditions of the target yields the remaining parameters b_i [5].

2.3 Direct Approximation in the Time Domain

Applying Parseval's formula to the objective function of the direct approximation (4) yields an equivalent expression in the time domain.

$$(7) \qquad J^\omega(a_i, b_i) = 2\pi \int_{+0}^{\infty} [R_y(\tau) - R^v(\tau)]^2 \, d\tau \equiv minimum$$

Assuming m conjugate complex eigenvalues $\lambda_i = \alpha_i \pm \beta_i$ and (n-2m) real eigenvalues α_i for the homogenuous correlation operator (5), the following form, with arbitrary constants C_i and S_i, is a general solution of (5).

$$(8) \qquad R_y(\tau) = \sum_{i=0}^{m} e^{\alpha_i \tau} (C_i \cos \beta_i \tau + S_i \sin \beta_i \tau) + \sum_{i=2m+1}^{n} C_i e^{\alpha_i \tau} \equiv$$
$$R_y(\alpha_i, \beta_i, C_i, S_i; \tau); \qquad \tau > 0$$

The real parts α_i and the imaginary parts β_i of the eigenvalues and the constants C_i and S_i are a complete parametric representation of the system (1). Using this set of parameters, we obtain a new optimization problem $J^\omega(\vec{x}) = J^\omega(\alpha_i, \beta_i, C_i, S_i) \equiv minimum$ which is equivalent to the original formulation (4). The corresponding restrictions can now be formulated in a very simple way.

3 Evolution Procedures

Evolution strategies have been successfully applied to optimization problems in Biological Sciences, Chemical Technics, Operations Research et. al. [3,2]. The procedures are based on the idea of mutation and selection.

Let $J(\vec{x}_k)$ be a objective function corresponding to a set of parameters $\vec{x}_k$. In a mutation step a new vector $\vec{x}_{k+1}$ results by a combination of $\vec{x}_k$ and a random component $\vec{\xi}_k$. In the selection step we compare the new value $J(\vec{x}_{k+1})$ with the previous one. If the new value is 'better' than $J(\vec{x}_k)$, the mutation was successful and the procedure can continue with the new parameters $\vec{x}_{k+1}$. Otherwise we have to perform an additional mutation step (with a different random vector $\vec{\xi}_k$). Usually heuristic components are integrated in evolution procedures to improve the convergence properties.

Basic concepts for the application of evolution processes have been suggested by Box [2], Rechenberg [3] et. al.. In the following we concentrate on 'pure' evolution procedures and describe a sightly modified concept of step size control which is based on the consideration of uniquely distributed mutation processes.

3.1 Controlled One Parameter Optimization

Let ξ be a normalized, uniquely distributed random process. The sum of the previous value x_k and the random component $\xi = \xi_k$ yields the mutated parameter.

$$(9) \qquad x_{k+1} = x_k + \sigma \xi,$$

where the 'random amplitude' σ is going to be determined. The success ratio s of a mutation can be defined as follows.

(10) $$s = \frac{number\ of\ successful\ mutations}{total\ number\ of\ mutations} = P\{J(x_k + \sigma\xi) \le J(x_k)\}$$

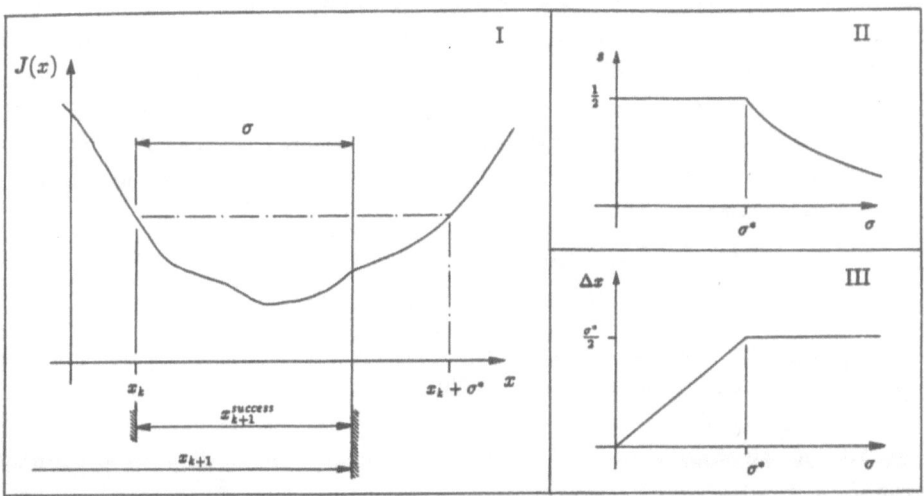

Figure 1: I. Relation between the interval of successful mutations (x_{k+1}) and the random amplitude σ. II.,III. The dependence of the success ratio s (II.) and of the average successful step size Δx (III.) on the random amplitude σ.

In the sense of a statistical mean the success ratio s is related to the random amplitude σ as shown in figure 1.

(11) $$s(\sigma) = \begin{cases} \frac{1}{2} & 0 < \sigma \le \sigma^*, \\ \frac{\sigma^*}{2\sigma} & \sigma^* \le \sigma \end{cases}$$

The characteristic amplitude σ^* is defined to be the minimal distance $|\ x_{k+1} - x_k\ |$ at which $J(x_{k+1})$ equals the previous value $J(x_k)$ for $x_{k+1} \ne x_k$. Obviously $\sigma = \sigma^*$ is the 'optimal' random amplitude. In this case the success ratio s and the average successful step size Δx are simultanuously maximal ($s = \frac{1}{2}$, $\Delta x = \frac{1}{2\sigma^*}$).

(12) $$\Delta x = E\left\{|\ x_{k+1}^{success} - x_k\ |\right\} = \begin{cases} \frac{\sigma}{2} & 0 \le \sigma \le \sigma^*, \\ \frac{\sigma^*}{2} & \sigma^* \le \sigma \end{cases}$$

Considering the running evolution procedure as a (discrete) dynamical process x_k of k, we should observe a monotonic decreasing optimal random amplitude $\sigma = \sigma^*$ 'on the way' to the optimum in most cases.

Let $\hat{s}_k$ and $\Delta \hat{x}_k$ be observations (estimates) of the success ratio s and the average (successful) step size Δx of the evolution step k. Replacing the actual values in (11) and

(12) by these estimates, we obtain two different concepts of step size control.

$$(13) \qquad \sigma_s = \begin{cases} constant & \hat{s}_k \geq \frac{1}{2}, \\ 2\hat{s}_k\sigma_s & \hat{s}_k < \frac{1}{2} \end{cases} ; \qquad \sigma_{\Delta x} = 2\,\Delta\hat{x}_k$$

Actually the evolution process is not stationary (i. e. σ^* is decreasing). Consequently the success ratio control is stable and forces σ_s to be a little bit larger than the optimum σ^*. In the other case (Δx control) the behaviour of the whole system depends on the initial (and intermediate) conditions. If $\sigma_{\Delta x} > \sigma^*$ is satisfied in the beginning and σ^* really decreases during the evolution process, $\sigma_{\Delta x}$ is going to be very close to σ^* ($\sigma_{\Delta x} > \sigma^*$). However, the Δx control may be unstable if one of the above conditions is violated.

Both concepts can be combined in such a manner that we obtain a stable procedure and the high accuracy of the Δx control, too.

$$(14) \qquad \sigma = \begin{cases} constant & \hat{s}_k \geq \frac{1}{2}, \\ minimum\ \{2\hat{s}_k\sigma, 2\Delta\hat{x}_k\} & otherwise \end{cases}$$

Assuming $s = \frac{1}{2}$ (which is approximately satisfied by the above control concept), one evolution step requires about 2 mutations. With the simplifying assumption that (on an average) every second evolution step increases the accuracy of the solution by factor 2 we obtain the following estimation[1] of the required number of mutations m to reach an error ε.

$$(15) \qquad m \leq -\frac{4}{\ln 2}\, ln\varepsilon; \qquad \varepsilon = \left|\frac{x_k - x^{min}}{x_1 - x^{min}}\right|; \qquad k = \frac{m}{2}$$

3.2 Multi-Dimensional Controlled Procedure

In the N-dimensional case the characteristics s and Δx are no longer independent of the local properties of the objective function $J(\vec{x})$. However, the described concept can be applied in almost the same way. The mutation process $\vec{\xi}$ is assumed to be uniquely distributed and independent. Similar to the one-dimensional case the mutation $\vec{x}_{k+1}$ is generated additively. Consequently the individual step size of the parameter x_i (of $\vec{x}$) is determined by the corresponding leading element σ_i of the amplification matrix σ. The success ratio s has already been defined in (10).

Assuming $J(\vec{x})$ to be a separable function of $(x_1, x_2, \ldots, x_N)$ we obtain a characteristic success ratio $s = \frac{1}{2N}$. In this case the individual average step sizes Δx_i of the parameters x_i have just reached their maximum (or almost reached it in the non-separable case respectively). Analogous observations (estimates) of the success ratio $\hat{s}_k$ and the vector of the average step sizes $\Delta\vec{\hat{x}}_k$ can be used to define the corresponding 'optimal' step size control.

$$(16) \qquad \vec{\sigma} = \begin{cases} constant & \hat{s}_k \geq \frac{1}{2N}, \\ 2\Delta\vec{\hat{x}}_k & otherwise \end{cases} ; \qquad \vec{\sigma}^T = (\sigma_1, \sigma_2, \ldots, \sigma_N)$$

---

[1]Taking the whole improvement of all mutations into account, a more precise estimate seems to be: $m \approx 2(1 - \ln 2\varepsilon)$.

Evolution procedures of that kind have been applied to the considered optimization problem for up to 16 parameters (system order $n = 8$). In most of the cases a preceding linear approximation was used to determine a starting vector $\vec{x}_0$. After a certain number of steps, depending on N and on $S^v(\omega)$, the evolution procedure was stopped. Finally a Newton iteration was started with the evolution results to evaluate the exact solution.

4 Some Typical Approximation Results

In general, the differences concerning the approximation errors (objective function J^ω) are small in relation to the dependence on the system order n. The results of the nonlinear approach are slightly better than the linear approximation, as was to be expected. However, if we increase the system order by $\Delta n = 1 \cdots 3$ in the linear case, the linear adaptation yields better approximations than the nonlinear procedure ($n \leq 8$). This is especially valid for more complex target spectra.

Regarding simple target spectra, the approximation error of the linear method is minimal, if the integration order j satisfies $j \approx -\frac{1}{2}$. In that case the weighting of the higher frequencies by derivation is equalized by the weighting of the low frequency range by integration. However, other examples show that the 'optimal' integration order j depends on the properties of the target spectrum. From a theoretical point of view the optimal integration order j^* should satisfy $1 - n \leq j^* \leq 0$, which is consistent with our experience.

References

[1] Ammon, D., *Generierung von diskreten zweidimensionalen stochastischen Prozessen mit definierten Leistungsdichtespektren*, Z. Angew. Math. u. Mech., 68(1988), T133-135.

[2] Box, G. E. P. and Draper, N. R., *Evolutionary Operation, a Statistical Method for Process Improvement*, Wiley, New York, 1969.

[3] Rechenberg, I., *Evolutionsstrategie: Optimierung technischer Systeme nach Prinzipien der biologischen Evolution*, Frommann-Holzboog, Stuttgart, 1973.

[4] Shinozuka, M., *Simulation of Multivibrate and Multidimensional Random Prozesses*, J. Acoust. Soc. Am., 49(1971), 357-368.

[5] Wedig, W., *Linear and Nonlinear Identification - One-dimensional Road Spectra and Nonlinear Oscillators*, in: Schiehlen, W., and Wedig, W. (eds.), *Analysis and Estimation of Mechanical Systems*, CISM Vol. No. 303, Springer, Berlin, 1987.

MODELLING SOFTWARE RELIABILITY FROM RUN-TIME DATA

A W ANDREW, R J COLE & J GOMATAM

DEPARTMENT OF MATHEMATICS, GLASGOW COLLEGE

ABSTRACT

A Software reliability model is presented here based on the
relationship between bug occurrence rates and program unit
execution frequencies. A set of testable reliability measures is
presented.

1. Introduction

Software products are increasing in size and complexity and
are now penetrating safety critical areas. Reliability models
are used to predict reliability from data obtained during the
software development cycle. The literature on the models is
extensive. Reviews can be found in [1 (pp 950-955)] and [2].

Suppose that a software program has been running from time 0
up to the present time, time τ (program execution time). During
this time n failures have occurred. It is assumed that each time
a failure occurs the failure-causing bug is removed before
execution of the program is restarted. The occurrence rate of a
bug is the rate at which failures attributable to that bug occur.
A software reliability model makes predictions about the future
failure behaviour of the program given the past failure history
of the program. The data set of most existing software
reliability models consists of only the past failure times of the
program, τ_1, τ_2,..., τ_n. Here τ_i corresponds to the occurrence

J. Manley et al. (eds.), Proceedings of the Third European Conference on Mathematics in Industry, 225–231.
© 1990 Kluwer Academic Publishers and B. G. Teubner Stuttgart.

time of the i th occurring bug.

2. The concept of exercise frequency

A major factor affecting a bug's occurrence rate is its location in the program code. Bugs in frequently exercised parts of the program tend to have higher occurrence rates than bugs in less frequently exercised code.

The exercise frequency of a bug is a measure of the frequency with which the program units (eg, subroutines, modules, statements) associated with the bug are executed. The program units associated with a bug are those units which have to be repaired in order to remove the bug. Each time a program unit associated with a particular bug is executed the bug may (or may not) cause the program unit to function incorrectly.

Definition. The exercise frequency of a bug is the sum of the execution frequencies of the bug's associated program units.

The failure rate of the program at any time is determined by the occurrence rates of the bugs remaining in the program at that time. By monitoring program execution it is possible to observe the frequencies with which different units of the program are executed. From this execution frequency data we can obtain the exercise frequencies, $f_1, \ldots f_n$, of the first n occurring bugs. We can use these exercise frequencies to help us to estimate the distribution of bug occurrence rates and to predict the reliability of the program more accurately than models which utilise only the occurrence times.

3 Reliability model based on exercise frequencies

The following three assumptions are common to existing
software reliability models: the environment in which the
program is used does not change over time; each bug contained in
the program causes failure independently of all other bugs
contained in the program (so that the times to occurrence of any
two bugs contained in the program are independent random
variables); the occurrence time of a bug whose occurrence rate
is x has the exponential distribution with parameter x,

$$(3.1) \qquad\qquad g(\phi) = xe^{-x\phi}.$$

The occurrence rates of bugs with given exercise frequency f
are assumed to be drawn from a probability distribution (gamma)
whose mean is proportional to f, with a coefficient of variation
independent of f. A simplified version of this model, in which a
deterministic relationship between exercise frequencies and
occurrence rates is assumed, is described in [3].

Denote by N the number of bugs initially present in the
program at time 0. Denote by pdf(x|f) the probability density
that a bug chosen at random from among the N initially present in
the program at time 0 has occurrence rate x given that the bug
has exercise frequency f, and let

$$(3.2) \ \ pdf(x|f) \ = \ gamma(x;\omega,\nu f) = x^{\omega-1}e^{-x/\nu f}/\Gamma(\omega)(\nu f)^{\omega}, \ \ \nu>0, \omega>0.$$

The gamma parametric form is assumed here because of its
mathematical tractability and flexibility. The mean and
coefficient of variation of this distribution are $\nu\omega f \propto f$ and

$1/\sqrt{\omega}$ respectively.

Let the probability density that a bug has exercise frequency f given that it occurs in $[0,\tau]$ be denoted by $pdf(f|[0,\tau])$. Then $f_1, f_2,...,f_n$ are drawn from the distribution with this probability density function.

Assume that $pdf(f|[0,\tau])$ is the gamma probability density function with parameters $a > 1$ and $b > 0$, that is;

$$(3.3) \qquad pdf(f|[0,\tau]) = gamma\ (f;\ a,b) = f^{a-1}e^{-f/b}/\Gamma(a)b^a.$$

Again the gamma parametric form is chosen here because of its mathematical tractability and flexibility. The parameters of the model a, b will be estimated from the data $f_1, f_2,...,f_n$.

Denote by $pdf_{init}(f)$ the probability density that a bug chosen at random from among the N initially present in the program has exercise frequency f. Using equations (3.1)-(3.3) it can be shown that

$$(3.4) \qquad pdf_{init}\ (f) \propto \frac{f^{a-1}e^{-f/b}}{1-(\nu f\tau+1)^{-\omega}}$$

(Full details can be found in [3].)

Denote by $pdf_{xinit}(x)$ the probability density that a bug chosen at random from among the N initially present in the program has occurrence rate x. Then by (3.2),

$$(3.5) \qquad pdf_{xinit}(x) \quad = \quad \int_0^\infty gamma(x;\omega,\nu f)\ pdf_{init}(f)\ df.$$

4. Measures of program reliability

Let Λ_τ be the program failure rate at the present time (time τ).

We assume that

(4.1) Λ_τ = program failure rate at time 0 –

$-\Sigma$ (occurrence rates of the bugs which have

occurred in $[0,\tau]$),

$$= \sum_{j=1}^{N} X_j - \sum_{i=1}^{n} Y_i,$$

where $X_1, \ldots, X_N$ are independent and identically distributed with probability density function pdf_{xinit}, and Y_i represents the occurence rate of the ith occurring bug (the bug which occurred at time τ_i and whose exercise frequency is f_i).

Using (3.1) and (3.2) it can be shown that the probability density function of Y_i is $gamma(y;\omega+1,\nu f_i/(\nu f_i \tau_i+1))$. Hence the expected value,

(4.2) $$E(Y_i) = \frac{\nu f_i(\omega+1)}{\nu f_i \tau_i+1}.$$

Notice that this is a decreasing function of τ_i. Of bugs with the same given exercise frequency, the later occurring bugs tend on average to have lower occurrence rates than the early occurring bugs. Thus, removal of the later occurring bugs tends on average to result in larger drops in the failure rate of the program.

Let T be the time between now (time τ) and the next failure. We assume that T has the exponential distribution with parameter λ,

(4.3) $$pdf_T(t) = \lambda e^{-\lambda t}, \quad \text{with} \quad \lambda = E(\Lambda_\tau).$$

Using (3.4), (3.5), (4.1), and (4.2) it can be shown that

$$(4.4) \qquad \lambda = \frac{NaZ(a+1,(\nu\omega\tau b)^{-1},\omega)}{\tau Z(a,(\nu\omega\tau b)^{-1},\omega)} - \sum_{j=1}^{n} \frac{\nu f_j (\omega+1)}{\nu f_j \tau_j + 1} .$$

where

$$(4.5) \qquad Z(p,q,r) \equiv \frac{1}{\Gamma(p)} \int_0^\infty \frac{\eta^{p-1} e^{-q\eta}}{1 - \left(1 + \frac{\eta}{r}\right)^{-1}} d\eta ,$$

is a generalisation of the Hurwitz zeta function.

Various measures of the reliability of the program can now be obtained, for example: current failure rate of the program = λ; mean time to failure = $E(T) = 1/\lambda$; and the probability of failure-free operation during the next t time units = $e^{-t\lambda}$.

5. Estimation of the model parameters a, b, N, ν, ω

Estimates of a and b ($\hat{a}$ and $\hat{b}$) are obtained are obtained by maximising the likelihood function,

$$(5.1) \qquad L_1(a,b) = \prod_{i=1}^{n} pdf(f_i|[0,\tau]) = \prod_{i=1}^{n} gamma(f_i|[0,\tau].$$

Define a second likelihood function

$$(5.2) \qquad L_2(N,\nu,\omega) = \text{Joint pdf } \{t_1, t_2, \ldots t_n\}, \quad t_i = \tau_i - \tau_{i-1}.$$

Using equ.(4.3) with $\tau = 0$, $\tau_1, \tau_2, \ldots, \tau_{n-1}$,

$$(5.3) \qquad L_2(N,\nu,\omega) = \prod_{i=1}^{n} \lambda_i e^{-\lambda_i t_i},$$

where, from (4.4),

$$(5.4) \qquad \lambda_i = \frac{N\hat{a}Z(\hat{a}+1,(\nu\omega\tau\hat{b})^{-1},\omega)}{\tau Z(\hat{a},(\nu\omega\tau\hat{b})^{-1},\omega)} - \sum_{j=1}^{i-1} \frac{\nu f_j (\omega+1)}{\nu f_j \tau_j + 1} .$$

Likelihood function L_2 is maximised with respect to N, ν, and ω to obtain estimates $\hat{N}$, $\hat{\nu}$, and $\hat{w}$.

6. Conclusions and outlook

The central assumption of the software reliability models presented here is the postulated relationship between bug occurrence rates and the frequency with which different parts of the code are executed. This relationship has been modelled stochastically and includes a deterministic result as a limiting process [3]. It is more realistic to start with the stochastic assumption on the tendency for bugs with high exercise frequencies to have high occurrence rates than to assume a deterministic relationship. The validity and accuracy of the models presented here await testing on real software for which the inter-failure times and execution frequencies are recorded. Preliminary simulation studies are being carried out at present.

References

[1] Abdel-Ghaly, A A; Chan, P Y; Littlewood, B: Evaluation of competing software reliability predictions. IEEE Transactions on Software Engineering vol SE-12 (1986), 950-967.

[2] Bendell, A; Mellor, P (Editors): State of the art report on software reliability. Pergamon Infotech (1986).

[3] Andrew, A W; Cole, R J; Gomatam, J: Modelling software reliability from run-time data. Technical Report TR/MAT/AWA,RJC&JG/88/1, Department of Mathematics, Glasgow College, (September 1988), pp 27.

SOFTWARE SIMULATION OF MODEL REFERENCE ADAPTIVE CONTROL SYSTEMS

Mr. M. Bakr & Dr. D. Bell

1 INTRODUCTION

Simulation of Model Reference Parameter Tracking Identification Techniques and Model Reference Adaptive Control Systems for design and stability analysis was implemented in the past using a complete analogue simulation and in more recent research, digital computers were introduced for the overall interface of the analogue components of the simulator and controller.

In this work the advances in Computer Aided Systems Engineering (C.A.C.S.E.) software enabled the simulation and design analysis to be confined to digital simulation of both plant and control mechanisms.

2 C.A.C.S.E. - The Simulation Tool

Most of the software packages for (CACSE) were developed on mainframe computers since they required large Random Access Memory (RAM) and storage media be it magnetic tape or disc, sophisticated and fast calculating capabilities, and in some cases high quality graphics.

The restricted availability of the software to large specialised establishments, ceased when a new marketing prospective was generated through the production of enhanced capability microcomputers, better known as personal computers. The reduced cost enabled small firms and establishments to buy and use these computers for low cost system simulation. At this stage manufactures began to to modify their software for personal computers in order to take advantage of this market sector.

The General Electric Company (GEC) developed a program they referred to as the Control Engineers Workbench (CEWB) for use by their engineers on VAX minicomputers, and until this date it is not known if is has been adapted for wider use in microcomputers.

Other firms hurriedly modified their programs and Intergrated Systems Inc. have now released Matrix X as Matrix X/PC with options System-Build/PC and System-ID/PC, even though other options and support facilities available on minicomputers are not yet modified for microcomputers such as (Real Time) RT-Build.

Two CACSE packages for the IBM-PC AT have been investigated for use in this work,

i) PC Matlab /The Mathworks Inc.

ii) Matrix X /Intergrated Systems Inc.

The final selection of Matrix X for its enhanced capabilities and a number of other features, will be discussed in this paper.

Both packages provide advanced mathematical analysis, matrix algebra, signal processing, system identification and block diagram design in two modes, first; direct input interactive mode and second; macro program execution.

The facility of block diagram design analysis provided by both Matrix X and Matlab is the subject of investigation as a tool of simulation for Model Reference Adaptive Control.

2.1 MATLAB's BLKBUILD & CONNECT

Systems comprising of of a plant and a dynamic controller are often represented in block-diagram form. For systems of even moderate complexity, it can be quite difficult to find the state- space model required, in order to bring analysis and design tools into use.

233

J. Manley et al. (eds.), Proceedings of the Third European Conference on Mathematics in Industry, 233–240.
© 1990 *Kluwer Academic Publishers and B. G. Teubner Stuttgart.*

Starting with systems block-diagrams, CONNECT is a function that can help form state-space models.

This feature in PC Matlab is based on the idea that having obtained a block diagram representation of the system to be simulated, each component will have to be defined only in a Laplace numerator-denominator form, hence restricted to single input / single output individual blocks. Interconnections between the blocks are defined as summation points.

Although PC Matlab provided a good software simulation package, it did not have the flexibility required for Model Reference Control simulation, since essential components of the technique are parameters adjusted via the feedback loop, which in turn requires the ability to multiply the output of two blocks.

2.2 Matrix X's SYSTEM-BUILD

The System-Build capability in Matrix X provides an interactive, menu-driven graphical environment for building, modifying and editing computer simulation modes.

Systems can be modelled by dividing them into individual components, with each component being described by a specific type of function block.

Super-blocks can be used to represent collections of individual blocks. This hierarchical structure is suitable for organising and documenting the modelling process.

Any super-block or set of nested super-blocks can be simulated, linearised and analysed. This provides a number of integration algorithms suitable for simulating a variety of systems.

3 MATRIX X SIMULATION

3.1 The creation of COMMAND FILES

Command files are stored a data files of a sequence of Matrix X commands and executed by Matrix X in a similar manner that batch files are executed under PC-DOS. They can be executed independently or nested with other command files.

They can be created by two methods as described by the User's manual:

A - Outside Matrix X by using any text editor or wordprocessor capable of storing text is ASCII format.

B - Inside Matrix X, this second method involves building a command file while entering commands by using the DIARY facility.

A very affective method when using System-Build commands is to create the file inside Matrix X, while building a simulation design, using the command DIARY. This makes more sense, as response to the commands are seen on the monitor. Then a text editor or in this case a wordprocessor is used to include commands of interactive nature such as INQUIRE which prompts the user for an input. This command was used to enter the plant and model gains and time constants and the sampling period by the user every time the command file is executed.

3.2 THE ADAPTIVE SYSTEM

The complete system is simulated in super-block MRAC which is composed of four blocks as shown in figure 3.3. This is built in Matrix X format be execution of macros;

MACRO	SUPER-BLOCK	COMMENT
SIM2	MRAC	Invokes SIM1
SIM1	ADJUST & ADAPT	Invokes SIMLTN
SIMLTN	PLANT	

The two super-blocks within the MRAC define the plant and the adaptive algorithm as follows;

A) Plant: A first order plant with time varying parameters (both gain and time constant) is simulated by Matrix X in super-block PLANT, figure 3.1a & 3.1b. The plant's transfer function is given by:

$$H(s) = \frac{1}{(G1 + g1)s + (Go + go)}$$

The original plant parameters G1 & Go are constants while g1 & go in the feedback loop are adjusted by extra inputs and can take any form e.g. random, pseudo random or sinsoidal and hence will define the parameters variations.

The detailed block diagram figure 3.2 shows how sections were combined in super-blocks for Matrix X since the maximum number of blocks allowed in each is only six while figure 3.3 shows it in Matrix X representation.

B) The reference model (MODEL) and the adjustment blocks are included in the second super-block (ADAPT) figure 3.4 and it's sub super-block (ADJUST) figure 3.5.

The hierarchical structure of super-block is in the form;

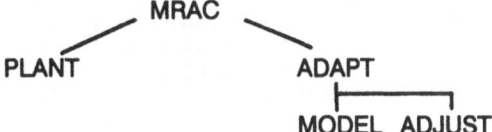

Adjustment blocks: define adaptation based on the minimisation of an error function f(e) = 1/2.q.e2 using the steepest decent error minimisation method.

The error signal   e = Yo-Ym   where   Yo is the plant output

                                                                                                Ym is the model output

The parameter influence forcing function defined by the equations;

$$a1\dot{u}1 + a0u1 = -\dot{Y}m$$

and

$$a1\dot{u}0 + a0u0 = -Ym$$

represent the parameter adjusting mechanism of the adaptive system and is defined in the simulation by super-block ADJUST figure 3.5.

This technique for the first order model investigated requires, in addition to the model output Ym, the first derivative with respect to time Ym. This is obtained applying the properties of the Laplace transform since;

$$L\{\frac{df(t)}{dt}\} = sF(s)-f(0+)$$

where L{f(t)} = F(s) and f(0+) is the initial value of f(t).

In this case, the transfer function of the Model:

$$\frac{1}{J1s + Jo}$$ is paralleled by another model $$\frac{s}{J1s + Jo}$$

is subjected to the same input to obtain Ym.

4 ANALYSING AND SIMULATING THE SYSTEM

The Analysed option prepares a super for simulation. When selected, the system is prepared for simulation, performing error checks and creating the state and output vectors.

When simulating the system, the variable step 'Implicit Stiff Solver' Integration Algorithm was selected. The choice was made because the solutions to the ordinary differential equations has rapidly decaying transient terms when compared with the slow varying steady state terms.

The 'Implicit Stiff System Solver' is a muti-step integration algorithm which intergrated past the desired time point and interpolates the solution at T + dT until the local error is within the tolerance.

5 SIMULATION RESULTS

At first a simple adaptation problem was considered with a small deviation in plant parameters to those of the reference model, simulating an analogue system, the result in figure 5.1 was that feedback compensation, through the adjustment mechanism, was for one parameter to adjust the plant gain to follow that of the reference model. MOre important was the elimination of the overshoot noticed in conventional control of heating systems by a thermostat (On-Off) figure 5.2; and optimisation methods. Then a perturbation to one plant parameter was injected figure 5.3 and the feedback compensation to adapt to it was very close.

A heating system's set the temperature needs to be adjusted for different requirements during one switch-on time, so the reference model gain was corrected during the simulation with satisfactory results shown in figure 5.4 showing that it is not rigid to fixed reference model parameters.

The real test that would show how applicable to heating system is this adaptive technique, was by simulating adaptation to a state space model of a single zone other than the approximated Laplace model. In figure 5.5 perturbations are automatically injected through the outside temperature which was assumed to follow the equation (h1 + g1) d0o/dt + (ho + go) = 0i defined in figure 5.6 and 5.7 and through pseudo random noise at the output. The result shown in figure 5.8 proves that although the feedback adjustment compensation oscillates for a considerable time after reaching the reference model steady state, oscillations on the plant are barely visible.

A final simulation test was performed by transforming the reference model and adjustment mechanism blocks into discrete section while keeping the plant simulation analogue, which is the case when using a digital controller. The sampling period was 30 seconds giving the simulation result of figure 5.9.

6 CONCLUSIONS

The digital simulation of MOdel Reference adaptive control using Matrix X/PC and it's option System-Build/PC eliminates the need for analogue simulation. This approach is more time efficient and avoids hardware design at simulation analysis stage.

The simulation procedure discussed in this chapter is flexible for different applications of Model Reference Techniques and applicable to higher order systems.

The results proved that for a heating system, a simplified adjustment mechanism can be used, with a single adaptive feedback loop. It is also possible to change the reference model parameters during adaptation with satisfactory results.

This adaptive method would still be stable for the control of systems defined by a different transfer function so far as the reference model parameters are close to a reduced or an approximate transfer function (as in the multi-input state space system).

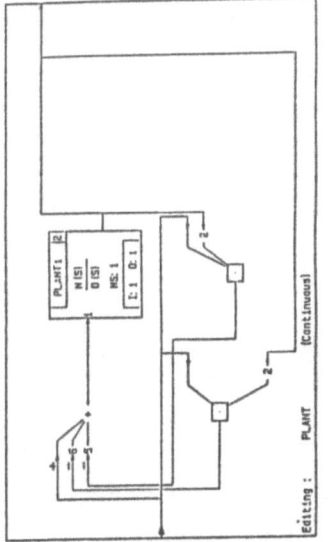

Figure (3.1a): Simulation of first order time varying parameters

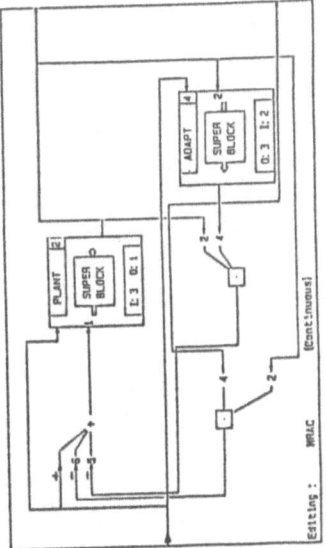

Editing :   PLANT   [Continuous]

Figure (3.1b) : MATRIXx Superblock (PLANT)

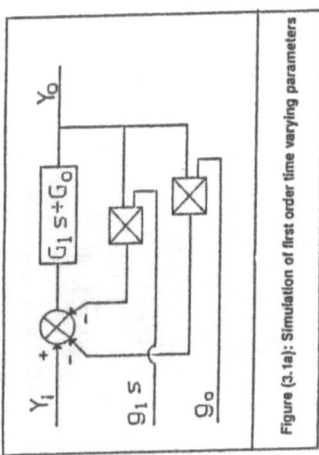

Figure (3.2): Simulated Model Reference Adaptive Control

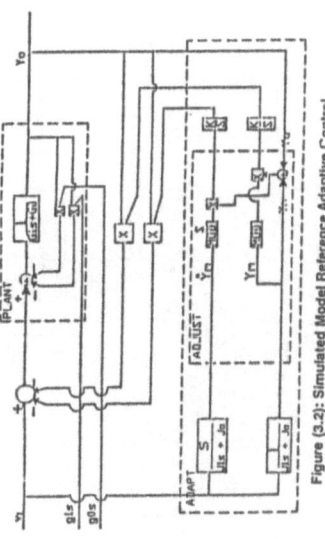

Editing :   MRAC   [Continuous]

Figure (3.3) : MATRIXx Superblock (MRAC)

238

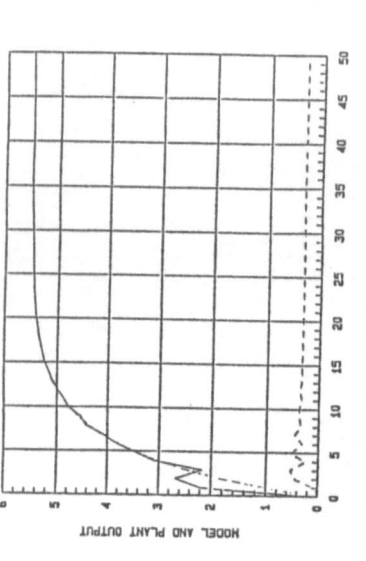

Figure (5.1) : Adaptation To Model Without Injected Perturbation To Plant Parameters

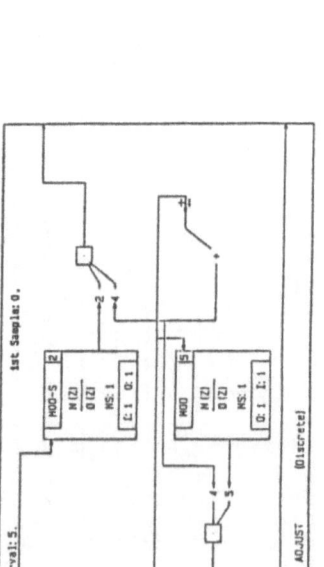

Figure (5.2) : Conventional Heating Control

Figure (3.4) : MATRIXx Superblock (ADAPT)

Editing :   ADAPT   (Continuous)

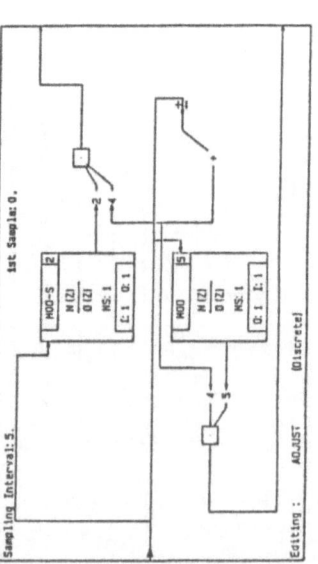

Figure (3.5) : MATRIXx Superblock (ADJUST)

Editing :   ADJUST   (Discrete)

239

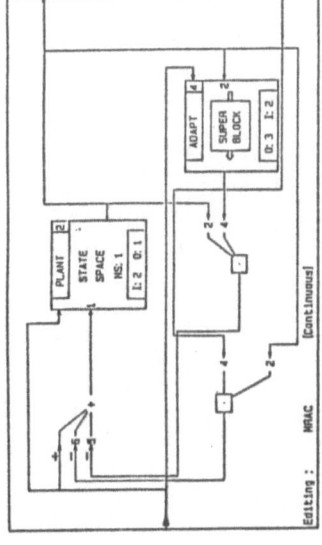

Editing :   MRAC   [Continuous]

Figure (5.5) : Adaptive System With State-Space Plant

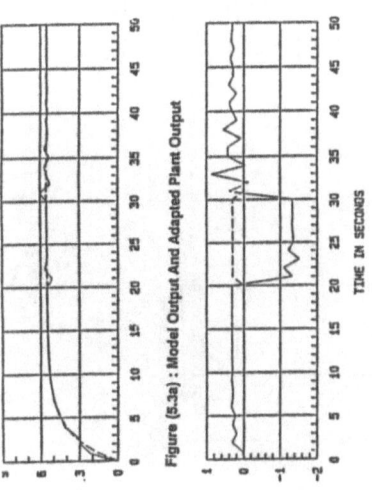

Figure (5.3a) : Model Output And Adapted Plant Output

Figure (5.3b) : Perturbation To And Adjustment Of Parameter

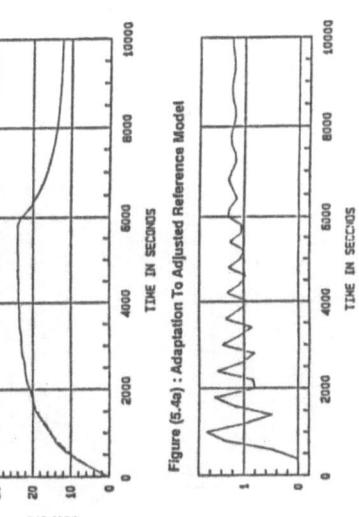

Figure (5.4a) : Adaptation To Adjusted Reference Model

Figure (5.4b) : Parameter Adjustment

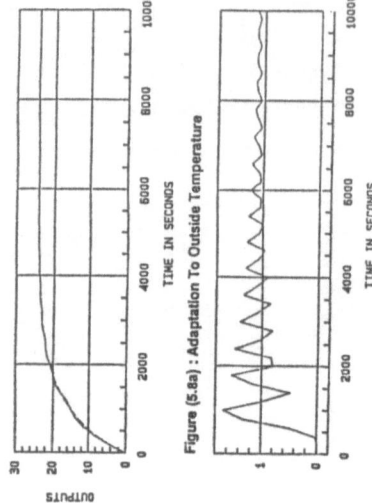

Figure (5.8a) : Adaptation To Outside Temperature

Figure (5.8b) : Parameter Adjustment

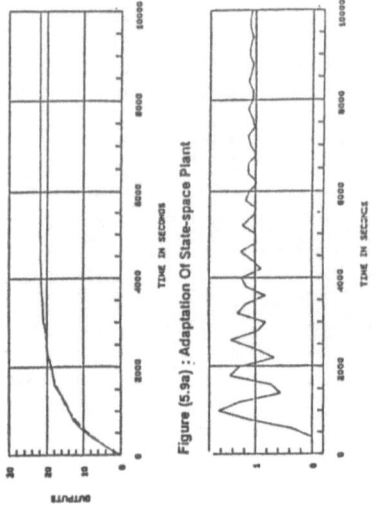

Figure (5.9a) : Adaptation Of State-space Plant

Figure (5.9b) : Parameter Adjustment

Figure (5.6): First order feedback system

Figure (5.7): Model Reference Adaptive Control

EFFECTIVE LENGTH OF AN ULTRAFILTRATION DEVICE

N.G. Barton, Australia

Summary: Industrial mathematics ventures have operated in Australia in a systematic way for the last half dozen years, and a notable feature has been the annual Mathematics-in-Industry Study Group at which 6-8 industrial problems are collaboratively investigated by academic and government mathematicians.[1] This paper is concerned with the effective length of an ultrafiltration device, a problem which was studied at the 1986 Study Group.[6]

1. Introduction

The centrepiece of the ultrafiltration device is a cartridge, typically about 0.3 m in length and 0.08 m in diameter, containing approximately 3000 small tubes. These tubes are manufactured using a stiff foam and can be considered rigid to radial stress at the typical operating pressures of about 100 kPa above the exit pressure. The tubes have inner and outer radii of about 100 μm and 300 μm respectively (1 μm $= 10^{-6}$ m), and the foam in the walls is fairly homogeneous with holes of typical dimensions 0.1 μm.

The dirty liquid is circulated under pressure around the tubes inside the cartridge, and clean liquid then percolates radially inwards through the walls of the tubes and out the lumina. The model presented here is a better treatment of exterior-driven flow through a porous tube than one which is widely available.[4] It is also noted in passing that it is more convenient to use the exterior of the tubes as the filtration surface than the interior of the tubes as in the common practice (see $e.g.$ [5]). In the case of exterior-driven flow, the lumina of the tubes do not become blocked during filtration, and it is possible to clean or 'back-wash' the filtration surfaces by blowing air at typical pressures of 500-700 kPa into the lumina of the tubes.

It is intuitively obvious that there is no point in making the fine tubes too long. The goal is to determine the effective length of the fine tubes as a function of the most important parameters of the apparatus - the inner and outer tube radii (r_i, r_o), the viscosity (μ) of the liquid, the pressure drop $(\Delta p = p_o - p_e)$ between the liquid to be filtered and the ends

241

J. Manley et al. (eds.), Proceedings of the Third European Conference on Mathematics in Industry, 241–247.

of the lumina, and the permeability (k) of the tube walls.

The basic fact which is exploited in the analysis which follows is that the ratio [*length* : *outer radius*] of the tubes is typically about 1000. It follows that the flow which percolates through the tube walls is almost radial. Moreover, the flow in the lumen of each tube in almost Poiseuille flow provided that the parameters of the apparatus satisfy a crucial condition (equation (9)) which is subsequently confirmed for realistic values of the tube parameters. Locally, therefore, the flow is well-understood. A slow variation of properties in the axial direction is then determined by considering the volume flux swept past any section of the tube.

2. Flow through the tube walls matched to flow in the lumen

The following time-independent theory using cylindrical co-ordinates (r, θ, z) is applicable to slow flow through stiff foam walls. The flow velocity $\mathbf{v}$ in the porous tube walls is given by Darcy's law (see *e.g.* [2]) pp. 223-224)

(1)
$$\nabla p = -\mu \mathbf{v}/k$$

where μ is the permeate viscosity, k is the permeability of the foam, and p, $\mathbf{v}$ are the pressure and velocity averaged over a volume whose dimensions are large compared with the individual holes in the foam (typically 0.1 μ) and small compared with the thickness of the tube walls (typically 200 μ).

The permeate flow in the tube walls is incompressible, and so the continuity equation $\nabla . \mathbf{v} = 0$ yields

(2)
$$\frac{\partial}{\partial z}\left(\frac{k}{\mu}\frac{\partial p}{\partial z}\right) + \frac{1}{r}\frac{\partial}{\partial r}\left(r\frac{k}{\mu}\frac{\partial p}{\partial r}\right) = 0$$

for flow in which p is independent of the azimuthal co-ordinate θ. The first term in equation (2) is completely negligible in comparison with the second term. If ℓ is the effective length of the tubes, the ratio of the terms is $O((r_0/\ell)^2)$, where it is anticipated that ℓ is comparable to L ($2L$ is the total length of individual tubes). Therefore $(r_0/\ell)^2$ is very small indeed, typically about 4×10^{-6}, so that equation (2) can be integrated to give

(3)
$$p(r, z) = p_o - C(z) \int_r^{r_0} \frac{\mu dr'}{r' k(r')} \, ,$$

$$v_{rad}(r, z) = -C(z)/r.$$

(4)

Here, $C(z)$ has dimensions of

$$[\text{pressure.permeability/viscosity}] = [ML^{-1}T^{-2}.L^2/ML^{-1}T^{-1}] = [L^2T^{-1}]$$

and is a function of the axial co-ordinate z, and p_o is the fluid pressure at the outer radius $r = r_o$. Suppose now that the almost radial flow holds for $z > z_0$, then the steady volume flux $Q(z)$ of permeate through the inside wall of the tube up to station z is

$$Q(z) = Q(z_0) + \int_{z_0}^{z} \int_0^{2\pi} \frac{C(z')}{r_i} r_i \, d\theta \, dz'$$

or

(5)
$$Q(z) = Q(z_0) + 2\pi \int_{z_0}^{z} C(z') \, dz'.$$

Now consider the flow in the lumen of each tube. This flow is pressure driven and almost parallel since the ratio $r_i/2L$ is very small. The axisymmetric axial momentum equation is

(6)
$$\rho u \frac{\partial u}{\partial z} + \rho v \frac{\partial u}{\partial r} = -\frac{\partial p}{\partial z} + \mu[\frac{\partial^2 u}{\partial z^2} + \frac{1}{r}\frac{\partial}{\partial r}(r\frac{\partial u}{\partial r})]$$

in which the radial velocity component v is of $O(r_i U/\ell)$ where U is a typical length scale for the axial velocity. It is now anticipated that the dominant form of equation (6) is

(7)
$$0 = -\frac{\partial p}{\partial z} + \frac{\mu}{r}\frac{\partial}{\partial r}(r\frac{\partial u}{\partial r})$$

in which case the steady axial velocity $u(r)$ is given by Poiseuille's Law

(8)
$$u(r) = -\frac{1}{4\mu}\frac{dp}{dz}(r_i^2 - r^2).$$

To confirm the approximation (7), observe that the axial derivatives in the viscous term of (6) are of $O((r_i/\ell)^2)$ compared to the radial derivatives, and may therefore be ignored. Moreover, the neglect of the non-linear inertia terms in equation (6) in comparison with the retained pressure and viscous terms is valid provided that

$$\rho U^2/\ell \ll \mu U/r_i^2.$$

Now U is of $O(Gr_i^2/\mu)$ where G is a typical value of the axial pressure gradient $\partial p/\partial z$. Hence the linearisation of equation (6) to give (7) is valid provided that

(9)
$$\frac{\rho G r_i^4}{\mu^2 \ell} \ll 1.$$

This property (9) is confirmed when numerical values are introduced into an example in Section 3.

Using equation (8), the local volume flux swept past station z is

$$Q(z) = \int_0^{r_i} \int_0^{2\pi} u(r) \, r \, dr d\theta$$

or

(10)
$$Q(z) = -\frac{\pi r_i^4}{8\mu} \frac{dp}{dz}.$$

The pressure gradient inside the lumen is seen from equation (3) to be

(11)
$$\frac{dp}{dz} = -\alpha \frac{dC}{dz}$$

where α is defined by

(12)
$$\alpha = \int_{r_i}^{r_o} \frac{\mu}{r' k(r')} dr'$$

and has dimensions $[L.ML^{-1}T^{-1}/L.L^2] = [ML^{-3}T^{-1}]$. Equation (10) therefore becomes

(13)
$$Q(z) = \frac{\alpha \pi r_i^4}{8\mu} \frac{dC}{dz}$$

and, if it assumed that this expression for the flux is true for $z > z_0$, the two steady-state flux expressions (5) and (13) yield

(14)
$$Q(z_0) + 2\pi \int_{z_0}^z C(z') dz' = \frac{\alpha \pi r_i^4}{8\mu} \frac{dC}{dz}.$$

This expression holds provided that condition (9) is satisfied.

Differentiation of equation (14) with respect to z gives the second order ordinary differential equation

(15)
$$\frac{d^2C}{dz^2} = \gamma^2 C$$

where the constant γ depends on the permeate and tube properties and is given by

(16)
$$\gamma^2 = \frac{16\mu}{\alpha r_i^4}.$$

The constant γ has dimensions $[ML^{-1}T^{-1}/ML^{-3}T^{-1}L^4]^{1/2} = L^{-1}$ and is, in fact, independent of μ. The general solution of equation (14) is

(17)
$$C(z) = c_1 e^{\gamma z} + c_2 e^{-\gamma z}$$

and the corresponding expression for the pressure in the lumen is (from equations 3, 12, 17)

$$(18) \qquad p = p_0 - \alpha[c_1 e^{\gamma z} + c_2 e^{-\gamma z}]$$

The constants c_1 and c_2 are now determined for the case of a tube of total length $2L$. If the origin $z = 0$ is chosen as the mid point of the tube, the pressure will be even in z so that $c_1 = c_2$ in (18). Also, the pressure at the end $z = L$ of the lumen is p_e which gives

$$p_e = p_0 - 2\alpha c_1 \cosh \gamma z,$$

so that

$$(19) \qquad c_1 = \frac{p_0 - p_e}{2\alpha \cosh \gamma L}.$$

Thus the solution for $-L < z < L$ is

pressure: $\quad p(z) = p_0 - (p_0 - p_e)(\cosh \gamma z)/\cosh \gamma L,$

fluid speed: $u(r, z) = [\gamma(p_0 - p_e)(\sinh \gamma z)/4\mu (\cosh \gamma L)] (r_i^2 - r^2).$

If the volume flux Q is taken to be zero at the mid point $z = 0$, it follows that

$$(20) \qquad Q(z) = \frac{\pi r_i^4 \gamma(p_0 - p_e) \sinh \gamma z}{8\mu \cosh \gamma L}$$

and the volume flux out of either end of the tube is

$$(21) \qquad Q(L) = \frac{\pi r_i^4 \gamma(p_0 - p_e)}{8\mu} \tanh \gamma L$$

Note that the dimensions of this expression are

$$[L^4.L^{-1}ML^{-1}T^{-2}/ML^{-1}T^{-1}] = [L^3 T^{-1}]$$

(as expected) and that $Q(L)$ tends to the finite value $\pi r_i^4 \gamma(p_0 - p_e)/8\mu$ as γL becomes large. In this limit, the pressure-driven flow through the lumen is limited by the viscous stress at the walls $r = r_i$ and, consequently, the tube is wastefully long. *The optimum length of the tube occurs when L is in the range $1/\gamma$ to $3/\gamma$.*

3. Numerical values and justification of equation (9)

Some numerical values are applied to illustrate the use of the above analysis. Suppose the fine tubes of the device have inner and outer radii of 100 μm and 300 μm, and the viscosity μ of the dirty liquid is about 1 mPa.s (appropriate to water at 20 ° C). If the fine holes in the tube walls are of diameter 0.1 μm, then the permeability k is about 10^{-17} m (see [3] §§ 5.5, 5.10). Also $\alpha = (\mu/k) \log 3$ and, from equation (16), γ is approximately 1.2 m^{-1}. Thus the theory suggests that the device would deliver no additional filtering capacity if its length was increased beyond the range 0.8 m to 2.4 m. [These figures are based on an estimate for the volume fraction of pores in the tube walls. Actual volume fractions are confidential, and the implementation of result (21) has been left to the company.]

The property (9) may also be confirmed. The maximum value for the pressure gradient is

$$G = -(p_o - p_e) \gamma \tanh \gamma L,$$

and using the above values for γ, μ and r_i together with $\rho = 10^3$ kg.m^{-3}, $\ell = L = 0.15$ m, $p_o - p_e = 10^5$ Pa, it follows that

$$\frac{\rho G r_i^4}{\mu^2 \ell} \sim 0.01$$

thus confirming the applicability of the theory.

4. Conclusions

Further details on the mathematical work on the ultrafiltration device are given in the Proceedings of the 1986 Study Group [6]. Additional work which was performed but has not been described here includes the ability of the fine tubes to withstand crushing by pressure of the dirty liquid and an optimisation calculation to indicate when the filter should be back-washed. Although the present problem has been simple in nature, the mathematical work was of undoubted economic benefit since the optimisation calculation mentioned just above resulted in the company taking out an international patent. Other problems identified by the Study Group have led to major mathematical investigations[7,8], and thus have been of benefit to applied mathematics itself.

5. <u>References</u>

[1] Barton, N.G.: A comparison of some Australian and European mathematics-in-industry ventures. *Austral. Math. Soc. Gazette* 14 (1987), 25-35.

[2] Batchelor, G.K.: *An Introduction to Fluid Dynamics* (Cambridge University Press, 1967).

[3] Bear, J.: *Dynamics of Fluids in Porous Media* (American Elsevier, N.Y., 1972).

[4] Bird, R.B.; Stewart, W.E.; Lightfoot, E.N.: *Transport Phenomena* (Wiley, N.Y., 1960).

[5] Breslau, B.R.; Tresta, A.J.; Milne, B.A.; Medjanis, G.: Advances in hollow fiber ultrafiltration technology" in A.R. Cooper (ed.), *Ultrafiltration Membranes and Applications in Polymer Science and technology* 13 (Plenum Press, N.Y., 1980).

[6] de Hoog, F.R.: *Proceedings of the 1986 Mathematics-in-Industry Study Group* (CSIRO Division of Mathematics & Statistics, 1987).

[7] Gates, D.J.: A microscopic model for progressive failure of brittle materials under load. *Int. J. Rock Mech. & Mining Sciences* (to appear).

[8] Please, C.P.; Wheeler, A.A.; Wilmott, P.: A mathematical model of cliff blasting. *SIAM J. Appl. Math.* 47 (1987), 117-127.

CSIRO Division of Mathematics and Statistics,
P.O. Box 218, Lindfield, N.S.W. 2070 Australia

<u>TRUNCATED SEQUENTIAL TESTS FOR MATERIAL CONTROL PROBLEMS</u>

Rainer Beedgen
c/o Berufsakademie Mannheim
Coblitzweg 7
6800 Mannheim 1
Federal Republik of Germany

Key Words: material accountancy, multivariate statistics,
parametric test, Neyman-Pearson Lemma,
sequential testing

<u>Abstract:</u>

Industrial facilities that handle valuable or hazardous materials are often forced to apply
material control procedures that effectively detect losses of the special material under
consideration. In many cases the material under consideration can only be measured with random
errors that means statistical procedures have to be considered. The statistical tests that are
presented in the paper have been developed to control Plutonium in nuclear facilities. The basic
principle of material control is accountancy, where material input, output and inventories are
measured and the book inventory ist compared with the measured inventory. The requirement of early
detection of possible material losses lead to a sequential evaluation of material balance data,
where the boundary condition of a final decision at the end of a given reference time has to be
observed. Truncated sequential test procedures based on the material accountancy statistic have
been developed and applied to detect losses of material with sufficient high probability.

1. Introduction:

Industrial facilities that handle valuable or hazardous materials are often forced to apply
material control procedures that effectively detect losses of the special material under
consideration. In many cases the material under consideration can only be measured with random
errors that means statistical control procedures have to be considered. The statistical test
procedures that are presented in the paper haven been developed to control Plutonium in nuclear
facilities. The basic principle of material control is accountancy [9], where material input,
output and inventories are measured and the book inventory is compared with the measured inventory
[1]. The requirement of early detection of possible material losses lead to a sequential
evaluation of material balance data, where the boundary condition of a final decision at the end
of a given reference time has to be observed. Truncated sequential test procedures based on the
material accountancy statistic have been developed and applied to detect losses of material with
sufficient high probability and in a timely manner [2].

2. Multiple Balances Model

2.1 Description of Material Balances Vector

In the following the statistical model for n material balances is given. For each balance period
i = 1, 2, ..., n in the reference time the balance equation

(2.1) $MUF_i = I_{i-1} + R_i - S_i - I_i$

(MUF = Material Unaccounted For) holds, where I_{i-1}: beginning inveniory of period i (I_0 is the
beginning inventory of period 1)

249

J. Manley et al. (eds.), Proceedings of the Third European Conference on Mathematics in Industry, 249–255.
© 1990 Kluwer Academic Publishers and B. G. Teubner Stuttgart.

I_i: ending inventory of period i
R_i: receipts during period i
S_i: shipments during period i.

The balances MUF_i are the result of numerous measurements and are assumed to be normally distributed. Furthermore, the random vector $\underline{MUF} = (MUF_1, MUF_2, ..., MUF_n)$ is assumed to be normally distributed, i.e. $\underline{MUF} \sim N(\mu, \Sigma)$.

2.2 Statistical Hypotheses

It may be assumed that the measurements of the material balance components (I, R, S) are unbiased estimates of the true amounts. It is assumed that the measurement error model, i.e. the variance-covariance matrix Σ, is known. The question of a possible material loss during the n balance periods results in an uncertainty about the parameter μ as mean of the random vector $\underline{MUF}$. In the case of no loss the equation $E(MUF_i) = 0$ holds for all $i = 1, 2, ..., n$ balances. If there is a material loss during some balance period i_0, the inequality $E(MUF_i) = l_{i_0} > 0$ has to be assumed. Following these arguments the parameter space under consideration is

(2.2) $\Theta = \{\mu = (l_1, l_2, ..., l_n | l_i \geq 0\}$.

The two situations "no loss of material" and "loss of material" are now translated into assertions about the unkown parameter μ. The situation of no loss is defined as the null hypothesis:

H_0: (no loss of material)

(2.3) $\mu\epsilon\Theta_0 = \{(0, 0, ..., 0)\}$.

Formula (2.3) means that in the non-loss case the mean values for all n material balances are zero. The alternative hypothesis is the situation that during the n balance periods a loss of amount L happened. This may be described as:

H_1: (loss of amount L)

(2.4) $\mu\epsilon\Theta_1 = \{(l_1, l_2, ..., l_n) | l_i \geq 0$ with $\Sigma l_i > 0\}$.

3. Considerations in the Neighbourhood of the Neyman-Pearson Lemma

The probability of Type I error is called "false alarm probability" because in our model this means a detection of a loss if there is indeed no loss. The probability to reject H_0 if H_1 is true will be called "detection probability".
One objective in selecting a statistical test was to find a test with highest guaranteed detection probability for some fixed amount of loss L [3]. The simple alternative hypothesis of a special fixed loss pattern is written as:

H_1': (special fixed distribution of loss L)

(3.1) $\Theta_1' = \{l_1', ..., l_n') | \Sigma l_i' = L\}$.

The Neyman-Pearson test for H_0 against H_1 has the following form:

$$(3.2) \quad (l_1', \ldots, l_n') \cdot \Sigma^{-1} \cdot (muf_1, \ldots, muf_n)^t$$

$$> s_\alpha, \text{ reject } H_0$$

$$\leq s_\alpha, \text{ reject } H_1$$

where muf_i is a realization of MUF_i and t means the transposed vector. The test threshold s_α depends on the selected false alarm probability α. The detection probability of the Neyman-Pearson test can be written as:

$$(3.3) \quad 1 - \beta[(l_1, \ldots, l_n)]$$
$$= \Phi([(l_1, \ldots, l_n)\Sigma^{-1}(l_1, \ldots, l_n)^t]^{\frac{1}{2}} - \Phi^{-1}(1-\alpha))$$

where $\Phi(..)$ is the standard normal distribution function. The test with the highest guaranteed detection probability is:

$$(3.5)$$
$$muf_1 + \ldots + muf_n \quad \begin{array}{l} > s_\alpha, \text{ reject } H_0 \\ \leq s_\alpha, \text{ reject } H_1 \end{array}$$

4. Truncated Sequential Tests

The objective of timely detection of a possible loss calls for a sequential testing of the material balance results. The statistical tests presented in the following are called Near-Real-Time Accountancy (NRTA) procedures, because they are intended to be used for a running process. At the end of the reference time the inspection authority has to make a final decision between "loss" and "non-loss". What means the test are truncated. The statistical problem may be described to find a test for the mean of a sequence of correlated normal distributed random variables where only one observation for each random variable is given. To approach the problem three tests have been suggested. There is no "optimal test" whatsoever available.

4.1. Truncated Sequential CUMUF Test

Based on the analysis in chap. 3 and on results in the area of quality control the cumulative sum of the material balance results is defined:

$$(4.1) \quad CUMUF_i = MUF_1 + MUF_2 + \ldots + MUF_i$$

for $i=1, 2, \ldots, n$. The random vector $(CUMUF_1, CUMUF_2, \ldots, CUMUF_n)$ is multivariate normally distributed with known variance-covariance matrix Γ. Under H_0 we have $E(CUMUF_i) = 0$ and under H_1 $E(CUMUF_i) = 1_1 + 1_2 + \ldots + 1_i$. If it is assumed that $cumuf_i$ is a realization of $CUMUF_i$ we define the following test for H_0 against H_1 [4]:

a) for $i=1, 2, \ldots, n-1$:

$$(4.2a)$$
$$cumuf_i \quad \begin{array}{l} > k_\alpha[var(CUMUF_i)]^{\frac{1}{2}}, \text{ reject } H_0 \\ \leq k_\alpha[var(CUMUF_i)]^{\frac{1}{2}}, \text{ go to the} \\ \qquad\qquad\qquad\qquad \text{next period} \end{array}$$

b) for i=n:

$$(4.2b) \quad cumuf_n \quad \begin{cases} > k_\alpha [var(CUMUF_n)]^{\frac{1}{2}}, \text{ reject } H_0. \\ \leq k_\alpha [var(CUMUF_n)]^{\frac{1}{2}}, \text{ reject } H_1. \end{cases}$$

4.2 GEMUF Test

This test is closely related to the idea of the Neyman-Pearson Lemma explained in chap. 3. The generally unknown loss for period i is estimated by the unbiased estimate MUF_i [11]. That leads to the following test statistic:

$$GEMUF_i = (MUF_1, \ldots, MUF_i) \Sigma_i^{-1} (MUF_1, \ldots, MUF_i)^t$$

for i=1, 2, ..., n, where Σ_i^{-1} is the inverse of the variance-convariance matrix for the first i balances. Under H_0 $GEMUF_i$ is $x^2(i)$ distributed with mean i and under H_1 non-central $x^2(i)$ distributed.
The test may be described as follows:

a) For i=1, 2, ..., n-1:

$$(4.3a) \quad gemuf_i \quad \begin{cases} > F_{x^2(1)}^{-1}((1-\alpha)^{n/2}) + m \cdot (4i-1), \text{ reject } H_0 \\ \leq F_{x^2(1)}^{-1}((1-\alpha)^{n/2}) + m \cdot (i-1), \text{ go to next} \\ \qquad\qquad\qquad\qquad\qquad\qquad\qquad \text{period} \end{cases}$$

b) For i=n:

$$(4.3b) \quad gemuf_n \quad \begin{cases} > F_{x^2(1)}^{-1}((1-\alpha)^{n/2}) + m \cdot (n-1), \text{ reject } H_0 \\ \leq F_{x^2(1)}^{-1}((1-\alpha)^{n/2}) + m \cdot (n-1), \text{ reject } H_1 \end{cases}$$

4.3 Truncated Sequential CUSUM Test

A third statistical test which has been applied to material accountancy data is the widely known CUSUM test based on an idea of Page [1o]. For the CUSUM test a linear transformation of the MUF vector is used which may be described as:

$$(4.4) \quad MUFR_i = E(MUF_i) - E(MUF_i | MUF_i, \ldots, MUF_{i-1})$$

for i = 2, ..., n with $MUFR_1 = MUF_1$. The random variables $MUFR_i$ are called MUF-residuals and are independent normally distributed variables. Under H_0 $E(MUFR_i) = 0$ holds for all i and under H_1 $E(MUFR_i) \neq 0$ for some i_0. That means, the CUSUM test has to be two-sided. Based on $MUFR_i$ the following test statistics are defined:

$$(4.5) \quad P_0 = 0, \quad P_i = max\{0, P_{i-1} + MUFR_i\}$$

$$(4.6) \quad N_0 = 0, \quad N_i = min\{0, P_{i-1} + MUFR_i\}$$

for i = 1, 2, ..., n. The statistical test is defined as follows:

for I = 1, 2, ..., n-1:

(4.7) I. for P_i > h or N_i < -h, reject H_0

  II. for P_i ≤ h and N_i ≥ -h, go to the next period

for i = n:

(4.8) I. for P_n > h or N_n < -h, reject H_0

  II. for P_n ≥ h and N_n ≥ -h, reject H_1

5. Example

The three different tests are demonstrated with Plutonium data of the reprocessing facility Wiederaufarbeitungsanlage Karlsruhe [8]. In Fig. 5.1 2o Plutonium material balances of the reprocessing campaign 3/83 are plotted.

Figure 5.1: Plutonium material balances in kg for 20

balance periods of a reprocessing facility

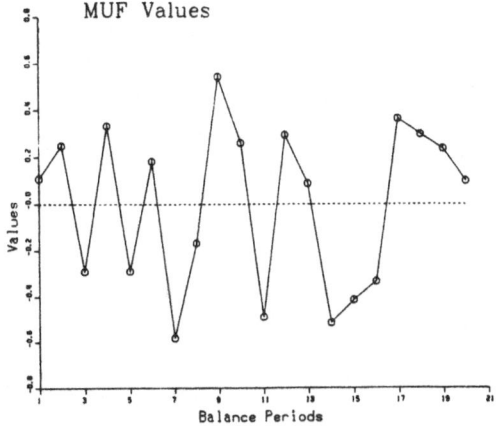

254

<u>Figure 5.2:</u> Test result of the Truncated Sequential CUMUF

test for the material balance reults in Fig. 5.1

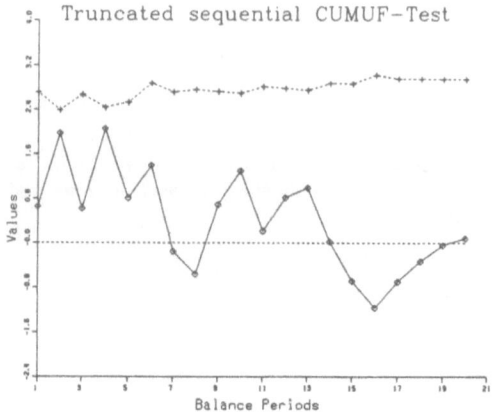

Truncated sequential CUMUF-Test

<u>Figure 5.3:</u> Test result of the GEMUF test

for the material balance reults in Fig. 5.1

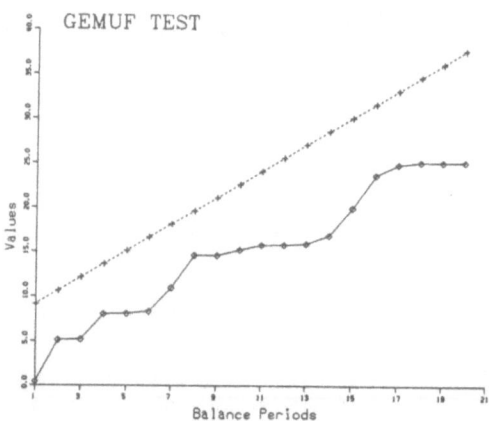

GEMUF TEST

From the data evaluation point of view the results can be interpreted in the following ways. Under
the assumption that the measurement model is correct, it can be assumed that a loss of Plutonium
during the last 2o balance periods has not taken place. But from the process control point of view
another aspect may be considered. For example, if we can assume that no loss of material has
happened, the results of the tests can indicate that the measurement model of the accountancy data
is correct and that all material is completely accounted for. For example, if the actual precision
of the measurements is lower than in the model assumed this may result in an alarm of one or more
tests. This means that a process control authority may also benefit from the NRTA evaluation.

References:

[1] R. Avenhaus:
Material Accountability: Theory, Verification and Applications.
Chichester: John Wiley & Sons 1977.

[2] R. Avenhaus, R. Beedgen, D. Sellinschegg:
Test Procedures to Detect a Loss of Material in a Sequence
of Balance Periods.
KfK 3539 (1985), Karlsruhe

[3] R. Beedgen:
Statistical Considerations Concerning Multiple Materials Balance Models.
Los Alamos National Laboratory. LA-9645-MS (1983), Los Alamos.

[4] R. Beedgen:
Truncated Sequential Test Procedure Using the CUMUF Statistic for a Timely Detection of Diversion.
In "Nuclear Safeguards Technology 1982", Vol. II. Vienna: International Atomic Energy Agency 1983.
pp. 383 - 392

[5] R. Beedgen, U. Bicking:
PROSA - A Statistical Computer Tool for Near Real Time Accountancy.
9th ESARDA Symp., London (1987) 241 - 247.

[6] R. Beedgen, U. Bicking:
PROSA: A computer Program for Statistical Analysis of Near-real-Time Accountancy (NRTA) Data.
KFK 4216 (1987), Karlsruhe.

[7] R. Beedgen:
Statistical Near-Real-Time Accountancy Procedures Applied to AGNS Minirun Data Using PROSA.
Los Alamos National Laboratory LA-11196-MS (1988), Los Alamos.

[8] R. Beedgen, E. Kugele, H. Orth, J. Lausch:
Development of an Integrated Process Information System for a Reprocessing Plant.
Proc. of the 3rd Int. Conference on Facility Operation - Safeguards Interface, San Diego Nov. 29 -
Dec. 4 1987. American Nuclear Society, La Grange Park (1988), pp. 23o - 236.

[9] International Atomic Energy Agency:
The Structure and Content of Agreements between the Agency and States Required in Connection with
the Treaty on the Non-proliferation of Nuclear Weapons. IAEA-Document INFCIRC/153 (1971), Vienna.

[1o] E. S. Page:
Continuous Inspection Schemes.
Biometrika (41) (1954), pp. 1oo - 115.

[11] R. Seifert:
Der GEMUF-Test und Fragen zur Near-Real-Time Accountancy.
KfK 4326 (1987), Karlsruhe.

ON THE APPROXIMATION OF FREE VIBRATION MODES OF A GENERAL THIN SHELL
APPLICATION TO TURBINE BLADES

Michel BERNADOU

INRIA, B.P. 105   and

78153 Le Chesnay Cedex

Bernard LALANNE

TURBOMECA, Bizanos

64320 Bordes

ABSTRACT

The computation of free vibration modes of a general thin shell is essential for engineering purpose. Thus, there exists many numerical methods to perform such a computation. By contrast, in our knowledge, there are very few *mathematical studies* of such approximations.

In this paper we point out some results obtained in this way, particularly, general modelization, existence of solutions, approximation by conforming finite element methods, convergence of discrete solutions to continuous solutions and error estimates. We conclude by giving some numerical results obtained in the study of turbine blades.

1 - FREE VIBRATION EQUATIONS OF A THIN SHELL WITH ARBITRARY SHAPE

These equations are deduced from the dynamic equations of shells in the absence of loading ; they are themselves obtained from the 3D-elasticity theory.

1.1. Dynamic equations of a general shell as a 3D elastic body :

Subsequently we use a regular representation of the 3D-shell $\mathscr{C}$ through a set of curvilinear coordinates (ξ^1, ξ^2, ξ^3), i.e.,

$$\begin{cases} \mathscr{C} = \{M \in \mathscr{E}^3 \ ; \ \vec{OM} = \vec{\phi}(\xi^1, \xi^2) + \xi^3 \vec{a}_3 \ , \ (\xi^1, \xi^2) \in \bar{\Omega} \ , \\ \qquad -\frac{1}{2} e(\xi^1, \xi^2) \leq \xi^3 \leq \frac{1}{2} e(\xi^1, \xi^2)\} \ . \end{cases}$$

The mapping $\vec{\phi}$ associates to the plane reference domain Ω the middle surface of the shell while ξ^3 denotes the normal coordinate through the thickness e of the shell. Here and subsequently, we use the notations of BERNADOU-BOISSERIE [1982] or KOITER [1966].

Then, according to DUVAUT-LIONS [1972, pp 124-125] or SCHOUTEN [1959, (7.26)], the displacement field $\vec{U}$ of the 3D shell $\mathscr{C}$ is solution of the problem

257

J. Manley et al. (eds.), Proceedings of the Third European Conference on Mathematics in Industry, 257–264.
© 1990 *Kluwer Academic Publishers and B. G. Teubner Stuttgart.*

$$\begin{cases} \textit{Find a function } t \in [0,T] \to \vec{U}(\xi,t) \in \mathcal{V} \textit{ such that} \\[6pt] \int_{\mathcal{C}} \rho \, \frac{\partial^2 \vec{U}}{\partial t^2} \, \vec{v} \, d\mathcal{C} + \int_{\mathcal{C}} E^{*ijk\ell} \, \gamma_{ij}^*(\vec{U}) \, \gamma_{k\ell}^*(\vec{v}) \, d\mathcal{C} = 0 \;, \; \forall \vec{v} \in \mathcal{V} \;, \\[6pt] \vec{U}(\xi;0) = \vec{U}_o(\xi) \;; \; \frac{d\vec{U}}{dt}(\xi;0) = \vec{U}_1(\xi) \;, \end{cases} \tag{1.1}$$

where $\mathcal{V} = \{\vec{v} \in (H^1(\overset{\circ}{\mathcal{C}}))^3 \;, \; \vec{v}|_{\partial\mathcal{C}_o} = \vec{0}\}$, $\partial\mathcal{C}_o$ — clamped part of the boundary $\partial\mathcal{C}$, γ_{ij}^* — covariant components of the spatial strain tensor referred to the local basis $(\vec{g}^1, \vec{g}^2, \vec{g}^3)$ and $E^{*ijk\ell}$ — contravariant tensor of elastic moduli in $\mathcal{C}^3$.

1.2. Corresponding two-dimensional thin shell equations :

By assuming that i) the normal to the undeformed middle surface remains normal to the deformed middle surface and ii) the stresses are approximatively plane and parallel to the tangent plane to the middle surface, KOITER [1966] has obtained

$$\int_{\mathcal{C}} E^{*ijk\ell} \, \gamma_{ij}^*(\vec{U}) \, \gamma_{k\ell}^*(\vec{v}) \, d\mathcal{C} \approx a(\vec{u},\vec{v})$$

with $a(\vec{u},\vec{v}) = \int_\Omega eE^{\alpha\beta\lambda\mu} \, \{\gamma_{\alpha\beta}(\vec{u}) \, \gamma_{\lambda\mu}(\vec{v}) + \frac{e^2}{12} \, \bar{\rho}_{\alpha\beta}(\vec{u}) \, \bar{\rho}_{\lambda\mu}(\vec{v})\} \, \sqrt{a} \; d\xi^1 d\xi^2$ and where $E^{\alpha\beta\lambda\mu}$, $\gamma_{\alpha\beta}$ and $\bar{\rho}_{\alpha\beta}$ denote respectively the surface contravariant tensor of elastic moduli, the middle surface strain tensor and the tensor of change of curvature, while $\vec{u}$ denotes the displacement field of the particles located upon the middle surface of the shell.

Similarly, by using the approximation (see BERNADOU-BOISSERIE [1982, (1.3.20)]) $\vec{U} = \vec{u} - \xi^3(u_{3|\alpha} + b_\alpha^\lambda u_\lambda) \, \vec{a}^\alpha$, we obtain $\int_{\mathcal{C}} \rho \, \frac{\partial^2 \vec{U}}{\partial t^2} \, \vec{v} \, d\mathcal{C} \approx b(\vec{\ddot{u}},\vec{v})$, $\vec{\ddot{u}} = \partial^2\vec{u}/\partial t^2$, with

$$\begin{cases} b(\vec{\ddot{u}},\vec{v}) = \int_\Omega \rho e\{[1 + \frac{e^2}{12}(b_1^1 b_2^2 - b_1^2 b_2^1)] \, [a^{\alpha\beta} u_\alpha v_\beta + u_3 v_3] \\[6pt] \qquad + \frac{e^2}{12} \, a^{\alpha\beta}[(u_{3|\alpha} + b_\alpha^\lambda u_\lambda)(v_{3|\beta} + b_\beta^\mu v_\mu) \\[6pt] \qquad + (u_\alpha v_{3|\beta} + u_{3|\alpha} v_\beta + 2b_\alpha^\lambda u_\lambda v_\beta)b_\eta^\eta]\} \, \sqrt{a} \; d\xi^1 d\xi^2 \;. \end{cases} \tag{1.2}$$

Thus, the problem (1.1) can be approximated by :

$$\begin{cases} \textit{Find a function } t \in [0,T] \to \vec{u}(\xi^1,\xi^2;t) \in \mathcal{W} \textit{ such that} \\[6pt] a(\vec{u},\vec{v}) + b(\vec{\ddot{u}},\vec{v}) = 0 \;, \; \forall \vec{v} \in \mathcal{W} \end{cases} \tag{1.3}$$

with $\mathcal{W} = \{\vec{w} \in (H^1(\Omega))^2 \times H^2(\Omega) \;; \; \vec{w}|_{\Gamma_o} = \vec{0} \;; \; \frac{\partial w_3}{\partial n}\big|_{\Gamma_o} = 0\}$.

1.3. Free vibration equations :

Now we restrict our attention to free vibrations which may take place in a shell free of the effect of any external loads at any time, and

subjected to time-independent kinematic boundary conditions. According to
KRAUS [1967, page 291], they can be represented by expressions of type

$$(1.4) \qquad u_j(\xi^1,\xi^2;t) = \bar{u}_j(\xi^1,\xi^2) \cos \omega t \ ; \ j = 1,2,3$$

where ω is the frequency of the vibration. By substituting (1.4) into (1.3),
the free vibration problem can be specified as :

$$(1.5) \quad \begin{cases} \textit{Find couples } (\lambda,\vec{u}) \in \mathbb{R}^+ \times \vec{\mathcal{W}} \quad \textit{such that} \\[2ex] a(\vec{u},\vec{v}) = \lambda \, b(\vec{u},\vec{v}) \ , \ \forall \vec{v} \in \vec{\mathcal{W}} \quad \textit{with } \lambda = \omega^2 \ . \end{cases}$$

2 - SOME RESULTS CONCERNING THE EIGENVALUE PROBLEM

2.1. Abstract setting :

Consider the following problem :

Problem 2.1 : Let V and H two real Hilbert spaces such that

i) $V \subsetneq H$, the inclusion is dense ;

*ii) the canonical injection of V into H is compact. The scalar
products (resp. the norms) on the spaces V and H are denoted $((.,.))_V$ and
$(.,.)_H$ (resp. $\|.\|_V$ and $|.|_H$).*

*Let $a(.,.)$ a continuous, symmetric, V-elliptic bilinear form on $V \times V$.
Then, the problem consists in finding couples (λ,u), $\lambda \in \mathbb{R}$, $u \in V - \{0\}$ such
that*

$$(2.1) \qquad a(u,v) = \lambda(u,v)_H \ , \ \forall v \in V \ .$$

$\square$

Then, we have the following theorem :

*Theorem 2.1 (RIESZ-NAGY [1952]) : The eigenvalues of problem 2.1 form an
increasing sequence*

$$(2.2) \qquad 0 < \lambda_1 \le \lambda_2 \le \ldots \le \lambda_m \le \ldots \ ,$$

*growing to $+\infty$ when the space V is of infinite dimension, each of these
eigenvalues having a finite multiplicity.*

*Moreover, there exists an orthonormal basis of the space H
constituted by the eigenvectors associated to the eigenvalues λ_j, i.e.,*
$a(u_j,v) = \lambda_j(u_j v)_H \ , \ \forall v \in V \ ; \ (u_j,u_i)_H = \delta_{ij} \ .$

$\square$

2.2. Application to the general thin shell problem :

Now we give the main result of this work :

Theorem 2.2 : If the mapping $\vec{\phi}$ is sufficiently regular, the problem (1.10) enters in the class of problems 2.1. In particular the eigenvalues satisfy (2.2) and the eigenvectors satisfy

$$(2.3) \quad \begin{cases} a(\vec{u}_j,\vec{v}) = \lambda_j \, b(\vec{u}_j,\vec{v}) \ , \ \forall \vec{v} \in \mathcal{W} \ , \\[2mm] b(\vec{u}_j,\vec{u}_i) = \delta_{ij} \ . \end{cases}$$

Proof : With notations of section 2.1, we take $V = "\mathcal{W}"$, $H = "\mathcal{H} = (L^2(\Omega))^2 \times H^1(\Omega)"$, $(\vec{u},\vec{v})_{\mathcal{H}} = b(\vec{u},\vec{v})$ for any $\vec{u},\vec{v} \in \mathcal{H}$. Then properties i) and ii) of problem 2.1 are satisfied as a consequence of Sobolev and Kondrasov's theorems (see ADAMS [1975]) on the one hand and of the following lemma on the other hand :

Lemma 2.1 : The expression (1.2) defines a scalar product on the space $\mathcal{H}$. Corresponding norm is equivalent to the usual ones.

It is generally admitted that a shell is *thin* when its thickness and normal curvatures R_N satisfy the inequality $e(\xi^1,\xi^2) \le \frac{1}{10} \min|R_N(\xi^1,\xi^2)|$ where $\min|R_N|$ is understood for all possible normal sections. In this proof we have only assumed that $e(\xi^1,\xi^2) \le \min|R_N(\xi^1,\xi^2)| /3$. Then, after several technical developments, we get for any $\vec{v} \in \mathcal{H}$:

$$b(\vec{v},\vec{v}) \ge \int_\Omega \rho e\{ \frac{11}{16} v_i v^i + \frac{e^2}{36} v_{3|\alpha} v_3|^\alpha \} \sqrt{a} \, d\xi^1 d\xi^2$$

from which we can deduce the existence of a constant $C > 0$ such that :

$$b(\vec{v},\vec{v}) \ge c_1 \{ |v_1|^2_{0,\Omega} + |v_2|^2_{0,\Omega} + \|v_3\|^2_{0,\Omega} \} \ , \ \forall \vec{v} \in \mathcal{H} \ . \qquad \square \text{ (Lemma 2.1)}$$

It remains to check the properties concerning the bilinear form $a(.,.)$. These properties were proved in BERNADOU-CIARLET [1976].

$$\square \text{ (Theorem 2.2)}$$

3 - APPROXIMATION BY CONFORMING FINITE ELEMENT METHODS

By using conforming finite element methods, we can construct finite dimensional subspaces $\vec{V}_h$ of space $\mathcal{W}$. The corresponding discrete problems can be stated :

Problem 3.1 : Find couples $(\lambda_h,\vec{u}_h) \in \mathbb{R}^+ \times \vec{V}_h$ such that

$$a(\vec{u}_h, \vec{v}_h) = \lambda_h \, b(\vec{u}_h, \vec{v}_h) \, , \quad \forall \vec{v}_h \in \vec{V}_h \, .$$

□

By assuming some regularity results, it is possible to extend to general thin shells the results of STRANG-FIX [1973, theorems 6.1 and 6.2]. For brievety, we summarize the results which we have obtained on Figure 3.1 for different choices of finite elements.

Finite element used to construct a subspace of $H^2(\Omega)$ / Finite element used to construct a subspace of $H^1(\Omega)$	ARGYRIS triangle (P_5) 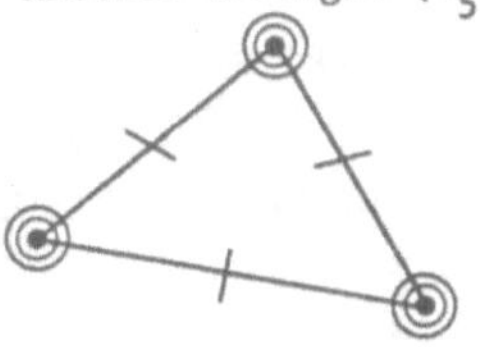	Reduced H.C.T.-triangle 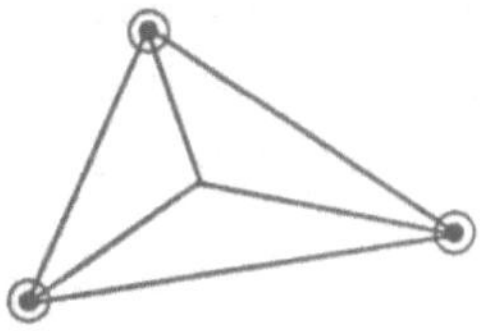
ARGYRIS triangle (P_5)	$\lambda_j \le \lambda_{hj} \le \lambda_j + 2Ch^8\lambda_j^5$ $\|\vec{u}_j - \vec{u}_{hj}\|_{\mathcal{H}} \le Ch^5\lambda_j^{5/2}$ $a(\vec{u}_j - \vec{u}_{hj}, \vec{u}_j - \vec{u}_{hj}) \le Ch^8\lambda_j^5$	✕
GANEV triangle (P_4) 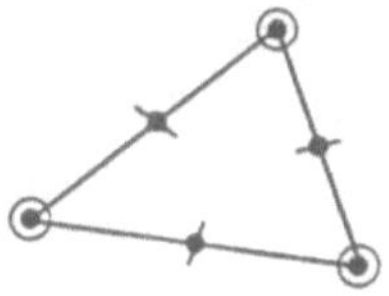	$\lambda_j \le \lambda_{hj} \le \lambda_j + 2Ch^8\lambda_j^5$ $\|\vec{u}_j - \vec{u}_{hj}\|_{\mathcal{H}} \le Ch^5\lambda_j^{5/2}$ $a(\vec{u}_j - \vec{u}_{hj}, \vec{u}_j - \vec{u}_{hj}) \le Ch^8\lambda_j^5$	✕
Reduced H.C.T.-triangle	✕	$\lambda_j \le \lambda_{hj} \le \lambda_j + 2Ch^2\lambda_j^2$ $\|\vec{u}_j - \vec{u}_{hj}\|_{\mathcal{H}} \le Ch^2\lambda_j$ $a(\vec{u}_j - \vec{u}_{hj}, \vec{u}_j - \vec{u}_{hj}) \le Ch^2\lambda_j^2$
Triangle (P_1) 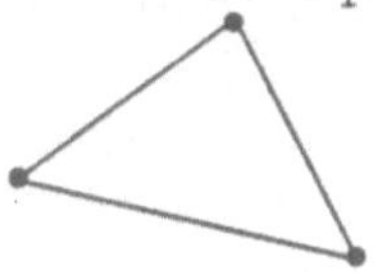	✕	$\lambda_j \le \lambda_{hj} \le \lambda_j + 2Ch^2\lambda_j^2$ $\|\vec{u}_j - \vec{u}_{hj}\|_{\mathcal{H}} \le Ch^2\lambda_j$ $a(\vec{u}_j - \vec{u}_{hj}, \vec{u}_j - \vec{u}_{hj}) \le Ch^2\lambda_j^2$

Figure 3.1 :

Error estimates between continuous and approximate eigenvalues and eigenvectors for $j = 1,\ldots,M_h$ where $M_h = \dim \vec{V}_h$

4 - APPLICATION TO THE COMPUTATION OF FREE VIBRATION MODES OF ROTATING TURBINE BLADES

This is a typical example of an *industrial problem* which has been studied in collaboration with TURBOMECA.

The first work we had to develop was to approximate the mappings $\vec{\phi}$ and e which define the geometry of the shell. In this study, the analytical definition of these mappings is unknown. We had only access to the values of these mappings at a finite number of points. By using B-spline approximations we were able to get very regular interpolate functions, say for instance $\vec{\phi}_{h\delta} \in (\mathcal{C}^3(\Omega))^3$, $e \in \mathcal{C}^1(\Omega)$. These additional approximations are analyzed in BERNADOU-LALANNE [1986].

Then, we apply all these results to *rotating turbine blades*. These blades are loaded by centrifugal forces and aerodynamical pressure. Then we look for free vibration modes around the corresponding loaded equilibrium configuration. The combined use of B-splines (to approximate the middle surface and the thickness of the shell) and finite elements (to approximate the displacement), both of *high degree of precision*, lead to an *excellent approximation of the free vibration modes*. These results are illustrated by Figures 4.1 and 4.2. They give respectively the first and fourth free vibration modes (flexion u_3, Von Mises stresses upon concave and convex faces and the points of maximum stresses upon concave (I) and convex (E) faces). The Figure 4.3 shows the evolution of the first eight free vibration modes according to the rotation speed of the pale.

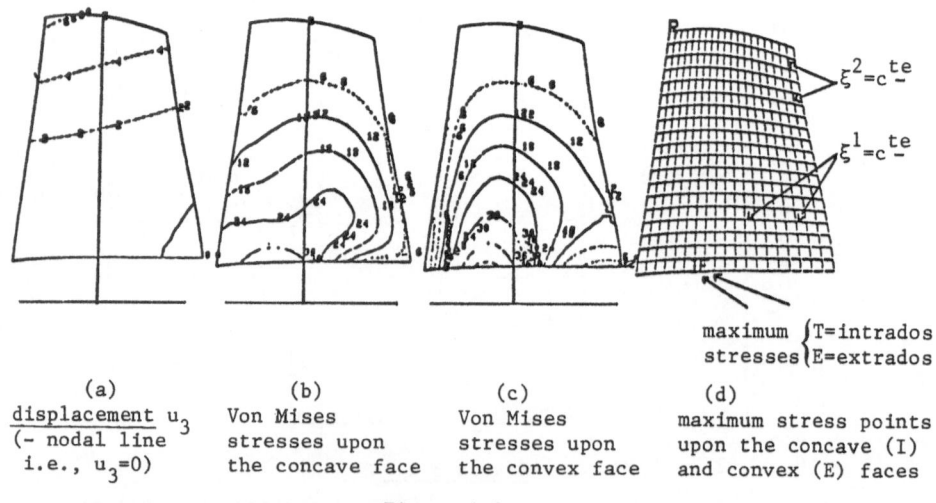

maximum {T=intrados
stresses {E=extrados

(a)	(b)	(c)	(d)
displacement u_3 (- nodal line i.e., $u_3=0$)	Von Mises stresses upon the concave face	Von Mises stresses upon the convex face	maximum stress points upon the concave (I) and convex (E) faces

Figure 4.1 :

First free vibration mode around the loaded equilibrium configuration

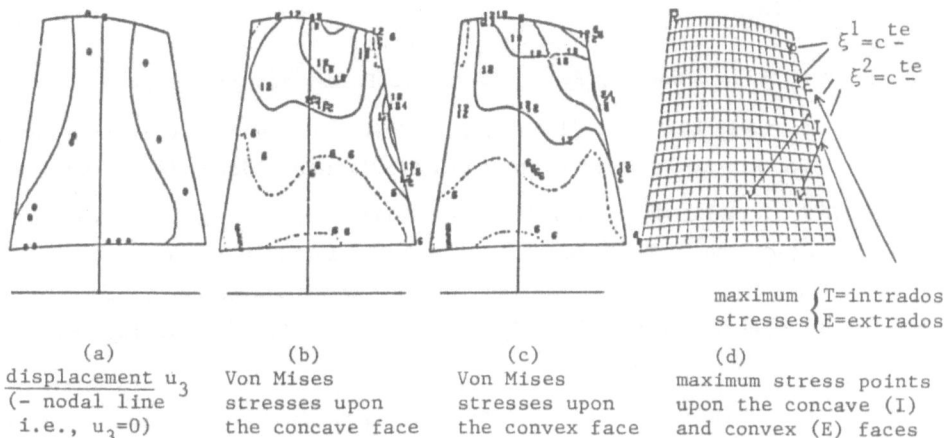

maximum { T=intrados
stresses { E=extrados

(a)	(b)	(c)	(d)
displacement u_3 (- nodal line i.e., $u_3=0$)	Von Mises stresses upon the concave face	Von Mises stresses upon the convex face	maximum stress points upon the concave (I) and convex (E) faces

Figure 4.2 :

Fourth free vibration mode around the loaded equilibrium configuration

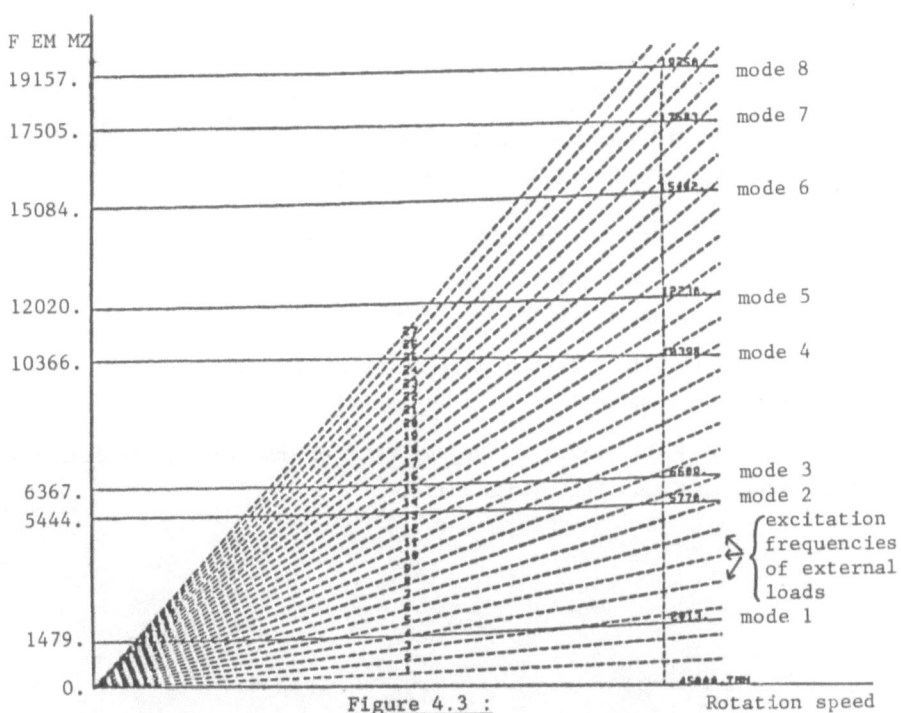

Figure 4.3 :

Variation of free vibration modes in terms of the rotation speed

5 - REFERENCES

ADAMS R.A. [1975] : *Sobolev Spaces*, Academic Press, New-York.

BERNADOU M., BOISSERIE J.M. [1982] : *The Finite Element Method in Thin Shell Theory : Application to Arch Dam Simulations*, Birkhaüser, Boston.

BERNADOU M., CIARLET P.G. [1976] : Sur l'ellipticité du modèle linéaire de coques de W.T. Koiter in : R. Glowinski and J.L. Lions ed., *Computing Methods in Applied Sciences and Engineering*, Lecture Notes in Economics and Mathematical Systems, *134* (Springer-Verlag, Berlin, 1976), pp. 89-136.

BERNADOU M., LALANNE B. [1986] : On the approximation of thin shells by "B-spline and finite element" methods, in *Innovative Numerical Methods in Engineering*, Edited by R.P. Shaw, J. Périaux, A. Chaudouet, J. Wu, C. Marino, C.A. Brebbia, pp. 585-592, Springer-Verlag, Berlin.

DUVAUT G., LIONS J.L. [1972] : *Les Inéquations en Mécanique et en Physique*, Dunod, Paris (English translation : Springer-Verlag, Berlin, 1976).

KOITER W.T. [1966] : On the nonlinear theory of thin elastic shells, Proc. Kon. Ned. Akad. Wetensch., *B69*, pp. 1-54.

KRAUS H. [1967] : *Thin Elastic Shells*, J. Wiley, New-York.

RIESZ F., NAGY B. Sz. [1952] : *Leçons d'Analyse Fonctionnelle*, Akadémiai Kiadó, Budapest.

SCHOUTEN J.A. [1959] : *Tensor Analysis for Physicists*, Oxford University Press (second edition).

STRANG G., FIX G.J. [1973] : *An Analysis of the Finite Element Method*, Prentice Hall, Englewood Cliffs.

INTEGRATION ALGORITHMS FOR THE DYNAMIC
SIMULATION OF PRODUCTION PROCESSES.

M. Berzins , P. M. Dew , A. J. Preston.

ABSTRACT

The essential features needed to construct an efficient integrator for dynamic simulation problems arising in the gas/ oil industry are examined. A series of numerical experiments are performed using two widely used codes: DASSL [8] and SPRINT [1], that are designed to solve differential-algebraic equations. From these experiments, a number of improvements to the codes have been identified and incorporated in a new integrator.

1. Introduction.

The development of general purpose codes, such as SPRINT [1] and DASSL [8], for the numerical solution of differential-algebraic equations (d.a.e.s), has made it possible to efficiently solve many of the routine equations that arise in modelling dynamic gas and oil networks [3]. Nevertheless, there still exists a number of important networks that cannot be adequately handled by existing integrators. These problems can be classified as Index two (or higher) d.a.e.s (see below). A number of numerical experiments have been performed using both DASSL [8] and SPRINT [1], and from these the following areas where problems may arise, have been identified:

(i)   the solution of the non-linear equations that arise when using an implicit time-integration scheme;

(ii)   the accurate estimation of the local error for the fixed or variable stepsize form of the backward differentiation formulae (b.d.f.) of Gear [4];

(iii) the stepsize and order selection mechanism used by the integrators, which are based on b.d.f. formulae.

This paper provides a summary of the work that has been undertaken in the above and related areas. For further details, the interested reader is referred to the companion report, [2].

The general class of differential-algebraic equations is defined by

(1) $\underline{f}(t , \underline{y}(t) , \underline{\dot{y}}(t)) = \underline{0}$ given constants $\underline{y}(0) , \underline{\dot{y}}(0)$ and where $\underline{f} : [t_0 , T] \times \mathbf{R}^n \times \mathbf{R}^n \to \mathbf{R}^n$

In the dynamic simulation problems considered in this paper, the function $\underline{f}$ can be written in the form :

J. Manley et al. (eds.), Proceedings of the Third European Conference on Mathematics in Industry, 265–271.
© 1990 *Kluwer Academic Publishers and B. G. Teubner Stuttgart.*

(2) $\underline{f} = A(t, \underline{y}) \, \underline{\dot{y}} - \underline{g}(t, \underline{y})$,   where A is a square matrix and can be singular.

A classification of d.a.e.s, the <u>index</u> of the system of equations, has been introduced by Gear and Petzold [5]. For details on the definition of the <u>index</u> and its implications see [2] and [6]. This paper is concerned with the *standard index two* problem [6] pp. 24-27, defined by

$$y_1 = \phi(t) \qquad t \, \varepsilon \, [\, t_0 \, , T \,] \, , \qquad \phi \, \varepsilon \, C^2 \, [\, t_0 \, , T \,] \, \to \, \mathbf{R}$$

(3) $$y_2 = \dot{y}_1 \qquad \underline{y} = \begin{bmatrix} y_1 \\ y_2 \end{bmatrix} \; : \; [\, t_0 \, , T \,] \to \mathbf{R}^2 \; , \quad \underline{y}(0) = \begin{bmatrix} \phi(0) \\ \dot{\phi}(0) \end{bmatrix}$$

where we have chosen $\phi(t) = \sin(2 \pi t)$, $t \, \varepsilon \, [0, 3]$.

Practical experiments conducted by the authors have shown that DASSL [8] is one of the most robust codes for solving index two problems, although it is far from being efficient. The purpose of this paper is to carry out a series of numerical experiments which compare the performance of DASSL [7] with that of SPRINT [1] when solving the *standard index two* problem. A number of recommendations are given to improve the performance of SPRINT [1] for this class of problems.

2. An <u>Investigation</u> <u>of</u> the <u>Factors</u> Affecting the <u>Performance</u> <u>of</u> <u>DASSL</u>.

In order that the particular features which make DASSL [8] successful in solving d.a.e.s. are understood, the following aspects are investigated:
(i) the form of the non-linear equations to be solved, and the scaling of these equations;
(ii) the error estimates based on [6] and the stepsize and order selection algorithm.

The results were obtained using finite differencing to compute the partial derivatives of the Jacobian matrix. Details of the statistics used to measure code performance are given in the Appendix. The number of evaluations of the residual vector $\underline{r}$, not counting those used in decomposing the Jacobian, is defined (in the case of *the standard index two d.a.e*) at the nth accepted step by

$$\underline{r}_n = \begin{bmatrix} y_{1,n} - \phi(t_n) \\ y_{2,n} - \dot{y}_{1,n} \end{bmatrix}$$

where $y_{1,n}$ denotes the computational approximation to the exact value $y_1(t_n)$ and similarly for $y_{2,n}$ and $\dot{y}_{1,n}$. The weighted vector norm of estimated local errors in $\underline{y}$, $\underline{le}$, is controlled at each step by means of a standard mixed error test and using the *averaged l_2 norm*, [2].

2.1. The <u>Form</u> <u>of</u> the <u>Non-Linear</u> <u>Equations</u>.

DASSL [8] uses the kth order b.d.f. of Gear ($k = 1,2,3,4,5$) to approximate $\underline{\dot{y}}_n$ by $\underline{\dot{y}}_n = \dfrac{y_n}{h_o \gamma_o} + \sum\limits_{i=1}^{k} \alpha_i \, \underline{y}_{n-i}$, where h_o is the stepsize used on the previous step and γ_0 , α_1 , α_2 , ... , α_k are the method dependent parameters.

The rate of convergence of the non-linear equations can be accelerated by using relaxation techniques. In DASSL [8] the relaxation factor used is ρ (see below), so that the $j+1$th iterative value of Newton's method, $y_n^{(j+1)}$, is found by solving for the correction $\Delta y_n^{(j)}$:

$$\left[\frac{\partial f}{\partial y} + \frac{1}{h_o \gamma_o} \frac{\partial f}{\partial \dot{y}}\right] \left[\Delta y_n^{(j)}\right] = -f\left[t_n , y_n^{(j)} , \frac{y_n^{(j)}}{h\gamma} + \sum_{i=1}^{k}\alpha_i y_{n-i}\right] \times \rho$$

$$y_n^{(j+1)} = y_n^{(j)} + \Delta y_n^{(j)} , \quad j = 1,2,3,...$$

where $\rho = \left[\dfrac{2 h_c \gamma_c}{h_o \gamma_o + h_c \gamma_c}\right]$ and the subscript c denotes current values, while the subscript o denotes old values. A further modification to the way in which the non-linear equations are solved is given by Petzold and Löstedt [9]. In this case, if the jth equation contains derivatives, then the factor $2 h_c \gamma_c$ in ρ is replaced by $h_o \gamma_o$ and a scaling of $\rho = 1$ is applied to the algebraic constraints.

An improvement to the above is to adopt the approach used in SPRINT[1], but select the relaxation factors by the following procedure. The row(s) of the iteration matrix representing the differential equation(s) are multiplied by $h_o \gamma_o$ and the corresponding residual entry is multiplied by $h_c \gamma_c$. This is expected to lead to a faster convergence of the iterates. The system of equations to be solved is then given by

$$\begin{bmatrix} -1 & 0 \\ 1 & -h_o \gamma_o \end{bmatrix} \left[\Delta^* y_n^{(j)}\right] = \begin{bmatrix} y_{1,n}^{(j)} - \phi(t_n) \\[2mm] \dfrac{2 h_o \gamma_o \, h_c \gamma_c \, (y_{2,n}^{(j)} - \dot{y}_{1,n}^{(j)})}{h_o \gamma_o + h_c \gamma_c} \end{bmatrix}$$

$$y_n^{(j+1)} = y_n^{(j)} + \Delta^* y_n^{(j)} \quad j = 1,2,3,...$$

$$y_n^{(1)} = y_n^{(p)} + \Delta^* y_n^{(0)}$$

By comparing the results in Tables 1 and 2a, a large improvement to the errors in y_1 for the *standard index two* problem can be observed. Indeed, the algebraic equation is now satisfied to within unit round-off error. It should be noted that although there is little improvement to the errors in $\dot{y}_1$, there is some improvement for the errors in $\dot{y}_2$ and also a considerable reduction in the amount of work performed, compared with DASSL [8]. However, the results show that the maximum global error in $\dot{y}_2$ is still well beyond accepted bounds.

2.2. Estimating the Local Error.

In DASSL [8], after convergence of the iterates has been achieved, the error estimates at orders k , $k-1$, $k-2$ are given by $e = \alpha^* (y_n - y_n^{(p)})$ where the superscript p denotes the predicted value and where $\alpha^* = \alpha^* (h , h_1 , h_2 , \cdots , h_k)$ depends on the current stepsize, h, and on the previous k stepsizes at order k, In particular, at order one, the value of α^* is given by

$$\alpha^* = \frac{h}{h + h_1} .$$

Clearly, if $[t_{n-1} , t_n]$ spans a discontinuity or sudden change in gradient the error test can always be satisfied in DASSL [8] at order 1, since the code can always select a small enough step size. However α^* does not have a convenient form for any of the higher orders. In contrast, the error estimate used in the codes of Hindmarsh [7] and in the b.d.f. SPGEAR module of SPRINT [1] is given by $\underline{E} = \alpha \, (\, y_n - y_n^{(p)})$ where α is a constant which, at order one, has the value $\alpha = \frac{1}{2}$. In this case, if $[t_{n-1} , t_n]$ spans a discontinuity in $\dot{y}$, it is not always possible to satisfy the error test at order one.

The results in Table 2b show that a strong relationship exists between the number of error test failures, the work done by the code, and the order selected. When the number of error test failures is high, the code spends a long time at order one, taking advantage of the stability properties, but in order to achieve the required accuracy, the code takes many steps.

An alternative is to use only the error estimates that arise from the differential equations in the system. The local error estimate, $\underline{le}$, is now defined by $\underline{le} = M^{-1} A \, \underline{e}$ rather than $\underline{le} = \underline{e}$ as in SPRINT [1], where M is the iteration matrix used in solving the non-linear equations, and A is the matrix defined by equation (2). These equations are then solved by performing an LU decomposition on M, and using back-substitution.

Tables 2b and 3b show that with this modification to the local error estimate , the new code takes advantage of the efficient high order methods. The number of error test failures has been reduced considerably and there is a drastic reduction in the global error.

2.3. Stepsize and Order Selection.

Suppose that at order k, the estimates of the error at the orders one above and below are denoted by $\underline{e}^{k+1}$ and $\underline{e}^{k-1}$. Then in the order selection algorithm of DASSL [8], $\underline{e}^{k-1}$, $\underline{e}^k$, and $\underline{e}^{k+1}$, are replaced by $A \, \underline{e}^{k-1}$, $A \, \underline{e}^k$, and $A \, \underline{e}^{k+1}$ respectively. This is expensive as it requires three matrix-vector products, each of which costs "half" an o.d.e. residual evaluation in the SPRINT [1] software. However, this appears to be one of the few ways of ensuring that the code selects the correct order. DASSL [8] also tends to make sure that $||\underline{e}^{k-2}|| > ||\underline{e}^k||$, to ensure that the order selection algorithm does not force the code to stay at an unnecessarily high order. In the case when this condition has been violated, the order is reduced whenever $||A \, \underline{e}^{k-2}|| < ||A \, \underline{e}^k||$.

3. Conclusions and Recommendations.

In this paper and also in [2], a number of important issues concerning the design of an efficient integrator for index two d.a.e.s. have been discussed. In particular, the strategies used in both DASSL [8] and SPRINT [1] concerning the method of solution and scaling of the non-linear equations have been mentioned. The experiments have shown that the form of the non-linear equations used in SPRINT [1] works well, and that scaling of the equations is an important issue. For this reason, the form of the non-linear equations as solved in SPRINT [1] is adopted with the DASSL [8] convergence strategy. The numerical results have also shown that the formulation of the local error estimate can have a major impact on the efficiency of the d.a.e. integration. Therefore the local error estimate based only on the differential variables is adopted. To ensure that the code selects the correct order, the strategy adopted is identical to that used in DASSL [8], except for an additional criterion based on the comparison of two matrix-vector products.

Applying these modifications has enabled an index two module to be constructed, SPDASL, which can be used within the SPRINT [1] software. The results from this new index two solver are given in Tables 4a and 4b. In fact, a comparison of the results in Table 4a with those in Table 1 shows an extremely large reduction in the work performed by the new code. Table 4b shows that the order selection strategy has worked extremely well, allowing a large proportion of time to be spent at a high order. The errors in $\dot{y}_2$ have been reduced considerably, and the algebraic equations are better satisfied since the errors in $\dot{y}_1$ have been reduced. This indicates a very large improvement over the original code.

We have also tested our module on a dynamic simulation problem supplied by Shell Research Ltd, and the preliminary results are encouraging [2].

Acknowledgements.

We wish to thank Shell Research Limited for permission to publish this paper and for funding the SERC CASE studentship for Andrew Preston. We are also grateful for the help that has been given by S Frost and L Scales of Shell Research Ltd.

REFERENCES.

[1] Berzins M., Dew P.M., and Furzeland R.M. (1986), Developing P.D.E. Software Using The Method of Lines and Differential Algebraic Integrators. High-lighted Talk presented at 1986 O.D.E. Conference, Albuquerque, New Mexico, July 1986 (to appear in Appl. Numer. Math.).

[2] Berzins M., Dew P.M., Preston A.J., (1988), Integration Algorithms for the Dynamic Simulation of Production Processes, Report 88.20, School of Computer Studies, Leeds University, Leeds LS2 9JT.

[3] Capstick M.A., (1987), On Improving the Performance of Dynamic Process Simulators, Ph. D. Thesis, Departments of Chem. Eng. and Comp. Studies, Leeds University, Leeds LS2 9JT.

[4] Gear C.W., (1971), Numerical Initial Value Problems in Ordinary Differential Equations, Prentice Hall, Englewood Cliffs, NJ, U.S.A.

[5] Gear C.W., Petzold L.R., (1984), ODE Methods for the solution of Differential/Algebraic Systems, SIAM Journal on Numerical Analysis, 21, pp. 716-728.

[6] Gupta G.K., Gear C.W., Leimkuhler B., (1985), Implementing Linear Multistep Formulas for solving D.A.E.s, Report UIUCDCS-R-85-1205, Department of Computer Science, University of Illinois, Urbana IL61801.

[7] Hindmarsh A.C., (1981), O.D.E Solvers for use with the method of lines, Advances in Computer Methods for Partial Differential Equations IV, R. Vichnevetsky and R.S. Stepleman, eds., IMACS, New Brunswick, NJ, pp. 312-316.

270

[8] Petzold L.R., (1982), A Description of DASSL : A Differential-Algebraic System Solver, SAND82-8637, Applied maths division 8331, Sandia National Laboratories, Livermore, California C.A 94550.

[9] Petzold L.R., Löstedt P., (1983), Numerical solution of Nonlinear Differential Equations with Algebraic Constraints, Report SAND 83-8877, SSISC, Sandia National Laboratories, Livermore, California C.A 94550.

Authors' Address.

School of Computer Studies, Leeds University, LEEDS LS2 9JT.

Appendix.

The tables below illustrate the results obtained for the numerical experiments. For the *standard index two* problem, the initial conditions for the test problem are consistent with the analytic solution. The column headings denoted by err y_1, err y_2 etc., denote the maximum global error (in y_1, y_2 etc.) that occurred over the interval $[t_0, T]$ for all accepted steps using a tolerance value TOL. The column *EF* denotes the number of error test failures and *STEP* the number of steps taken by the integrator. *CALL* records the number of function evaluations for the problem, not counting those used in the decomposition of the Jacobian. *JAC* is the number of decompositions of the Jacobian and *SOLVE* denotes the total number of iterations incurred by the non-linear equations solver. The number of steps taken at each of the five orders is also recorded.

Tab. 1 The standard index two problem using DASSL [8].

TOL	err y_1	err y_2	err $\dot{y}_1$	err $\dot{y}_2$	STEP	CALL	JAC	SOLVE
0.5D-02	0.21D-03	0.92D-01	0.92D-01	0.90D+03	462	1504	521	1504
0.2D-02	0.31D-04	0.17D+00	0.17D+00	0.87D+02	608	2115	644	2115
0.1D-02	0.60D-05	0.28D-01	0.28D-01	0.12D+04	908	3179	1376	3179
0.5D-03	0.26D-05	0.16D-01	0.16D-01	0.13D+03	4936	19381	7638	19381

Tabs. 2a,b The standard index two problem after modifications to the non-linear equation solver.

TOL	err y_1	err y_2	err $\dot{y}_1$	err $\dot{y}_2$	STEP	CALL	JAC	SOLVE
0.5D-02	0.00D+00	0.81D-01	0.81D-01	0.10D+04	331	1042	344	1042
0.2D-02	0.00D+00	0.33D-01	0.33D-01	0.59D+01	369	1278	509	1278
0.1D-02	0.14D-16	0.26D-01	0.26D-01	0.12D+03	1296	4841	2173	4841
0.5D-03	0.00D+00	0.95D-01	0.95D-01	0.27D+02	469	1298	399	1298

TOL	ORD 1	ORD 2	ORD 3	ORD 4	ORD 5	EF
0.5D-02	259	32	28	12	0	176
0.2D-02	290	21	18	22	18	256
0.1D-02	1167	60	36	31	2	1108
0.5D-03	311	66	45	31	16	167

Tabs. 3a,b The standard index two problem after changes made to the estimation of local error.

TOL	err y_1	err y_2	err $\dot{y}_1$	err $\dot{y}_2$	STEP	CALL	JAC	SOLVE
0.5D-02	0.00D+00	0.10D-01	0.10D-01	0.17D+01	66	141	11	344
0.2D-02	0.14D-16	0.67D-02	0.67D-02	0.62D+00	85	198	7	482
0.1D-02	0.14D-16	0.48D-02	0.47D-02	0.62D+00	95	218	7	532
0.5D-03	0.00D+00	0.22D-02	0.14D-02	0.40D+02	123	289	20	676

TOL	ORD 1	ORD 2	ORD 3	ORD 4	ORD 5	EF
0.5D-02	1	1	7	20	37	2
0.2D-02	1	1	15	7	61	10
0.1D-02	1	1	17	7	69	10
0.5D-03	6	12	11	14	80	14

Tabs. 4a,b The standard index two problem modifying the order selection strategy, convergence strategy, and error test in the SPDASL module of SPRINT [1].

TOL	err y_1	err y_2	err $\dot{y}_1$	err $\dot{y}_2$	STEP	CALL	JAC	SOLVE
0.5D-02	0.31D-12	0.51D-01	0.18D-01	0.46D+01	73	291	10	201
0.2D-02	0.12D-11	0.27D-01	0.73D-02	0.43D+01	86	319	11	217
0.1D-02	0.88D-13	0.11D-01	0.54D-02	0.70D+00	99	347	11	232
0.5D-03	0.40D-13	0.41D-02	0.24D-02	0.21D+01	118	425	9	286

RTOL=ATOL	ORD 1	ORD 2	ORD 3	ORD 4	ORD 5	EF
0.5D-02	1	11	15	13	33	3
0.2D-02	1	1	17	24	43	2
0.1D-02	1	1	22	22	53	1
0.5D-03	1	12	23	19	63	5

MULTICOMPONENT FLOW COMPUTATION

WITH APPLICATION TO STEAM CONDENSERS

A W Bush*, G S Marshall* and T S Wilkinson**

* Dept of Mathematics and Statistics, Teesside Polytechnic.

** NEI Parsons Ltd., Newcastle Upon Tyne.

1 INTRODUCTION

This paper describes the collaborative research undertaken between NEI Parsons Ltd and Teesside Polytechnic to predict the flow behaviour of the three component mixture of air, water vapour and liquid water occurring in power station steam condensers. This paper describes an algorithm, based on the MAC method of Harlow and Welch [1], for the numerical solution of the coupled governing equations.

NEI Parsons Ltd have monitored an experimental condenser rig under a range of operating conditions. The rig has been discussed previously, principally by Al-Sanea et al [2],[4] and Carlucci and Cheung [3]. Papers [2] and [3] used the SIMPLE algorithm of Patankar and Spalding [5] but applied different empirical correlations to the source terms, while paper [4] extended the model used in [2] to consider two components.

2 GOVERNING EQUATIONS

The conservation of mass and linear momentum of the i-th component, in two dimensional cartesians, is given by equations (2.1),(2.2) and (2.3).

(2.1) $\dfrac{\partial}{\partial t}(\alpha_i \rho_i) + \dfrac{\partial}{\partial x}(\alpha_i \rho_i u_i) + \dfrac{\partial}{\partial y}(\alpha_i \rho_i v_i) = \Gamma_i$

(2.2) $\dfrac{\partial}{\partial t}(\alpha_i \rho_i u_i) + \dfrac{\partial}{\partial x}(\alpha_i \rho_i u_i u_i) + \dfrac{\partial}{\partial y}(\alpha_i \rho_i u_i v_i) = -\alpha_i \dfrac{\partial p}{\partial x} + \alpha_i \rho_i b_x$

$$+ \dfrac{\partial}{\partial x}(\alpha_i \mu_i \dfrac{\partial u_i}{\partial x}) + \dfrac{\partial}{\partial y}(\alpha_i \mu_i \dfrac{\partial u_i}{\partial y})$$

273

J. Manley et al. (eds.), Proceedings of the Third European Conference on Mathematics in Industry, 273–281.
© 1990 Kluwer Academic Publishers and B. G. Teubner Stuttgart.

$$+ \sum_{j \neq i} [\tau_{ij}(u_j - u_i) + S_{ij}^x]$$

$$(2.3) \quad \frac{\partial}{\partial t}(\alpha_i \rho_i v_i) + \frac{\partial}{\partial x}(\alpha_i \rho_i v_i u_i) + \frac{\partial}{\partial y}(\alpha_i \rho_i v_i v_i) = -\alpha_i \frac{\partial p}{\partial y} + \alpha_i \rho_i b_y$$

$$+ \frac{\partial}{\partial x}(\alpha_i \mu_i \frac{\partial v_i}{\partial x}) + \frac{\partial}{\partial y}(\alpha_i \mu_i \frac{\partial v_i}{\partial y})$$

$$+ \sum_{j \neq i} [\tau_{ij}(v_j - v_i) + S_{ij}^y]$$

where α_i is the volume fraction occupied by component i, ρ_i is the density, $\underline{v}_i = (u_i, v_i)$ is the velocity vector, Γ_i is the net mass source rate due to interactions with the other components, p is the common pressure, μ_i is the effective viscosity, $\underline{b} = (b_x, b_y)$ is a body force per unit mass, τ_{ij} are intercomponent drag coefficients and $\underline{S}_{ij}$ is the momentum source associated with a transfer of mass between components i and j.

The need to solve an enthalpy equation is avoided by assuming that the vapour is at its saturation temperature throughout the condenser. (This assumption is based on the observation that at temperatures greater than saturation temperature the heat transfer is hardly affected). By assuming that the steam and air are perfect gases the temperature of the vapour mixture is calculated from the local partial pressure of the steam.

There are two auxiliary relations which must be satisfied. The first concerns the volume fractions and is given by equation (2.4).

$$(2.4) \qquad \sum_i \alpha_i = \epsilon \qquad \text{the fraction of volume available to the fluid mixture}$$

Usually ϵ will be unity. However, in steam condensers part of the volume is taken up by cooling tubes and in this situation $\epsilon \approx 0.52$. Since there can be no net source of mass, the mass source rates must satisfy equation (2.5) everywhere in the calculation domain.

$$(2.5) \qquad \sum_i \Gamma_i = 0$$

3 SOLUTION ALGORITHM

A brief summary of the algorithm is now given, followed by a description of steps (ii), (iii) and (iv). Bush and Marshall [6] describe the algorithm in more detail.

(i)   Prescribe initial values for all variables.

(ii) Calculate intermediate velocity fields using the momentum equations (2.2) and (2.3).

(iii) Construct a pressure correction equation to impose overall mass conservation.

(iv) Solve equations (2.1) for the new volume fractions.

(v)   Store these new values as old values and repeat steps (ii) to (iv) until some convergence criteria has been satisfied.

Steps (iii) and (iv) are repeated several times in each time step since changes made in one will affect the other. Also required are appropriate boundary conditions and empirical relations for the various source terms.

The momentum equations may be written as (3.1) and (3.2).

$$(3.1) \qquad \alpha_i \rho_i \left[\frac{\partial u_i}{\partial t} + u_i \frac{\partial u_i}{\partial x} + v_i \frac{\partial u_i}{\partial y} \right] = -\alpha_i \frac{\partial p}{\partial x} + \text{other terms}$$

$$(3.2) \qquad \alpha_i \rho_i \left[\frac{\partial v_i}{\partial t} + u_i \frac{\partial v_i}{\partial x} + v_i \frac{\partial v_i}{\partial y} \right] = -\alpha_i \frac{\partial p}{\partial y} + \text{other terms}$$

Intermediate velocity fields $\underline{v}_i^*$ are obtained from equations (3.1) and (3.2) using the old time values, i.e.

$$(3.3) \qquad \alpha_i \rho_i \frac{\underline{v}_i^* - \underline{v}_i^o}{\delta t} = -\alpha_i \nabla p^o + \text{all other terms.}$$

The new velocity fields, $\underline{v}_i^n$, are obtained using the gradient of the pressure correction, q, i.e.

$$(3.4) \qquad \alpha_i \rho_i \frac{\underline{v}_i^n - \underline{v}_i^*}{\delta t} = -\alpha_i \nabla q$$

Once the pressure correction has been determined, equation (3.4) gives $\underline{v}_i^n$ explicitly. Note that adding equations (3.3) and (3.4) yields the updated momentum equations with the corrected pressure p^0+q .

An equation for q is obtained by taking the divergence of equation
(3.4) and enforcing overall mass conservation to give (3.5).

(3.5) $$\frac{1}{\delta t} \sum_i \left[\underline{\nabla} \cdot (\alpha_i \underline{v}_i^*) - \frac{\Gamma_i}{\rho_i} \right] = \underline{\nabla} \left[\sum_i \frac{\alpha_i}{\rho_i} \underline{\nabla}q \right]$$

Equation (3.5) is a Poisson like equation which is solved for q using
standard iterative techniques.

The preceding paragraphs describe the algorithm in its simplest
explicit form. If the intercomponent drag is dominant it is beneficial, in
terms of economy of computing effort, to make these terms implicit and thus
use a larger time step. For problems which are known to have steady state
solutions greater degrees of implicitness are advantageous. Whether one
uses the fully explicit algorithm or a more implicit version the
underlying idea of splitting each time step into two steps and the ability
to accommodate an arbitrary number of components are retained.

4 EXPERIMENTAL CONDENSER

The experimental condenser at NEI Parsons was originally designed to
demonstrate the effects of air pockets and so is not considered to be a
good example of condenser design. However, this does not preclude it from
being a good test of a model, indeed it is hoped that such models will
ultimately predict the shortcomings of particular condenser designs.

The test rig shown in figure 1 uses an internal vent two-thirds of the
way into a tube bank with 20 rows of cooling tubes, with 20 tubes in each
row. Details of the test rig and operating conditions are given in
Marshall [7].

Momentum Sources

The source terms in the momentum equations which require modelling are
the resistance to the flow past the cooling tubes and the intercomponent
drag between the gaseous and liquid components. The tube resistance term

is estimated using the Jakob correlation, described in McAdams [8], for flow across staggered tubes. Using this correlation the resistance to the flow may be written as

(4.1) $$-f\rho_i |\underline{v}_i| \underline{v}_i N'$$

where f is the friction factor and N' is the number of tube rows in the flow direction per unit length.

The intercomponent drag between the gaseous and liquid components is estimated by assuming that the condensate exists as droplets of a single size. Thus, by computing the projected area of the droplets and applying the formula for the friction factor of spherical objects described in Clift et al [9], the following estimate of the τ term in (2.2) is obtained, (the subscripts g and l refer to the gaseous and liquid phases respectively)

$$\tau_{g,l} = 1/2 \, \rho_g \, f \, A \, |u_g - u_l|$$

where A is the projected area of the droplets and f is the friction factor. A similar expression will appear in (2.3)

Heat and Mass Transfer

The method used to calculate the local heat and mass transfer is an adaptation of that described by Chisholm [10]. The sink term for the water vapour is given by equation (4.2).

(4.2) $$m = \frac{-Q}{\lambda}$$

where m is the condensation rate per unit area, Q is the heat flux and λ is the latent heat.

The local value of the heat flux, Q, is determined by the local temperature gradients and heat transfer coefficients. The assumption that the vapour is saturated throughout the condenser means the local saturation temperature is deduced from the local steam concentration and the operating pressure of the condenser. Having determined the local values of the heat flux the condensation rates are calculated and stored as sink and source terms respectively for the water vapour and liquid condensate continuity equations.

Simulation Results

The previously described model has been used to predict the flow behaviour in the test rig shown in figure 1. Figures 2(a-c) show the volume fractions of the three components. Figures 3(a-d) give a direct comparison between prediction and observation by comparing the variations of mean heat flux along four of the tube rows. As can be seen from these figures there is good agreement between the results in the lower half of the condenser but rather greater discrepancies between the results near the top of the tube nest. This is a consequence of the underestimation of the size of the air pocket in the upper half of the tube nest and suggests that the assumptions and correlations should be reconsidered. However, as an initial test of the solution algorithm and the condenser model the results are encouraging and can hopefully be improved upon.

5 REFERENCES

[1]   Harlow F.H. and Welch J.E.
          Numerical Calculation Of Time Dependent Incompressible Flow Of A
          Fluid With A Free Surface
          Phys. Fluids, Vol 8(12), pp2182-2189, (1965)
[2]   Al-Sanea S., Rhodes N., Tatchell D.G. and Wilkinson T.S.
          A Computer Model For Detailed Calculation Of The Flow In Power
          Station Condensers.
          I.Chem.E. Symposium Series No.75, p70, (1983)
[3]   Carlucci L.N. and Cheung I.
          Numerical Predictions Of Flow And Heat Transfer In Power Plant
          Condensers.
          pp569-582,May 1985.
[4]   Al-Sanea S., Rhodes N. and Wilkinson T.S.
          Mathematical Modelling Of Two-Phase Condenser Flows.
          2nd International Conference On Multi-Phase Flow, (1985)
[5]   Patankar S.V. and Spalding D.B.
          A Calculation Procedure For Heat, Mass And Momentum Transfer In
          3-D Parabolic Flows.
          Int. Journal Of Heat And Mass Transfer, Vol 15, p1787, (1972)
[6]   Bush A.W. and Marshall G.S.
          "The SMAC Code For Multi-component Flow: A First Approach" in
          Industrial Fluid Flow Computation. Editors A.W. Bush and M.J.
          O'Carrol. EMJOC Press (1986).
[7]   Marshall G.S.
          Multicomponent Fluid Flow Computation.

Ph.D Thesis, submitted October 1988.

[8]   McAdams W.H.
        Heat Transmission.
        McGraw-Hill, pp162-164 (1954).

[9]   Clift R., Grace J.R. and Weber M.E.
        Bubbles, Drops and Particles.
        Academic Press (1978).

[10]   Chisholm D.
        "Modern Developments In Marine Condensers: Non-Condensable Gases:
        An Overview" in Power Condenser Heat Transfer For Technology.
        Editors P.J.Marto and R.H. Nunn. Hemisphere (1981).

FIGURES

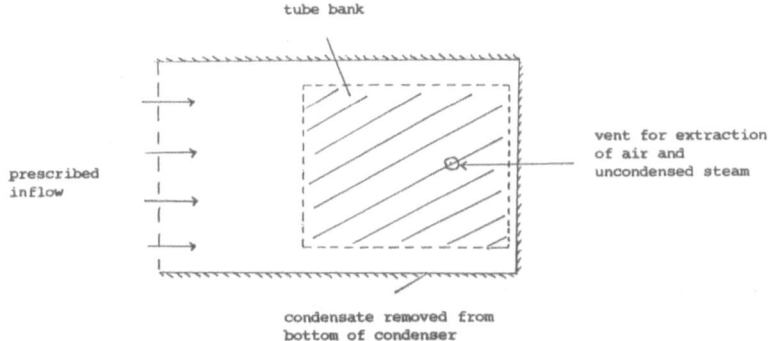

Figure 1 Condenser setup.

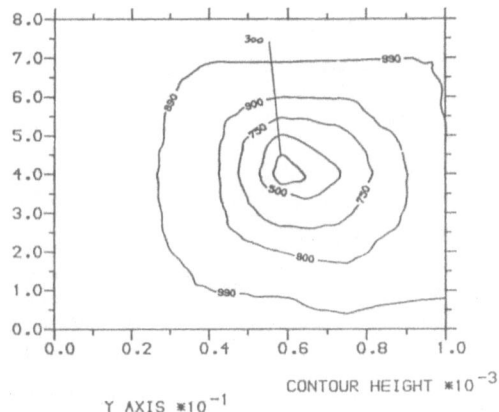

Figure 2(a) Volume fraction of steam.

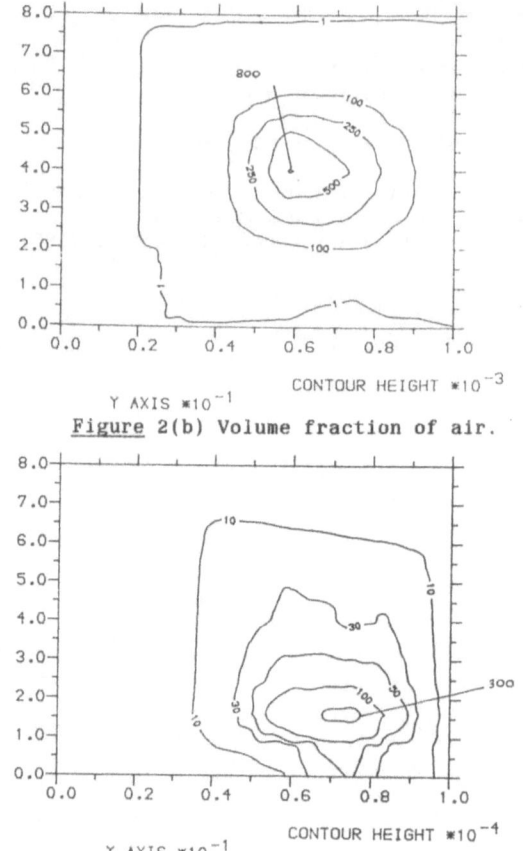

CONTOUR HEIGHT *10^{-3}

Y AXIS *10^{-1}

Figure 2(b) Volume fraction of air.

CONTOUR HEIGHT *10^{-4}

Y AXIS *10^{-1}

Figure 2(c) Volume fraction of condensate.

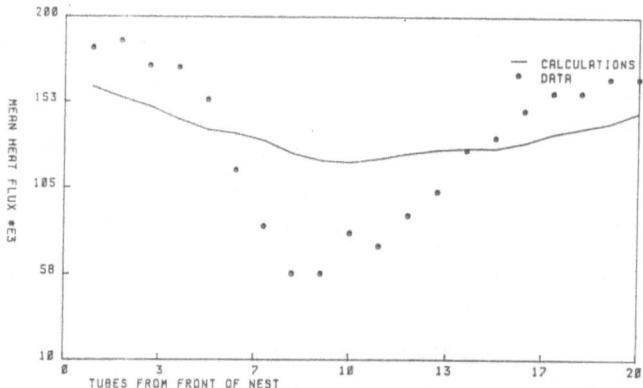

Figure 3(a) Heat flux comparison along 3rd row from the top.

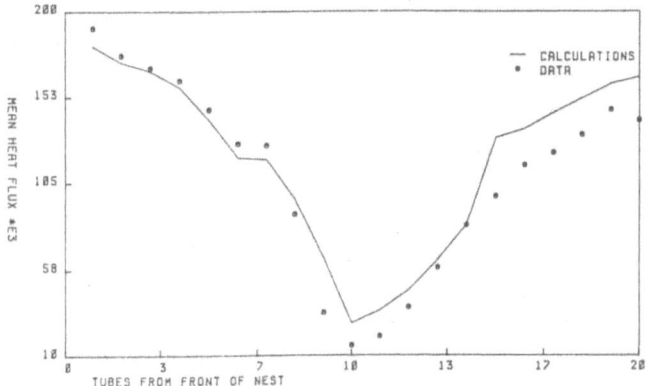

Figure 3(b) Heat flux comparison along 8th row from the top.

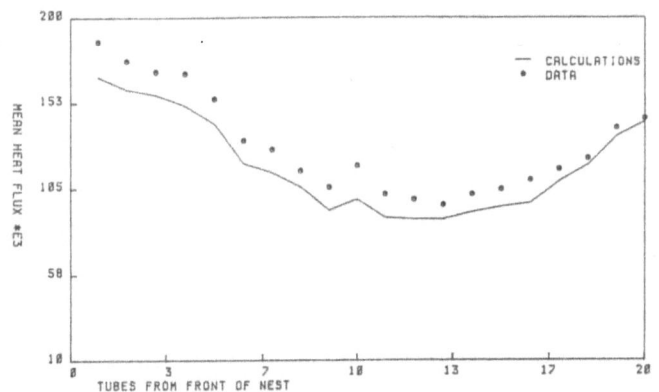

Figure 3(c) Heat flux comparison along 13th row from the top.

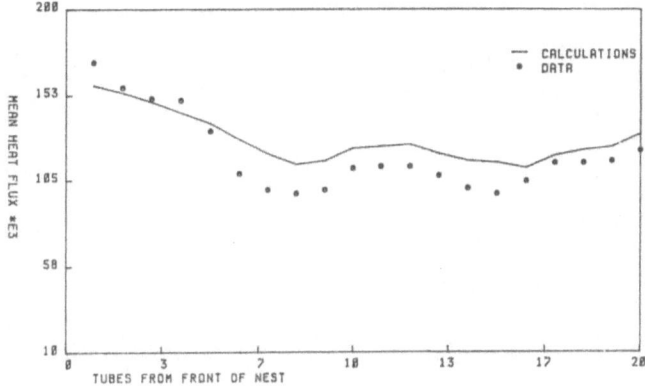

Figure 3(d) Heat flux comparison along 18th row from the top.

NATURAL CONVECTION WITHIN A DROPLET
AS A RESULT OF A CHEMICAL REACTION ON ITS SURFACE

E.H. de Groot

Fakultät für Physik

Universität Bielefeld

D-4800 Bielefeld 1, FRG

Abstract

The problem of natural convection in a spherical droplet is considered resulting from an exothermic reaction on its surface of a substance, dissolved within the droplet.

I. Introduction

We have a spherical drop of liquid in which a substance is dissolved which reacts on the surface of the droplet with a homogeneous medium outside. In the absence of convection the replenishing of the solute on the surface (and therefore the chem. reaction) would solely rely on the diffusion of the solute, which is often quite a slow process. If on the other hand the reaction is exothermic the heat produced on the surface will induce a lower local density which acted upon by gravity will produce convection, which in turn will tend to enhance the reaction rate, giving more heat leading to convection etc. How effective such a convection will be in enhancing the reaction rate does not only depend on the strength of the velocity field but also on how effective it mixes i.e., how complicated the field is. For example: a Hill's vertex as produced within a droplet by falling through another liquid is rather ineffective as it tends to only mix the liquid immediately around its central vertical axis with the liquid on the surface)[1], and is therefore not capable of explaining the high reaction rates found experimentally)[2]. In principle natural convection stands a better chance of explaining these high reaction rates as although the magnitude of the velocities involved are not likely to be so great, the form of the field tends to be more complicated and has therefore a better chance of being able to produce the required "mixing" effect.

A major restriction of this paper is that we confine ourselves to the convection resulting from the temperature gradient and not due to the change in density as a result of the depletion of the solute i.e., we assume that the specific weight of solute and solvent are equal.

In the following we will not try to depict a realistic situation but rather simplify and approximate in such a way as to keep the essential features of the problem intact, e.g. the medium outside the droplet is treated in a rather cavalier fashion

283

J. Manley et al. (eds.), Proceedings of the Third European Conference on Mathematics in Industry, 283–291.

© 1990 *Kluwer Academic Publishers and B. G. Teubner Stuttgart.*

only described by boundary conditions on the surface of the droplet; likewise the chemical reaction only finds place on the surface of the droplet and is considered to be intantaneous. In this way the concentration of solute within the droplet is ruled by the equation:

$$(\partial_t + \underline{u} \cdot \underline{\nabla}) C = D \Delta C \qquad\qquad t = 0 \qquad\qquad C \approx C_0 \qquad\qquad (1.1)$$
$$t > 0, r = R \quad C = 0$$

where D is the diffusion constant of the solute in the solvent and where we have written $C \approx C_0$ at $t = 0$ to allow for small deviations from isotropy in the initial state, which may possibly be amplified if the nonlinear term $(\underline{u} \cdot \underline{\nabla})C$ becomes dominant.

Here we do not want to solve (1.1) (but hopefully in a future paper) but rather determine the velocity field $\underline{u}$ which of course depends on the concentration C itself, or rather on the outgoing flux $\partial_r C|_{r=R}$ as this outgoing flux induces an inward flux of heat which in turn creates the velocity field $\underline{u}$. The result will then be a non-linear eq. (1.1) · which we would like to solve (probably numerically) in the near future, but for the moment we only want to determine the velocity field $\underline{u}$, for which we in principle have to solve a temperature eq. of the same type as (1.1) and use the solution as the input of the Navier-Stokes eq. to give us our $\underline{u}$. In general one would then expect $\underline{u}$ to depend on the whole past history of C making (1.1) into a complicated integro-differential equation in t. On the other hand, the diffusion (conduction) of heat and momentum (viscosity) are in practice mostly very much larger then the diffusion of the solute. We can therefore think of applying a sort of adiabatic approximation where the timescale of the depletion of the solute is quite large compared to the timescale at which the temp. gradient and velocity field establish themselves, so that we can assume the temperature and velocity field to instantaneously reach their steady state values after each change in C. This approximation would then clearly make $\underline{u}(t)$ only dependent on the value of $\partial_r C|_{r=R}$ on the surface of the droplet at the same moment t and not on the past, resulting in (1.1) becoming a differential eq. in t and not an integro-diff. eq. So the equation will be local in t but cannot possibly be local in space, of course, as the velocity at the point $\underline{r}$ will depend on $\partial_r C|_{r=R}$ all over the surface of the droplet (only the surface values count of course). So we expect to obtain an expression like:

$$\underline{u}(\underline{r},t) = \int\limits_{r'=R} d\,\Omega'\,\underline{\underline{K}}\,(\underline{r},\underline{r}')\,\frac{\underline{r}'}{C_0}\,\partial_{r'}\,C\,(\underline{r}',t) \qquad\qquad (1.2)$$

where the integration is over the surface of the droplet, and our task is to find the kernel $\underline{\underline{K}}(\underline{r},\underline{r}')$. In general $\underline{\underline{K}}(\underline{r},\underline{r}')$ will still depend on $\partial_r C|_{r=R}$ itself as otherwise (1.2) would describe a linear dependence of $\underline{u}$ on C which cannot be correct as the Navier-Stokes eq. is non-linear in $\underline{u}$. On the other hand, as we have already assumed that the diffusion of heat and momentum (i.e. viscosity) is very large compared with the diffusion of solute it must be a good approximation to neglect the convection terms (i.e. the non-linear terms) in the temp. and Navier-Stokes eqs. as the essential nonlinearity is then given by the concentration eq. (1.1), which will make (1.2) into a linear relation where $\underline{K}(\underline{r},\underline{r}')$ is independent of $\partial_r C|_{r=R}$.

II. The equations for the temperature and the velocity field

We first rewrite eq. (1.1) as:

$$(\partial_t + \underline{u}\cdot\underline{\nabla})\Gamma = D\Delta\Gamma \quad \text{with} \quad \Gamma := \frac{C_0 - C}{C_0} \qquad t = 0 \qquad \Gamma \approx 0 \tag{2.1}$$
$$t > 0 \quad r = R \quad \Gamma = 1$$

We will now, for the sake of simplicity, assume that a constant fraction of the heat produced on the surface diffuses into the droplet, which then remains there forever, which explains our boundery condition on the current below. Defining the normalized temperature as

$$\Theta = \frac{T - T_0}{T_\infty - T_0} \quad \text{where} \quad T_0 = T(t=0) , \quad T_\infty = T(t=\infty) \tag{2.2}$$

we have for the temperature

$$(\partial_t + \underline{u} \cdot \underline{\nabla}) \Theta = k \Delta \Theta \qquad t = 0 \qquad \Theta = 0 \tag{2.3}$$
$$t > 0 \quad r = R \quad k \partial_r\Theta = D \partial_r\Gamma$$

where k is the coefficient of heat conduction and $T_\infty - T_0$ is a measure of the total amount of heat which flows into the drop during the whole process. For the velocity field we take the Navier-Stokes eq. in the Boussinesq approximation)[3-5] where the effect of the variation in density is limited to the external force.

$$(\partial_t + \underline{u}\cdot\underline{\nabla})\underline{u} = - \frac{1}{\rho_0} \underline{\nabla} p - \Theta\underline{g}^* + \nu \Delta \underline{u} \; ; \; \underline{\nabla}\cdot\underline{u} = 0 \tag{2.4}$$
$$\text{with} \quad \underline{g}^* : = \alpha \; (T_\infty - T_0)\underline{g}$$

where α is the expansion coeff. of the solvent, $\underline{g}$ the accaleration of gravity and ρ_0 the density at $t = 0$. The boundary conditions are

$$t = 0 \qquad \underline{u} = 0$$
$$t > 0 \quad r = R \quad u_r = 0.$$

Free boundary conditions: $\partial_r \, r^{-1} u_\theta = \partial_r \, r^{-1} u_\varphi = 0$ (No tangential stress)

Rigid boundary conditions: $u_\theta = u_\varphi = 0$ (No slip)

Note that $\Theta\underline{g}^*$ is not a conservative force (as Θ will not only depend on the vertical coordinate) so that $\underline{u} = 0$ is not a solution.

As mentioned before we assume

$$k, \nu \gg D \tag{2.5}$$

which garantees that the essential non-linearity is given by the concentration equation (2.1), so that we can drop the convection terms $(\underline{u}\cdot\underline{\nabla})\Theta$ in (2.3) and

$(\underline{u} \cdot \underline{\nabla})\underline{u}$ in (2.4) but not in (2.1) of course.

We observe that the only connection between temp. and velocity on the one hand and the concentration on the other hand is by way of the boundary condition (2.3)

$$k \ \partial_r \ \Theta \ = \ D \ \partial_r \ \Gamma \quad \text{on the surface.} \tag{2.6}$$

In calculating the dependence of $\underline{u}$, Θ on $\partial_r \Gamma|_{r=R}$ we cannot a priori assume azimuthal independence of $\partial_r \Gamma|_{r=R}$ (although the eqs. are invariant for rotations around the direction of gravity) in view of the possible instability, mentioned under (1.1), where small deviations from isotropy in the initial state are amplified in the course of time. Taking into account such a possible $\varphi-$ dependence of $\partial_r \Gamma|_{r=R}$ complicates our task considerably, as we will see. As the time scale of the concentration eq. (2.1) is so much longer than the time scale of the diffusion of heat and momentum we can in solving $\underline{u}$, Θ apply an adiabatic approximation where $\underline{u}$, Θ fully establish themselves instantaneously after every change in the concentration Γ i.e., we take

$$\nu t, \ kt \ \rightarrow \ \infty \ ; \ \frac{\nu}{k} \ , \ Dt \ \text{fixed.} \tag{2.7}$$

We call the expressions for Θ, $\underline{u}$ obtained in this way the steady state temp. and velocity resp.

III. The steady state temperature

The steady state solution of (2.1), neglecting the convective term, is as one can check easily:

$$\Theta \ (\underline{r},t) \ = \ \int\limits_0^t \ dt' \ \frac{1}{4\pi} \int\limits_{r'=R} \ d \ \Omega' \ \tilde{\Theta} \ (\underline{r},\underline{r}',t,t') \ r' \ \partial_{r'} \ \Gamma(\underline{r}',t') \tag{3.1}$$

with

$$\tilde{\Theta} \ = \ \frac{3D}{r'^2} \ + \ \frac{D}{k} \ \delta \ (t \ - \ t') \ [\tfrac{1}{2}\big(\tfrac{r}{r'}\big)^2 \ + \ \sum_{l=1}^{\infty} \frac{\big(\tfrac{r}{r'}\big)^l}{l} \ (2l \ + \ 1) \ P_l \ (\hat{r} \cdot \hat{r}') \ =$$

$$= \ \frac{3D}{r'^2} \ + \ \frac{D}{k} \ \delta \ (t \ - \ t') \ [\tfrac{1}{2} \big(\tfrac{r}{r'}\big)^2 \ + \ 2 \ \{|\underline{r} \ - \ \underline{r}'|^{-1} \ r \ - \ 1\} \ - \tag{3.2}$$

$$- \ ln\Big\{\frac{(|\underline{r}-\underline{r}'|+ \ r')^2 - \ r^2}{4r'^2}\Big\}]$$

where $P_l(\hat{r} \cdot \hat{r}')$ is the usual Legendre polynomial and

$$(\mu'':=) \ \hat{r} \cdot \hat{r}' = \ \mu\mu' \ + \ (1-\mu^2)^{1/2}(1-\mu'^2)^{1/2}\cos(\varphi-\varphi') \tag{3.3}$$

is the cosine of the angle between $\underline{r}$ and $\underline{r}'$.

In summing the series in (3.2) we made use of the well known formula:

$$r'|\underline{r} \ - \ \underline{r}'|^{-1} \ = \ \sum_{l=0}^{\infty} \ \big(\tfrac{r}{r'}\big)^l \ P_l\big(\hat{r} \ \cdot \ \hat{r}'\big) \quad \text{if} \ r \ < \ r'$$

In (3.2) the first term $\frac{3D}{r'^2}$ describes the overall temperature rise of the droplet which plays no role in creating the velocity field $\underline{u}$ as $\underline{u}$ only depends on $\nabla\Theta$ of course, which is instanteneous as, clearly, $\nabla\Theta \sim \delta(t-t')$ due to our adiabatic approximation ($kt \rightarrow \infty$, Dt fixed). Having obtained the temperature we are ready to insert this into (2.4) and calculate $\underline{u}$.

IV. The steady state velocity field

Instead of calculating $\underline{u}$ directly we calculate only the radial component of the velocity field u_r and of the vorticity ω_r. These then suffice (as $\nabla \cdot \underline{u} = 0$) to contruct the entire $\underline{u}$ field (see ref. 4). (As mentioned we neglect the convective term in (2.4)).

$\nabla \times$ (2.4) gives us the vorticity eq.

$$\partial_t \, \underline{\omega} = g^* \, \nabla \times \Theta \, \underline{e}_3 + \nu \, \Delta \, \omega \qquad \underline{\omega} := \nabla \times \underline{u} \qquad \underline{g}^* = - g^* \underline{e}_3 \qquad (4.1)$$

$\underline{r} \cdot$ (4.1) gives us for the radial vorticity

$$\partial_t \, r\omega_r = g^* \, \partial_\varphi \, \Theta + \nu \, \Delta \, r\omega_r \qquad t = 0 \qquad\qquad \omega_r = 0 \qquad (4.2)$$

$$t > 0 \quad r = R \quad \text{Rigid:}\omega_{,r} = 0 \text{ Free: } \partial_r \omega_r = 0.$$

N.B. These boundary conditions follow from those in (2.4) as

$$\omega_r = \frac{1}{\sin\theta} \, \partial_\theta \, \sin\theta \frac{u_\varphi}{r} - \frac{1}{\sin\theta} \, \partial_\varphi \, \frac{u_\theta}{r}.$$

To obtain the equation for u_r we take

$\nabla \times$ (4.1)   and as $\nabla \times (\nabla \times \underline{u}) = - \Delta\underline{u}$ (as $\nabla \cdot \underline{u} = 0$)

we obtain

$$\partial_t \, \Delta \, \underline{u} = g^*(\Delta \, \Theta \, \underline{e}_3 - \nabla \, \partial_3 \, \Theta) + \nu \, \Delta^2 \, \underline{u} = \text{ (by (2.3))} \qquad (4.3)$$

$$= g^* \Big[\frac{1}{k} \, \partial_t \, \Theta \, \underline{e}_3 - \nabla \, \partial_3 \, \Theta \Big] + \nu \, \Delta^2 \, \underline{u}.$$

Taking $\underline{r} \cdot$ (4.3) gives us then for the radial velocity

$$\partial_t \, \Delta \, r \, u_r = \Big(\frac{1}{k} \, \partial_t \, \Theta \, x_3 - r \, \partial_r \, \partial_3 \, \Theta \Big) + \nu \, \Delta^2 \, r \, u_r \qquad (4.4)$$

where the boundary conditions are:

$t = 0 \qquad\qquad u_r = 0$

$t > 0 \quad r = R \quad u_r = 0 \qquad\qquad$ Rigid: $\partial_r r u_r = 0 \qquad$ Free: $\partial_r^2 r u_r = 0$

where as we work in polar coordinates we have:

$$x_3 = r\mu = r \cos\theta \qquad \partial_3 = \mu \, \partial_r + r^{-1}(1 - \mu^2)\partial_\mu.$$

The boundary conditions follow immediately from (2.4) by considering:

$$\nabla \cdot \underline{u} = r^{-2} \, \partial_r \, r^2 \, u_r + \frac{1}{\sin\theta} \, \partial_\theta \, \sin\theta \, \frac{u_\theta}{r} + \frac{1}{\sin\theta} \, \partial_\varphi \, \frac{u_\varphi}{r} = 0 \qquad (4.5)$$

as for a rigid boundary we have for $r = R$, $u_\varphi = u_\theta = 0$ we have:

$$\partial_r r^2 u_r \underset{u_r=0}{=} \partial_r r u_r = 0$$

while for a free boundary we differentiate:

$$\partial_r \nabla \cdot \underline{u} = 0$$

and obtain as $\partial_r r^{-1} u_\theta = \partial_r r^{-1} u_\varphi = 0$ for $r = R$

$\partial_r \, r^{-2} \, \partial_r \, r^2 \, u_r \underset{u_r=0}{=} \partial_r^2 \, r \, u_r = 0.$

Before we write down the solution for ω_r of (4.2) we should observe that in the case where the concentration Γ and thus also Θ has azimuthal symmetry, which will certainly be the case as long as $(\underline{u} \cdot \nabla)\Gamma$ in (2.1) is small, the driving term in (4.2) i.e., $\partial_\varphi \Theta$ will be 0 and thus $\omega_r = 0$. Indeed, we then should have:

$$\underline{\omega} = \omega \, \underline{e}_\varphi \qquad \text{in case of azimuthal symmetry.}$$

Writing down the solution of (4.2) makes this apparent:

$$\omega_r \, (\underline{r},t) = \int_0^t dt' \, \frac{1}{4\pi} \int_{r'=R} d\,\Omega' \,\, \widetilde{\omega}_r \,(\underline{r},\underline{r}',t,t') \,\, r' \,\partial_{r'} \,\Gamma \,(\underline{r}',t') \tag{4.6}$$

with $\widetilde{\omega}_r = \partial_\varphi \, \widetilde{\widetilde{\omega}}_r$ \hfill (4.7)

with for free boundary conditions

$$\widetilde{\widetilde{\omega}}_r = 3 \, r'^{-1} \, g^\bullet \, \frac{D}{k} \, P_1(\hat{\underline{r}},\hat{\underline{r}}') + \tfrac{1}{2} \, g^\bullet \, \frac{D}{k\nu} \, r'^{-1} \, \delta(t - t') \, *$$

$$* \sum_{l=2}^\infty \frac{1}{l} \, \frac{2l+1}{2l+3} \, \left(\frac{l+1}{l-1} \, r'^2 - r^2\right)\left(\frac{r}{r'}\right)^{l-1} \, P_l(\hat{\underline{r}} \cdot \hat{\underline{r}}) \tag{4.8}$$

and for rigid boundary conditions

$$\widetilde{\widetilde{\omega}}_r = \tfrac{1}{2} \, g^\bullet \, \frac{D}{k\nu} \, r'^{-1} \, \delta \, (t - t')(r'^2 - r^2) \sum_{l=1}^\infty \frac{1}{l} \, \frac{2l+1}{2l+3} \, \left(\frac{r}{r'}\right)^{l-1} \, P_l \, (\hat{\underline{r}} \cdot \hat{\underline{r}}'). \tag{4.9}$$

We now see immediately what happens in case $\partial_r \Gamma|_{r'=R}$ is independent of φ', we then have $\widetilde{\widetilde{\omega}}_r$ only depending on $\varphi - \varphi'$ by virtue of (3.3) and therefore $\int d\Omega' \widetilde{\widetilde{\omega}}$ is independent of φ and thus $\omega_r = 0$ by (4.7).

We also observe that in the case of a free boundary we have a non-instanteneous term: $3r'^{-1} g^\bullet \frac{D}{K} \hat{\underline{r}} \cdot \hat{\underline{r}}'$ ($P_1(\mu^-) = \mu^-$) which simply describes a rigid body rotation of the droplet (notice the independence of ν) around an axis in the horizontal plane in response to the azimuthally asymmetric weight distribution; a term which is of course absent for a rigid boundary. In practice such a rotation is of course always present in the case of azimuthal asymmetry but will be damped by the viscosity of the medium outside.

We now turn to the solution for the radial velocity i.e., eq. (4.4):

$$u_r(\underline{r},t) = \int dt' \frac{1}{4\pi} \int_{r'=R} d\Omega' \widetilde{u}_r(\underline{r}, \, \underline{r}',t,t') r' \partial_{r'}\Gamma(\underline{r}',t') \tag{4.11}$$

with for a free boundary

$$\widetilde{u}_r = \frac{D}{8k\nu} \, g^\bullet \, \delta \, (t - t') \, (r'^2 - r^2) \, [- \tfrac{2}{5} \left(\tfrac{1}{3} \, r'^2 - \tfrac{1}{7} \, r^2\right) r'^{-2} \mu \, +$$

$$+ \, (r \, r')^{-1} \sum_{l=2}^\infty \frac{l-1}{l} \, \left(\frac{r'^2}{2l-1} - \frac{r^2}{2l+3}\right) \left(\frac{r}{r'}\right)^{l-1} (l\mu' - (1 - \mu'^2) \, \partial_{\mu'}) \, P_{l-1} \, (\hat{\underline{r}} \cdot \hat{\underline{r}}')] \tag{4.12}$$

while for a rigid boundary

$$\widetilde{u}_r = \frac{D}{8k\nu} \, g^\bullet \, \delta \, (t - t') \, (r'^2 - r^2)^2 \, [- \tfrac{2}{35} \, r'^2 \, \mu \, +$$

$$+ \, (r \, r')^{-1} \sum_{l=2}^\infty \frac{l-1}{l} \, \left(\frac{r}{r'}\right)^{l-1} (l \, \mu' - (1 - \mu'^2)\partial_{\mu'}) \, P_{l-1}(\hat{\underline{r}} \cdot \hat{\underline{r}}') \tag{4.13}$$

Having obtained u_r and ω_r we now proceed to fully determine the velocity distribution from these functions (see for an elaborate explanation ref. 4).

Our first step is to construct the functions $\widetilde{\widetilde{\Phi}}$ and $\widetilde{\widetilde{\Psi}}$ from $\widetilde{u}_r$ and $\widetilde{\widetilde{\omega}}_r$ in such a way that:

$$\widetilde{u}_r = \frac{1}{r^2} \, L^2 \, \widetilde{\widetilde{\Phi}} \qquad\qquad \widetilde{\widetilde{\omega}}_r = \frac{1}{r^2} \, L^2 \, \widetilde{\widetilde{\Psi}} \tag{4.15}$$

where L^2 is the squared angular momentum operator:

$$L^2 = - \, (\partial_\mu(1 - \mu^2)\partial_\mu + (1 - \mu^2)^{-1} \, \partial_\varphi^2) \tag{4.16}$$

as $L^2 P_l(\mu'') = - \, \partial_{\mu''}(1 - \mu''^2)\partial_{\mu''} \, P_l(\mu'') = l(l + 1) \, P_l(\mu'')$ \qquad $(\mu'' := \hat{\underline{r}} \cdot \hat{\underline{r}}')$

and \qquad\qquad $L^2\mu = L^2P_1(\mu) = 2 \, P_1(\mu) = 2\mu$ \hfill (4.17)

we only have to multiply each summand in $\widetilde{u}_r$ by $\frac{r^2}{l(l-1)}$ and in $\widetilde{\widetilde{\omega}}_r$ by $\frac{r^2}{l(l+1)}$ to obtain our functions $\widetilde{\Phi}$ and $\widetilde{\widetilde{\Psi}}$ respectively. We only do this in the case of free boundary conditions, for rigid boundary conditions it goes exactly the same way and is even simpler.

So specialising to the most interesting case of a free boundary we obtain:

$$\widetilde{\Phi} = \frac{D}{8k\nu} \, g^* \, \delta \, (t - t') \, (r'^2 - r^2) \, [\, - \frac{1}{5} \left(\frac{1}{3} \, r'^2 - \frac{1}{7} \, r^2 \right) \left(\frac{r}{r'} \right)^l \mu \, +$$

$$+ \sum_{l=2}^{\infty} \frac{1}{l^2} \left(\frac{r'^2}{2l-1} - \frac{r^2}{2l+3} \right) \left(\frac{r}{r'} \right)^l (l \, \mu' - (1 - \mu'^2) \, \partial_{\mu'}) \, P_{l-1}(\hat{r} \cdot \hat{r}')] =$$

$$= - \frac{D}{8k\nu} \, g^* \, \delta \, (t - t') \, (R^2 - r^2) \, [\frac{1}{5} \left(\frac{1}{3} \, R^2 - \frac{1}{7} \, r^2 \right) \left(\frac{r}{R} \right)^2 \mu \, +$$

$$+ \partial_{3'} \Big|_{r'=R} \sum_{l=2} \frac{1}{l^2} \left(\frac{r}{r'} \right)^l \left(\frac{R^2}{2l-1} - \frac{r^2}{2l+3} \right) P_{l-1} \left(\hat{r} \cdot \hat{r}' \right)] \tag{4.18}$$

which after summing gives

$$\tilde{\Phi} = \frac{D}{8k\nu} \, g^* \, \delta \, (t - t') \, (R^2 - r^2) \, [\, - \frac{1}{5} \left(\frac{1}{3} \, R^2 - \frac{1}{7} \, r^2 \right) \left(\frac{r}{R} \right)^2 \mu \, + \tag{4.19}$$

$$+ \partial_{3'} \Big|_{r'=R} \int_0^r d \rho \, K_\Phi(r,\rho) \, \{\frac{1}{|\rho - r'|} - \frac{1}{r'}\}]$$

where $\underline{\rho} := \rho \frac{\underline{r}}{r}$ and $\partial_{3'} := \frac{\partial}{\partial x'_3} = \mu' \partial_{r'} + (1 - \mu'^2) r'^{-1} \partial_{\mu'}$,

and $K_\Phi(r,\rho) = (1 + \frac{1}{3}(\frac{r}{R})^2) \ln \frac{r}{\rho} + 2(1 - (\frac{\rho}{r})^{1/2}) - \frac{2}{9} (\frac{r}{R})^2(1 - (\frac{\rho}{r})^{3/2})$ \hfill (4.20)

and where Ra is the dimensionless Rayleigh number

$$Ra = \frac{g^* R^3}{k\nu} = \frac{\alpha(T_\infty - T_0) R^3 g}{k\nu} \tag{4.21}$$

Analogously:

$$\widetilde{\widetilde{\Psi}} = \frac{1}{2} \, Ra \, D \, [3 \left(\frac{r}{R} \right)^2 \frac{\nu}{R^2} \, \mu'' + \delta \, (t - t') \, r \int_0^r \frac{d\rho}{\rho} \, K_\Psi(r,\rho) \, \{\frac{1}{|\rho - r'|} - \frac{1}{r'}\}] \tag{4.22}$$

with

$$K_\Psi = \frac{1}{3} [1 + \left(\frac{r}{R} \right)^2] \ln \frac{r}{\rho} + \frac{8}{9} [(\frac{\rho}{r})^{3/2} - 1] [(\frac{r}{R})^2 - \frac{1}{5}] + [\frac{r}{\rho} - 1] (\frac{r\rho}{R^2} - \frac{3}{5})$$

with the help of the functions $\widetilde{\Phi}$ and $\widetilde{\widetilde{\Psi}}$ we can now determine the complete velocity field (ref. 4) $\underline{u} = \underline{\nabla} \wedge (\underline{\nabla} \wedge \Phi \frac{\underline{r}}{r}) + \underline{\nabla} \wedge \Psi \frac{\underline{r}}{r}$ \hfill (4.23)

where: $\Phi = \int_0^t dt' \frac{1}{4\pi} \int_{r'=R} d \, \Omega' \, \tilde{\Phi} \, (r, r', t, t') \, r' \, \partial_{r'} \Gamma \, (r' \cdot t')$ \hfill (4.24)

$$\Psi = \int_0^t dt' \frac{1}{4\pi} \int_{r'=R} d \, \Omega' \, \tilde{\Psi} \, (r, r', t, t') \, r' \, \partial_{r'} \Gamma \, (r', t') \tag{4.25}$$

$$\tilde{\Psi} = \partial_\varphi \, \widetilde{\widetilde{\Psi}} \tag{4.26}$$

or working out (4.22)

$$u_r = r^{-2} L^2 \Phi = - r^{-2} (\partial_\mu (1 - \mu^2) \partial_\mu + (1 - \mu^2)^{-1} \partial_\varphi^2) \, \Phi$$

$$u_\theta = r^{-1} (1 - \mu^2)^{-1/2} (\partial_\varphi \Psi - (1 - \mu^2) \partial_r \partial_\mu \Phi) \tag{4.27}$$

$$u_\varphi = r^{-1}(1 - \mu^2)^{-1/2} ((1 - \mu^2) \partial_\mu \Psi + \partial_r \partial_\varphi \Phi).$$

In the case of azimuthal symmetry $\Psi = 0$ as $\widetilde{\widetilde{\Psi}}$ only depends on $\varphi - \varphi'$ while in the case of spherical symmetry (which is only realized for $0 < t \ll \frac{R^2}{(1.5\pi)^2}\left(\frac{1}{\nu}+\frac{1}{k}\right)$) not only $\Psi = 0$ but also the second term of Φ in (4.19) (i.e., the term $\sim K_\Phi$) does not contribute and we obtain only a simple Hill's vortex.

V. The resulting eq. for the concentration

Having obtained the $\underline{u}$ field in terms of the concentration flux on the surface we are in a position to discuss the resulting eq. for the concentration (2.1).

Introducing dimensionless variables we can write (2.1) as

$$(\partial_{t^\bullet} + \underline{u}^\bullet \cdot \underline{\nabla}^\bullet)\Gamma^\bullet = \Delta^\bullet \Gamma^\bullet \qquad \Gamma^\bullet = \frac{C_0 - C}{C_0} \tag{5.1}$$

with $\underline{u}^\bullet = \frac{R}{D}\underline{u}$; $t^\bullet = \frac{Dt}{R^2}$; $\underline{r}^\bullet = \frac{\underline{r}}{R}$

where, as we have seen, we can split up $\underline{u}^\bullet$ into a term describing the internal circulation and a rigid body rotation, which is absent for a rigid boundary or in the case of azimuthal symmetry.

$$\underline{u}^\bullet = Ra\ \underline{u}^\bullet_{inst(anteneous)} + \frac{\nu}{D}\ Ra\ \underline{u}^\bullet_{rot(ation)} \tag{5.2}$$

$$\text{with } Ra = \frac{g^\bullet R^3}{k\nu} = \frac{\alpha g(T_\infty - T_0)}{k\nu}$$

and where

$$\underline{u}^\bullet_{inst} = \frac{1}{4\pi}\int_{r'^\bullet=1} d\Omega'\ \underline{K}_{inst}(\underline{r}^\bullet, \underline{r}'^\bullet)\ \partial_{r'^\bullet}\ \Gamma^\bullet(\underline{r}'^\bullet, t)$$

$$\underline{u}^\bullet_{rot} = \frac{1}{4\pi}\int_{r'^\bullet=1} d\Omega'\ \underline{K}_{rot}(\underline{r}^\bullet, \underline{r}'^\bullet)\partial_{\varphi'}\int_0^{t^\bullet}\partial_{r'^\bullet}\ \Gamma^\bullet(\underline{r}'^\bullet, t'^\bullet)_{dt'^\bullet} \tag{5.3}$$

where $\underline{K}_{inst}$, $\underline{K}_{rot}$ are known, universal functions of $\underline{r}^\bullet$, $\underline{r}'^\bullet$ (actually only $r^\bullet$, $\cos\theta$, $\cos\theta'$, $\cos(\varphi - \varphi')$) i.e., independent of all other parameters (C_0, ν, k, ...) so that one only has to calculate them once as indicated in the previous chapter.

Strictly speaking, our expression for $\underline{u}$ in (5.3) is not valid for very small times i.e., for times smaller than the response time of velocity and temperature to a change in concentration which we in our adiabatic approximation have taken to be negligibly small. So eq. (5.3) is only valid for

$$t^\bullet \gtrsim t^\bullet_R = \frac{1}{(\frac{3}{2}\pi)^2}\left(\frac{D}{\nu} + \frac{D}{k}\right) \tag{5.4}$$

which has to be compared with the total time of depletion of the solute $t^\ast_D$

$$t^\bullet_R \ll t^\bullet_D = \frac{1}{\pi^2} \text{ (due to (2.5)).} \tag{5.5}$$

That this is not a real restriction can be seen by realising that for such small times ($t^\bullet < t^\bullet_R$) the diffusion term in (5.1) is dominant over the convection term.

Conclusion

We derived under the condition that ν, k $\gg$ D (i.e., viscosity and heat conduction of the solvent very large compared to the diffusion of the solute) the dependence of $\underline{u}$ on $\partial_r C|_{surface}$ in the concentration of solute C within the drop

$$(\partial_t + \underline{u}\cdot\underline{\nabla})\Gamma = D\ \Delta\ \Gamma \qquad \Gamma = \frac{C_0 - C}{C_0}.$$

The essential control parameter of the non-linear term turns out to be $Ra = \frac{\alpha g(T_\infty - T_0)}{k\nu}$ where $T_\infty - T_0$ is a measure of the chemical energy contained within the droplet at the start.

$$\underline{u} \sim Ra \int_{surface} \underline{K} \, \partial_r \Gamma$$

where $\underline{K}$ is a known universal kernel i.e. independent of all parameters.

An equation of this type might be very worthwhile to study within the context of synergetics)[6].

Let us e.g. start with an initial state which is isotropic (apart from small deviations) $C \approx C_0$ i.e., $\Gamma \approx 0$ at $t = 0$. It is clear that for $t = \infty$ all solute will be gone so we have $\Gamma = 1$ at $t = \infty$, for all Ra. If $Ra = 0$ we have stable spherical symmetry. If $Ra \neq 0$ we have cylindrical symmetry, but this solution might be unstable i.e., if we start with a very small anisotropy $C \approx C_0 + \varepsilon(\varphi)$ at $t = 0$ will u_φ stay small or will there be a critical value Ra_c above which such an initial anisotropy gets amplified to macroscopic proportions? Other questions are, if we will just obtain a somewhat modified Hill's vortex or more complicated multi-cell patterns etc. In several ways the problem is more complicated than usual (like e.g. the Rayleigh-Benard problem) in that the problem is essentially parabolic and in that (for $Ra > 0$) $\underline{u} = 0$ is not a solution or more to the point: one does not actually have the cylindrically symmetric solution which makes it difficult to determine if it is stable or not at some value of Ra. Nevertheless, or perhaps because of this, we think it to be a very interesting problem. We believe to have cast it in more tangible form and will persue the answers to the above questions.

Acknowledgements

I am indebted to Profs. P. Stichel and D. Meinköhn for many stimulating discussions.

REFERENCES

1. *L.E. Johns, Jr.,* and *R.B. Beckmann*
   A.I.Ch.E. Journal 1966, V12, 11
   R.J. Brunson and *R.M. Wellek*
   A.I.Ch.E. Journal 1971, V17, 1123
2. *R.M. Wellek* and *R.J. Brunson*
   Can. J. Chem. Engngn. 1975, V53, 150
   M. Morbidelli, A. Servida, G. Storti and *S. Carra*
   Chem. Eng. Sc. 1982, V37, 1653
   J.D. Thornton and *T.J. Anderson*
   Int. J. Heat Mass Transfer 1981, V24, 1847
3. *G.K. Batchelor*
   Fluid Dynamics, Cambridge University Press
4. *S. Chandrasekhar*
   Hydrodynamic and Hydromagnetic Stability, Dover Public. N.Y.
5. *P.G. Drazin* and *W.H. Reid*
   Cambridge University Press
6. *H. Haken*, Advanced Synergetics, Springer Verlag 1987

ZONE MODELLING THE ONSET OF HAZARDOUS CONDITIONS

H. A. DONEGAN - DEPT MATHEMATICS
T. J. SHIELDS - DEPT BUILDING
G. W. SILCOCK - DEPT BUILDING

UNIVERSITY OF ULSTER, JORDANSTOWN, N. IRELAND.

ABSTRACT The ability to predict the onset of hazardous conditions within a
fire compartment is essential in fire safety engineering. To this end,
various mathematical models have been devised. These include deterministic
zone and field models and also stochastic models which rely on a wide
statistical data base. Fire zone models which are popular in the fire
engineering profession use the pragmatic lumped parameter approach and can
be described using differential equations which can be solved analytically
or by numerical techniques depending on the level of sophistication.
Strategies of differing complexity are discussed in some detail to enable
the reader to gain an understanding not only of the mathematics of the
modelling but also of the associated fire engineering objectives.

1. INTRODUCTION

A building is essentially an assemblage of compartments and for the past
decade the fire engineering professionals have been concentrating their
efforts on modelling a typical compartment fire. As an introduction it is
necessary to describe the physical processes which contribute to the growth
and development of a fire within a compartment.

The complexity of the problem is readily illustrated by means of a simple
correspondence diagram (fig 1).

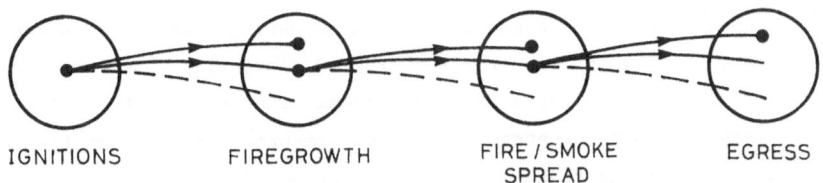

| IGNITIONS | FIREGROWTH | FIRE / SMOKE SPREAD | EGRESS |

FIG.1. MANY VALUED CONSEQUENCES OF UNWANTED IGNITIONS

Lack of image uniqueness characterises the lack of absolute repeatability
over time. It is clear that given an ignition which by the nature of heat
release develops into unwanted fire, there is no unique fire growth,

J. Manley et al. (eds.), Proceedings of the Third European Conference on Mathematics in Industry, 293–301.
© 1990 Kluwer Academic Publishers and B. G. Teubner Stuttgart.

although with more advanced testing the predictability range is narrowing.
The corresponding fire and smoke spread is difficult to predict although
building geometry and pressure differentials can act as global constraints
to assist technologists in the assessment of variations. Any building on
fire puts its occupants under threat and their safety depends on the time
required for their escape being less than the time available. The
significance of this inequality is developed in the model descriptions.
The consequence of non-egress is either no survivors or some survivors.

The present state of fire modelling, though conceptually the same as it was
in the early seventies, now has the advantages acruing from computer
facilities with the result that more complex problems can be addressed.
Deterministic and stochastic modelling are both used and each has its
advocates. Generally stochastic modelling has advantages when statistics
are readily available and particularly in the environment of the multi-
compartment-occupier interface. Deterministic modelling on the other hand
relies on an understanding of the physical laws concerning aspects of mass
balance, energy and momentum conservation. Such modelling is described
under two headings - Field modelling and Zone modelling. The former utilises
CFD (computational fluid dynamics) whereas the latter uses the much more
pragmatic lumped parameter approach. Field modelling is complex and
requires extensive computing power which is beyond the reach of the average
fire engineering professional whereas the zone modelling approach is at a
state where it is becoming sufficiently user friendly and significantly
meaningful.

The remainder of this short paper is devoted to illustrating the development
of simple zone models from a state where no computing power is required to
a state where coupled non linear first order differential equations are used
to describe the time available for evacuation. Such a model has been used
by this research group to determine escape time profiles for a multi-storey
office building and the results will be presented.

2. ZONE MODELLING

A compartment with an unwanted fire has essentially three zones or lumped
environments. These are the hot upper gas layer, the buoyant plume from
the fire source and the cooler lower air layer. Often the buoyant plume is
subsumed in the hot gas layer to simplify the calculations and for the
purpose of these illustrations, this approach is adopted.

Chitty[1]   defines zone modelling as:

The deterministic analysis of a system by reduction to a network of quantifiable components.

Such an approach has been described [4] as reasonable provided the compartment size is not of the proportions of a large electricity generating hall or a covered football stadium.   The principal features [6] which characterise a zone model are:

(i)   the assumption of uniform (but not constant) conditions over entire areas and volumes

(ii)   the   approximations are sufficiently close to reality, and

(iii) the geometry of the situation is well represented.

The following paragraphs will give the reader the necessary background to the simple lumped parameter zone models, with worked examples where feasible.

2.1 DESCENDING SMOKE LAYER (Negligible Ventilation)

The physical equation relating mass smoke production rate $\dot{m}$ to the height z (fig 2) of the smoke layer above the floor level is given [7] by:

$$(2.1.1) \qquad \dot{m} = 0.096 \ P \ \rho_o \ z^{3/2} \ g^{1/2} \ \left(\frac{T_o}{T_f}\right)^{1/2} \qquad \text{where:}$$

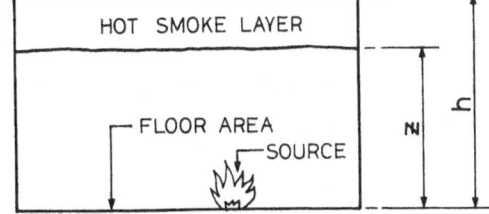

P = perimeter (m) of fire plan

ρ_o = ambient air density (kgm^{-3})

g = gravitational constant (ms^{-2})

T_o = ambient air temperature (K)

T_f = source temperature (1200K)

FIG. 2. COMPARTMENT WITH DECENDING SMOKE LAYER.

Assuming that the rate of smoke production equates with the rate of air entrainment then it is clear that:

$$(2.1.2) \qquad \dot{m} = \frac{d}{dt} [\rho'A(h - z)] = -\rho' \ A \ \frac{dz}{dt}$$

where $\rho' = \rho_o (T_o/T)$ the density of the smoke at ceiling level and T is bounded thus; $400 \le T^o C \le 500$.   Hence from (2.1.1)

$$(2.1.3) \qquad -\rho'A \ \frac{dz}{dt} = 0.096 \ P\rho_o \ z^{3/2} \ g^{1/2} \ (T_o/T_f)^{1/2}$$

$$(2.1.4) \quad \Rightarrow t = \frac{20A}{pg^{\frac{1}{2}}} [z^{-\frac{1}{2}} - h^{-\frac{1}{2}}].$$

From (2.1.4) it is a simple process to determine the time for conditions in a poorly ventilated room to become untenable. For example consider the following:

A = 40 m² (about the size of a small classroom)
P = 6 m (equivalent to a 1.25 MW output fire [7])
g = 9.81ms⁻²
h = 3m

From (2.1.4) t is approximately 10 seconds whenever z is taken to 1.5 m (approximately eyelevel). Hence with this as a stipulated hazard condition the time available for evacuation is 10 seconds. This must be less than the time required to physically evacuate such a compartment if no casualty is to result.

2.2 DESCENDING SMOKE LAYER (Good Ventilation at Ceiling Level)

This calculation is demonstrated only for the steady state case. Using the previous data the mass flow rate from the plume to the smoke layer reduces to:

$$(2.2.1) \qquad \dot{m}_i = 0.18 \ P \ z^{3/2}.$$

Given that the constant extraction rate at ceiling level is v m³s⁻¹ then the mass flow rate at extraction is given by:

$$(2.2.2) \qquad \dot{m}_o = v\rho_o (T_o/T).$$

On the assumption that no heat is lost from the hot layer then:

$$(2.2.3) \qquad T - T_o = \dot{Q}/\dot{m}_i C_p$$

where $\dot{Q}$ is the rate of heat release from the fire having perimeter P, and C_p is the specific heat of the ambient air at constant pressure.

For the steady state condition:

$$(2.2.4) \qquad \dot{m}_i = \dot{m}_o \quad \text{hence:}$$

$$(2.2.5) \qquad T - T_o = \dot{Q}T/v\rho_o \ T_o \ C_p$$

which when T/T_o is expressed as ϕ, reduces to:

$$(2.2.6) \qquad T_o(\phi - 1) = \dot{Q}\phi/v\rho_o C_p.$$

Also from (2.2.1) the following relationship emerges:

$$(2.2.7) \qquad 0.18Pz^{3/2} = v\rho_o/\phi.$$

Solving equations (3.2.6/7) for z yields:

$$(2.2.8) \qquad z = [\frac{v\rho_o - \dot{Q}/C_p T_o}{0.18P}]^{2/3}.$$

This shows that the steady state height of the smoke layer above floor level is a function of the ventilation rate for a given rate of heat release. Furthermore the layer temperature T can also be derived from (2.2.6/7) to give:

(2.2.9) $$T = T_o + \dot{Q}/[0.18\ PC_p\ z^{3/2}].$$

2.3 DESCENDING SMOKE LAYER (Good Ventilation at Ceiling level)
Time dependent solution.

With notation as before, the volumetric increase in smoke within the compartment over unit time is given by:

(2.3.1) $$\dot{V} = \frac{d}{dt}\ [A(h - z)] = -A\ \frac{dz}{dt}\ .$$

In this case:

(2.3.2) $\dot{V} = \dfrac{\dot{m}_i}{\rho'} - E$ where E is the volume extraction rate at ceiling level and ρ' is as before.

Equating equations (2.3.1) and (2.3.2) yields:

(2.3.3) $$-A\ \frac{dz}{dt} = \frac{\dot{m}_i}{\rho'} - E$$

(2.3.4) $$\Rightarrow \frac{dz}{dt} = \frac{E}{A} - c\ z^{3/2} \quad \text{where}$$

(2.3.5) $$c = 0.18P/A\rho'.$$

Equation (2.3.4) is solved numerically using small time steps with the initial condition that $z(0) = h$.

For example consider the situation in a modern atrium building where the central void is 17 m high having a plan area of 120m². Smoke at ambient temperature is extracted from the top of the atrium at 25m³s⁻¹. From equation (2.2.7) the height of the smoke layer is determined as 12.3 m. In this calculation P is assumed to be 4 m corresponding to a small 0.5 MW fire and ρ' is taken as 1.2 kg m⁻³. The time for the equilibrium state to occur is derived numerically as follows:

Equation (2.3.4) reduces to:

(2.3.6) $$\frac{dz}{dt} = 0.21 - 5.2 \times 10^{-3}\ z^{3/2}$$

(2.3.7) $$\Rightarrow \Delta z_{n+1} = \Delta t\ [0.21 - 5.2 \times 10^{-3}\ z_n^{3/2}\ ]$$

starting with z_o = 17 m and using Δt = 10 seconds the following results emerge:

TAB. 1 Time To Equilibrium Position of Smoke Layer

z_n	17.00	15.45	14.39	13.65	13.13	12.74	12.48	12.29	12.14
Δz_{n+1}		-1.55	-1.06	-0.74	-0.52	-0.37	-0.26	-0.19	-0.14
t	0	10	20	30	40	50	60	70	80

Thus in approximately 80 seconds the smoke layer will reach its equilibrium
state. (The rate of heat release from the fire is assumed constant).

2.4 INTRODUCTION TO THE ASET [2] MODEL

In this approach there is no restriction on the fire perimeter. The rate
of heat release is used directly and as such introduces a further complexity
with respect to time. In general there is a more precise evaluation of the
rate of air entrainment and in this case the average plume temperature is
used. There is no ventilation at ceiling level and any leakage from the
compartment resulting from the fire driven gas expansion is assumed to
occur at floor level. Using the same notation as before:

(2.4.1) $\qquad \dot{m}_i = 0.21 \, \rho g^{1/2} \, z^{5/2} \, [\dot{Q}*]^{1/3}$

where z in this case is the distance above the source and:

(2.4.2) $\qquad \dot{Q}* = (1 - \lambda_r)\dot{Q}/\rho_o \, C_p \, T_o \, g^{1/2} \, z^{2/5}$.

Here λ_r is the fraction of $\dot{Q}$ lost by radiation from the plume.

The origin of these relationships is part analytical and part experimental.

In this model the heat source is at a well defined height Δ above the
floor level, see fig. 3 and the facility exists therefore, for handling
the smoke layer below the source level, although for this introduction,
$0 < z(t) \leq h$ and $-\Delta < z_i(t)$.

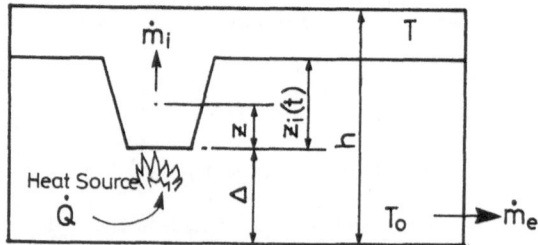

FIG. 3. ASET CONFIGURATION — LOW LEVEL OF LEAKAGE AT FLOOR LEVEL

A mass balance for the lower shrinking volume yields:

(2.4.3) $\qquad \rho_o A \dfrac{dz_i}{dt} = \dot{m}_e - \dot{m}_i \quad (z = z_i) \quad$ where:

(2.4.4) $\dot{m}_e = (1 - \lambda_c) \dot{Q}/C_p T_o$ with λ_c the fraction of $\dot{Q}$ lost to the bounding surfaces of the compartment.

(2.4.5) Hence:

$$\frac{dz_i}{dt} = \frac{-(1 - \lambda_c)\dot{Q}}{\rho_o AC_p T_o} - \frac{0.21 \ g^{1/3} \ z^{5/3}(1 - \lambda_r)^{1/3} \ \dot{Q}^{1/3}}{A(\rho_o C_p T_o)^{1/3}}$$

This differential equation is now set in dimensionless form by letting:

$\tau = t/t_c$; $\zeta = z_i/L_c$ and $\dot{q} = \dot{Q}/\dot{Q}_o$ where t_c and L_c are the characteristic time and length respectively and $\dot{Q}_o$ is taken as $\dot{Q}$ (t = 0). Hence

(2.4.6) $\frac{d\zeta}{d\tau} = -c_1\dot{q} - c_2\dot{q}^{1/3} \ \zeta^{5/3}$; $0 < \zeta \leq \zeta_o$

where $\zeta_o = h/L_c$, $c_1 = \frac{(1 - \lambda_c)\dot{Q}_o \ t_c}{\rho_o C_p T_o AL_c}$ and

$c_2 = \frac{0.21 t_c}{A} \ [\frac{(1 - \lambda_r) \ \dot{Q}_o \ g \ L_c^2}{\rho_o C_p T_o}]^{1/3}$.

By performing an energy balance for the hot layer it can be shown in dimensionless form that:

(2.4.7) $\frac{d\phi}{d\tau} = \phi[c_1\dot{q} - (\phi - 1)c_2\dot{q}^{1/3} \ \zeta^{5/3}]/(\zeta_o - \zeta)$; $0 < \zeta < \zeta_o$.

where $\phi = \bar{T}/T_o$, $\bar{T}$ being the average upper layer temperature.

In order to solve equations (2.4.6/7) for ζ and ϕ, initial conditions must be specified. One set of such conditions assume that the fire starts with a small heat release $\dot{Q}_o$ at time t = 0. Using this, the initial conditions in dimensionless form become:

$\zeta(\tau = 0) = \zeta_o$ and $\phi(\tau = 0) = \phi_o = 1 + \zeta_o^{-5/3} c_1/c_2$. Although $d\phi/d\tau$ in (2.4.7) is inderterminate at $\tau = 0$, it is possible with considerable analysis of this apparent zero time singularity to deduce that for small time estimates:

(2.4.8) $\frac{\Delta\phi}{\Delta\tau} = \frac{c_1}{c_2} \ [2\dot{q}_o' \ \zeta_o + 5(c_1 + c_2\zeta_o^{5/3}] \ /6 \ \zeta_o^{8/3} + O(\tau).$

Equations (2.4.6/7) have been solved using the improved Euler method and software similar to that available from NBS [3] for such calculations. The solution is terminated in a given situation at the hazard time τ_{haz} whenever:

(2.4.9) $\phi \geq \phi_{haz}$ i.e. when the layer temperature reaches a hazardous value. The hazard criterion used in the application of this technique occurs when the upper layer temperature reaches 183°C.

3. APPLICATION OF ASET

The method has been applied by the authors to determine residual populations at risk associated with office building design. The following table summarises results for a four storey office building having two stairways and designed to comply with current codes of practice.

DETECTION CRITERIA	HAZARD CRITERIA	TIME (SECS) AVAILABLE FOR ESCAPE, T_A	TIME (SECS) REQUIRED TO ESCAPE, T_R	RESIDUAL POPULATION
IMMEDIATE DETECTION	UPPER LAYER TEMPERATURE = 183°C	329	220	0
SMOKE DETECTOR	UPPER LAYER TEMPERATURE = 183°C	255	220	0
THERMAL	UPPER LAYER TEMPERATURE = 183°C	183	220	139
SPRINKLER ACTIVATION	UPPER LAYER TEMPERATURE = 183°C	151	220	206
DETECTION IN ABSENCE OF AUTOMATIC DETECTION SYSTEM	UPPER LAYER TEMPERATURE = 183°C	179	220	129

TAB. 2   Residual population for deemed to satisfy design - 4 storey office building (two stairways). The time required for escape in the above table was determined using an optimal network model [5] which determines the evacuation time for a fully occupied building not exposed to any detection criterion (See Tab. 3).

TAB. 3

DETECTION CRITERIA

(1) Immediate detection
   This may be characterised in the program by specifying the rate of rise of upper gas layer temperature to be 0°C/ min.

(2) Smoke detector
   An upper layer temperature of 35°C was taken to correspond to conditions sufficient to activate a smoke detector

(3) Thermal sensor
   Activation characterised by an upper layer temperature of 57°C.

(4) Sprinkler Activation
   Characterised by an upper layer temperature of 74°C.

(5) Absence of Automation
   Average human detection time 2.5 minutes.

4. CONCLUSION

Simple mathematical models have been used to determine the onset of
hazardous conditions within fire compartments. The results of the simple
models yield times to a critical event and these have to be matched to
various egress senarios to gain some estimate of risk.

The more sophisticated model ASET which is non restrictive in boundary
conditions has been used to determine estimates of time available for
escape. These times have been compared with the time required estimated
from the Evacnet network model.

Zone models can be used with good effect to compare levels of fire safety
in buildings that comply with codes of practice to those of an innovative
design.

REFERENCES

[1] Chitty, R.: Zone Models Theory - Fire Modelling and Its Impact on
    Design. Outline lecture notes - 3 day course. 3-5 October 1988.
    IMPERIAL COLL. OF SCI. AND TECH., LONDON.

[2] Cooper, L.Y.: A Matheamtical Model for Estimating Available Safe Egress
    Time in Fires. Fire and Materials. Vol 6 (1982) 135-144.

[3] Cooper, L.Y. (National Bureau of Standards) Private Communication 1989.

[4] Cox, G.; Kumer, S.: Field Modelling of Fire in Forced Ventilated
    Enclosures. Combust. Sci. and Tech. 52 (1987), 7-23.

[5] Kisko, T.M.; Francis, R.K.: EVACNET[+] A Computer Program to Determine
    Optimal Building Evacuation Plans. Fire Safety Journal Vol 9, (1985),
    7-16.

[6] Mitler, H.E.: The Harvard Fire Model. Fire Safety Journal.
    Vol 9 (1985), 7-16.

[7] Shields, T.J., Silcock, G.W.H.: Buildings and Fire. Longman (1987).

A TREE SEARCH APPROACH BASED ON AN ASSIGNMENT RELAXATION FOR THE SOLUTION OF SET COVERING PROBLEMS

by

Elia El-Darzi and Gautam Mitra

ABSTRACT. An assignment relaxation for the set covering problem (SCP) is introduced and discussed. A tree search method is then developed which makes use of this relaxation. Computational experience of processing a collection of test problems is reported. The work reported here constitutes a part of a generalised tree search method for the solution of the SCP which is described in [11]

Key words: integer programming, discrete optimisation, set covering, scheduling, assignment, branch and bound.

1. Introduction to the set covering problem (SCP)

Integer programming represents a large class of discrete optimisation problems [7,21]. In this paper we are interested in a special class of 0-1 integer programming problems, namely the set covering problem (SCP). The SCP is a well known problem in the field of graph theory and combinatorial optimisation and represents a wide range of industrial scheduling and planning problems. These include bus crew scheduling [18], air crew scheduling [1], vehicle routing [6], steiner problem [14], facility location [8] and others. For a comprehensive survey on the application of the SCP see Balas [2], Balas et al [4] and El-Darzi et al [10].

Consider a set $R=\{1,2,...,m\}$ and a class H of subsets of R, such that $H=\{H_1,H_2,...,H_n\}$. Let $J=\{1,2,...,n\}$ be the set of indices for the subsets which make up the class H. A cover for R is a subclass of H defined as $\{H_j \mid j \in J_c\}$ where $J_c \subseteq J$, satisfying $\bigcup_{j \in J_c} H_j = R$. Let c_j be the cost of including H_j in the cover. Thus the minimum cost set covering problem is that of finding a cover $\{H_j \mid j \in J_c^*\}$ as above such that $\sum_{j \in J_c^*} c_j$ is a minimum.

The SCP may be also posed as a zero-one integer programming problem.

$$\text{Min} \quad \sum_{j=1}^{n} c_j x_j \tag{1.1}$$

subject to
$$\sum_{j=1}^{n} a_{ij} x_j \geqslant 1, \quad i = 1,...,m \tag{1.2}$$

$$x_j \in \{0,1\}, \quad j = 1,...,n \tag{1.3}$$

303

J. Manley et al. (eds.), Proceedings of the Third European Conference on Mathematics in Industry, 303–311.
© 1990 Kluwer Academic Publishers and B. G. Teubner Stuttgart.

where
$$x_j = \begin{cases} 1 \text{ if } H_j \text{ is included in the cover} \\ 0 \text{ otherwise} \end{cases}$$

and
$$a_{ij} = \begin{cases} 1 \text{ if } i \in H_j \\ 0 \text{ otherwise} \end{cases}$$

It is convenient to introduce the index sets R_i, $i = 1,2,...m$, such that for a row i, R_i denotes the indices of the columns with unit entry. Similarly, the index sets H_j, $j=1,2,...,n$, denote the indices of the rows with unit entry in column j. Thus H_j and R_i are related to a_{ij} as set out in (1.4) and (1.5)

$$H_j = \{i | a_{ij} = 1, \; i=1,...,m\} \text{ for all } j \qquad (1.4)$$

and
$$R_i = \{j | a_{ij} = 1, j=1,...,n\} \text{ for all } i \qquad (1.5)$$

where $\quad |H_j| = \sum_{i=1}^{m} a_{ij} \quad$ and $\quad |R_i| = \sum_{j=1}^{n} a_{ij}$.

A feasible solution to the SCP is called a cover. A prime cover x^*, is a cover for which x_j currently taking the value one cannot be reduced to zero without violating a constraint. An optimal solution to the SCP is a prime cover if all the costs are positive.

The contents of this paper is organised as follows. The assignment relaxation which we have introduced constitutes a new approach to solving the SCP and is described in section 2. The tree search algorithm is described in section 3 and the computational results are presented in section 4.

2. A lower bound derived by an assignment relaxation (ASP1)

Let $a_j = (a_{1j},...,a_{mj})^T$ denote the column a_j of the SCP problem and let this column be decomposed into a set of k_j columns $(a_j^1, \; ..., \; a_j^{k_j})$ each with at least one unit entry and at most two unit entries. It follows from this decomposition that

$a_j = \sum_{p=1}^{k_j} a_j^p$. The column (nonzero) count of a_j^p can only be 1 or 2, that is

$$\sum_{i=1}^{m} a_{ij}^p \epsilon \{1,2\}, \; p = 1,...,k_j \qquad (2.1)$$

where $\quad a_{ij}^p \; \epsilon \{0,1\}$.

Let q_j be the largest positive integer such that $q_j \leqslant (|H_j|+1)/2$. The allowable range for k_j is easily seen to be $q_j \leqslant k_j \leqslant |H_j|$. Let the set R (set of rows) be partitioned heuristically [9] into two disjoint sets R' and R" such that R'∪R"=R and R'∩R" =Φ. Introduce two (redundant) constraints indexed m+1 and m+2 such that

$$\sum_{j=1}^{n} a_{m+1,j}\, x_j \geqslant 0 \quad \text{and} \quad \sum_{j=1}^{n} a_{m+2,j}\, x_j \geqslant 0$$

where
$$a_{m+1,j} = \begin{cases} 1 & \text{if} \quad |\sum_{i \in R'} a_{ij}| < |\sum_{i \in R"} a_{ij}| \\[2mm] 0 & \text{otherwise} \end{cases}$$

and
$$a_{m+2,j} = \begin{cases} 1 & \text{if} \quad |\sum_{i \in R"} a_{ij}| < |\sum_{i \in R'} a_{ij}| \\[2mm] 0 & \text{otherwise} \end{cases}$$

Thus the SCP (1.1-1.3) can be rewritten as follows

$$\text{Min} \sum_{j=1}^{n} c_j x_j \tag{2.2}$$

subject to
$$\sum_{j=1}^{n} a_{ij} x_j \geqslant 1, \qquad i \in R' \tag{2.3}$$

$$\sum_{j=1}^{n} a_{ij} x_j \geqslant 1, \qquad i \in R" \tag{2.4}$$

$$\sum_{j=1}^{n} a_{m+1,j}\, x_j \geqslant 0 \tag{2.5}$$

$$\sum_{j=1}^{n} a_{m+2,j}\, x_j \geqslant 0 \tag{2.6}$$

$$x_j \in \{0,1\} \text{ for all } j. \tag{2.7}$$

Let $R^+=R'\cup\{m+1\}$ and $R^{++} = R"\cup\{m+2\}$. From the definitions set out in (2.1) it follows that it is always possible to derive a decomposition of a_j to $a_j^p (p=1,..,k_j)$ where a_j^p takes one of the three alterative forms.

a) $a_j^P = (..a_{rj}^P..o|..a_{sj}^P..o)^T$ where $r \epsilon R' \cap H_j$ and $s \epsilon R'' \cap H_j$ (2.8)

b) $a_j^P = (..a_{rj}^P..o|o..o \ a_{m+2,j}^P)^T$ where $r \epsilon R' \cap H_j$ (2.9)

c) $a_j^P = (o..o \ a_{m+1,j}^P|..a_{sj}^P..o)^T$ where $s \epsilon R'' \cap H_j$ (2.10)

In (2.8-2.10) all $a_{ij}^P = 1$ values and the rest of the components are zero.

The SCP as presented in (2.2-2.7) can be relaxed as an assignment problem by introducing a bipartite graph with two sets of vertices and arcs as shown below.

Set of vertices. For the m+2 rows introduce m+2 corresponding vertices $v_1,...,v_{m+2}$ which are used in the (assignment) graph representation of the problem.

The set of arcs and associated costs. Introduce three sets of arcs A_B, $A_{D'}$ and $A_{D''}$ defined as

(i) $A_B = \{(v_r,v_s)|r\epsilon R' \text{ and } s\epsilon R''\}$. There is an arc from v_r to v_s if there exists a column a_j^P which satisfies (2.8), with the associated cost $d_{jp} = 2c_j/|H_j|$.

(ii) $A_{D''} = \{(v_r,v_{m+2})|r\epsilon R'\}$. There is an arc from v_r to v_{m+2} if there exists a column a_j^P which satisfies (2.9) with the associated cost $d_{jp} = c_j/|H_j|$.

(iii) $A_{D'} = \{(v_{m+1},v_s)|s\epsilon R''\}$. There is an arc from v_{m+1} to v_s if there exists a column a_j^P which satisfies (2.10). The associated cost d_{jp} is given as $d_{jp} = c_j/|H_j|$.

Finally introduce an arc set A_{n+1} with a dummy arc from v_{m+1} to v_{m+2}, $A_{n+1} = \{(v_{m+1},v_{m+2})\}$ with zero associated cost ($d_{n+1,1} = 0$), whereby the flow requirements ($\geqslant 1$) may be imposed on v_{m+1} and v_{m+2}. Let a_{n+1} denote the corresponding dummy column. Let A denote the directed arcs of the resulting graph such that $A = \{A_B \cup A_{D'} \cup A_{D''} \cup A_{n+1}\}$. Let A_j denote the set of arcs obtained by decomposing the column j in the manner indicated earlier. Thus there are k_j arcs in A_j, where for $(v_r,v_s)\epsilon A_j$, $v_r \epsilon R^+$ and $v_s \epsilon R^{++}$ as explained in (2.8-2.10). It is easy to see that

$$\overset{n+1}{\underset{j=1}{U}} A_j = A .$$

A statement of the relaxed problem [13]. With each column a_j associate a variable y_{jp}, $y_{jp} \epsilon \{0,1\}$ and a cost coefficient d_{jp} which are defined for $j=1,...,n+1$ and $p=1,...,k_j$. Then the relaxed problem ASP1 is stated as

$$\text{Min} \quad \sum_{j=1}^{n+1} \sum_{p=1}^{k_j} d_{jp} y_{jp} \tag{2.11}$$

subject to

$$\sum_{j=1}^{n+1} \sum_{p=1}^{k_j} \alpha_{rsj}^p y_{jp} \geqslant 1, \ r \epsilon R^+ \tag{2.12}$$

$$\sum_{j=1}^{n+1} \sum_{p=1}^{k_j} \alpha_{rsj}^p y_{jp} \geqslant 1, \ s \epsilon R^{++} \tag{2.13}$$

$$y_{jp} \epsilon \{0,1\} \quad \text{for all } j \text{ and } p \tag{2.14}$$

$$\text{where} \quad \alpha_{rsj}^p = \begin{Bmatrix} 1 \text{ if } (v_r, v_s) \ \epsilon \ A_j \\ 0 \text{ otherwise} \end{Bmatrix} j=1,\dots,n+1 \tag{2.15}$$

ASP1 set out in (2.11–2.14) is a proper relaxation of the SCP [9]. The ASP1 can be made equivalent to the SCP by introducing the following relations for those columns a_j which are decomposed such that $k_j \geqslant 2$, $j=1,\dots,n$

$$y_{jp} = y_{jp+1} \ , \ p=1,\dots,k_j. \tag{2.16}$$

3. A tree search algorithm

Branch and bound is one of the most successful techniques for solving combinatorial optimisation problems, covering discrete optimisation models in general and integer programming problems in particular [19]. For an overview of the branch and bound techniques see Breu et al [5], Shapiro [20] and Mitra [17]

An outline of the tree search strategy

The alternative strategies which are adopted in the design of the branch and bound procedure are outlined in the statement of the algorithm set out below. At each step of the solution process it is necessary to know whether or not a feasible solution to the SCP is obtained. It is also necessary to know the value of the best objective function which is denoted by z_{min}. Also we use z_ℓ to denote the optimal solution value of the relaxed problem ASP1, that is, a lower bound on the objective function value of the SCP. Before stating the algorithm we explain how the search is terminated at each branch of the tree depending on the outcome of the subproblem investigation. A node is fathomed if after the solution of the subproblem one of the following conditions hold. (1) The lower bound is greater than the upper bound. (2) The subproblem has no feasible solution. (3) The optimal solution to the relaxed problem is found and this also satisfies the side contraints. In the tree search the subgradient optimisation [16] is used to solve the lagrangean relaxation of the SCP (2.11-2.14) and 2.16 [9]. Reduction tests which are reported in [9] are also used in this algorithm.

Step(1) Preprocessing procedure. In this step a number of heuristic procedures [2], [9] and [15] are applied to reduce the model size, to derive an upper bound (z_{min}) and a lower bound (z_ϱ) on $\bar{z}$ and to derive the cost vector of the relaxed problem. If $z_\varrho > z_{min}-1$ go to Step(11), otherwise go to Step(2)

Step(2) Solution at the root node. The relaxed SCP problem is solved at the root node of the tree. If $z_\varrho > z_{min} - 1$ or the side constraints are satisfied go to Step(11) else initialise the lagrangean multipliers and go to Step(8).

Step(3) Choice of the branching variable(s). Select a group of network variables y_{jk} corresponding to the arcs in the arc set A_j for branching. Add the two subproblems with ($y_{jk}=1, k=1,...,k_j$) and ($y_{jk}=0, k=1,...,k_j$) to the list of subproblems, store their positions in the list and go to Step(4b).

Step(4) Subproblem selection. (a) If the list of subproblems is empty go to Step(11).
(b) Choose a subproblem from the list and go to Step(5).

Step(5) Subproblem preparation. Prepare the subproblem to be optimised and go to Step(6).

Step(6) Subproblem solution. Solve the subproblem using a network optimiser [12] and go to Step(7).

Step(7) Subproblem solution analysis. If the subproblem has no feasible solution or the objective solution value is greater than $z_{min}-1$ then go to Step(4). If the side constraints are satisfied go to Step(10). Otherwise go to Step(8).

Step(8) Solution improvement and model reduction. Test for optimal network solution improvement. Temporarily remove the redundant columns and go to Step(9).

Step(9) Lagrangean and subgradient procedure. If the number of subgradient iterations exceed an iteration counter (LMAX) go to Step(3). Otherwise compute the lagrangean multipliers, update the corresponding costs and go to step(5).

Step(10) Update the best SCP solution. A feasible solution to the SCP has been found. Update z_{min} and the corresponding solution vector and go to Step(4).

Step(11) SCP optimal solution. Output the best SCP feasible solution and the corresponding optimal objective value (z_{min}).

4. Computational results

The computational results in the branch and bound algorithm designed for the ASP1 are given in tables 4.1. This algorithm is applied to the test problems [11] which have not been optimally solved using the preprocessing procedures. The abbreviations used in these tables headings are explained below.

PROBLEM NAME	z	REDUCED DIMENSIONS m	n	z_u	PREPRO- CESSING TIME	INITIAL NETWORK SOLUTION	TIME	NO OF NODES	NO OF LAG ITER	B&B TIME	TOTAL TIME
AIR1	16635	120	297	16660	40.9	16577.2	6.1	1136	601	954.1	995.0
AIR2	18880	129	436	18885	269.3	18833.1	7.3	26	11	67.0	336.3
AIR3	18195	104	499	18195	175.4	18124.0	6.9	145	113	528.3	703.7
AIR4	19715	138	1089	19795	100.9	19462.7	16.6	225	144	197.9	298.8
AIR5	21560	131	1003	21800	104.5	21377.3	14.7	357	173	342.7	347.2
AIR6	16925	124	998	17000	124.3	16795.0	14.6	31	0	44.1	168.4
BUS2	41051	26	90	41537	1.5	41036.4	0.04	16	2	0.9	2.4
BUS3	64749*	55	977	64749	62.3	63000.2	7.6	487	539	>1000	>1000
BUS4	74787*	53	547	75212	26.6	72086.1	2.6	818	966	>1000	>1000
RDM4	97	99	74	97	11.5	93.2	0.21	75	30	45.4	56.9
RDM5	96	31	54	96	10.7	94.4	0.12	37	30	34.8	45.5
RDM7	87	57	40	87	15.2	85.6	0.59	70	31	16.3	31.5

Table 4.1

All times are in cpu seconds of Honeywell
Multics DP6840 processing

$\bar{z}$ is the optimal solution value if found. A value with a "*" represents the best feasible solution found within the time limit. m, n are the reduced dimensions of the test problems, where m denotes the number of rows and n denotes the number of columns. z_u is the best upper bound derived using the preprocessing procedures. The computing time taken of the preprocessing procedures is given in the next column. The bound derived by the ASP1 at the root node of the tree and the corresponding computing time are also reported in this table, followed by the total number of nodes developed in the process of searching for the optimal solution. The number of lagrangean iterations applied within the framwork of the tree search is also tabulated. The last two columns in this table represent the total time for the branch and bound algorithm (not including Step(1)) and the total execution time. The maximum running time permitted was set to 1000 cpu seconds. The ASP1 solved nearly all the test problems within the time limit, except for BUS3 and BUS4. The lagrangean and subgradient procedures managed to improve the bounds at the lower levels of the tree, that is, when a large number of variables were fixed to either one or zero. But this was achieved at the expense of very high computing time.

Our overall conclusion is that more experiments are needed to "successfully" incorporate the lagrangean and subgradient procedures within the tree search scheme. Different branching strategies also may be worth investigating and of course a faster network optimiser will improve the performance of this algorithm.

REFERENCES

[1] Baker, E., and Fisher, M.: Computational results for very large air crew scheduling problems. Omega, 9, 613-618, (1981)

[2] Balas, E.: A class of location, distribution and scheduling problems: modelling and solution methods. Rev. Belge Stat. Inform. Recherche Oplle. 22, 36-57, (1983)

[3] Balas, E., and Ho, A.: Set covering algorithms using cutting planes, heuristics and subgradient optimization:A new computational study. Math. Prog. Stud. 12, 37-60, (1980)

[4] Balas, E., and Padberg, M.W.: Set partitioning: A survey. SIAM Review, 18, 710-760, (1976)

[5] Breu, R., and Burdet, C.A.: Branch and bound experiments in 0-1 programming. Math. Prog. Stud. 2, 1-50, (1974)

[6] Christofides, N.: Vehicle routing. in Travelling salesman problem. Lawler, E.L., Lenstra, J.K., Rinooy Khan, A.H.G., and Shmays, D.S., eds, Academic Press, (1985)

[7] Dantzig, G.: Linear programming and extensions. Princeton University Press, Princeton,

New Jersey, (1963)

[8] Daskin, M.S., and Stern, E.D.: A hierarchical objective set covering model for emergency medical service vehicle deployment. Transp. Sci. 15, 137–152, (1981)

[9] E, El–Darzi.: Methods for solving the set covering and set partitioning problems using graph theoretic (relaxation) algorithms. Brunel University, PhD thesis, (1988)

[10] E, El–Darzi., and G, Mitra.: Set covering and set partitioning: A collection of test problems. Brunel University, Internal Report, (1988). To appear in Omega

[11] E, El–Darzi., and G, Mitra.: A tree search approach for the solution of set problems using alternative relaxations. Brunel University, Internal Report, (1988)

[12] Ford, L., and Fulkerson, D.: flows in networks. Princeton University Press, (1962)

[13] Frieze, A.: Private communication. (1985)

[14] Fulkerson, D.R., Nemhauser, G.L., and Trotter, L.E.: Two computationally difficult set covering problems that arises in computing the 1–Width of incidence matrices of steiner triple systems. Math. Prog. Stud. 2, 72–81, (1974)

[15] Garfinkel, R.S., and Nemhauser, G.L.: Integer programming. Wiley, (1972)

[16] Geoffrion, A.M.: Lagrangean relaxation for integer programming. Math. Prog. Stud. 2, 82–114, (1974)

[17] Mitra, G.: Investigation of some branch and bound strategies for the solution of mixed integer linear programs. Math. Prog. 4, 150–170, (1973)

[18] Mitra, G., and Darby–Dowman, K.: CRU–SCHED A computer based bus crew scheduling system using integer programming. in Computer scheduling of public transport. Rousseau, J.-M., Ed., North Holland, (1985)

[19] Murty, K.G.: Linear and combinatorial programming. John Wiley and Sons, (1976)

[20] Shapiro, J.F.: A survey of lagrangean techniques for discrete optimization. Ann. Discrete Math. 5, 113–138, (1979)

[21] Williams, H.P.: Model building in mathematical programming. John Wiley and Sons, (1985)

Authors address: Brunel University, Dept of Math., Uxbridge, Middlx, UB8 3PH, England.

Making a Workpiece with Spiral Turns
by Means of Forming Cutters*

Heinz W. Engl and Thomas Langthaler, Linz

Abstract:

Workpieces with spiral turns (e.g. milling cutters, borers) are produced by means of forming cutters, generators or abrasive wheels. For this purpose, a cylinder of the desired raw material is passing the rotating tool with the correct twist. Through this process, parts of the slug are milled away and one cog of the workpiece is formed. Our problem is, given the desired shape of the workpiece, to construct (compute) a tool which forms a cog of this workpiece. It is of course important to select shape and position of the forming tool in such a way that already correctly made parts of the workpiece are not damaged in a later stage of the production process.

First, we describe the traditional construction of a solution on the drawing board and formulate it mathematically. We show that the problem is ill-posed in the sense that small data changes can lead to an unsolvable problem. The discretization of the traditional approach has the property that local errors in the data result in global changes in the solution, which is not desirable.

Then we describe a new solution technique, which is computationally more efficient and whose discretization has the property that data errors propagate into the solution only locally. We can show continuous dependence of the solution on the data and give error bounds for a technically important special case. Numerical results will be presented.

* This work has been partially supported by the Austrian Fonds zur Förderung der wissenschaftlichen Forschung (project S32/03) and the Austrian Federal Ministry for Science and Research. The practical part of this project was performed for VOEST-ALPINE Werkzeuge- und Präzisionstechnik GmbH, Ferlach, Austria, while both authors were at the University of Klagenfurt

J. Manley et al. (eds.), Proceedings of the Third European Conference on Mathematics in Industry, 313–332.
© 1990 *Kluwer Academic Publishers and B. G. Teubner Stuttgart.*

314

1. The Traditional Method

The practical problem to be solved is described in the Abstract. In this Section, we start with a description of the traditional construction method for solving this problem on the drawing board, which we will then formulate in mathematical terms: Given a workpiece with spiral turns to be produced, we look at its projections into two orthogonal planes, namely its "ground plane" and its "front view", where we fix these planes such that the axis of the workpiece lies parallel to the ground plane and orthogonal to the front view plane. Given the frontal intersection and the height of one spiral turn, the workpiece is determined uniquely. Here, the frontal intersection is defined as the figure of intersection of the workpiece with a plane orthogonal to its axis. The value of the height of one spiral turn represents the ratio between shifting and rotating the frontal intersection to get the desired workpiece. Thus, the workpiece can be thought of as created by shifting and rotating the frontal intersection along its axis according to the height of one spiral turn. Following this procedure, the two-dimensional frontal intersection forms the three-dimensional workpiece.

Other input variables are the distance between the axis of the workpiece and the axis of the desired forming cutter, which should be able to produce the given workpiece and the "turn of axes", i.e., the angle between the axis of the workpiece and the axis of the forming cutter projected onto the ground plane. In the front view we always see that frontal intersection which corresponds to the intersection of the axis of the workpiece and the axis of the forming cutter in the ground plane. That frontAal intersection is determined by its "screw-in angle", i.e., the angle between the radius connecting the center and the first point of the frontal intersection and the radius orthogonal to the ground plane and pointing away from it. Fig. 1 shows a typical example.

The basis for the traditional method is the concept of spiral turns. By a spiral turn we mean that helical curve which we get following one point of the frontal intersection during the construction of the workpiece. In the front view, such a spiral turn is always a circle with center in the axis of the workpiece. The totality of all spiral turns, corresponding to all points of the frontal intersection, determines the surface of the workpiece in $I\!R^3$ uniquely.

Knowing all spiral turns, we can formulate our problem in geometrical terms:

The desired forming cutter is the tool with the property that it is tangent to all spiral turns and that it does not intersect any of those spiral turns.

This can be seen as follows: Being tangent to a specific spiral turn, means that the forming cutter produces exactly the given point of the frontal intersection throughout the whole workpiece; if it were not tangent to a spiral turn, this would mean that the forming cutter would not produce the given point of the frontal intersection. If it were intersecting a spiral turn, this would mean that the forming cutter would mill away the corresponding point of the frontal intersection.

For the construction of that forming cutter we need two profile views, the first one being orthogonal to the ground plane and to the axis of the forming cutter. We project our spiral turns onto that plane. The next construction step is a circular projection around the forming cutter axis as center onto the plane orthogonal to the ground plane and containing the forming cutter axis. In the next step we need the second profile view orthogonal to the ground plane and parallel to the axis of the forming cutter. We project our circularly projected spiral turns onto that plane.

Let us pick one of those spiral turns: Any forming cutter being tangent to this curve will produce the correct spiral turn in $I\!\!R^3$. Thus, considering all (projected) spiral turns, the problem of finding the desired forming cutter is equivalent to the problem of finding a curve being tangent to all projected spiral turns and not intersecting any of them. That curve is called the "envelope".

Fig. 2 shows the construction of one projected spiral turn according to the steps described above.

In the sequel we distinguish between two problems: The "continuous (infinite-dimensional) problem", where we consider all points of the given frontal intersection and all resulting projected spiral turns, and the "discrete (finite-dimensional) problem", where we consider only a finite number of points of the frontal intersection and their corresponding projected spiral turns.

There exist values of the control variables "turn of axes", "screw-in angle" and "distance of axes" for which one or more projected spiral turns are hidden by others. In such a situation it is impossible to find a curve being tangent to all spiral turns without intersecting any of them. We call such a problem "unsolvable". On the other hand, if there exists a curve being tangent to all spiral turns without intersecting any of them, we call the problem "solvable".

It can be seen easily that the values of the control variables have a great influence on the solvability of the problem. The smaller the distance of the axes, the better. The turn of axes should be chosen near the value of the twist angle (which is the angle naturally connected with the height of one spiral turn; see Proposition 2.4). The screw-in angle should be chosen in such a way that the absolute value of the slope of all tangentials of the frontal intersection is as small as possible. Reasons for these suggestions are given in [1].

316

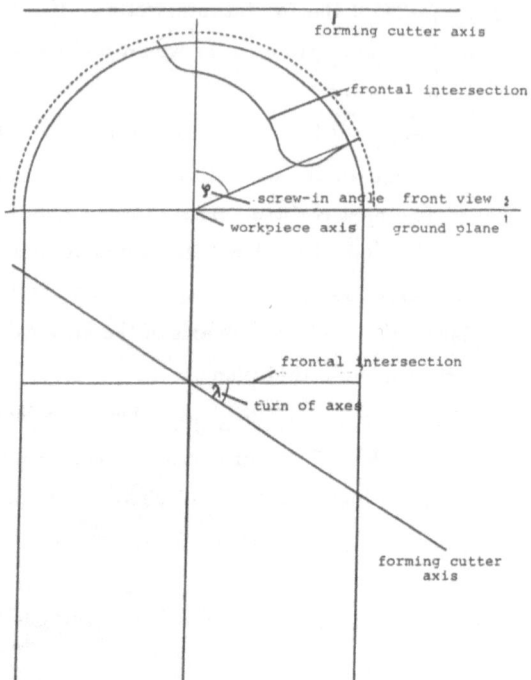

Figure 1

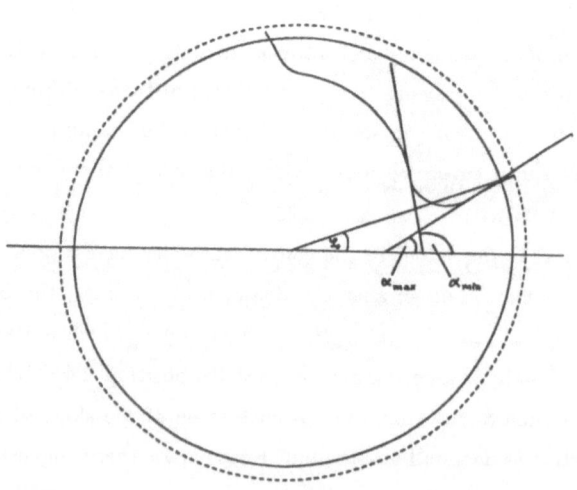

Figure 3

317

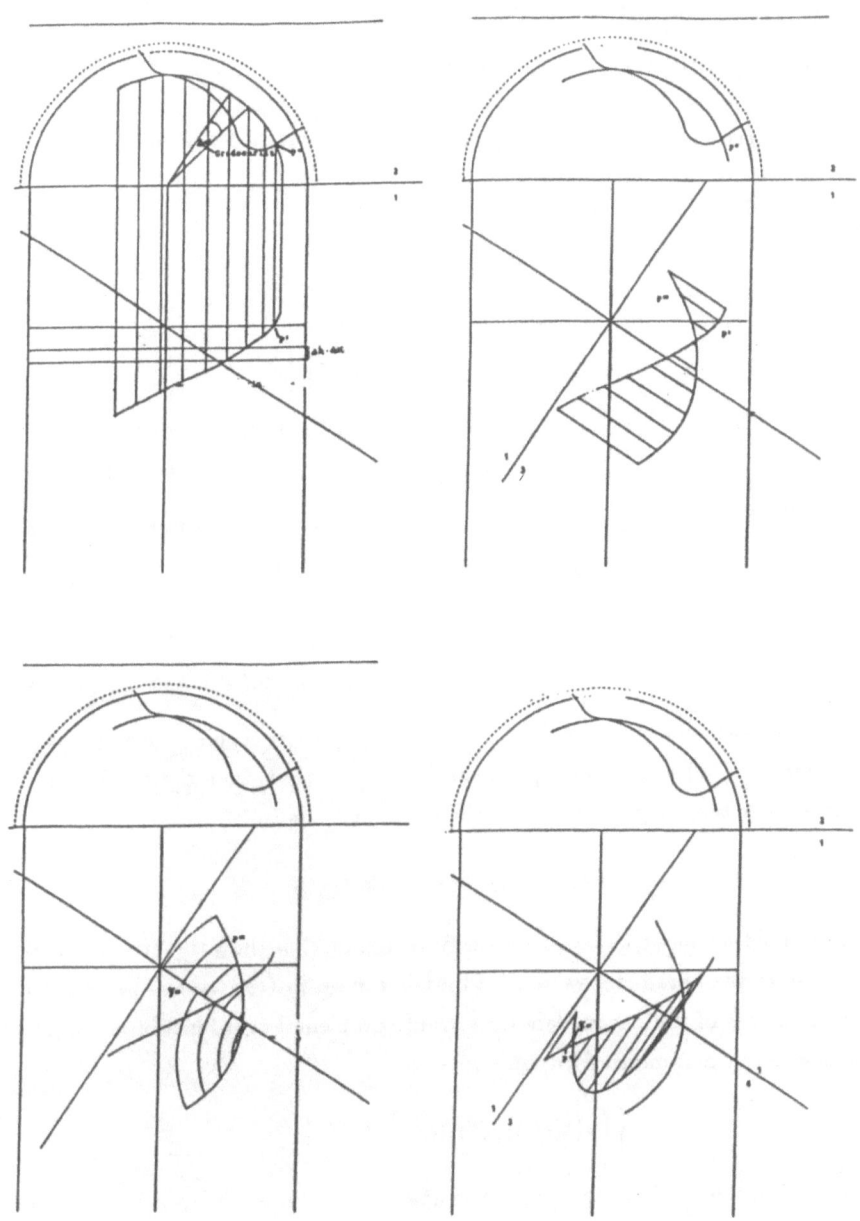

Figure 2

between the frontal intersection and the projected spiral turns:

$$x(s,t) = r(t)\cos\big(\varphi(t) + s\big)\cos\lambda + s\triangle h\sin\lambda$$

$$y(s,t) = fa - \sqrt{\Big(fa - r(t)\sin\big(\varphi(t) + s\big)\Big)^2 + \Big((x(s,t)\sin\lambda - s\triangle h)/\cos\lambda\Big)^2}$$

with

fa	...	distance of axes
λ	...	turn of axes
$\triangle h$	...	height of one spiral turn
$\{r(t), \varphi(t) \mid t \in T\}$	...	parametric form of the frontal intersection in polar coordinates (t = curve parameter)
$\{x(s,t), y(s,t) \mid t \in T, s \in \mathbb{R}\}$	...	parametric form of the set of all projected spiral turns, given in cartesian coordinates corresponding to the second profile view; t parameterizes the different points of the frontal intersection. For each fixed t, s is the curve parameter for this specific spiral turn.

Using the above formulas, we are able to construct any projected spiral turn for any given frontal intersection. The desired shape of the forming cutter is then the envelope of that set of curves. The tangency condition for the envelope leads to

$$x_t(s,t)y_s(s,t) = x_s(s,t)y_t(s,t),$$

which has to hold for all (s,t) such that the envelope touches the t-th curve at the point corresponding to the parameter value s. If these s can be written (as can be achieved by appropriate choices of the parameters for solvable problems) as a function $s = s(t)$, the envelope is given in parametric form by

$$\left\{ \Big(x\big(s(t), t\big), y\big(s(t), t\big) \Big) / t \in T \right\}.$$

For a practical problem of constructing a forming cutter, this condition is only necessary, but not sufficient. There may be more than one envelope, which all satisfy the above equation. Out of this set, we have to select that curve which is closest to the axis of the forming cutter, since this is the only one that will not destroy any part of the workpiece. A difficulty that we encounter in solving this problem is its ill–posedness in the sense that arbitrarily small changes in the "data", i.e., the frontal intersection, can change a intersection that causes a change in the shape of the projected spiral turns (depending on t) exceeding the change in the shape of the projected spiral turns depending on s. In this situation an envelope does no longer exist. The practical construction of such a perturbation runs as follows: Make a slot in the frontal intersection such that its width is small in relation to its depth. The resulting set of projected spiral turns will contain one or more hidden by others, depending on the ratio between width and depth of the slot. Thus, no envelope exists.

Now we turn our attention to the finite-dimensional problem. We are now looking for an envelope of finitely many projected spiral turns. To force uniqueness, we have to pick our envelope from a set of curves described by finitely many parameters. It is natural to take as this set the set of cubic splines with variable nodes, since this class of functions has some practically desirable properties (e.g. smoothness, small total curvature). We need variable nodes to fulfill the tangency conditions of an envelope, which arise in addition to the usual conditions for a cubic spline. We decide to use the nodes as the points where the envelope touches the curves.

Let $\{t_1, \ldots, t_n\}$ be a set of n parameters (with $t_1 < \ldots < t_n$) and $(r_i := r(t_i), \varphi_i := \varphi(t_i))_{i=1,\ldots,n}$ be the corresponding points of the frontal intersection, $x_i(s) := x(s, t_i)$, $y_i(s) := y(s, t_i)$ the resulting projected spiral turns according to the above formulas.

Considering the practical background, we use natural boundary conditions for the spline. If the spline envelope P is described by

$$P(x) = a_i(x - x_i)^3 + b_i(x - x_i)^2 + c_i(x - x_i) + d_i, \quad x_i \leq x < x_{i+1}$$

then the nodes $x_i = x_i(s_i)$ and the coefficients are described by

(1.1)
$$d_i = y_i(s_i) \qquad i = 1, \ldots, n$$

(1.2)
$$c_i \quad = \frac{y_{is}(s_i)}{x_{is}(s_i)} \qquad i = 1, \ldots, n$$

(1.3)
$$a_i h_i^3 + b_i h_i^2 + c_i h_i + d_i = y_{i+1}(s_{i+1}) \quad i = 1, \ldots, n-1$$

(1.4)
$$3a_i h_i^2 + 2b_i h_i + c_i \quad = c_{i+1} \qquad i = 1, \ldots, n-1$$

(1.5)
$$3a_i h_i + b_i \quad = b_{i+1} \qquad i = 1, \ldots, n-1$$

(1.6)
$$a_n = b_1 = b_n = 0$$

with $h_i := x_{i+1}(s_{i+1}) - x_i(s_i)$. This is a nonlinear system of $5n$ equations in the $5n$

of the nodes and the computation of the resulting spline. Knowing the values of $\bar{s} :=$ $(s_1,\ldots,s_n)$, i.e., the points of tangency, which are also the nodes, the cubic spline is uniquely determined by equations (1.1), (1.3), (1.4), (1.5) and (1.6). Since as usual one can write all other coefficients in terms of the c_i, one can view (1.1), (1.3), (1.4), (1.5), (1.6) as a system of equations in $s = (s_1,\ldots,s_n)$ and $c = (c_1,\ldots,c_n)$, which has the form

(1.7) $$Ac = z$$

with

$$z^T := 3\left(\frac{y_2 - y_1}{h_1}, \frac{y_3 - y_2}{h_2}, \frac{y_4 - y_3}{h_3}, \ldots, \frac{y_n - y_{n-1}}{h_{n-1}}, 0\right), y_i = y_i(s_i)$$

and

$$A := (a_{ij})_{i,j=1}^n \text{ with } a_{ij} = \left.\begin{cases} 0 & j < i \\ 1 & j = i \\ 2 + \frac{h_i}{h_j} & j = i+1 \\ (-1)^{n-2-j} h_i \left(\frac{1}{h_{j-1}} + \frac{1}{h_j}\right) & i+1 < j < n \\ (-1)^{n-i-1} \frac{1}{h_{j-1}} & j = n \end{cases}\right\} 1 \le i \le n-1$$

$$\text{and } a_{nj} = \begin{cases} \frac{1}{h_j} & j = 1 \\ (-1)^{j-1}\left(\frac{1}{h_{j-1}} + \frac{1}{h_j}\right) & 2 \le j \le n-1 \\ (-1)^{n-2} \frac{1}{h_{j-1}} & j = n. \end{cases}$$

Not only the coefficients of A, but also those of z depend on s via h_i and y_i. For $\bar{s}$ fixed, (1.7) is linear. We define

$$T : T^n \times \mathbb{R}^n \to \mathbb{R}^n$$

$$T(t,\bar{s}) := \left(c_i(\bar{s}) - \frac{y_{is}}{x_{is}}(\bar{s}_i)\right)_{i=1,\ldots,n},$$

where c is a solution to (1.7) for $\bar{s}$ and

$$T^n := \{(t_1,\ldots,t_n)/t_1 < t_2 < \ldots < t_n; \quad t_i \in T, \quad i = 1,\ldots,n\}.$$

With this definition of the function T, solving (1.1)–(1.6) is equivalent to solving the system of nonlinear equations

$$T(t,s) = 0, \qquad t \in T_n,$$

for $s \in \mathbb{R}^n$. Note that this equation comes from (1.2) (if $x_{is}(\bar{s}_i) \ne 0$, which is fulfilled in practically relevant situations), while the remaining equations are incorporated into the definition of the function T. Suppose we know for given t_0 such a solution $\bar{s}_0$ of $T(t_0,\bar{s}) = 0$, then the following Proposition is a direct consequence of the Implicit Function Theorem and the Strong Column-Sum Criterion for linear equations:

Proposition 1·1. *Let $\hat{s}$ such that $T(t, \hat{s}) = 0$ and for all $i \in \{1, \ldots, n\}$, let*

(1.8)
$$\left| \frac{y_{iss}(\hat{s}_i)x_{is}(\hat{s}_i) - y_{is}(\hat{s}_i)x_{iss}(\hat{s}_i)}{\left(x_{is}(\hat{s})\right)^2 \left(1 + \left(\frac{y_{is}(\hat{s}_i)}{x_{is}(\hat{s}_i)}\right)^2\right)^{3/2}} \right| > \| \nabla_{s_i} c(\hat{s}) \|_1.$$

Then $\det \left(\frac{\partial T}{\partial \bar{s}}(t, \hat{s}) \right) \neq 0$, i.e., T can be locally solved for $\bar{s}$, $\bar{s}$ is differentiable in t.

Remark 1.2. For a dual result to Proposition 1.1 see [1]. The above result is a stability result for changing the discretization of the frontal intersection. Condition (1.8) can be interpreted in the following way: We consider the solution spline being tangent to all n spiral turns in the tangential points $\left(x_i(\bar{s}_i), y_i(\bar{s}_i)\right)$. Changing the tangential point for one spiral turn causes a change in the interpolating spline (with the tangency conditions omitted) even if all other tangential points are kept fixed. Now, the sum of all absolute values of all changes in the slopes of these interpolating spline in the tangential points kept fixed has to be smaller than the absolute value of the curvature of that spiral turn (in the original contact point) of which the tangential point has been modified. This criterion has to be fulfilled for all spiral turns.

Summing up, we see that fulfilling the requirements of Proposition 1.1 enables us to change the discretization in t, to compute the resulting changes in the curve parameters s, which fix the contact points, and thus we can estimate the change in the spline coefficients using equation (1.7) in combination with the Perturbation Lemma. In this sense, the problem of computing our spline envelope is stable with respect to changes in the discretization of the frontal intersection.

Remark 1.3. Similar methods can be used to analyse the consequences of a perturbation in the frontal intersection itself, i.e., if for fixed t the values of the corresponding points of the frontal intersection $(r(t), \varphi(t))$ are perturbed. By the same arguments also the combined effect of changes in the discretization and in the frontal intersection itself can be estimated (see [1]).

We now turn to the question if, using the traditional method, data errors propagate locally or globally. A data error which is non-zero only in a small subinterval for the parameter t parameterizing the frontal intersection, the extremal case being that where only one point of the frontal intersection is changed, will be called "local". We will call the resulting change of the spline envelope "global", if at least one coefficient in each polynomial describing the spline is changed. Using these definitions, it can be shown that for each solvable, discrete problem there exists a local data error in the frontal intersection that leads to a global change in the corresponding spline envelope; for details see [1]. In our opinion this is the major drawback of the traditional method in its suggested implementation.

Remark 1.4.   To illustrate another drawback, we consider the problem of finding an envelope for the set of identical parabolas of the form $y = (x - m)^2$, $m \in I\!N$, with the natural requirement that the spline envelope should be symmetric with respect to 0. It is clear that the solution is the x-axis. Elevating the central parabola $y = x^2$ by an arbitrarily small distance, which is a symmetric change, leads to a problem that does not have a symmetric solution any longer. Thus we see that a symmetrically perturbed symmetric problem with symmetric solution need not have a symmetric solution, which is certainly strange.

Maybe the problems outlined in Remarks 1.3 and 1.4 are caused by our selection of cubic splines for the envelopes; other choices of functions might avoid these problems. However, no such choice seems to be readily available.

Summing up, we see that the traditional method works well on the drawing board, but has disadvantages in its numerical implementation using cubic spline functions. For this reason we develop a new method, which is in a certain sense equivalent to the traditional method without having its disadvantages. This new method will be described in Section 2.

2. The New Method

Let us consider the original situation, where we have a given workpiece, defined by its frontal intersection and its height of one spiral turn as explained in Section 1. In the distance fa, there is the axis of the desired forming cutter forming the angle λ (turn of axes) with the axis of the workpiece. The forming cutter itself can be thought of as being created by a rotation of its shape around its axis; it is rotationally symmetric with respect to its own axis.

Our problem is to find a forming cutter that is tangent to all spiral turns of the workpiece without intersecting any of them. In $I\!R^3$, this characterization of the shape of the forming cutter can be formulated as follows: The rotationally symmetric forming cutter must be tangent to the given workpiece with all points of its shape, but is must not cut through it.

These considerations lead to the new construction principle:

For computing one distinct point of the shape of the forming cutter, which is fixed by its distance to the axis of the forming cutter, we take a plane orthogonal to the axis of the cutter and containing that point. We construct the figure of intersection of that plane with the given workpiece. The requirement that the cutter has to be tangent to the workpiece results in the desired radius of the cutter as the minimum distance between the axis of the cutter and the above figure of intersection. Repeating this process along the whole axis of the cutter gives the desired shape of the forming cutter.

Based on these considerations, the problem of finding a point of the shape of the cutter reduces to computing the minimum distance between a point and a curve, namely the figure of intersection of a plane with the workpiece, which can be computed without much computational effort. Note that we only need to compute the value of the minimum distance, but not the point where it is achieved.

In contrast to the traditional method described in Section 1, we need not calculate an envelope. The results of this new method are points of the shape of the forming cutter itself. This new method can also be used for constructing a solution on the drawing board.

We describe the new method in a more algorithmic way:

New Method (NV):

i Put a plane orthogonal to the axis of the cutter through a point xfr of the axis of the cutter

ii Compute the points of intersection of all spiral turns with that orthogonal plane (i.e.,

iii Compute the minimum distance between the axis of the cutter and all points of intersection; the result is the radius yfr of the cutter for the point xfr of the axis of the cutter

iv Repeat (i)—(iii) for all points xfr of the axis of the cutter to get the shape $yfr(xfr)$ of the cutter.

Using this definition of (NV) and the above notation we can show:

Proposition 2·1. *For a solvable continuous problem, method (NV) has the same solution as the traditional method, i.e., the envelope of the set of projected spiral turns as described in Section 1.*

Sketch of the proof (for details see [1]):

The surface of the workpiece is determined by the set of all spiral turns:

$$W := \{(x,y,z)/x = r(t)\cos\big(\varphi(t) + s\big), y = r(t)\sin\big(\varphi(t) + s\big), z = s\triangle h,$$

$$\text{for} \quad s \in I\!R \quad \text{and} \quad t \in T\}.$$

The axis of the cutter is given by

$$F := \{(x, fa, z)/z = x\tan\lambda\}.$$

Next we compute for an arbitrary point $P(x,y,z)$ of an arbitrary spiral turn the distance $\tilde{x}$ of the ground projectional plane, orthogonal to the cutter axis and containing that point, to the origin. We also compute its distance to the cutter axis.

Now we compute all points of all spiral turns having the same value $\tilde{x}$ and take the minimum over their distances to the cutter axis. Thus we get the radius of the cutter for $\tilde{x}$. Let Q be that point of the cutter axis with distance $\tilde{x}$ to the origin. Using the formulas of Section 1 and the following figure, we obtain

$$\tilde{x}(s,t) = s\triangle h \sin\lambda + r(t)\cos\big(\varphi(t) + s\big)\cos\lambda$$

and

$$dist(P,Q) = \sqrt{\Big(fa - r(t)\sin\big(\varphi(t) + s\big)\Big)^2 + \left(\frac{\tilde{x}(s,t)\sin\lambda - s\triangle h}{\cos\lambda}\right)^2}$$

Thus we get the radius yfr of the cutter via

$$yfr(xfr) = \min_{\{(s,t)/\tilde{x}(s,t)=xfr\}} dist(P,Q).$$

Comparing the formula for $dist(P,Q)$ and for $\big(x(s,t),y(s,t)\big)$ in Section 1, this is equivalent to

$$yfr(xfr) = \min_{\{(s,t)/x(s,t)=xfr\}} \big(fa - y(s,t)\big).$$

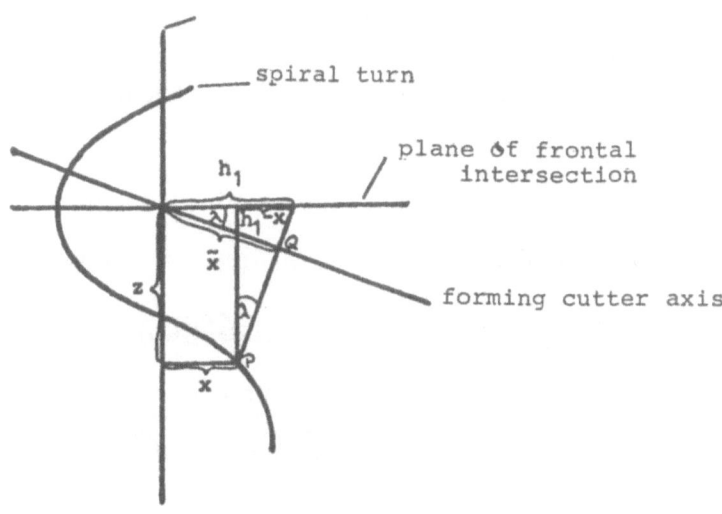

Using the Lagrange function for this problem, we see that $x_s y_t - x_t y_s = 0$ is a necessary condition for that minimum problem. Thus we can conclude that the result of our new method (NV) is always an envelope to the set of all spiral turns. This condition is sufficient, too, when we restrict ourselves to solutions that make practical sense (i.e., the envelope with smallest distance to the cutter axis), as we already did in the last section.

$\Diamond$

Remark 2.2. As indicated in Section 1, for the traditional method, it is possible to construct small modifications of the frontal intersection that result in an unsolvable problem. In contrast, (NV) has always a solution, though it does not change an unsolvable problem into a solvable one. The "solution" generated by (NV) for an unsolvable problem is that shape of a forming cutter which mills as many spiral turns correctly as possible, without intersecting any spiral turn. For solvable problems we get the desired solution, for unsolvable ones we get a good approximation that does not mill away any desired parts of the workpiece. Already this feature represents a major advantage of (NV) over the traditional method.

Next we consider the influence of errors in the frontal intersection on the resulting shape of the cutter. For this purpose, let E be a set of feasible error parameters η and $\bigl(r(t,\eta), \varphi(t,\eta)\bigr)$ the corresponding set of frontal intersections. For $\eta = 0$ we get the unperturbed frontal intersection.

Proposition 2·3. *Let* $r(t,\eta)$, $\varphi(t,\eta)$ *be continuous in both arguments and the turn of axes* $\lambda > 0$, *then we have for every point* xfr *of the cutter axis that the radius* $yfr(xfr,\eta)$ *of the cutter of the perturbed problem converges for* $\eta \to 0$ *to the exact radius* $yfr_0(xfr)$ *of the cutter.*

Proof:

Similar to the proof of Proposition 2.1 we describe the surface of the workpiece by

$$W(s,t,\eta) := \{(x,y,z)/x = r(t,\eta)\cos(\varphi(t,\eta)+s),$$

$$y = r(t,\eta)\sin(\varphi(t,\eta)+s),$$

$$z = s\triangle h, s \in \mathbb{R}, t \in T, \eta \in E\}.$$

Now we substitute s by xfr via the relation

$$s\triangle h = xfr \sin \lambda,$$

which is always uniquely determined, since $\lambda > 0$. Thus we get the surface of the workpiece equivalently by

$$\widetilde{W}(xfr,t,\eta) := \{(x,y,z)/x = r(t,\eta)\cos(\varphi(t,\eta)+\frac{\sin\lambda}{\triangle h}xfr),$$

$$y = r(t,\eta)\sin(\varphi(t,\eta)+\frac{\sin\lambda}{\triangle h}xfr),$$

$$z = xfr \sin \lambda, xfr \in \mathbb{R}, t \in T, \eta \in E\}.$$

Now, let Q be an arbitrary, but fixed, point of the cutter axis with distance xfr to the origin. The resulting radius of the cutter can be computed by

$$yfr(xfr_0,\eta) = \inf_{t\in T} dist(\widetilde{W}(xfr_0,t,\eta),Q).$$

According to the above assumptions, $dist(\widetilde{W}(xfr_0,t,\eta),Q)$ is continuous in η and hence, as the infimum of continuous functions, so is $yfr(xfr_0,\eta)$.

$\diamond$

This result is only a qualitative statement, but in the following proposition we get error bounds for the technically relevant special case of "over-milling" (i.e, the turn of axes is greater than the twist angle):

Proposition 2·4. *Let $(\tilde{r}, \tilde{\varphi})$ be a perturbed version of the exact frontal intersection (r, φ) with the property that for all $t \in T$, $dist\left((\tilde{r}(t), \tilde{\varphi}(t)), (r(t), \varphi(t))\right) < \epsilon$. Then we have for the case of over-milling (i.e., $\lambda > \arctan\left(\frac{R}{\Delta h}\right)$, R = radius of the workpiece) that*

$$|\widetilde{yfr}(xfr) - yfr(xfr)| \leq C \cdot \epsilon \quad \text{for all} \quad xfr,$$

where the value of C comes out of the proof; it depends on the height of one spiral turn, the radius of the workpiece, the turn of axes and the twist angle, but not on ϵ.

Sketch of the proof (for details see [1]):

Every point of the perturbed frontal intersection is at most at a distance of ϵ away from the corresponding point of the exact frontal intersection. We choose one arbitrary point and fix it for the sequel. Taking all points closer to the exact point than ϵ and twisting this set according to the height of one spiral turn results in an "ϵ-tube" around the exact spiral turn, which contains all spiral turns resulting from points fulfilling our assumptions.

We try to get bounds for the dimensions of the figure of intersection between this ϵ-tube and a ground projectional plane orthogonal to the cutter axis. Using these bounds for all spiral turns intersecting that plane, we get bounds for the error in the radius of the cutter, which is the minimum distance of the cutter axis to the figure of intersection with the ground projectional plane orthogonal to the cutter axis.

The following figure illustrates our situation:

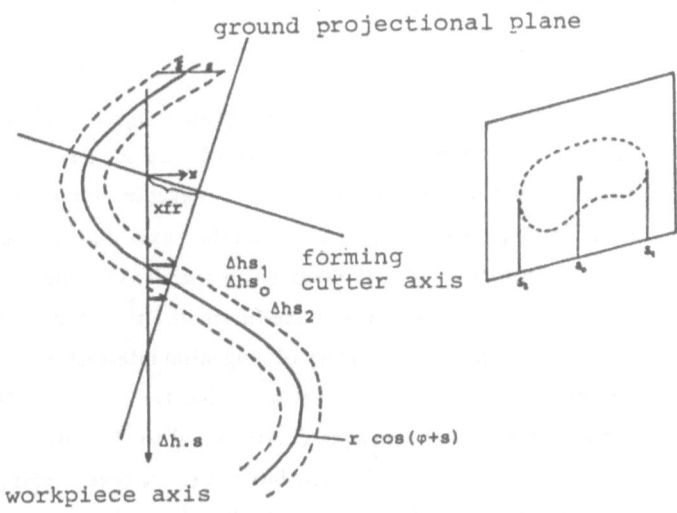

First we treat the exact spiral turn. By s_o, we denote the parameter which characterizes its intersection with the ground projectional plane, fulfilling the equation

$$\cos(\varphi + s) = - \underbrace{\frac{\triangle h \tan \lambda}{r}}_{=:k} s + \underbrace{\frac{xfr}{r \cos \lambda}}_{=:d} .$$

According to our assumptions this equation has always exactly one solution $\left(1 < |\frac{\triangle h \tan \lambda}{R}| \le |\frac{\triangle h \tan \lambda}{r}| = |k|\right)$.

Further computations lead to values s_1 and s_2 for the largest possible distance of the figure of intersection of the ϵ- tube from its center s_0 in direction of s:

$$s_1 = s_0 - \frac{\epsilon}{|k+1|}, \qquad s_2 = s_0 + \frac{\epsilon}{|k+1|} .$$

These values can be used to find the corresponding bounds for the x- coordinates in the normal representation of the spiral turns. Yet what we need are bounds for the y-coordinates of that figure of intersection which we get via a continuity argument by

$$x(s_2) > x > x(s_1) \Rightarrow y(x) \in \left[r \sin(\varphi + s_0) - \epsilon - r \frac{\triangle h \epsilon}{|k+1|}, r \sin(\varphi + s_0) + \epsilon + r \frac{\triangle h \epsilon}{|k+1|}\right] .$$

Summing up, we can bound the error, caused by one perturbed spiral turn, in the radius of the cutter by $C(r)\epsilon$, where $C(r) = \sqrt{|\frac{\triangle h \tan \lambda}{|k+1|}|^2 + |1 + \frac{r \triangle h}{|k+1|}|^2}$.

Considering errors in all points of the frontal intersection, fulfilling our assumptions, we have to replace r by R (since if $r \le R$, then $C(r) \le C(R)$) and get with $C := C(R)$ that $|\widetilde{yfr} - yfr| \le C \cdot \epsilon$.

$\diamond$

Remark 2.5. For $|k| \to 1$ the above bounds are getting larger, which can be interpreted in the following way: $|k| \to 1$ means that the turn of axes converges to the twist angle. If both are identical, there will be the possibility of tangential intersections of an ϵ-tube with ground projectional planes orthogonal to the cutter axis. In that case, contraction of the ϵ-tube for $\epsilon \to 0$ can happen at arbitrarily small speeds. One has to expect the same troubles for the situation where the turn of axes is smaller than the twist angle ("under-milling"), because there is also the possibility of tangential intersections. In the latter case another problem may arise. It is possible that one spiral turn has more than one point of intersection with a ground projectional plane orthogonal to the cutter axis. Thus it can happen that parts out of an ϵ-tube intersect, but the exact central spiral turn does not. This may influence the value of the radius of the forming cutter, because the minimum distance is taken for a point of this figure. With ϵ getting smaller, this additional figure of intersection vanishes, the value of the minimum distance may have a discontinuity, and bounds like in Proposition 2.4 cannot be found.

In Proposition 2.3 and 2.4 we get pointwise bounds for the shape of the cutter from uniform bounds for the frontal intersection. Thus we have stability, but only in a weak form.

It is obvious from the local character of (NV) that by using (NV), local erros in the data (frontal intersection) lead only to local errors in the result (shape of the forming cutter), a fact that holds also for the finite-dimensional problem, which we will treat now.

According to the above, using (NV) for the discrete problem results in solving one minimum problem over a finite number of spiral turns for every point of the shape of the forming cutter. The resulting envelope runs always along that projected spiral turn that lies nearest to the cutter axis. This procedure gives poor results for coarse discretization, but the computations are inexpensive and thus one can take a very fine discretization, which results in a higher quality of the envelope.

There are other possibilities to use (NV), which result in even higher quality of the resulting envelope. For example, interpolation of the discretized frontal intersection by splines gives as many spiral turns as needed for a smooth solution and exact computation of the minimum distances. It is obvious that for an increasing number of spiral turns (i.e., given points of the frontal intersection) the quality of the result gets better, because the solution of the discrete problem converges to the solution of the continuous one.

The conclusion is that our method (NV) has the following advantages over the traditional method:

- stability with respect to perturbations in the data (frontal intersection)
- local errors in the data cause only local errors in the solution
- computation much easier and less expensive
- unique solution for the continuous problem without further assumptions
- computation of projected spiral turns not necessary; in the numerical examples, we computed them only for judging the quality of the results from the envelope property of the solution.

3. Numerical Realization

The basic setup of our numerical method is the following: We interpolate the discretized frontal intersection by a cubic spline. Then we discretize the cutter axis equidistantly and put ground projectional planes orthogonal to the cutter axis through each of these points. For each such plane, we compute the figure of intersection with the workpiece by computing points of intersection between that plane and spiral turns. By means of the cubic spline we are able to compute this figure of intersection very accurately around the point with minimum distance to the cutter axis. That procedure enhances our results.

Using continuity arguments for choosing starting values for the minimization, we could reduce the computational effort by a factor of 40; for details see [1]. Considering practical requirements such as lower and upper bounds for the value of the screw-in angle reduces the effort even further.

A practically important characterization of the desired forming cutter is the shaft slot it causes in the workpiece. By "shaft slot" we mean that part of the workpiece which is between the correctly milled part and the cylindrical shaft. Since it is useless for practical purposes, it is clear that the height of this shaft slot should be as small as possible.

Our computer program based on (NV) achieves this aim by computing for a given frontal intersection that forming cutter, which causes the smallest shaft slot. This optimization uses the turn of axes and the screw-in angle as control variables.

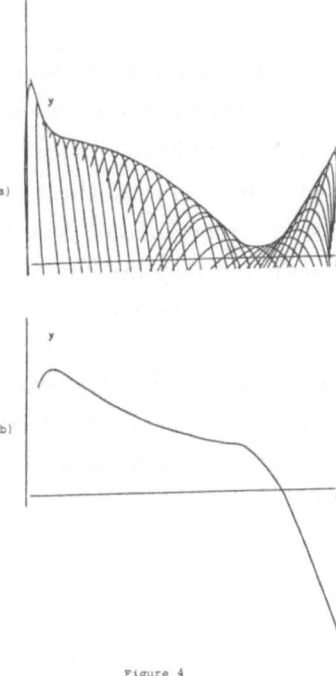

a)

b)

Figure 4

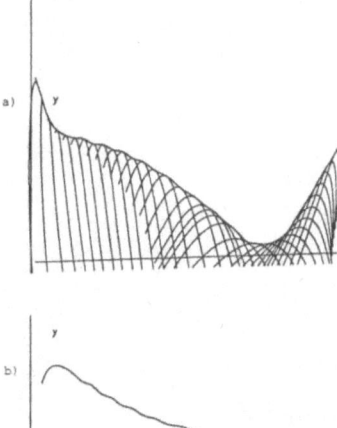

a)

b)

4. Numerical Results

We present two examples, both based on the same frontal intersection, computed with (NV). Fig. 3 shows that frontal intersection which was taken as data for our first example. For the second example, some points of that frontal intersection have been perturbed by changing the second coordinate by $\pm$ 0.5 mm in an alternating way. Fig. 4a shows the resulting shape of the forming cutter for unperturbed data, Fig. 5a for perturbed data. Comparing these figures we can easily recognize the strong influence of small changes in the data on the resulting shape of the forming cutter.

In both figures, the set of projected spiral turns is only drawn for the purpose of checking the result by the quality of the envelope; note that these curves were not required for the computation of the forming cutter, since (NV) does not use the envelope property. Fig. 4b and Fig. 5b show the shaft slots for both examples.

Reference:

[1] Langthaler, T.: CAD von Formfräsern, Dissertation, Universität Linz, 1988

Authors' addresses:

Prof.Dr. Heinz W. Engl
Institut für Mathematik
Johannes-Kepler-Universität
A-4040   Linz, Austria
Dr. Thomas Langthaler
Technologietransfer Linz
Kammer der gewerblichen Wirtschaft
Wiener Straße 131 (LIG)
A-4020   Linz, Austria

State variables feedback control of stepping motors with flexible shaft

R. FIROOZIAN, M.Sc., Ph.D. and J. G. BAJIR, B.Eng.

Summary: In this paper the authors present an active control algorithm for position control of stepping servo motors with flexible transmission shaft. A non-linear mathematical model is derived to describe the dynamic behaviour of the system. To implement the full state variables active control technique, the non-linear model is first linearised. It is shown that the full state variables could be used in a closed loop controller to introduce damping in all eigenvalues of the system. Furthermore the effectiveness of linear control algorithm on the non-linear system is studied.

Nomenclature

A , B	the plant matrix (5x5) and the input vector (5x1)
$\bar{c}_m$, $\bar{c}_\ell$	effective damping coefficients of motor and load
i	stator current at each phase
J_ℓ, J_m	load and rotor inertia
K_ϑ, K_t, K_θ	the torque constant of the motor and the coefficients of linearised equation of the torque
K_S	the stiffness of flexible shaft
L, R	the inductance and resistance of stator winding
r	number of teeth on rotor
s	the Laplace operator
T_m, T_f, T_ℓ	torque on rotor, shaft and load respectively
u_i	the ith eigenvector (5x1)
$\bar{u}$	the matrix of eigenvectors (5x5)
$\bar{v}_i$	input voltage to motor
w	the system controllability vector (5x1)
$\bar{x}_i$	the ith state variable
x	the vector of state variables (5x1)
$\bar{\theta}_m$, θ_0	angular rotation of rotor and load
$\wedge_i$	the ith eigenvalue of the system
ρ_i	desired value of the ith eigenvalue
$(^T)$	represents the transpose of a matrix
$(\cdot)$	represents the derivative with respect to time
F	the input force

Introduction

The trend in automation technology is for lightweight servo position control systems with increased demand on reliability and performance. Traditionally the position control is achieved by monitoring the angular position of servo motor for feedback with assumption that the link between the motor and load is rigid. This assumption is no longer valid when the design is based on optimised strength and weight. There is a need,

333

J. Manley et al. (eds.), Proceedings of the Third European Conference on Mathematics in Industry, 333–341.
© 1990 Kluwer Academic Publishers and B. G. Teubner Stuttgart.

therefore, to develop a control strategy where the position control is achieved with minimum vibration and inaccuracy due to the flexibility of the transmission mechanism and with maximum speed of response.

In the present work the application of active control technique for hybrid type stepping motors with flexible transmission shaft is studied. Stepping motors of hybrid type (PM) have become predominant in many position control application that require incremental motion control. A non-linear mathematical model was presented by Pollack [1], for the PM step motor to verify the explanation of resonances and sudden loss of damping. He concluded that resonance rather than being a sudden loss of torque, is actually a very sudden loss of dynamic damping. Various linear/non-linear mathematical models to predict the dynamic behaviour of stepping motors, have been proposed by research workers in this field [2-3]. The aim has been to explain the relationships between the torque and the applied current rather than to obtain an overall control strategy. The optimum performance of servo motors using position, velocity and acceleration feedback employing classical control technique was studied by Firoozian et al [4-5]. It was shown that with flexible transmission shaft a compromise between speed of response, accuracy and stability has to be made. Such a position control systems would have two sets of eigenvalues which correspond to closed loop and structures respectively. The classical feedback control theory provides an easy method of introducing damping into the closed loop eigenvalues. The method, however, fails to provide sufficient damping on the eigenvalues due to the flexibility of the structures. The active control method provides the means of assigning an arbitrary damping to all eigenvalues of the system. The state variables are fed in a controller, with predefined gains, to provide a single control force.

Theoretical model

A schematic diagram of position control systems with flexible shaft is shown in Fig. 1. The system consists of a servo motor (stepping motor), shaft, an equivalent load inertia, position transducer and a controller. To describe the dynamic characteristic of the system, the governing differential equation for each component is first derived. The torque equation of the motor and load using the Laplace operator (s) may be written as:

$$T_m = J_m s^2 \theta_m + c_m s \theta_m + T_f \tag{1}$$

$$T_f = K_s (\theta_m - \theta_0) \tag{2}$$

$$T_f = J_\ell s^2 \theta_0 + c_\ell s \theta_0 + T_\ell \tag{3}$$

The torque developed by stepping motor with a current i in the phase winding can be derived by equation [1]:

$$T_m = K_a i \sin (r \theta_m) \tag{4}$$

Equation (4) shows that there is an internal position feedback in stepping motors. The voltage equation of each phase of the stepping motor can be simplified as:

$$V_i = Ri + L s i \tag{5}$$

Equations (1) to (5) describes the overall dynamic characteristics of stepping servo motors in general form. The system is non-linear because of the term $\sin(r\theta_m)$ in equation (4). The application of active control technique requires a set of linear equations which describes the system. To implement this, equation (4) is linearised about a steady state point as:

$$T_m = K_f i - K_\theta \theta_m \text{ where: } K_t = \frac{\partial T_m}{\partial i}\Big|_{\theta = \text{const.}} \text{ and } K_\theta = -\frac{\partial T_m}{\partial \theta_i}\Big|_{i = \text{const.}} \tag{6}$$

The matrix equation of the linearised model can be obtained by using equation (6) instead of equation (4), i.e:

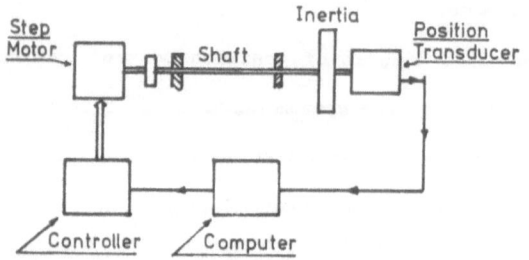

Fig. 1.   A Schematic diagram of a position control
system using a stepping motor

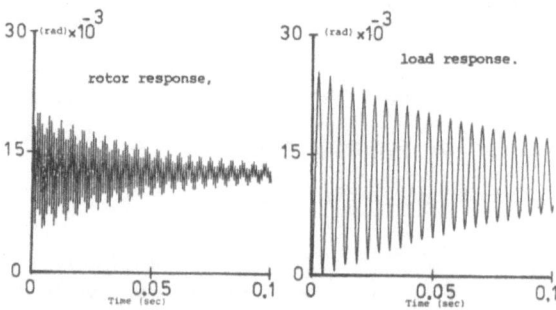

Fig. 2.   System response to a step input of 0.72° rotation (non-linear
model).

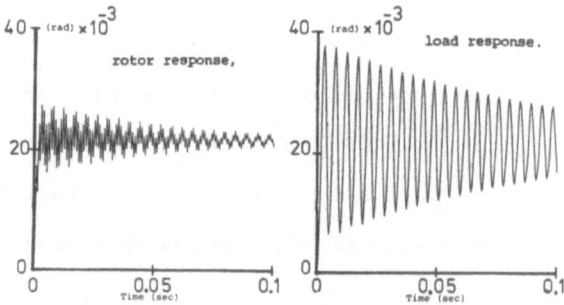

Fig. 3.   System response to a step input of 0.72° rotation (linear
model).

$\dot{\underline{x}} = \underline{A} \, \underline{x} + \underline{B} \, \underline{F}$ where, $x_1 = \theta_o$, $x_2 = S\theta_o$, $x_3 = \theta_m$, $x_4 = S\theta_m$ and $x_5 = i$ (7)

System equation (7) has 5 state variables, an input voltage (v_i) and external torque input (T_ℓ). Since the matrix equation (7) is linear the principal of superposition can be used to obtain a solution for an input function of v_i by setting T_ℓ to zero. Matrix equation (7) can now be written in standard single input state space form as:

$$\underline{x} = \underline{A} \, \underline{x} + \underline{B} \, v_i' \tag{8}$$

The eigenvalues of the system λ_i ($i = 1, 5$) have two components of real and imaginary parts. The value of the real part in relation to the imaginary part represent the stability and damping ratio of each mode of oscillation. The objective is to construct a control signal (v_i) in such a way to shift the eigenvalues of the plant matrix '$\underline{A}$' to a desired value 'ρ' where the system becomes stable with damping ratio of at least 0.7. Porter [6] has shown that the control signal (v_i) which shifts all eigenvalues λ_i to a desired value ρ_i can be obtained as:

$$v_i' = \left[\sum_{J=1}^{5} K_j \, \underline{u}_j^T \right] \underline{x} \tag{9}$$

where K_j is the controller gain and may be obtained by [6]

$$K_j = \frac{\prod\limits_{K=1}^{5} (\rho_K - \lambda_j)}{W_j \prod\limits_{\substack{K \neq j \\ K = 1}}^{5} (\lambda_K - \lambda_j)} \qquad j = 1, .., 5 \tag{10}$$

W_j in equation (10) is the element of the controllability vector and $\underline{U}$ is the matrix of eigenvectors. If W_j in equation (10) is zero, then it represents an uncontrollable system with respect to the control signal v_i. Equations (9) and (10) are the active control law to shift all eigenvalues of the plant matrix $\underline{A}$ to a desired value. Replacing for (v_j) in equation (8) from equation (9), the matrix equation of the closed loop system becomes:

$$\underline{\dot{x}} = \left\{ \underline{A} + \underline{B} \sum_{j=1}^{5} K_j \underline{u}^T_j \right\} \underline{x} \tag{11}$$

To implement the closed loop control law of equation (11), all the state variables of the system ($\underline{x}$) must be measured experimentally.

Numerical Results

To illustrate the suitability of active control strategy a numerical study on the performance of stepping motor is carried out. A 1 kW 5-phase stepping motor is connected to the load by a continuous shaft which has a stiffness of 50 Nm/rad. The rotor and load have an equivalent inertia of 1.2×10^{-3} and 2.4×10^{-3} (kg.m^2). The stepping motor has a resolution of 0.72^0 step angle with 50 teeth on the rotor. The resistance and inductance of each phase are 0.37 (Ω) and 0.25×10^{-3} (henry) with a torque constant of 2 (Nm/amps). The internal damping coefficients (C_s and C_g) were assumed to be negligible. The non-linear model of the system (equation (1) to (5)) is first solved using the numerical integration technique of Runge-Kutta[7]. The response of the rotor and load for a step input of v_j is shown in Fig. 2. The transient response shows two distinct oscillations corresponding to structural and magnetic resonance frequencies. This is the major problem with flexible shaft where the internal damping is negligible. The linearized model (equation (8)) is also solved for a similar input function (Fig. 3). It shows that the linear model predicts a similar response to that of non-linear model. There is, however, a

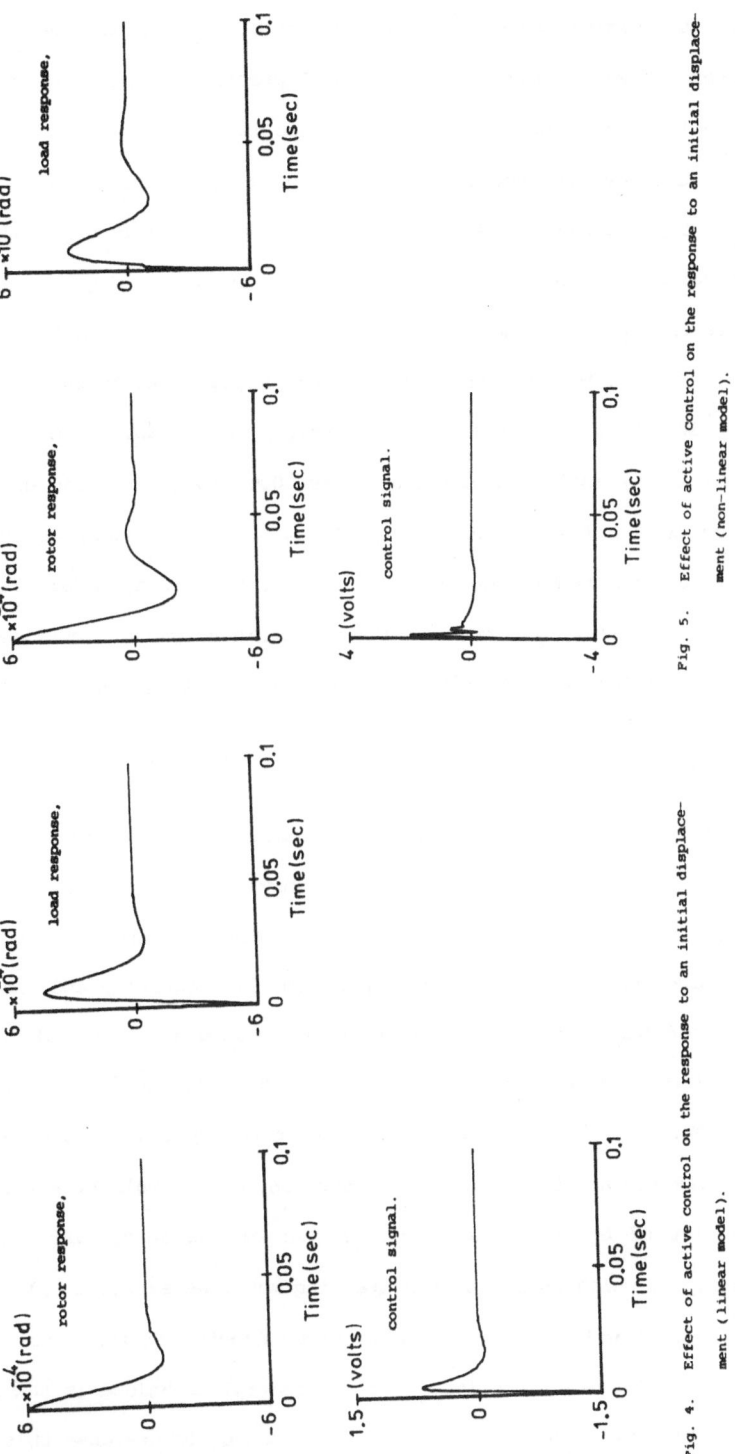

Fig. 4. Effect of active control on the response to an initial displacement (linear model).

Fig. 5. Effect of active control on the response to an initial displacement (non-linear model).

considerable difference on the amplitude of oscillation on the response of the rotor. Nevertheless the linear model predict the trend of oscillation which could be used in the design of controller.

To eliminate the undesirable oscillation the control law of equation (11) is implemented on the linear model. The desired eigenvalues were chosen arbitrary to have a damping ratio of 0.7. Fig. 4. shows the response of the rotor and load inertia together with the required control signal. It can be seen that the control signal introduces the required damping in all eigenvalues of the system. The response of Fig. 4. is obtained to an initial condition of +0.0006 (rad) on rotor and −0.0006 (rad) on load of the motor. Fig. 5. shows the response of the system when the active control law is applied to a non-linear model. This shows that although the active control law is obtained from a linearised model, it is also effective on non-linear models representing the real system.

Discussion and conclusion

A non-linear mathematical model is obtained to describe the dynamic characteristic of stepping motors. To apply the active control technique the non-linear model is linearised about a steady state operating point. It is shown that the linear model provides an accurate prediction of the response of the system. The model is also used to construct an active control technique to introduce sufficient damping in all eigenvalues of the system. If all the state variables of the system are available from direct measurement, then the gain of the controller could be set using the control law of equation (10). The output of the controller would be a single signal which is fed to the stepping motor as an input voltage. This control signal will introduce the required damping in the eigenvalues of the system. The major problem of active control technique is the require- ment of state variables from direct measurements. To overcome this problem an estimator could be used to predict the state variables from a single measurement. Initial work has shown that the estimator can effectively predict the five state variables of the system from the measurement on the angular rotation of the load inertia.

Further work is in progress to study the real time application of observer based active control strategy. A test rig is being set up consisting of a stepping motor, load inertia and a flexible shaft. The active control technique is particularly suitable where the parameters of the system remain constant through the operations. The applicability of the active control when the parameters such as the load inertia (specifically for robotic application) is being studied.

References

[1] Pollack, S. H., "On stability characteristic of PM step motors", 7th annual symposium incremental motion control systems and devices. University of Illinois, May 1978.

[2] Johnson, R. C. and Justice, M., "Mathematical model of a hybrid stepper motor and drive circuit", Proc. 4th annual symposium incremental motion control systems and devices, April 1-3, 1975.

[3] Singh, G. Leenhouts, A. C. and Kaplan, M. "Accuracy consideration in step motor systems", 7th annual symposium incremental motion control systems and devices. University of Illinois, May 1978.

[4] Firoozian, R.,Brimson, J. R. and Foster, K., "The comparison of performance of servo feed-drive systems", Proc. I.Mech.E. Conf. on electric versus hydraulics drives, Mechanical Engineering Publications Ltd., London, 27 October 1983, pp. 41-50.

[5] Firoozian, R. and Foster, K., "The choice of a servo motor for a specific application". Proc. I.Mech.E. Conf. on electric versus hydraulics versus pneumatics, Mechanical Engineering Publications Ltd., London, 22 October 1985, pp. 67-76.

[6] Porter, B. and Crossley, T. R., "Modal control, theory and applications", Taylor and Francis Ltd., 10-14 Macklin Street, London, 1972.

[7] NAG, Numerical Algorithm Group Version 11, IBM Corporation, Oxford University, Nov. 1983.

Department of Mechanical and Process Engineering,
The University of Sheffield,
Mappin Street, Sheffield, S1 3JD

Application of Mathematics to Heat Processing in the Meat Industry

Fulton, G.S., Burfoot, D., James, S.J. and Bailey, C.

INTRODUCTION

Food processing companies, equipment designers and manufacturers are increasingly using mathematical models in the design, optimisation, 'scale-up' and comparison of the many processing operations used throughout the food industry. A number of the models have been developed at the Institute of Food Research - Bristol Laboratory (IFR-BL) specifically to study heat and mass transfer processes in the meat industry.

When consumers purchase fresh meat or meat products from their butcher or supermarket, few think about the long, complicated chain of processing operations that are required to produce a high quality product at an economic price. Many of the complications arise because meat is an inhomogeneous biological material, ie made up of fat, lean and bone, and irregular in shape. Quality and microbiological changes, and the properties of all three components are very temperature dependent, which introduces further complications in attempts to model any of the processing operations that occur in the distribution system. Four of these operations are considered in this paper.

(1) Primary chilling - Immediately after slaughter, the carcass is hot (approximately 39°C) and the surface wet. Cooling is required to reduce both the rate of evaporative weight loss and bacterial growth and transform the soft musculature into a material which is suitable for boning or jointing. Primary chilling is therefore an operation where the principal requirement is to remove heat with minimum mass transfer.

(2) Retail display - After chilling, conditioning, storage and transport the majority of fresh meat is sold from retail display cabinets. In retail display, refrigeration is used to maintain the temperature of the pre-cooled meat and, with unpackaged products, to minimise surface drying.

(3) Convection cooking - Very little meat is consumed in a raw uncooked state and meat joints are usually cooked in hot air convection ovens. Heat is transferred into the meat to raise the internal temperature to the minimum required to meet microbial safety considerations and effect the protein denaturation and chemical changes required for development of flavour and tenderness. These objectives are confounded by the need to limit cooking losses which can be more than 40% of the uncooked weight.

(4) Pressure cooking and vacuum cooling - Most low quality meats present in fillings for pies and pasties and for ready meals are cooked in liquids under pressure to reduce cooking times. To limit overcooking and increase throughput, rapid cooling rates are achieved by a vacuum cooling process.

In these, and many other meat processing operations, the different systems available, the range of environmental conditions that can be used and the considerable variation in the properties of the raw material make it very difficult to optimise the process using a series of experimental trials.

The development of a mathematical model of the process can reduce the extent of experimental investigation, extend experimental results to produce process design data charts, produce a better understanding of the mechanisms underlying the process and suggest new/novel approaches. A general model for heat and mass transfer within and from a typical meat/meat product is described below together with particular examples of modifications carried out to meet the requirements of the four processes.

J. Manley et al. (eds.), Proceedings of the Third European Conference on Mathematics in Industry, 343–351.
© 1990 Kluwer Academic Publishers and B. G. Teubner Stuttgart.

MATHEMATICAL MODEL

The transfer of heat and mass within a material with positional and temperature dependent properties is described by the Fourier equations:

$$\rho\, Cp\, \frac{\partial T}{\partial t} = \frac{\partial}{\partial x}\left(k\, \frac{\partial T}{\partial x}\right) + \frac{\partial}{\partial y}\left(k\, \frac{\partial T}{\partial y}\right) + \frac{\partial}{\partial z}\left(k\, \frac{\partial T}{\partial z}\right) \text{ for heat transfer} \qquad [1]$$

$$\frac{\partial c}{\partial t} = \frac{\partial}{\partial x}\left(D\, \frac{\partial c}{\partial x}\right) + \frac{\partial}{\partial y}\left(D\, \frac{\partial c}{\partial y}\right) + \frac{\partial}{\partial z}\left(D\, \frac{\partial c}{\partial z}\right) \text{ for mass transfer} \qquad [2]$$

Explicit finite difference approximations can be used to solve these equations to determine the enthalpy, temperature and moisture distributions within the material when subjected to the following boundary conditions:

$$t = 0;\ x < X;\ y < Y;\ z < Z;\ T = T_i;\ c = c_i \qquad [3]$$

$$t > 0;\ x = 0;\ \frac{\partial T}{\partial x} = 0 \qquad [4]$$

$$y = 0;\ \frac{\partial T}{\partial y} = 0$$

$$z = 0;\ \frac{\partial T}{\partial z} = 0$$

$$t > 0;\ x = X;\ k\, \frac{\partial T}{\partial x} = h\, (T_a - T_s) + h_m\, (p_a - \alpha p_s)\, \lambda;\ D\, \frac{\partial c}{\partial x} = h_m\, (p_a - \alpha p_s) \qquad [5]$$

$$y = Y;\ k\, \frac{\partial T}{\partial y} = h\, (T_a - T_s) + h_m\, (p_a - \alpha p_s)\, \lambda;\ D\, \frac{\partial c}{\partial y} = h_m\, (p_a - \alpha p_s)$$

$$z = Z;\ k\, \frac{\partial T}{\partial z} = h\, (T_a - T_s) + h_m\, (p_a - \alpha p_s)\, \lambda;\ D\, \frac{\partial c}{\partial z} = h_m\, (p_a - \alpha p_s)$$

At IFR-BL this classical approach has been adopted to calculate temperature changes and weight losses during a range of meat processing operations eg. chilling /1/, freezing /2,3,4/, thawing /5,6,7/, retail display /8,9/ and cooking /11,12,13/. In all cases a series of major complications have had to be overcome which can be separated into four groups :

(i) Boundary conditions There are few published data on heat and mass transfer coefficients at the surface of meat carcasses, joints etc. In some cases coefficients have been determined experimentally /4,14,15,16,17/ for the particular operation, whilst in others the coefficients used in the models have been selected to obtain the required degree of agreement between measured and predicted product temperatures or weight losses. The latter approach is acceptable when design charts are being produced by interpolation between experimental data but extrapolation outside experimental values must be carried out with caution.

Problems also occur in simulating existing commercial operations, such as in chill rooms or cookers, where large variations are found in air temperature, velocity, humidity and the size and composition of products. This can produce substantial changes in transfer coefficients around the carcasses and joints /11/. Consequently the model has to be solved a number of times with a range of coefficients to determine the differences in product treatments resulting from such variations.

(ii) Material properties Some of the thermophysical properties, diffusion coefficients and water activities of the various red meats (beef, pork, lamb) and their components (lean, fat, skin) have been measured under specific conditions /18/. International cooperation under COST 90 has produced predictive equations that allow selected thermophysical properties to be calculated from the chemical composition of the products /19/. For some materials and temperature ranges no data are available. In processes such as cooking where considerable weight loss occurs the properties of the material have to be varied to take into account changes in composition during processing.

(iii) <u>Composition</u> The composition and conformation (shape) of a carcass or meat joint varies considerably within a single processing batch and the statistical distribution and mean of the properties can also differ markedly between batches. The complex shape of meat carcasses and joints also leads to variations in boundary conditions.

(iv) <u>Mechanisms</u> Many of the mechanisms of mass movement within meat have yet to be fully understood. Certainly during cooking, studies have shown that moisture loss is too rapid to be caused by diffusion and new theories for mechanisms of weight loss are being proposed at IFR-BL /20/. In freezing very high internal pressures are known to build up, due to a frozen surface shell resisting internal expansion, and then rapidly fall as the meat splits /21/. Under these conditions the Fourier laws of heat conduction do not strictly apply. Cleland /22/ has also suggested that in freezing and thawing of meat the thermal properties are a function of rate of temperature change as well as temperature itself.

Some of the practical problems could be minimised by better control in the meat processing and distribution chains and at least one meat processor in the UK is developing systems based on statistical process control to attain more product uniformity. As this approach is adopted more widely the use of models to predict and control processing times and weight losses will become even more prevalent. Several problems, for example estimating the rates of drip during thawing and cooking, will remain while research determines the mechanisms of weight loss, the structure of meat and interaction of individual components and the effect on meat quality. The following examples illustrate how mathematical modelling techniques in conjunction with pilot plant testing are being applied at IFR-BL in current investigations of four industry processes.

APPLICATIONS

Primary chilling of pig carcasses

The main aim of primary chilling is to reduce the temperature of the carcass to prevent the growth of food poisoning organisms and limit the growth of food spoilage bacteria. Reducing the internal temperature of the meat below 20°C within a few hours post mortem has also been shown to reduce protein denaturation which would otherwise produce 'drip' on cutting /23/. The rate of evaporative weight loss from the carcass is also a function of surface temperature and rapid chilling consequently reduces weight loss. Current EC legislation for intercommunity trade requires a maximum carcass temperature of 7°C before cutting or transport, whilst practical considerations have shown that the best temperature for cutting pork is approximately 4°C. To increase throughputs and maximum utilisation of expensive plant, processors would like to achieve these temperatures in the minimum time. All these considerations illustrate the advantages of rapid primary chilling. However, if any of the muscle (lean meat) in the carcass is frozen during the chilling process, drip loss during cutting will increase substantially. Rapid temperature reduction below 10°C within 3 to 5 h post mortem increases toughness in pork when finally cooked /24,25/. It is also physically difficult to cool rapidly large pig carcasses because of their high heat capacity and poor thermal conductivity. For conventional single stage air chilling operations, simple process design charts for pig chilling have been produced by a combination of large scale experimental trials and a basic linear model /26/. The number of combinations that have to be considered in optimising a multi-stage system where air temperature, velocity and relative humidity can be varied independently is too great to be covered experimentally. A model has therefore been developed to guide the experimental programme and provide data for industrial plants. The problems of modelling the irregular shape of a pig carcass are considerable, and in the current model infinite cylinders of different radii are used as analogues to carcasses of different weights. Each cylinder is built up from three concentric shells with different temperature dependent thermophysical properties that represent the skin, fat and lean of the pig. One main purpose of the model will be to identify conditions that will minimise the total chilling time without any toughening or freezing in the lean. However, it is currently being used to study the mechanism of mass transfer within the meat and from the meat to the air stream. The model has predicted that the percentage weight loss from pig carcasses after chilling for 24 h does not vary with fat or skin thickness. After 72 h, there was no still no effect of fat thickness on weight loss, but increasing the skin thickness from 1.5 to 3 mm increased the losses from a 50 mm radius cylinder from 3.5 to 4.4% (Figures 1a and 1b). The predictions of moisture profiles within the skin, fat and lean showed that most moisture was lost from the skin with the fat acting as a barrier to moisture loss from the lean. Consequently, when a thick layer of skin was present more moisture was lost. In conventional abattoir operations that produce pork with the rind in place, a model that predicts conditions to minimise weight loss can have important economic potential. However, some producers supply retail outlets with pork that has already been trimmed of the skin and most of the fat. In these cases the speed of chilling is likely to be more important than weight loss.

Retail display

Recent IFR-BL investigations have concentrated on display of chilled unwrapped foodstuffs because large amounts of fresh meat are still sold in this way and there is considerable growth in supermarket sales of both unpackaged fish and delicatessen products, eg sliced meats, paté and cheese. Discolouration is probably the most important factor controlling the display-life of pre-packaged meat /27/ and a relationship between weight loss and surface colour changes that occur during retail display of unwrapped meat has been established /8/. One major retailer alone has estimated losses of £1M/year due to weight loss and consequent deterioration in appearance.

Investigations have been carried out in a number of stages:
(1) measurements of air temperature, velocity and humidity and product temperature and weight loss were carried out on a range a commercial cabinets,
(2) weight loss from a range of products was determined during controlled wind tunnel tests, and
(3) a finite difference mathematical model was developed of heat and mass transfer under retail display conditions.

Initially high weight losses were thought to be due to the large variations in air conditions (fluctuations of up to 11°C and 47% relative humidity were recorded) and cabinet designers were developing improved control systems to reduce these. However, the model has predicted that the mean values of air

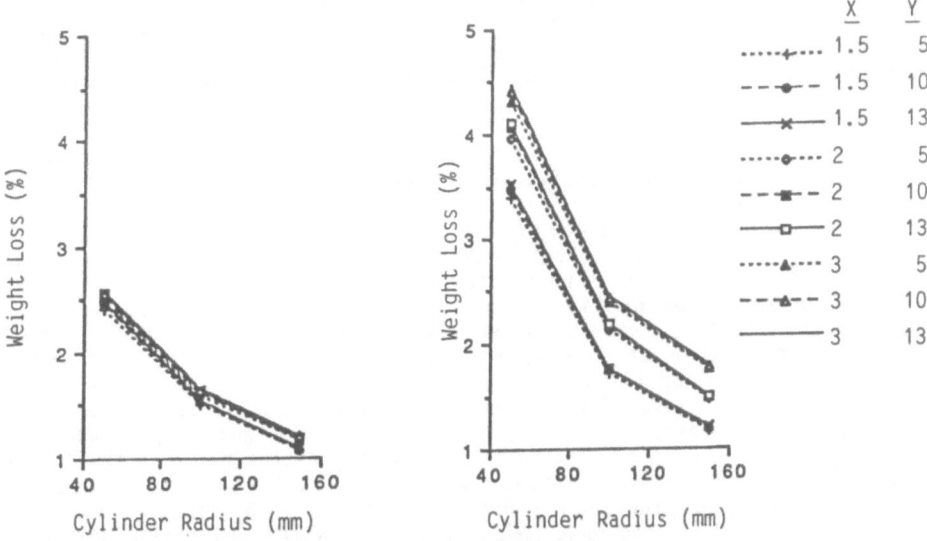

Figure 1: Effect of pig carcass composition on percentage weight loss during one stage of chilling in environmental conditions of 90% air relative humidity, air temperature of 2°C and air velocity of 1.4 m/s.
Key: X = skin thickness (mm)   Y = fat thickness (mm)

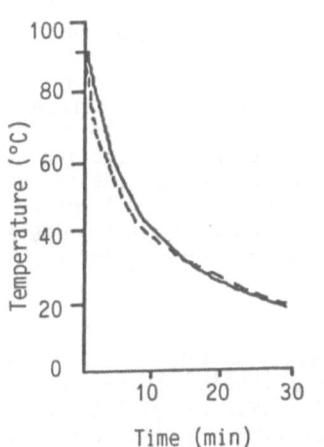 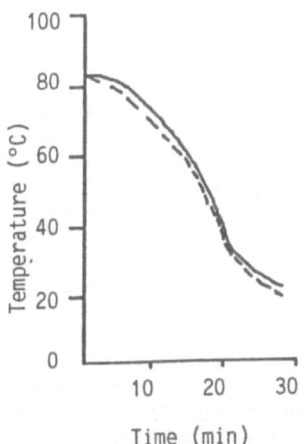

(a)   Water   (b)   Water & Diced Carrot

Figure 2: Measured (-) and predicted (----) temperatures of water during vacuum cooling of 250 kg batches of (a) water and (b) 50/50 (w/w) mixture of water and diced carrot.

conditions over the display period have a greater effect on weight loss than the degree of fluctuation /9/ provided that defrosting is not carried out during display. The model has also shown that at low air velocities mass transfer from the product surface into the air limits the rate of moisture loss from the product whilst diffusion of water within the product controls losses at high air velocities. The environmental conditions at which the rate controlling mechanism moves from surface to diffusion control are dependent on the diffusion coefficient of water through the product.

Pressure cooking and vacuum cooling

Most soups, pie fillings, sauces and components of ready-to-eat meals are prepared in jacketed vessels where steam is injected during cooking and water is circulated during cooling. Some of these vessels are sealed during heating to increase pressures above atmospheric and reduce cooking times. The product is then often evaporatively cooled by reducing the vessel pressure below atmospheric. Product manufacturers could determine the heating and cooling times for new recipes by carrying out practical tests but these can be expensive requiring considerable time and large amounts of product, often as much as 3 tonnes. Unless very extensive, such tests would not show the effects of operating variables, such as steam pressure or cooling water temperature, on processing times. A mathematical model was developed using heat and mass balances on the system and finite difference methods to predict the temperature of solids in the product and proved using mixtures of water and diced solids in pilot scale equipment (250 kg batch size) at IFR-BL /12/. It has also been used to simulate the heating of typical product compositions /28/ in both pilot scale tests and during commercial processing of batches up to 750 kg. During tests using 250 kg batches, the model showed good agreement with measured heating times which varied from 17 to 24 minutes with vacuum cooling times varying from 16 to 39 minutes depending on the product composition. The shape of the temperature-time curve obtained during vacuum cooling is also affected by the product composition, even with a simple example such as the addition of diced carrot to water (Figures 2a and 2b).

The differences in these temperature-time curves are due to the method of controlling foam formation during vacuum cooling. No foam is produced when water boils at reduced pressure, however, juices from carrot can lead to formation of a foam when water/carrot mixtures are vacuum cooled. To ensure this foam is not transferred to the vacuum equipment, a valve between the vessel and vacuum equipment closes when the foam reaches a certain height in the vessel. This valve action restricts the rate of evaporative cooling (cf Figures 2a and 2b). The alternative method of cooling by passing water through the jacket is strongly dependent on the water flowrate and temperature. This is a major reason for the variation found in cooling time in commercial operations which can range from 112 to 328 minutes for a 250 kg batch cooling from 95 to 20°C.

Convection cooking

Cooking causes larger losses in the weight of meat than any other processing operation, 20% being typical and 50% not exceptional. There should be considerable scope for the meat processor to improve yields in such operations without adversely affecting product quality. Simple calculations showed that weight losses during cooking were too rapid to result from diffusion alone but few data were available to produce a quantitative model of weight loss. In the first phase of the investigation, emphasis was concentrated on predicting heating times until further information on weight loss mechanisms was available. A four stage approach /29/ was adopted:

(i)   develop the mathematical model
(ii)   measure the missing values in the model
(iii) confirm the model using relatively ideal materials
(iv)   test the application of the model to meat cooking
The model, based on finite difference methods, has been developed to predict the temperature within a finite cylinder with thermophysical properties which can vary with position and temperature. The primary missing value in the model was the surface heat transfer coefficient which was measured at the surface of a finite cylinder during pilot scale studies. A PTFE cylinder of the same size as that used to determine the heat transfer coefficients was heated in a range of air conditions and the times to attain specific temperatures were measured and compared with predictions from the model. Agreement between measured and predicted heating times to 74°C, a commonly accepted minimum temperature for cooking meat, was within 5%. Clearly the model was reliable when using an ideal material. However, measured and predicted cooking time for meat cylinders differed by up to 65% primarily because the model neglected dimensional changes and weight losses. More sophisticated models are now being developed to allow for weight loss and the effects of temperature, time and different rates of transverse and longitudinal shrinkage of the meat fibre system.

Further models
All of the above systems have concentrated on predicting heating times or weight losses. However, other factors affect the profitabilty of an operation, the most important of which are microbial growth and quality. Since both of these depend on the temperature and moisture content of the meat, the finite difference models being developed to predict temperature and moisture profiles can be used simultaneously with models of bacterial growth and equations to predict rates of change of quality parameters. In most cases, laboratory experiments are needed to determine such factors as growth rate constants for bacteria which can be used in the models. However, once these factors have been determined the models can be used to estimate bacterial loads and quality within commercial operations.

CONCLUSIONS
Mathematical models have been developed to predict the temperature and weight changes that occur in a wide range of meat processing operations from the farm gate to storage/cooking by the consumer. These models are used in the design of new processes and in increasing the profitabilty of existing operations. In all cases, pilot scale or factory tests have been required to determine unknown factors in the models such as surface transfer coefficients or to confirm the application of the models. Once these factors are known, the models can be applied with confidence to predict the effect on the product of changing the processing conditions to those which either have not, or could not, be used in existing pilot plants or factory tests. A range of data is now available for the models which allows initial estimates to be made of meat temperatures and weight losses in a number of unit processes.
Many of the benefits of adopting modelling methods within the meat industry are difficult to quantify, for example, in the investigation of methods of increasing throughput or in preparing new equipment designs, both of which would otherwise need extensive and costly test work. However, during several confidential projects, savings of £100K/year have been achieved by using models to indicate how processing factors should be changed within existing equipment. With the annual turnover of the meat industry currently exceeding £8000M it is easy to see that potential savings of 1% in weight loss by using models to optimise systems could have a large effect on operating economics.

NOMENCLATURE

C = Specific heat (J/kg K)
c = Concentration of water (kg/m^3)
c_i = Initial concentration of water (kg/m^3)
D = Diffusivity (m^2/s)
h = Heat transfer coefficient (W/m^{2}K)
h_m = Mass transfer coefficient (kg/s m^2 Pa)
k = Thermal conductivity (W/m K)
P_s = Partial pressure of water at the surface (Pa)
P_a = Partial pressure of water vapour in the air (Pa)
t = Time (s)
T_s = Temperature of the surface (°C)
T_a = Temperature of the surrounding medium (°C)
T_i = Initial temperature (°C)
x, y and z = Rectangular co-ordinates
X, Y and Z = Surface of the product
ρ = Density (kg/m^3)
α = Water activity
λ = Latent heat of evaporation of water (J/kg)

REFERENCES

(1) James, S.J. and Swain, M.J. (1983) The effect of surface fat layers on the chilling time of meat. Proc. 16th Int. Cong. Refrig., Paris, 3, 417-423.

(2) James, S.J., Creed, P.G. and Bailey, C. (1979) The determination of the freezing time of boxed meat blocks. Proc. Inst. Refrig., 75, 73-83.

(3) Creed, P.G. and James, S.J. (1984) The prediction of freezing and thawing times of mutton carcasses. Proc. 30th Eur. Meeting Meat Res. Workers, Bristol, 59-60.

(4) Creed, P.G. and James, S.J. (1985) Heat transfer during the freezing of liver in a plate freezer. J. Food Sci., 50, 285-288, 294.

(5) Bailey, C., James, S.J., Kitchell, A.G. and Hudson, W.R. (1974) Air-, water- and vacuum-thawing of frozen pork legs. J. Sci. Food Agric., 25, 81-97.

(6) James, S.J. and Creed, P.G. (1980) Predicting thawing time of frozen beef fore and hind quarters. Int. J. Refrig., 3(4), 237-240.

(7) Creed, P.G. and James, S.J. (1981) Predicting thawing times of frozen boneless beef blocks. Int. J. Refrig., 4, 355-358.

(8) Fulton, G.S., Burfoot, D., Bailey, C. and James, S.J. (1987) Predicting weight loss from unwrapped chilled meat in retail displays. Proc 17th Int. Cong. Refrig. Vienna, 555-563.

(9) James, S.J., Fulton, G.S., Swain, M.V.L. and Burfoot, D. (1988) Modelling the effect of temperature and relative humidity fluctuations on weight loss in retail display. Proc. Int. Inst. Refrig. Refrigeration meeting 'Food and People', Brisbane.

(10) Burfoot, D. and James, S.J. (1988) The effect of spatial variations of heat transfer coefficient on meat processing times. J. Food Engineering, 7, 41-46.

(11) Burfoot, D. and James, S.J. (1984) Problems in mathematically modelling the cooking of a joint of meat. Proc. COST 91 Seminar, Athens, 1983. In 'Thermal processing and Quality of Foods', ed. P. Zeuthen et al. (1984), Elsevier Applied Science Publishers, London and New York, 467-472.

(12) Burfoot, D., Hayden, R. and Badran, R. (1987) Simulation of a pressure cook/water and vacuum cooled processing system. Poc. IChemE Symposium 'Engineering Innovation in the Food Industry', Bath, 21-28.

(13) Burfoot, D. (1984)   Predicting the effect of fat thickness and

distribution on the heating times of joints of rolled meat. Proc. 30th Euro. Meeting Meat Res. Workers, Bristol, 213-313.

(14) Arce, J.A. and Sweat, V.E. (1979) Survey of published heat transfer coefficients encountered in food refrigeration processes. Report on Project No. RP-228, Agricultural Engineering Department, Texas A&M University.

(15) James, S.J. and Bailey, C. (1982) Changes in the surface heat coefficient during meat thawing. Proc. 28th Eur. Meeting Meat Res. Wkrs., Madrid, 1, 3.16, 160-163.

(16) James, S.J., Swain, M.V.L. and Daudin, J.P. (1988) Mass transfer under retail display conditions. 34th Int. Cong. Meat Sci. Tech., Brisbane.

(17) Burfoot, D. and Self, K. Predicting the heating times of beef joints. Submitted to J. Food Engineering.

(18) Morley, M.J. (1972) Thermal properties of meat: Tabulated data. Meat Research Institute Special Report No 1.

(19) Jowitt, R., Escher, F., Hallstrom, B., Meffert, H.F.Th., Spiess, W.E.L. and Vos, G. (1983) Physical Properties of Foods. Applied Science Publishers. London.

(20) Offer, G., Restall, D. and Trinick, J. (1984) Water holding in meat. In 'Recent advances in the chemistry of meat', Ed. Bailey, A.J., Proc. Symp. organised by Food Chemistry Group of Royal Soc. Chem. and Food Group of Soc. Chem. Ind. held at AFRC Meat Research Institute, April. Proceedings produced by Royal Society of Chemistry.

(21) Miles, C.A. and Morley, M.J. (1977) Measurement of internal pressures and tensions in meat during freezing, frozen storage and thawing. J. Fd. Technol., 12, 387-402.

(22) Cleland, A.C., Earle, R.L. and Cleland, D.J. (1982) The effects of freezing rate on the accuracy of numerical freezing calculations. Int. J. Refrigeration, 5(5), 294-301.

(23) Gigiel, A.J., Swain, M.V.L. and James, S.J. (1985) Effects of chilling hot boned meat with solid carbon dioxide. J.Fd. Technol., 20, 615-622.

(24) James, S.J.,Gigiel, A.J. and Hudson, W.R. (1985) The ultra rapid chilling of pork. Meat Science 13, 19.

(25) Honikel, K.O. and Regan, J.O. (1987) Hot boning of pig carcasses: influence of chilling conditions on meat quality. In accelerated Processing of meat. Elsevier Applied Science Publishers Ltd, London and New York, 97.

(26) Brown, T. and James, S.J. (1988) Process design data for pig chilling. IFR-BL Meat chilling subject day, 1988.

(27) Hood, D.E. and Riordan, E.B. (1973) Discolouration in pre-packaged beef: measurement by reflectance spectrophotometry and shopper discriminatiom. J. Fd. Technol. 8, 333-343.

(28) Burfoot, D., Badran, R. and Fulton, G.S. Pressure cooking and vacuum/water cooling of liquid/solid mixtures. Submitted to J. Food Engineering.

(29) Burfoot, D. and Self, K.P. (1988) Prediction for heating rates of cubes of meat undergoing water cooking. Int. J. Food Sci. Technol., 23, 247-257

AFRC Institute of Food Research - Bristol Laboratory
Langford, Bristol BS18 7DY

Numerical modelling of conjugate heat transfer in an Advanced Gas-cooled
Reactor fuel standpipe

M T R Fung and R P Hornby, National Nuclear Corporation, UK

Abstract

This paper presents an application of a finite difference technique which
solves the fluid flow and energy equations to predict the conjugate natural
convection heat transfer behaviour in a fuel standpipe of an Advanced
Gas-cooled Reactor. An algebraic model for turbulence is used and the full
effect of the change in fluid density with temperature is taken into
account. Numerical results are compared with temperature measurements from
several reactor standpipes and the agreement is found to be generally
satisfactory.

Keywords

Numerical modelling, conjugate heat transfer, natural convection, reactor
fuel standpipe.

1   Introduction

In an Advanced Gas-cooled Reactor, carbon dioxide under high pressure is

driven in a closed circuit, transporting the heat produced in the reactor

core to boiler units to raise steam for electricity generation. Natural

convection flows may arise in the reactor under full load operation in

reactor components shielded from the main gas flow path, producing

temperature gradients in the components which have to be calculated for

subsequent stress analysis. A conjugate heat transfer analysis is

necessary owing to the inter-dependence of fluid and solid temperatures.

Recent examples of conjugate heat transfer analysis for natural convection

flows include [1] and [4].

A numerical method has been used to predict the conjugate heat transfer

behaviour in a fuel standpipe of an Advanced Gas-cooled Reactor. Results

obtained from the method are compared with reactor temperature

measurements.

J. Manley et al. (eds.), Proceedings of the Third European Conference on Mathematics in Industry, 353–361.
© 1990 Kluwer Academic Publishers and B. G. Teubner Stuttgart.

2   Heat transfer in a fuel standpipe

A fuel standpipe forms a vertical penetration through the top of the
reactor pressure vessel and is the access route for fuel to the reactor
core. A narrow gas annulus, open at the top and closed at the bottom,
exists between the externally cooled fuel standpipe and the fuel plug unit,
below which fuel elements are attached. A schematic sketch of a fuel
standpipe is shown in Fig 1. The fuel standpipe is about 6 m high and
270 mm in diameter with a gas annulus about 6 mm wide.

Gas in the annulus between the standpipe and the plug unit is cooled by the
liner wall at a rate determined by the combined thermal resistances of the
liner components and the gas-to-surface heat transfer coefficients. This
cooling raises the gas density and produces a buoyancy-driven flow. A
balance between the buoyancy force and wall friction controls the strength
of the flow and the amount of heat transfer in the standpipe.

3   The conjugate heat transfer model

The conjugate heat transfer model represents the components shown in Fig 1.
A finite difference technique is used to solve the partial differential
equations used in describing the flow and the temperature in the
standpipe, based on a cylindrical polar grid. A porosity model is used to
adjust the coefficients of the finite difference equations and represents
the non-uniform width of the standpipe gas annulus.

Flow and temperature in the standpipe gas annulus is represented mathematically using the time-averaged, steady-state form of the continuity,
momentum and energy equations. These equations can be represented as:

$$\text{div } (\rho U \phi) = \text{div } (\Gamma_\phi \text{ grad } \phi) + S_\phi / V_\phi \qquad (1)$$

where $\phi = 1$ for mass conservation, $\phi = u$, v or w for momentum conservation and $\phi = h$ for energy conservation.

ρ is modelled as ($\rho = 2.342 \times 10^7/h$ kg/m^3).
Γ_ϕ is taken to be ($\mu_t + \mu$) in the momentum equations with μ_t modelling the
Reynolds stress terms (with $\mu = 3.2 \times 10^{-5}$ Ns/m^2). In the energy equation
it is taken to be ($\mu_t/\sigma_t + \mu/\sigma$) where σ_t has a value close to unity.

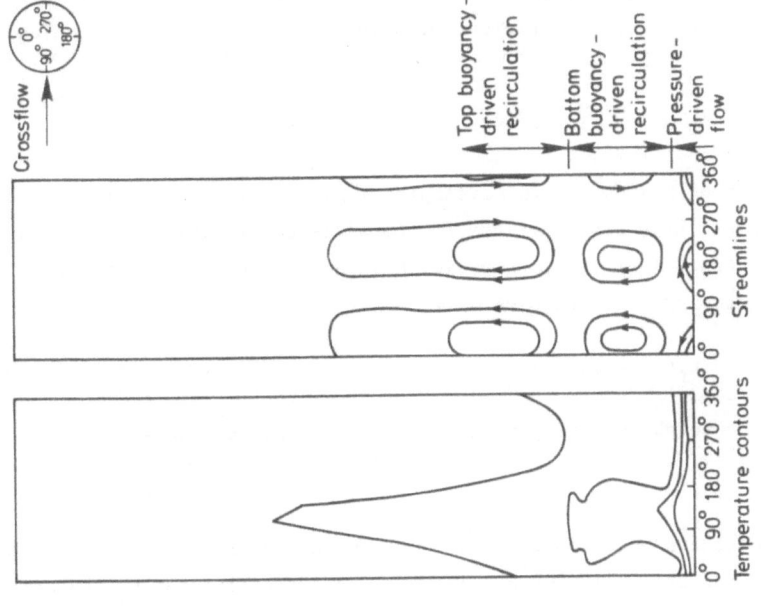

FIG. 2. A TYPICAL TEMPERATURE/FLOW PATTERN IN THE FUEL STANDPIPE GAS ANNULUS

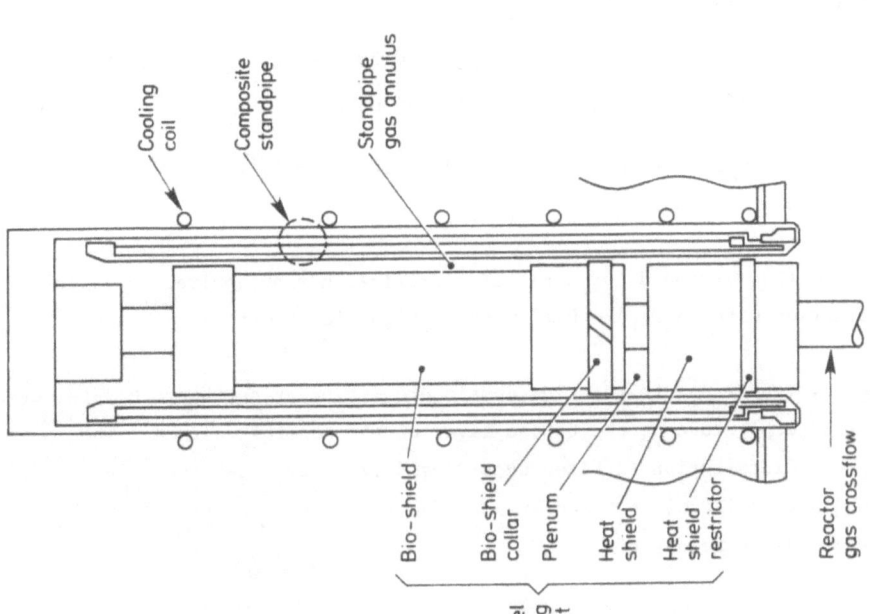

FIG.1. SCHEMATIC SKETCH OF A FUEL STANDPIPE

356

An algebraic model for turbulence based on the Prandtl free-shear-layer model [6], is considered sufficient with the relatively coarse calculation grid used:

$$\mu_t = 0.01 \; \rho \; \pi \; R \; | \; w_{max} - w_{min} \; | / 2 \qquad (2)$$

The promotion of mixing in the developing region of the mixing layer, the attenuation of large scale motions by the walls and thermal radiation are not represented. A similar algebraic turbulence model is used to describe mixing in the plenum of the plug unit:

$$\mu_t = 0.0044 \; \rho | w | z \qquad (3)$$

with z as the height of the plenum and the coefficient 0.0044 deduced from [6].

S_ϕ in the momentum equation is used to represent the friction between the gas and the annular walls ($-f \; \rho \; \phi \; |U| \; A/2$) or a constriction to the axial flow ($-Eu\rho \; w \; |U| \; a_b/2$), that is too short for the calculation grid.

f is taken to be the higher of f_l and f_t. [3]:

$$f_l = 24/Re \; ; \; f_t = 0.079/Re^{0.25} \; ; \; Re = \rho(u^2 + w^2)^{0.5} \; D/\mu \qquad (4)$$

where D is defined as twice the local annulus gap.

$$Eu = (1/\alpha - 1)^2 \; (a_b/a_c)^2 + (a_b/a_c - 1)^2; \qquad (5)$$
$$\alpha = 0.63 + 0.37 \; (a_c/a_b)^3$$

Equation (5), intended for a circular orifice, may be applied to an annular contraction with $(a_c/a_b) = 0.17$ without significant error.

Boundary conditions to the flow domain are set by prescribing the pressure and the temperature at the bottom layer of the control volumes. The pressure distribution is fixed according to measurements made in a test rig at the National Nuclear Corporation. The temperature is set to the hot reactor gas temperature.

The code also computes temperatures in the composite fuel standpipe and the top half of the plug unit in which heat conduction effects are significant. It employs the following anisotropic heat conduction equation:

$$\text{div } (k \text{ grad } T) + S_T/V_T = 0 \tag{6}$$

with harmonic averaging for k to handle heat transfer between dissimilar materials.

For the standpipe, the following source terms are used respectively to model heat transfer to the standpipe gas annulus on the inside and to the helical cooling water coil on the outside:

$$S_T = -H_g \ (T - h/c) \ A \ ; \ S_T = -H_a \ (T - T_a) \ A \tag{7}$$

H_a is deduced to be 300 W/m^{2}K and H_g is assumed as the sum of the following [2]:

$$H_n = \frac{k}{z} \ 0.13 \ (Gr.\sigma)^{0.33}; \ H_f = \frac{k}{D} \ 0.023 \ Re^{0.8} \ \sigma^{0.4} \tag{8}$$

Heat transfer between the standpipe gas annulus and the plug unit adopts an expression similar to equation (7). Source terms, represented by this equation, link the energy equation for the gas annulus to the conduction equation for the standpipe/plug unit, but carry opposite signs for the two equations to maintain energy conservation in the standpipe.

PHOENICS 81 was used to solve for field variables in the flow domain [5]. The partial differential equations for the fluid flow, represented by equation (1), are expressed in an analogous finite-difference form for a control volume. The equations are solved slab by slab except for the pressure-correction equation which is solved over the whole-field. Further details on the solution procedure may be found in [5].

The three-dimensional heat conduction equation for the solid domains is discretised with central differencing and the finite difference equations are solved using a tri-diagonal matrix algorithm. A solver with a higher convergence rate is considered unnecessary because the calculation time is dominated by the fluid flow solver.

The equations are solved iteratively. Each iteration involves first obtaining a solution of the equations for the flow domain followed by the solution of the equations for the solid domain.

4   Results and comparisons with reactor data

Results were obtained using a CRAY X-MP computer. Convergence was generally slow and typical of experience with computing natural convection flows. Slow convergence is most probably due to the strong coupling between the momentum and energy equations and the inability of the generalised PHOENICS 81 solver to take account of this.

A typical computed time-averaged flow and temperature field is shown in Fig 2, in which the horizontal scale of the figure is exaggerated by a factor of two. Reactor gas pressure is 41 bar absolute and the temperature at standpipe entry is 650°C. It can be seen in Fig 2 that the natural convection pattern along the standpipe is divided into two circulations. This is produced by the constriction at the bio-shield collar. The circumferential pressure difference at the standpipe entry produces a pressure driven flow which in turn sets up a temperature differential at the standpipe entry and maintains a buoyancy-driven hot upflow on the high pressure side of the annulus.

Temperature measurements are available from several reactor fuel standpipes for comparison. However, the flow resistance of the bio-shield collar varies from plug unit to plug unit and consequently the measured temperatures show considerable scatter. Therefore comparison has been made with code results based on the highest and the lowest bio-shield collar resistances possible. The computed maximum/minimum gas temperature distribution along the standpipe gas annulus is shown in Fig 3 and encompasses favourably the maximum/minimum temperatures measured on the surface of the heat shield. The comparison can be made because of the good thermal coupling between the gas and the adjacent heat shield surface in relation to other standpipe components. Over-prediction in azimuthal temperature range may be partly attributed to code modelling inaccuracies but may also be attributed to insufficient reactor measurements. A

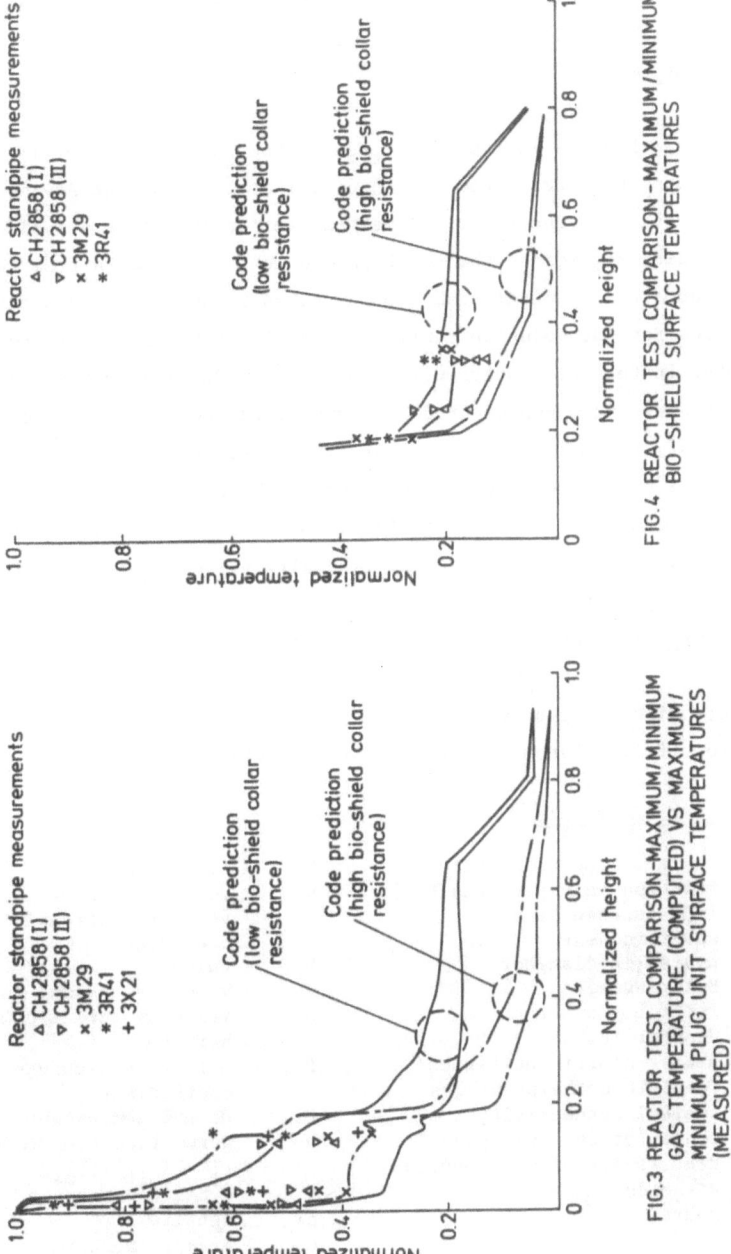

FIG.4 REACTOR TEST COMPARISON - MAXIMUM/MINIMUM
BIO-SHIELD SURFACE TEMPERATURES

FIG.3 REACTOR TEST COMPARISON-MAXIMUM/MINIMUM
GAS TEMPERATURE (COMPUTED) VS MAXIMUM/
MINIMUM PLUG UNIT SURFACE TEMPERATURES
(MEASURED)

360

favourable comparison of measurements on maximum/minimum surface
temperatures along the bio-shield with code results is shown in Fig 4.

5   Conclusions

A numerical method has been used to predict the conjugate natural
convection heat transfer behaviour in a fuel standpipe of an Advanced
Gas-cooled Reactor. The numerical method uses an iterative solution
procedure which segregates the equations for the flow domain from those for
the solid domain. Computational economy makes it necessary to use fairly
coarse grids for the solution domains. Details of the geometry are
represented in the calculation with a porosity model and surface friction
and heat transfer are represented by source terms based on empirical
correlations. A simple algebraic model for turbulence is used and the full
effect of gas density changes is included in the equations solved.
Temperature measurements are available from several reactor fuel standpipes
for comparison and the comparison is generally favourable.

6   Acknowledgement

The authors wish to acknowledge the kind permission of the National Nuclear
Corporation Limited for publishing the information in this paper.

7   List of symbols

A	surface area (for wall friction and heat transfer)	T	temperature
		U	velocity vector
a	flow passage area	u	velocity, circumferential component
cp	specific heat		
D	hydraulic diameter	V	volume
Eu	Euler Number	v	velocity, radial component
f	friction factor	w	velocity, axial component
Gr	Grashof Number	z	height
H	heat transfer coefficient	Γ	effective exchange coefficient
h	specific enthalpy of gas		
k	thermal conductivity	ξ	normalised height (Z/L)
L	length of the standpipe	θ	normalised temperature
R	mean radius of gas annulus		$(T - T_{min})/(T_{max} - T_{min})$
Re	Reynolds Number	μ	dynamic viscosity
S	source term	ρ	density
		σ	Prandtl Number
		ϕ	hydrodynamic variable

Subscripts

a	cooling coil	min	minimum
b	upstream	n	natural convection
c	constriction	T	temperature
f	forced convection	t	turbulent
g	gas	u	velocity, azimuthal
h	specific enthalpy		component
l	laminar	w	velocity, axial component
max	maximum	ϕ	hydrodynamic variable

8   References

[1] Kuehn T H and Balvanz J L. Conjugate heat transfer by natural convection from a horizontal heat exchanger tube. Proceeding of 7th International Heat Transfer Conference. $\underline{2}$ (1982) 317-322

[2] McAdams W H. Heat transmission, 3rd edition. McGraw Hill Book Company Inc, New York, USA (1954)

[3] Miller D A. Internal flow. BHRA, Bedford, UK (1971)

[4] Quarini G L. Buoyancy induced flows caused by inert solids immersed in Stratified fluids. Natural convection fundamentals and applications, edited by Kakac, S. et al (1985) 461-471. Hemisphere Publishing Corporation, New York, USA

[5] Spalding D B. A general purpose computer program for multi-dimensional one- and two-phase flow. Mathematics and computers in simulation. North Holland Press $\underline{23}$ (1981) 267-276

[6] Spalding D B. Turbulence models. A lecture course. Imperial College of Science and Technology, London, UK (1983)

Fung M-T R and Hornby R P. National Nuclear Corporation Limited, R & D Division, Warrington Road, Risley, Cheshire, WA3 6BZ, UK

RESTORATION OF NMR IMAGES

By

Fred Godtliebsen

The Norwegian Institute of Technology, Trondheim

Abstract:

Today, in addition to standard smoothing methods, averaging is the only statistical technique in use to reduce the noise in NMR images. The problem with this approach is that it is very time—consuming. The fact that averaging improves image quality indicates that the error in each pixel is distributed around the same mean. T1—weighted transversal NMR images of the head are used in our investigations. Several measurements are made of the same slice. Averaging these measurements gives us a very good approximation to the true image. By comparing the pixel values of the different measurements with the averaged image much information about the noise is gained. Two kinds of data are available; raw data and reconstructed data. We deal with both kinds of data. Row and column correlations in noise and signal are estimated and discussed. Possible dependencies between noise variance and signal magnitude are studied. Our aim is to use image restoration methods to recover the true scene from a single measurement.

Key words: Nuclear magnetic resonance, noise distribution, correlations,
                  image restoration, iterated conditional modes.

J. Manley et al. (eds.), Proceedings of the Third European Conference on Mathematics in Industry, 363–368.
© 1990 *Kluwer Academic Publishers and B. G. Teubner Stuttgart.*

1. Introduction.

Today, NMR image quality is improved by averaging. This means that the colour in each pixel is estimated by the corresponding pixels in several measurements. A problem with this approach is that each measurement is very time—consuming. As a result there is some chance that the patient will move during the image acquisition time and the number of patients treated each day is rather small.

The intention with this paper is to get information about the noise and to apply a simple version of a powerful restoration method based on probabilistic considerations.

In the results we deal with raw or reconstructed data, where raw data are the Fourier components of the composite signal, and reconstructed data are the result from an inverse 2D Fourier transform of the raw data.

2. Description of the approach.

Suppose that we have m measurements. By averaging the measurements a very good approximation of the true scene is found. Let's define $x^*(i,j)$ = true value in (i,j), $y(i,j)$ = averaged value in (i,j), $y_k(i,j)$ = value of measurement k in (i,j) where $k = 1,2,...,m$. The value of the error between the true scene and measurement k in (i,j) is given by $x^*(i,j) - y_k(i,j)$. By the triangle inequality we have

$$|x^*(i,j) - y_k(i,j)| \leq |x^*(i,j) - y(i,j)| + |y(i,j) - y_k(i,j)|$$

Let's denote $y(i,j) - y_k(i,j)$ and $x^*(i,j) - y(i,j)$ by error of type 1 and error of type 2, respectively. The value of $x^*(i,j)$ is unknown. However, by increasing m the error of type 2 will be of minor importance. We will therefore concentrate on reducing the error of type 1 as much as possible. We try to achieve this by using a simple version of ICM, see [1].

In our implementation of ICM, we assume that the image contain $c = 64$ colours and use the eight nearest neighbours as a neighbourhood. We compute

$$\underset{x_i}{\text{Min}} \left\{ \frac{1}{2\kappa} (y_i - x_i)^2 - \beta \hat{u}_i(x_i) \right\}$$

for each pixel in turn, where y_i is the observed pixel value, κ is the noise variance, β is a smoothing parameter and $\hat{u}_i(x_i)$ is the current number of neighbours of pixel i having colour x_i.

3. Results.

In our analysis we have used eight T1—weighted measurements of the same slice with repetition time and echo time equal to 450 ms and 30 ms, respectively. All results about the noise are given for the error of type 1. Real and imaginary parts of the data are examined separately.

3.1. Noise distributions.

For the error in the raw data it seems that a normal distribution gives a very good fit. The histogram in Figure 1 is based on 262144 error estimates.

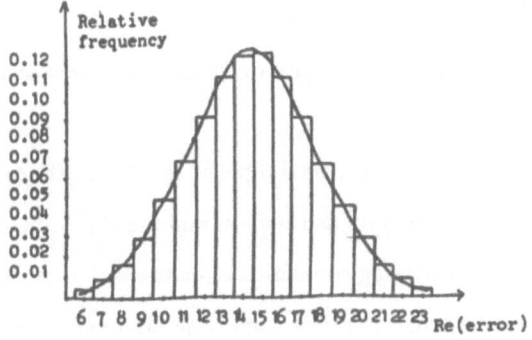

Figure 1: Noise distribution for the real values in the raw data, $\hat{\mu} = 14.4$, $\hat{\sigma}^2 = 10.2$.

About one half of the reconstructed data is outside the object (which is a head in this case). It is therefore of interest to see whether there are differences in the noise inside and outside the object. When this was done the histograms indicated that in both cases a normal distribution with mean zero would give a good fit. The estimated variances were found to be $(0.021)^2$ and $(0.018)^2$ inside and outside the object, respectively. These estimates indicate that the noise variance in tissue is greater than in the surrounding pixels. The estimates inside and outside the object are based on 18048 and 20480 values, respectively.

3.2. Correlations.

We have estimated row and column correlations for the noise and the signal in the images. The correlation estimates of the noise are based on a single measurement and the averaged data, and the signal correlations are based on the averaged image.

It seems that the noise in different points are essentially uncorrelated in the raw data. It is seen from the noise correlations in the reconstructed data that the magnitudes of the correlations decrease slowly as the distances between the pixels increase. From these results it also seems like the correlation signs alternate in the following way $-,+,-,+,...$ The results given for the reconstructed data indicate that there are no differences between the noise correlations in the surroundings and the object.

In the raw signal we observe small negative column correlations and positive row correlations for nearest neighbours.

It is observed that there are fundamental differences between the correlation structure in signals from the surroundings and signals from the object. In the surroundings there is a cyclic, decreasing correlation structure with signs $-,+,-,...$, while in the object there are large slowly decreasing and positive correlations.

3.3. Variance instability.

Restoration methods are often based on the assumption that the noise is independent of the signal magnitude. For this reason the estimated noise variances were plotted as a function of signal magnitude. In each point the noise variance is estimated from the eight measured values by the standard estimator for variance. Histograms of the noise variance indicated that the noise depends on the magnitude of the signal. It seems like the noise increases when the signal magnitude increases in the raw data. In the reconstructed data the opposite effect seems to take place.

3.4. Restorations.

We have applied the simple version of ICM on a single reconstructed measurement with several β–values. From the restorations it can be seen that β is very critical for the degree of smoothing in the image. Unfortunately, none of the restorations we made can be said to give a satisfactory (compared to prior knowledge and the averaged image) restoration of the true scene.

The main reason for the rather poor restorations lies probably in the choice of the prior distribution. It is very likely that a prior distribution of the form presented in [2] would give a much better restoration of the true scene. This will be studied in more detail in the future.

As a last point it is important mentioning that even though the restoration we tried here were not successful, the investigations have given much insight about what kind of noise a NMR image typically contains.

4. Conclusions.

The noise seems normally distributed in both the raw and the reconstructed data. Several interesting results are given about the correlation structure of the noise and the signal in NMR images. Plots of the noise variance as a function of the signal magnitude indicate that there are dependencies between noise and signal magnitude.

The simple version of ICM did not give satisfactory results. More realistic prior distributions should be used in the restorations.

References.

[1]   Besag, J.: On the Statistical Analyses of Dirty Pictures. J.R. Statist. Soc., B, Vol.48, 1986, 259–302.

[2]   German, S. and McClure, D.: Statistical Methods for Tomographic Image Restoration, Proceedings 46.th Session ISI, Bulletin of the ISI 52 (1987).

An on-line augmented price correction technique for

hierarchical control of interconnected industrial processes

Z. M. Hendawy and P. D. Roberts
Control Engineering Centre
City University
Northampton Square
EC1V OHB,
U.K.

Abstract

This paper presents an on-line steady-state optimisation technique of hierarchical nature, which is based on the augmented Lagrangian function and involves shifting the model outputs. The new algorithm is applicable to non convex problems with duality gap, but can also be used with advantage when applied to convex problems. Derivation and structure of the algorithm is presented, and simulation examples are given.

Keywords: Large scale systems; hierarchical control; interconnected subsystems; optimisation; Augmented Lagrangian.

Introduction

A multilayer control scheme for an industrial process consists of two main layers   . Firstly, an optimising control layer, and secondly, a regulating or follow-up control layer. The regulating control layer is responsible for keeping the selected set-points at their desired values inspite of rapid process disturbances, while an optimising control layer has the responsibility of determining and maintaining optimal values of the set

369

J. Manley et al. (eds.), Proceedings of the Third European Conference on Mathematics in Industry, 369–375.
© 1990 Kluwer Academic Publishers and B. G. Teubner Stuttgart.

points under slowly varying disturbances (Lefkowitz, 1977).

The presented algorithm in this paper describes and investigates an optimising control structure for large scale industrial process consisting of interconnected sub-processes. The price correction mechanism with global feedback by Findeisen et al (1980) has been extended by Shao and Roberts (1983) to provide a double iterative coordination strategy by incorporating a shift on model outputs. This algorithm has been successful in solving convex example problems and has shown that a significant reduction of on-line set-point changes are required in order to achieve the final converged solution of the optimisation problem, However, many steady state optimisation problems are often nonlinear and nonconvex, which can result in duality gaps. Hence, the technique by Shao and Roberts, (1983) cannot always be safely applied on such problems.

To cover the case of non-convex problems, another approach is proposed in this paper, which extend the applicability of the previous method. It also improves the effectiveness of the iterative correction process and further reduces on-line control correction time. Therefore, the method is also powerful for convex problems because of the reduced number of controller set-point changes required.

In addition to shifting the model outputs, the main idea of this novel algorithm is based on the application of the augmented Lagrangian technique. The resulting algorithm is of a hierarchical structure and contains three iterative loops. The real process measurements are required only within the outer loop while the other two inner loops involve model based computations only. By this means, on-line control correction time is significantly reduced and the method is applicable for both convex and nonconvex problems.

General model structure

The model output is shifted from

$$y = F(c,u) \text{ to}$$
$$Y_s = F(c,u) + s \tag{1}$$
$$u_s = HY_s = u + Hs \tag{2}$$

where $s \in \mathcal{Y}$ is a fixed vector provided by the coordinator. At some value of s, the model output $\hat{y}$ is expected to match the real output y_*.

The performance index and feasible set within the local problems are also modified as follows:

$$G(c,u_s,Y_s) \leq \dot{0} \tag{3}$$
$$Q(c,u,y) = Q(c,u_s,Y_s) \tag{4}$$

Hence, the model based problem is

$$\min_{c,u} \{Q(c,u_s,Y_s)$$
$$\text{s.t.} \tag{5}$$
$$u_s = HY_s$$
$$G(c,u_s,Y_s) \leq 0$$

The augmented Lagrangian of the overall problem is:

$$L_{as}(c,u,y,\lambda,\mathcal{P},s) = Q(c,u+Hs,y+s) +$$
$$\lambda^T(u-HF(c,u)) + \tfrac{1}{2}\mathcal{P}\|u-HF(c,u)\|^2 \tag{6}$$

Equation — (6) can be rewritten—in— a hierarchical decomposable form as :

$$L_{as}(c,u,y,\lambda,\mathcal{P},s) = \sum_{i=1}^{N} [Q_i(c_i,u_i+H_{ji}s,y_i+s) + \lambda_i^T u_i - \sum_{j=1}^{N}\lambda_j^T$$

$$H_{ji}(F_i(c_i,u_i)) + 0.5\mathcal{P}(\|u_i\|^2 + \|F_i(c_i,u_i)\|^2)] - \mathcal{P}\, u^T HF(c,u) \tag{7}$$

The last term in equation (7) is not separable. To overcome this difficulty a suitable linearization of this term around some point $(\underline{u}, HF(\underline{c},\underline{u}))$ was proposed by Stephanopoulos and Westerberg (1975), this results is

$$u^T HF(c,u) \tilde{=} -\underline{u}^T HF(\underline{c},\underline{u}) + u^T HF(\underline{c},\underline{u}) + \underline{u}^T HF(c,u) \tag{8}$$

Using (8), the augmented Lagrangian (7) can be transformed to the following separable form:

$$\Lambda_a(c,u,\lambda,\mathcal{P},s,\underline{c},\underline{u}) = \sum_{i=1}^{N} [Q_i(c_i,u_i+H_{ji}s,F_i(c_i,u_i)+s_i) + \lambda_i^T u_i +$$

$$-\sum_{j=1}^{N}\lambda_j^T H_{ji} F_i(c_i,u_i) + 0.5\mathcal{P}(\|u_i\|^2 + \|F_i(c_i,u_i)\|^2 - 2$$

$$u_i^T H_i F(\underline{c},\underline{u}) - 2\sum_{j=1}^{N} \underline{u}_j^T H_{ji} F_i(c_i,u_i))] + \mathcal{P}\underline{u}^T HF(\underline{c},\underline{u}) \tilde{=}$$

$$\sum_{i=1}^{N}\Lambda_{ai}(c_i,u_i,\lambda,\mathcal{P},s,\underline{c},\underline{u}) + \mathcal{P}\underline{u}^T HF(\underline{c},\underline{u}) \tag{9}$$

Structure of the Algorithm.

The proposed structure of the algorithm is shown in figure (1) .

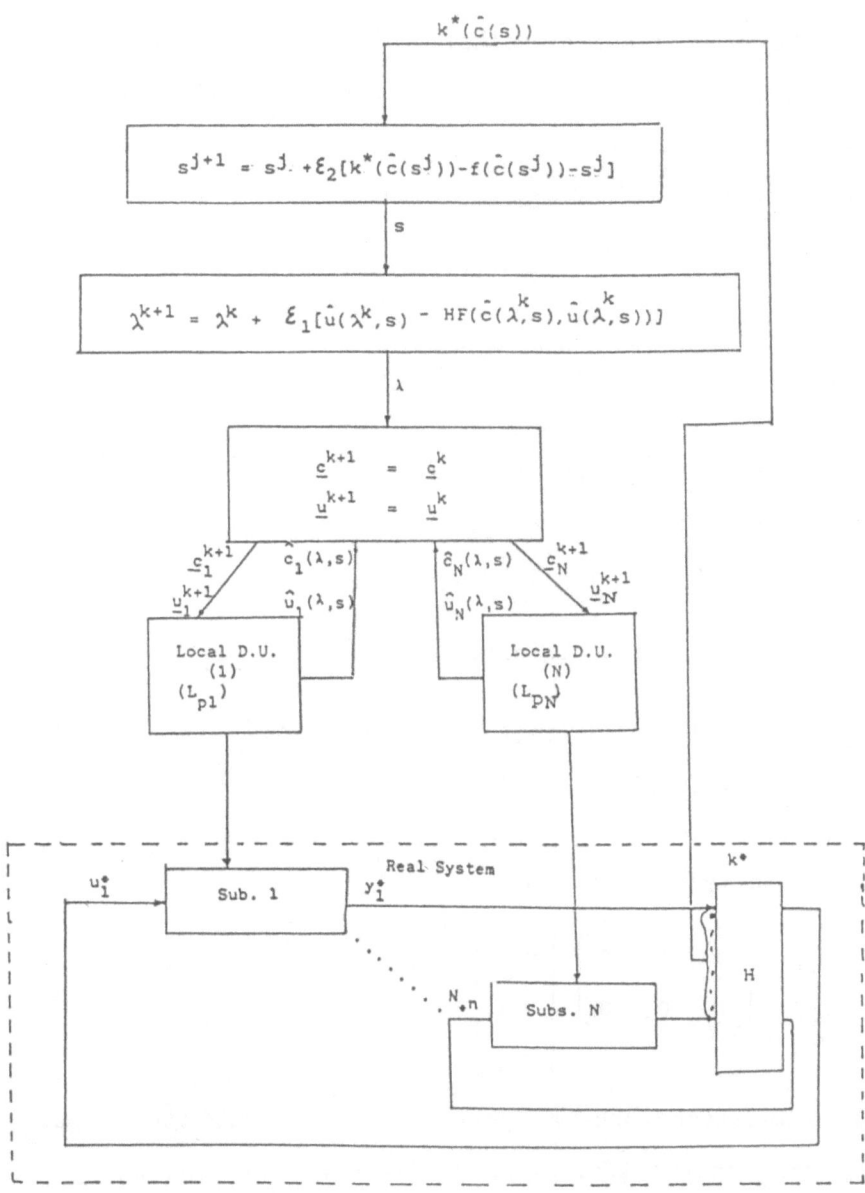

Fig. (1)   Structure of Algorithm

Simulation study

Example 1: This example consists of two interconnected subsystems. The real and model equations of the system are:

$$y_{*11} = 1.4 \ c_{11} - 0.6 \ c_{12} + 1.8 \ u_{11}$$

$$y_{*21} = 1.3 \ c_{21} - 1.1 \ c_{22} + 1.1 \ u_{21}$$

$$y_{*22} = 2.3 \ c_{22} - 0.7 \ c_{23} - 1.1 \ u_{21}$$

$$y_{11} = c_{11} - c_{12} + 2u_{11}$$

$$y_{21} = c_{21} - c_{22} + u_{21}$$

$$y_{22} = 2c_{22} - c_{23} - u_{21}$$

The system performance indices are:

$$Q_1 = (y_{11} - 1)^2 + c_{11}^2 + c_{12}^2$$

$$Q_2 = 2(y_{21}-2)^2 + (y_{22}-3)^2 + c_{21}^2 + c_{22}^2 + c_{23}^2$$

The system constraints are:

$$\{ |c_{11}| \leq 1, |c_{12}| \leq 1, \ y_{11} \geq 0 \text{ and } (0.8 - c_{12} - 0.6 \ u_{11}) \geq 0$$

$$|c_{21}| \leq 1, \ |c_{22}| \leq 1, |c_{23}| \leq 1, \ y_{21} \geq, \ 0, y_{22} \geq 0 \text{ and}$$

$$(2.04 + 1.05 \ u_{21}^2 - c_{21}^2 - c_{22}^2 - c_{23}^2 \geq 0) \}$$

and finally the coupling equation is:

$$\begin{bmatrix} u_{11} \\ u_{21} \end{bmatrix} = \begin{bmatrix} 0 & 1 & 0 \\ 1 & 0 & 0 \end{bmatrix} \begin{bmatrix} y_{11} \\ y_{21} \\ y_{22} \end{bmatrix}$$

The simulation work was performed on a Prime 750 computer and the local optimisation problems were solved using the NAG Library Subroutine E04UAF which uses the augmented sequential Lagrangian technique.

Algorithm	ε_1	ε_2	IS	IT	Real Performance	Real Optimum
present	1.0	1.0	120	4	91	6.0334	
Shao IV	1.0	1.0	-	6	106	5.9844	5.9826
Shao VII	1.0	1.0	-	5	144	6.0431	

Table 1. Comparison of Simulation Result, Example 1.

Conclusions

A novel price correction mechanism is presented in this paper, which is based on shifting model outputs as well as applying the augmented Lagrangian technique. It is hierarchical in structure with three iterative loops. These loops involve model computations only except in the outer loop where measurements from the system is required. Simulation results show that the algorithm is applicable to convex and non convex problems. The technique also reduces on-line correction time at the expense of the off line computations.

References

Lefkowitz, I. (1966): "Multilevel approach applied to control system design' Trans. ASME Series D.J. of Basic Eng, Vol. 88 No. 2, pp392-398

Shao, F. Q. and Roberts P.D. (1983). A price correction mechanism with global feedback for hierarchical control of steady state systems. Large Scale Systems 4, pp67-80

Stephanopoulos, G. and Westerberg, A.W. (1975): The use of Hestenes method of multipliers to resolve dual gaps in engineering system optimisation'. J Optimisation theory Appl. No. 15. pp 285-309

Findeisen, W., Bailey, F.N., Brdys, M., Malinowski, K., Tatjewski, P., and Wozniak, A., (1980). Control and Coordination in hierarchical system', Wiley, London

A LEAST-SQUARES FITTING TECHNIQUE FOR USE WITH LARGE NON-LINEAR PLANT MODELS

J. Hope, CEGB

Summary : Reported in this paper are details of a technique used to extend a power station performance model to allow its application to the fitting of parameters to measured plant data. The original Newtonian iteration scheme is generalised to least squares by use of Lagrange multipliers, and a method which defines the accuracy of the fitted parameters is indicated.

1.   Introduction

Plant models have been developed within the Central Electricity Generating Board as an aid to predicting and improving performance. These models typically employ Newtonian iteration to find the values of a set of unknowns in order to reduce an equal number of residuals to zero.

As application of these models has increased a need has arisen to match the model to a steady state plant condition, perhaps on a routine or automatic basis. In such a case there will be an excess of measurements and rather than discarding some of these, a least squares fitting option was considered. However the requirement on the set of residuals to be zero still remains since these reflect the links between various components in the plant model. An extension to the solution scheme, is therefore required, and is described below. Firstly, however, a summary of the engineering application is given.

2.   Engineering Application

Figure 1 shows the scope of the plant model which includes sub-models of boiler, turbine, feed heaters, valves, pumps, etc. The sub-models will each involve up to three residuals which represent, for example:

(a)   Mass balance of flow in and flow out.

(b)   Heat balances.

(c)   Pressure rise balances.

By economical program writing the number of these residuals can be minimised, but there will remain a number which are unavoidable. In addition a number of links

377

J. Manley et al. (eds.), Proceedings of the Third European Conference on Mathematics in Industry, 377–382.

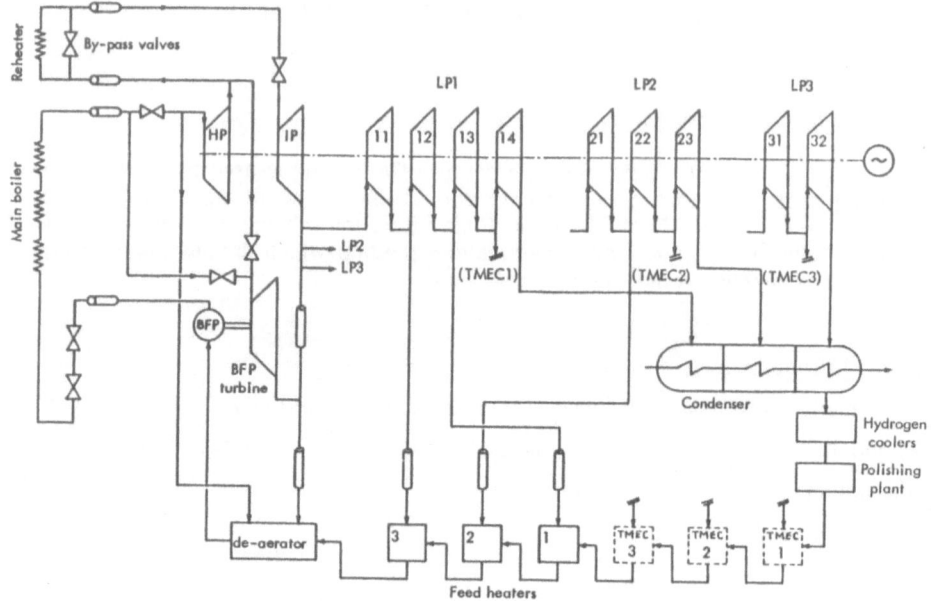

Figure 1. Flow Diagram for Typical Boiler and Turbine Plant

```
FEED HEATER 1
-------------

PRESSURE                0.16 BAR
WATER INLET FLOW      379.95 KG/SEC
WATER INLET TEMP       36.21 DEG C
STEAM INLET FLOW       12.72 KG/SEC
STEAM INLET TEMP       55.14 DEG C
EXIT FLOW             392.67 KG/SEC
EXIT TEMP              55.14 DEG C
```

Figure 2. Example of Printed Results for a Plant Item

between the sub-models will also require to be expressed as residuals and at the present level of modelling some 30 or 40 in all are required for the whole model.

It should be noted however that this is not the number of problem variables, which will number many hundred. A small section of the overall print-out of these variables is shown in Figure 2.

The model has been much used for investigating operating strategies, and for this purpose the present Newtonian solution method has been satisfactory. A new requirement arose in the context of plant monitoring whereby the model should be closely fitted to the present operating point, using a least squares criterion. The approach used is described next.

3. Problem Formulation

We require the values of the unknowns x_j, $(j = 1, m)$ which give a minimum value of the sum of squares:

$$C = G_1^2 + G_2^2 + \ldots G_p^2 \tag{1}$$

where G_i is a measure of model fit and could be defined for example as

$$G_i = \frac{y_i - m_i}{\sigma_i} \tag{2}$$

where

m_i = measured variable

y_i = variable as predicted by model

σ_i = standard deviation of measurement error

In addition the problem is constrained by the residuals:

$$F_i = 0, \quad i = 1, n \tag{3}$$

Available also are derivatives of F and G, evaluated numerically, so that

$$A_{ij} = \frac{\partial G_i}{\partial x_j} \quad \text{and} \quad B_{ij} = \frac{\partial F_i}{\partial x_j} \tag{4}$$

These derivatives will be evaluated at the current solution point x_N , and an improved solution point x_{N+1} will be required, where

$$x_{N+1} = x_N + \delta x \tag{5}$$

4.   Solution Scheme

By invoking linearity about the current solution point, and by use of the Lagrangian method for constrained optimisation, e.g. Fletcher, [1] the correction δx is given by

$$\begin{bmatrix} 2A^T A & B^T \\ B & 0 \end{bmatrix} \begin{bmatrix} \delta x \\ \lambda \end{bmatrix} = \begin{bmatrix} -2A^T G \\ -F \end{bmatrix} \tag{6}$$

where λ , G and F are the vectors of Lagrange multipliers, errors (2) and residuals (3).

To date the method used for the power plant model has used inversion to solve (6). However variations are possible. In particular if the inverse exists and is expressed as

$$\begin{bmatrix} 2A^T A & B^T \\ B & 0 \end{bmatrix}^{-1} = \begin{bmatrix} R & Q \\ Q^T & S \end{bmatrix} \tag{7}$$

then

$$\delta x = -2RA^T G - QF$$

or, from (5)

$$x_{N+1} = x_N - PG - QF \tag{8}$$

where $P = 2RA^T$

Relation (8) has the potential to improve efficiency when P and Q vary little and need to be updated only occasionally.

5.   Computer Implementation

The method has been successfully applied to a range of problems, each generated by a different choice of the errors G_i . In part the reason for the satisfactory implementation results from several years' use of the Newton scheme, from which the Lagrangian method was a natural extension.

It can be seen that information exists to define the confidence that can be placed on the fitted parameters, reflecting the accuracy of the original measurements. This information is expressed by a covariance matrix whose calculation is a straightforward extension and is now described.

6.   Covariance of Estimate

The errors to be fitted were defined in equation (2), and if we assume linearity around the fitted point, then from (8), with $F_i = 0$ (as we already have a solution), it follows that

$$\delta \underline{x} = P \delta G$$

What is required is the covariance matrix of $\delta \underline{x}$ i.e.

$$E\left(\delta \underline{x}.\delta \underline{x}^T\right) = E\left(P\delta G.\delta G^T P^T\right) = P\left[E\left(\delta G.\delta G^T\right)\right]P^T \tag{9}$$

From (2), taking measurement errors to be independent,

$$E(\delta G_i \delta G_j^T) = \frac{1}{\sigma_i \sigma_j} E(\delta m_i \delta m_j^T) = \delta_{ij}$$

Thus $E(\delta G.\delta G^T)$ is the unit matrix, and from (9) the covariance matrix of $\delta \underline{x}$ is therefore PP^T.

The co-variance properties of other variables in the model can now be directly obtained from the appropriate partial derivative matrix. For example this matrix is already available for the measured parameters in relation (4), though for other plant variables an additional numerical evaluation of the derivative is required.

7.   Conclusions

A practical route has been obtained for least squares fitting of a large non-linear model to plant measurements, without recourse to hill-climbing methods. The fitted parameters will form the basis from which studies of power station output optimisation can then proceed.

In practice there is considerable flexibility as to the choice of which variables are set to be free to achieve the fit, and a further benefit is that the structure of the original model is retained. A route by which covariance data may be obtained is also given, so allowing the accuracy of the fitted model parameters to be quantified.

8.   Acknowledgement

The author would like to thank the Central Electricity Generating Board for permission to publish this paper.

9.   Reference

Fletcher, R., Practical Methods of Optimisation, 1987, J. Wiley & Sons.

J. Hope
Generation Development & Construction Division
Central Electricity Generating Board
Barnett Way
Barnwood
Gloucester GL4 7RS

SPLINE APPROXIMATION OF OFFSET CURVES AND OFFSET SURFACES

Josef Hoschek, Franz-Josef Schneider

Abstract: This paper deals with the approximation of (regular) offset curves or offset surfaces (of given spline curves or spline sufaces of degree n) by spline curves or spline surfaces of arbitrarily chosen degree m. The approximating splines are determined by geometric continuity conditions and by parameter optimization for minimizing the range of the approximation error.

1. Introduction

Offset curves or offset surfaces have diverse potential application in geometric modelling, for example for generation of tool paths for milling machines, for description of the outer surface of a car body or a ship body by an offset with material thickness, for definition of tolerance zones.

We suppose that the given surface has the parametric representation $X = X(u,v)$, then the corresponding offset surface X_d at (oriented) distance d along the unit normal vector $N(u,v)$ is given by [3] [4] [7]

(1) $$X_d(u,v) = X(u,v) + N(u,v)\, d.$$

The normal vector N specifies a unique side of X on which the offset d is performed, the opposit side can be chosen by adopting a negative offset magnitude d. In the present paper we will use the Bézier technique therefore we can assume that the given surface $X(u,v)$ is a tensor product Bézier surface of degree (n,m) with the parametric representation

(2) $$X(u,v) = \sum_{i=o}^{n} \sum_{k=o}^{m} V_{ik}\, B_i^n(u)\, B_k^m(v),$$

where $B_i^n(t)$ are the Bernstein polynomials of degree n and V_{ik} are the Bézier points of $X(u,v)$.

The approximation of the offset surface X_d will be a tensor product Bézier surface $Y(u,v)$ of degree $p = 3$ or $p = 5$ and Bézier points W_{ik}.

The approximation of the offset surface X_d by the approximating surface Y can be performed

- *segmentwise*, that means that one segment of the given surface $X(u,v)$ can be approximated by one segment of the surface $Y(u,v)$ with respect to the offset surface X_d and a given error tolerance ε_0,

J. Manley et al. (eds.), Proceedings of the Third European Conference on Mathematics in Industry, 383–389.
© 1990 *Kluwer Academic Publishers and B. G. Teubner Stuttgart.*

- by *splitting* into further segments, that means that one segment of the given surface $X(u,v)$ leads to several surface patches of the approximation surface $Y(u,v)$ of the offset surface X_d with respect to a given error tolerance ε_0,

- by *merging* surface patches, that means that several patches of the given spline surface will be approximated by one patch of the approximation surface $Y(u,v)$ of the offset surface X_d with respect to a given error tolerance ε_0.

The present approach extends results for spline approximation of offset curves [9] to spline approximation of offset surfaces. The given results are extensions of offset spline approxmations for plane cubic splines [1] [14].

2.   Approximation Strategy for Spline Approximation

The key idea of the proposed method is to use the parametrization as a design parameter: The shape of an approximation surface of a set of points will be changed if the parameter values of the points are changed during the approximation process. If we want to change parametrization during the design process we must use spline conditions which are invariant to the parametrization. We can use osculating conditions which are well known in differential geometry [2], [5], [6] as *contact of order k* of two curves or surfaces.

We will discuss approximations of the offset surfaces by Bézier surfaces of polynomial degree $(3,3)$ or $(5,5)$. The approximation process in two steps:

I. the approximation of the boundary curves of the offset surface X_d,
II. the approximation of the interior of the offset surface X_d.

Step I: We assume that the offset surface X_d and the approximation surface Y may have the same corner points

$$(3) \qquad X_d(0,0) = Y(0,0), \quad X_d(1,0) = Y(1,0),$$
$$X_d(0,1) = Y(0,1), \quad X_d(1,1) = Y(1,1).$$

Now we start with the approximation procedure: the boundary curves of the offset surface X_d are the offset curves of given surface determined by (1), for example the boundary curve $v = 0$ is described by

$$(4) \qquad X_d(u,0) = X(u,0) + N(u,0)\, d.$$

For the approximation of the boundary curves we use the methods introduced in [8] combined with parameter-correction [10]. .

Step II approximates the interior of the given offset surface $X_d(u,v)$ while the approximation of the boundary curves remains unchanged. We assume that the given offset surface X_d and the approximation surface Y fulfill in the corner points (3) *contact of order one* for bicubic surfaces and *contact of order two* for biquintic surfaces [2]. Now we choose N points P_{ik} on the given offset surface $X_d(u,v)$ at the parameter values $u_i = \frac{i}{r}$ ($i = 0, \ldots, r$), $v_k = \frac{k}{s}$ ($k = 0, \ldots, s$) with $r > n$, $s > m$ and $N = (r+1)(s+1)$. The error vectors $\quad d_{ik} = P_{ik} - Y(u_i, v_k)$ lead to the total error sum

(5) $\qquad d = \sum_{i=0}^{r} \sum_{k=0}^{s} d_{ik}^2 \qquad\qquad$ which must be minimized.

In general the error vectors d_{ik} are not perpendicular to the approximation surface, thus the total error sum can be reduced again by parameter correction. In [10] for parameter correction the projection of the error vectors onto the corresponding tangents planes was used, a better convergence can be obtained by Newton-like methods introduced in [15].

After this approximation process the offset surface X_d and the approximation surfce Y differ with a maximal error ε_1. If ε_1 is greater than the given error tolerance ε_0, the offset surface X_d will be subdivided into more than one patch. The new boundary will be determined by a special segmentation algorithm which takes into account the curvature of the offset surface (see [11]).

3. Approximation via Bicubic Surface

The goal is to approximate the offset surface (1) by the Bézier surface

(6) $\qquad Y(\mu, v) = \sum_{i=0}^{3} \sum_{k=0}^{3} W_{ik} B_i^3(\mu) B_k^3(v)$

The boundary points W_{00}, W_{03}, W_{30}, W_{33} are determined by (3). In the first approximation step we approximate the boundary curves with the method introduced in [8]: For example for the boundary curve 1 [u = 0, v ϵ [0,1]] the new Bézier points are determined by contact of order one conditions about

(7) $\qquad W_{00} = V_{00} + N(0,0)d, \qquad\qquad W_{03} = V_{0m} + N(0,1)d,$

$\qquad W_{01} = W_{00} + \frac{\lambda_1}{3} \left[m(V_{01} - V_{00}) + N_v(0,0)d \right],$

$\qquad W_{02} = W_{03} + \frac{\lambda_2}{3} \left[m(V_{0,m} - V_{0,m-1}) + N_v(0,1)d \right].$

The parameters λ_1 and the parameter correction are used for minimization the total error of the boundary curves with the help of least square methods. After this step the Bézier points W_{0k}, W_{10}, W_{3k}, W_{13} are determined by the scalar parameter λ_1, the boundary curves of the given and of the approximation surface have common tangents in the corner points.

For evaluation of the unknown inner Bézier points W_{11}, W_{12}, W_{21}, W_{22} we further assume that the following conditions hold for the cross derivatives of the boundary curves

(8)
$$\text{edge 1:} \quad \frac{\partial Y}{\partial u}(0,v) = \alpha_1(v) \frac{\partial X_d}{\partial u}(0,v), \quad 2: \frac{\partial Y}{\partial v}(u,0) = \alpha_2(u) \frac{\partial X_d}{\partial v}(u,0).$$
$$\text{edge 3:} \quad \frac{\partial Y}{\partial u}(1,v) = \alpha_3(v) \frac{\partial X_d}{\partial u}(1,v), \quad 4: \frac{\partial Y}{\partial v}(u,1) = \alpha_4(u) \frac{\partial X_d}{\partial v}(u,1),$$

where the unknown functions α_1 are determined at the boundary curves by the (known) parameters λ_k. We set up the functions α_1 as quadratic Bézier functions with respect to (7) and ω_1 as parameters which shall be used for optimization. If we insert α_1 into the continuity conditions (8) we obtain the following vector-valued linear system for the unknown Bézier points W_{11}, W_{12}, W_{21}, W_{22}:

$$(9) \quad
\begin{pmatrix}
B_1^3(v) & B_2^3(v) & 0 & 0 \\
B_1^3(u) & 0 & B_2^3(u) & 0 \\
0 & 0 & B_1^3(v) & B_2^3(v) \\
0 & B_1^3(u) & 0 & B_2^3(u)
\end{pmatrix}
\begin{pmatrix}
W_{11} \\
W_{12} \\
W_{21} \\
W_{22}
\end{pmatrix}
=
\begin{pmatrix}
\frac{n}{3}\omega_1 N_1(v) + Q_1(v) \\
\frac{m}{3}\omega_2 N_2(u) + Q_2(u) \\
\frac{n}{3}\omega_3 N_3(v) + Q_3(v) \\
\frac{m}{3}\omega_4 N_4(u) + Q_4(u)
\end{pmatrix}$$

where the N_1, Q_1 are expressions with wellknown quantities. The linear system has rank 3, thus (9) describes for a (free chosen) point $P(u_0,v_0)$ the Bézier points W_{12}, W_{21}, W_{22} for example as functions of the parameters $\omega_1, \ldots, \omega_4$ and the Bézier point W_{11}. One of the vanishing determinants of system (9) leads to a (vectorvalued) condition determining $\omega_1 = \omega_1(\omega_1)$ ($i = 2,3,4$), thus the whole problem is reduced to four unknown linear variables: ω_1 and the components of W_{11}.

The total error d depends on these four variables, for the minimun of the total error the following conditions hold

(10) $\quad \dfrac{\partial d}{\partial W_{11}} = 0 \qquad \dfrac{\partial d}{\partial \omega_1} = 0$

These provide a linear system for ω_1 and the components of W_{11}.

4. Approximation by a Biquintic Surface

Now the approximation surface of the given offset surface may have the representation

(11) $\quad Y(\mu,v) = \sum\limits_{i=0}^{5} \sum\limits_{k=0}^{5} W_{ik} B_i^5(\mu)\, B_k^5(v)\,.$

Again the boundary points are determined by continuity conditions and the approximation of the boundary curves are evaluated first with methods developed in [8]. After this first step of the approximation process the Bézier points W_{0k}, W_{10}, W_{5k}, W_{15} are described by the scalar parameters λ_i, μ_j. In the boundary points of the boundary curves the given and the approximation surfaces have common tangents and the same curvature.

Now we choose a special continuity conditions at the corners of the given surface and the approximating surface: for example at the boundary curve 1 the following cross conditions shall hold

$\dfrac{\partial Y}{\partial u}(0,v) = v_{10}(v)\dfrac{\partial X_d}{\partial u}(0,v)\,,$

(12) $\quad \dfrac{\partial^2 Y}{\partial u^2}(0,v) = \left(v_{10}(v)\right)^2 \dfrac{\partial^2 X_d}{\partial u^2}(0,v) + {}^1v_{20}\dfrac{\partial X_d}{\partial u}(0,v)\,,$

$\dfrac{\partial^2 Y}{\partial u \partial v}(0,v) = v(v)\dfrac{\partial^2 X_d}{\partial u \partial v}(0,v) + \sigma(v)\dfrac{\partial X_d}{\partial v}(0,v)$

(the second derivatives $Y_{vv}(0,v)$ are formally determined by a contact of order two condition). The unknown functions ${}^k v_{10}$, ${}^k\overline{v}$, ${}^k\overline{\sigma}$ ($k=1,..,4$) can be set up as linear or as quadratic functions, these functions must fulfill the boundary values of μ_k, λ_k which are evaluated for the boundary curves. Therefore we we can introduce the following functions:

$${}^1v_{10}(v) = \frac{5}{n}\lambda_3 B_0^2(v) + {}^1\omega_8 B_1^2(v) + \frac{5}{n}\lambda_7 B_2^2(v)$$

$${}^1v_{20}(v) = \frac{20}{n}\left(\mu_3 + (\lambda_3)^2\frac{5(n-1)}{4n} - 2\lambda_3\right) B_0^2(v) + {}^1\omega_7 B_1^2(v) +$$

(13) $$\qquad + \frac{20}{n}\left(\mu_7 + (\lambda_7)^2\frac{5(n-1)}{4n} - 2\lambda_7\right) B_2^2(v)$$

$$^1\bar{v}(v) = \frac{25}{n}\left(\lambda_1\,\lambda_3\,B^1{}_0(v) + \lambda_2\,\lambda_7\,B^1_1(v)\right)$$

$$^1\sigma(v) = {}^1\omega_9\,B^2_1(v)$$

with the Bernstein polynomials $B^i_k(v)$ and the unknowns $^1\omega_8$, $^1\omega_7$, $^1\omega_9$.

Additionally we assume that the twist vectors $(X_d)_{uv}$ and Y_{uv} of the given and the required surfaces are parallel in the vertices. Out of this parallelism of the twist vectors the Bézier points W_{11}, W_{14}, W_{41}, W_{44} are determined by conditions like

$$W_{11} = \lambda_1\lambda_3(V_{00} - V_{01} - V_{10} + V_{11} + W_{01} + W_{10} - W_{00} + \tfrac{1}{nm}N_{uv}(0,0)\,d)$$

Now we have to record the still undetermined 12 Bézier points: From (12) and the analogous conditions for the other boundary curves we obtain a linear system for these 12 unknown Bézier points. Out of this system a linear system for the Bézier points W_{22}, W_{23}, W_{32}, W_{33} can be isolated. This system has a singular coefficient matrix analogously to (9). If we now choose (u_0, v_0) out of this small system the Bézier points W_{22}, W_{23}, W_{32} can be recorded, further the vanishing determinant of the system leads to three constraint conditions. From the big system for the unknown 12 Bézier points the points W_{12}, W_{13}, W_{21}, W_{31}, W_{42}, W_{43}, W_{24}, W_{34} can be evaluated as functions of the parameters $^1\omega_k$ and the components of W_{33}. Thus we can determine the total error sum as function of these unknowns: the components of W_{33} and the parameters $^1\omega_7$, $^1\omega_9$, $^2\omega_9$, $^3\omega_9$, $^4\omega_9$ are linear unknowns, while the parameters $^1\omega_1$, $^2\omega_1$, $^3\omega_1$, $^4\omega_1$ are nonlinear unknowns. For minimizing the total error sum nonlinear approximation methods are used combined with suitable parameter corrections as described in [10], [13]. At the end of this approximation process the given offset surface X_d and the approximation surface Y have the following geometric correspondence: they have common tangent planes and the same Dupin indicatrix in each corner point $X_d(0,0) = W_{00}$, $X_d(0,1) = W_{05}$, $X_d(1,0) = W_{50}$, $X_d(1,1) = W_{55}$.

5. Acknowledgement

The research reported here was done at the Centre for Applied Mathematics at the Technical University Darmstadt/ Kaiserlautern University and was sponsored by the Volkwagenwerk Foundation.

References

[1] *Arnold, R.*: Quadratische und kubische Offset-Bézierkurven. Dissertation, Universität Dortmund (1986)

[2] *Cohen, S.*: Beitrag zur steuerbaren Interpolation von Kurven und Flächen. Dissertation, Technische Universität Dresden (1982)

[3] *Farouki, R.T.*: The approximation of non-degenerated offset surface, Computer Aided Geometric Design **3**, 15-43 (1986)

[4] *Faux, I.D.* and *Pratt, M.J.*: Computational Geometry for Design and Manufacture. Ellis Horwood, Chichester. (1981)

[5] *Favard, J.*: Cours de Géométrie Différentielle Locale. Gauthier Paris (1953)

[6] *Geise, G.*: Über berührende Kegelschnitte einer ebenen Kurve, Zeitschr. Angew. Math. Mech. **42**, 297-304 (1962)

[7] *Hoschek, J.*: Offset curves in the plane. Computer-aided Design **17**, 77-82 (1985)

[8] *Hoschek, J.*: Approximate conversion of spline curves. Computer Aided Geometric Design **4**, 59-66 (1987)

[9] *Hoschek, J.*: Spline approximation of offset curves. Computer Aided Geometric Design **5**, 33-40 (1988)

[10] *Hoschek, J.*: Intrinsic parametrization for approximation. Computer Aided Geometric Design **5**, 27-31 (1988)

[11] *Hoschek, J. - Schneider, F.J. - Wassum, P.*: Optimal approximate conversion of spline surfaces. (sub. Computer Aided Geometric Design)

[12] *Hoschek, J.*: Approximate conversion of spline curves. Proceedings of the Second European Symposium on Mathematics in Industry (Ed. by H. Neunzert), 239-249, Teubner (1988)

[13] *Hoschek, J.-Wassum, P.-Schneider, F.J.*: Approximate conversion of splines. Proceedings PROLAMAT Dresden 1988 (Ed. by D. Kochan)

[14] *Klass, R.*: An offset spline approximation for plane cubic splines. Computer-aided Design **15**, 297-299 (1983)

[15] *Wassum, P.*: Approximative Basistransformation von Spline-Flächen mit beliebigem Polynomgrad. Preprint Fachbereich Mathematik Technische Hochschule Darmstadt (1988)

Fachbereich Mathematik
Technische Hochschule Darmstadt
Schlossgartenstrasse 7
6100 Darmstadt, Fed.Rep. of Germany

Diffusion flame ignition by a recirculating flow

M. Konczalla, DFVLR, Institute for chemical propulsion, D-7101 Hardthausen

Abstract

The ignition of a diffusion flame results from the interaction of transport processes and an exothermic chemical reaction. Diffusion flame ignition by a recirculating flow downstream of a backward-facing step is investigated. Special features of this flow are taken into account. The reaction rate is given by an Arrhenius expression, which introduces a nonlinear contribution to the conservation equations. The activation energy of the chemical reaction is used as large parameter in a singular perturbation approach. A well-defined singularity determines an ignition point y_{ig} on the dividing streamline. The influence of the inlet conditions on y_{ig} is investigated. It is shown, that y_{ig} is a decreasing function of the entrance velocity. This behaviour is experimentally observed.

1. Model

A diffusion flame close to a recirculating flow is ignited due to the strong exchange

of heat and mass between the wake and the surrounding flow. A fraction of the heat

released from the flame is transported by the reverse flow in regions upstream of the

flame, where it heats the entering cold flow. Thus a continuous process is maintained.

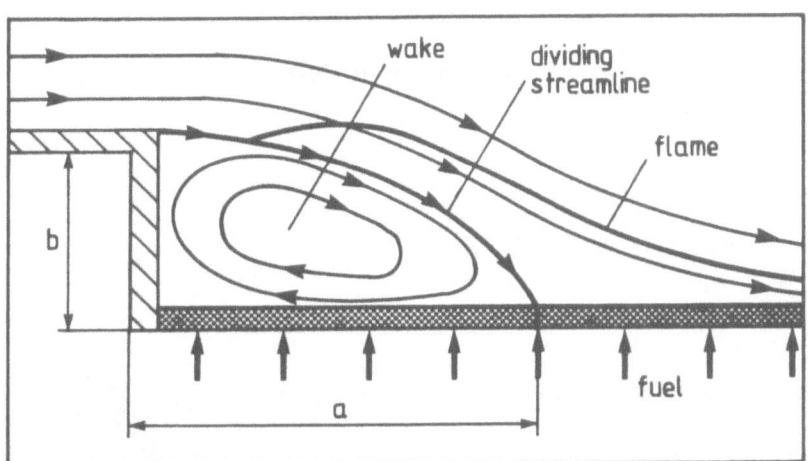

Fig. 1.1 Physical situation

A cold and fast flow (temperature $T'=T'_E$, oxidator mass fraction $Y'_{Ox}=Y'_{Ox,E}$,

vanishing fuel mass fraction $Y'_F=0$, characteristic velocity U_0) separates at the step and

a wake of well-defined length is established. Fuel is supplied through the chamber-

walls (sintermetal). The entering gas mixes with the hot, fuel-rich gas from the re-

J. Manley et al. (eds.), Proceedings of the Third European Conference on Mathematics in Industry, 391–400.
© 1990 *Kluwer Academic Publishers and B. G. Teubner Stuttgart.*

circulation zone in the vicinity of the dividing streamline. In this region ignition occurs. Downstream of the ignition point a stationary flame is observed. A fraction of the flames hot waste gases is entrained into the wake. Inside the stagnant wake conditions are homogenous ($T'=T_R'$, $Y_{Ox}'=0$, $Y_F'=Y_{F,R}'$). T_R' approximately coincides with the flame temperature. By heat and mass exchange an ignitible mixture is established in the surrounding flow. Further details are given in [1,4,5,11].

This description shows, that several processes effect ignition. In the surrounding flow we have to take into account convective transport of heat and species along the streamlines, crosswise diffusion and an exothermic, reactant consuming chemical reaction. Using several assumptions [6] a closed equation for a nondimensional temperature T is obtained:

(1.1) $$\left[\frac{\partial}{\partial y} - \frac{\partial^2}{\partial \eta^2}\right]T = \frac{\Gamma}{\chi^2} f(y)(S_F-T)(S_{Ox}-T)\exp(-T_a/T),$$

(1.2) $$\frac{\partial T}{\partial \eta}\bigg|_0 = \chi\,(T-1)\bigg|_0, \quad T(y=0,\eta) = \lim_{\eta\to\infty} T(y,\eta) = 1-\delta$$

T is given by

(1.3) $$T = T'/T_R'.$$

The normalized mass fractions Y_i (i = F, Ox)

(1.4) $$Y_i = \frac{m_F\,Q}{m_i\,c_p\,T_R'}\,Y_i'$$

are expressed in terms of T and coupling functions S_i [10]

(1.5) $$S_i = T + Y_i,$$

which are solutions of homogenous transient heat equations

(1.6) $$\left[\frac{\partial}{\partial y} - \frac{\partial^2}{\partial \eta^2}\right]S_i = 0$$

with appropriate linear boundary conditions. In (1.2) the nondimensional temperature difference δ is introduced. Γ is a Damköhler number and χ describes the exchange between the wake and its surrounding. $f(y)$ is defined in (1.8). Further m_i (i=F, Ox) refers to the molecular mass of fuel and oxidizer, Q is the specific heat of combustion per unit mass of fuel and c_p is the specific heat of the mixture. T_a is the activation temperature which is used as large parameter.

To derive these equations, the following assumptions are used:

- The flow is divided into the stagnant wake and an outer flow, which is approximated by the potential flow surrounding a quarter ellipse. This choice is suggested by experiments.

- For large Peclet-numbers Pe streamwise diffusion and heat conduction is negligible outside a small region close to the stagnation point.

- An Arrhenius-type, one-step irreversible reaction is used to desccribe the reaction rate. Important properties like reactant consumption and heat release are taken into account.

- Following Cheng and Kovitz [4] Newtonian boundary conditions are used to describe the exchange between the wake and the surrounding flow. It is assumed that the wake behaves as a well stirred reactor with constant values of temperature ($T_R=1$) and mass fractions ($Y_F=Y_{F,R}$, $Y_{Ox}=0$). Experiments suggest [1,11] that the exchange rate is proportional to U_o. This results in an exchange coefficient χ, which is coupled to U_o,

(1.7) $\chi \approx U_o^{\frac{1}{2}}$.

The coordinates y and η are related to the chosen flow. y specifies a streamwise distance measured from the entry plane where y=0 to the stagnation point where y=1. η is related to the streamlines in the vicinity of the dividing streamline. η ranges from 0 (dividing streamline) up to infinity (outer flow). **Fig. 1.2** shows some lines y=const. and η=const. over the wake. An ellipse with a/b=2 is selected.

The deceleration of the flow along η=0 is represented by f(y) appearing on the right-hand side of (1.1),

(1.8) $f(y) = \left[1 - K^2 y^2/a^2\right]/\left[1 - y^2\right]$.

f(y) is proportional to the local residence time of a fluid element at a position y on the dividing streamline.

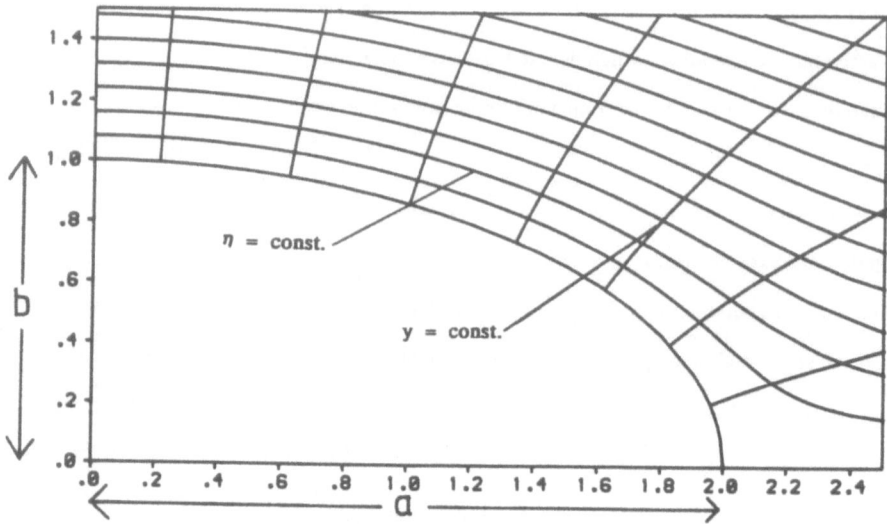

Fig. 1.2 Coordinates y = const. and η = const. over the wake

Eqs.(1.1) and (1.2) show, that chemistry and transport generate competing effects. Improving the exchange by increasing U_0 and χ reduces the local heat generation rate. Also, locally slower velocities given by larger values of $f(y)$ increase the reactive effects due to a larger ratio of a typical fluid dynamical and a characteristic chemical time.

2. Ignition points on the dividing streamline

2.1 Large activation temperatures T_a and large exchange coefficients χ

The initial and boundary value problem for T is analyzed for $T_a \gg 1$ [2]. For diffusion flames this is first done by Linan [7,8]. The asymptotic analysis exhibits a well-defined singularity, which defines an ignition point y_{ig} on the dividing streamline. For $T_a \gg 1$

$$(2.1) \qquad \Gamma = \gamma \exp(T_a), \quad T_a \to \infty,$$

is required. γ is of algebraic order with respect to T_a. (2.1) also implies, that the flame temperature is close to the wake temperature $T_R = 1$. This agrees with experiments [11]. A first result of the analysis is that large exchange coefficients χ

$$(2.2) \qquad \chi = \chi_1 T_a, \quad \chi_1 = O(1)$$

are necessary to obtain proper conditions for ignition. This is seen from analyzing the reactionless ($\Gamma \equiv 0$) temperature T_0 given by

$$(2.3) \qquad T_0 = 1-\delta \left\{ \mathrm{erf}\left(\frac{\eta}{2\sqrt{y}}\right) + \exp\,(x^2 y + x^\eta)(1-\mathrm{erf}\left(\frac{\eta}{2\sqrt{y}}+x\sqrt{y}\right)) \right\}.$$

T_0 describes the inert heating of the entering flow, which is important prior to ignition. Formal analysis shows, that the chemical reaction contributes terms of algebraic order with respect to T_a if

$$(2.4) \qquad 1-T = O(1/T_a);$$

otherwise the reaction term is exponentially small ($O(\exp(-T_a))$) or chemical equilibrium is observed. For $y=O(1)$ and $T=T_0$ condition (2.4) is fullfilled for $\eta=O(1/T_a)$ and $x=O(T_a)$. These conditions result from estimating the order of magnitude of the expression in the large brackets in (2.3).

2.2 Asymptotic expansions of the temperature

$\eta=O(1/T_a)$ suggests to use a stretched coordinate ξ

$$(2.5) \qquad \xi = x_1 T_a \eta, \quad \xi=O(1).$$

The temperature in this region is written as

$$(2.6) \qquad T=1- \frac{1}{T_a} \left[\frac{\delta}{x_1\sqrt{\pi y}}(\xi+1)-p_v \vartheta \right] + o(1/T_a).$$

$p_v\vartheta$ describes the temperature increase due to the chemical reaction, whereas the other terms represent the leading terms of an inner expansion of T_0.

Inserting (2.5) and (2.6) into (1.1) yields an equation for ϑ:

$$(2.7) \qquad \frac{d^2\vartheta}{d\xi^2} = \lambda_{ig} \left[\vartheta-(\xi+1)\right] \exp\left[p_v\left[\vartheta-c(\xi+1)\right]\right].$$

As boundary conditions for ϑ we use

$$(2.8) \qquad \frac{d\vartheta}{d\xi}\Big|_0 = \vartheta\,|_0, \quad \lim_{\xi\to\infty} \frac{d\vartheta}{d\xi} = 0$$

The first condition is the transformed Newtonian boundary condtion, whereas the second condition results from matching with an outer solution $\tilde{\vartheta}$. $\tilde{\vartheta}$ is expressed as Duhamel-integral [3] using the relation

(2.9) $\tilde{\vartheta}(y, \eta=0) = \lim_{\xi \to \infty} p_v(y) \vartheta(\xi; y)$.

In eq.(2.7) the following definitions are chosen:

(2.10) $p_v = \dfrac{Y_{Ox,E}}{x_1 \sqrt{\pi y}}$,

(2.11) $\lambda_{ig} = \dfrac{\gamma}{(x_1 T_a)^4} f(y) \, Y_{F,R}$,

(2.12) $c = \dfrac{\delta}{Y_{Ox,E}}$.

c is an important parameter [7] which describes the reaction's capability to produce maximal temperatures higher (c < 1) or lower (c > 1) as the wake temperature $T_R = 1$. This is seen from computing an adiabatic flame temperature in terms of $Y_{F,R}$, $Y_{Ox,E}$ and δ.

2.3 Analysis of the inner reaction layer

ϑ is determined by a nonlinear boundary value problem. The governing equations are similar to those given by Linan [7] and Linan and Crespo [8], who investigate the ignition of a diffusion flame at the hot system boundary. Here a half-space problem $(0 \le \eta < \infty)$ is treated, whereas the a.m. authors consider a full-space problem, which results in slight differences in the governing equations. More important is the use of Newtonian boundary conditions which introduces an additional parameter into the equations - this choice describes the physically important exchange between the wake and its surrounding.

Similar to Linan [7], multiple solutions for ϑ are expected. To obtain the full set of solutions ϑ_0

(2.13) $\vartheta_0 = \vartheta(\xi = 0)$, $0 \le \vartheta_0 < 1$

is used as independent parameter. This reduces the boundary value problem to an initial value problem with unique solutions. To satisfy the second boundary condition not every value λ_{ig} yields a solution; only eigenvalues λ

(2.14) $\lambda = \lambda(\vartheta_0; p_v, c)$

allow for a solution of the boundary value problem. These λ-values have to be determined numerically.

2.4 Ignition points

If $c=1$ ignition points y_{ig} are determined easily, because $\lambda(\vartheta_0; p_v, c=1)$ is given analytically by

$$(2.15) \quad \lambda(\vartheta_0, p_v, c=1) = \frac{p_v^2}{2} \frac{(2\vartheta_0 - \vartheta_0^2)}{1 - p_v\vartheta_0 + p_v^2} \exp(-p_v\vartheta_0 + p_v).$$

These functions show a maximum at $\vartheta_0^* < 1$ for $p_v > \sqrt{2}$, whereas for $p_v < \sqrt{2}$ the maximum appears at $\vartheta_0 = 1$. In addition for $p_v > \sqrt{2}$ a minimum is observed at $\vartheta_0 = 1$. The appearance of a maximum shows that only up to a well-defined value $\lambda^* = \lambda^*(\vartheta_0^*; p_v, c = 1)$ solutions of the boundary value problem exist. For $\lambda < \lambda^*$ two solutions are determined by different values of ϑ_0, whereas for $\lambda > \lambda^*$ no solution exists. For $p_v < \sqrt{2}$ unique solutions are observed.

Fixing $Y_{F,R}$, $Y_{Ox'E}$, γ/T_a^4, χ and a, b, an ignition point y_{ig} is determined by

$$(2.16) \quad \frac{\gamma}{(\chi_1 T_a)^4} f(y) Y_{F,R} = \lambda^*(\vartheta_0^*; p_v(y), c=1), \quad y=y_{ig}$$

This definition of y_{ig} results in a diverging derivative of ϑ_0 at y_{ig}: $\left. \dfrac{d\vartheta_0}{dy} \right|_{y_{ig}} \to \infty$.

Of special interest is the curve $y_{ig} = y_{ig}(U_0)$ or $y_{ig} = y_{ig}(\chi_1)$. **Fig.** 2.1 shows $y_{ig}(\chi_1)$ for special values of the a.m. parameters. In contradiction to simple physics, there is a decreasing segment of $y_{ig}(\chi_1)$, which agrees with experiments [5].

For $c \neq 1$ the same procedure is applicable. Obtaining an expression for $\lambda(\vartheta_0; p_v, c)$ is the main problem. This is solved by using a perturbation scheme [6] for $(1-c) \ll 1$. Fortunately $(1-c) \ll 1$ is the physically interesting situation, because $(1-c)$ is a measure for the temperature difference between the wake and the flame. A close examination of the resulting expressions shows, that for $c < 1$ ignition points are determined for unlimited values of χ_1 - the maximum of λ observed at $\vartheta_0 = 1$ for $c = 1$ and $p_v < \sqrt{2}$ moves to smaller values of ϑ_0.

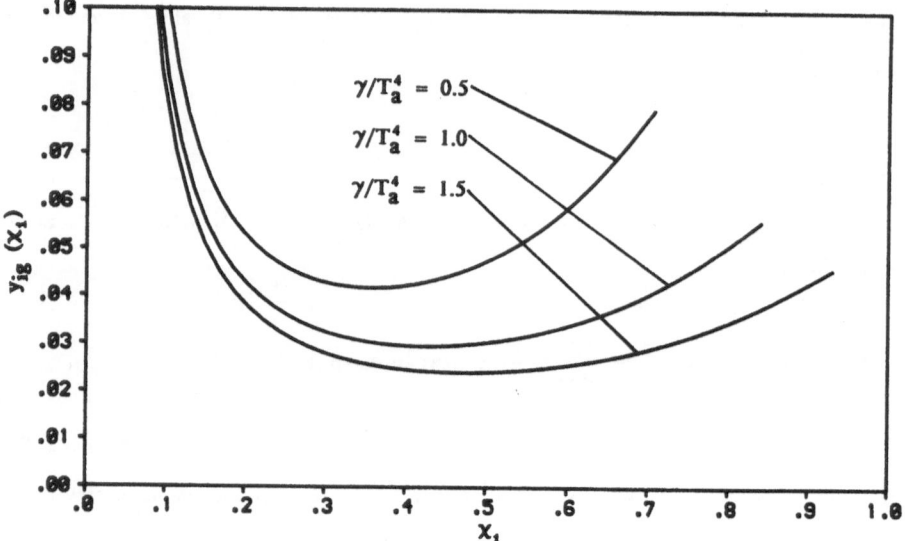

Fig. 2.1 Igniton points on the dividing streamline for c = 1.0
$Y_{F,R} = Y_{Ox,E} = = 0.5$, a/b =2

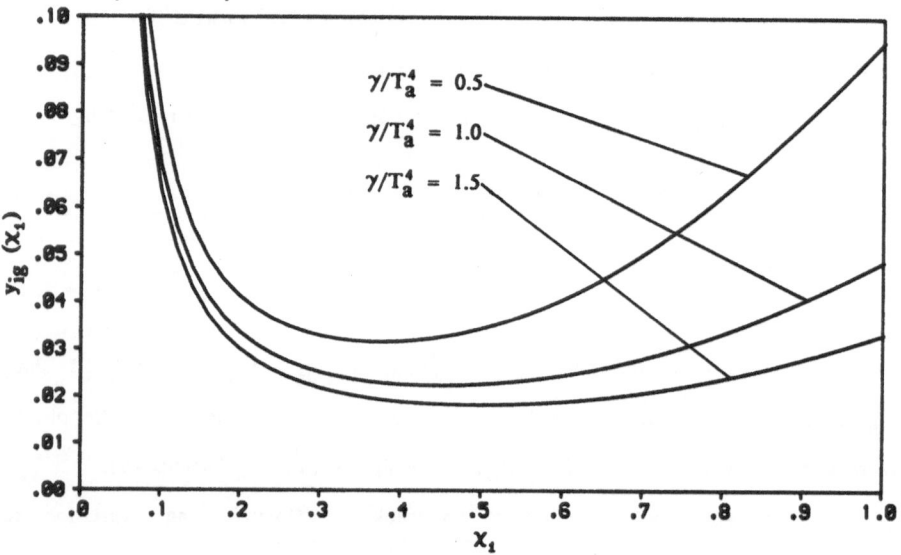

Fig. 2.2 Ignition points on the dividing streamline for c = 0.9.
$Y_{F,R} = Y_{Ox,E} = 0.5$, a/b = 2

For c = 0.9 resulting curves for y_{ig} (x_1) are shown in **Fig. 2.2**. If c > 1.0 critical behaviour is only observed for p_v-values exceeding a special value p_v^+ (c). Critical solutions disappear in a cusp in the λ-p_v-plane which is typical for a two-parameter-system [9].

3. Conclusions

Using a singular perturbation approach ignition points y_{ig} are determined. A decreasing segment of $y_{ig} = y_{ig}(X_1)$ is obtained. This agrees with experiments [5], where a decreasing ignition distance with increasing entrance velocity is observed. It is important to note that decreasing functions $y_{ig}(X_1)$ result from using Newtonian boundary conditions and the coupling of the exchange coefficients X to the entrance velocity U_0. Downstream from y_{ig} a flame is observed, which is described in terms of Linan's premixed flame analysis [8]. An asymptotic analysis of the corresponding equations [6] results in a free boundary value problem for the flame.

4. Literature

[1]   Beér, J.M.; Chigier, N.A.: Combustion Aerodynamics Reprint Edition. Malabar: Robert E. Krieger Publishing Company 1983

[2]   Buckmaster, J.D.; Ludford, G.S.S.: Theory of Laminar Flames, Cambridge: Cambridge University Press 1982

[3]   Carslaw, H.S.; Jaeger, J.C.: Conduction of Heat in Solids. Second Edition. Oxford: Oxford University Press 1959

[4]   Cheng, S.I.; Kovitz, A.A.: Theory of Flame Stabilization by a Bluff Body, 7th Symposium (Intern.) on Combustion 692-696 London: Butterworth 1959

[5]   Kawamura, T.: The Ignition Front of a Fuel Jet Flame stabilized by a Step. Comb. Flame 22 (1974) 283-288

[6]   Konczalla, M.: Ph. D. Thesis, University of Bielefeld, 1989

[7]   Linan, A.: The Asymptotic Structure of Counterflow Diffusion Flames for Large Activation Energies. Acta Astronautica, 1 (1974) 1007-1039

[8]   Linan, A.; Crespo, A.: An Asymptotic Analysis of Unsteady Diffusion Flames for Large Activation Energies. Comb. Sci. Techn. 14 (1976) 95-117

[9]   Meinköhn, D.: Das Zünden und Verlöschen einer autothermen Reaktion in einem porösen Partikel in oxidierender Atmosphäre. Ber. Bunsenges. Phys. Chem. 87 (1983) 992-996

[10]   Williams, F.A.: Combustion Theory. Second Edition. Menlo Park: Benjamin/ Cummings Publishing Company 1985

[11]   Winterfeld, G.: On Processes of Turbulent Exchange behind Flame Holders. 10th Symposium (Intern.) on Combustion 1265-1275 Pittsburgh: The Combustion Institute 1965

Matthias Konczalla

DFVLR Institute for chemical propulsion and engineering

D - 7101 Hardthausen a. K.

West - Germany

ON THE REAL–TIME SIMULATION AND CONTROL
OF THE CONTINUOUS CASTING PROCESS

E. LAITINEN[1], P. NEITTAANMÄKI[1] AND T. MÄNNIKKÖ[1]

Abstract. In this paper we present a method for controlling in real-time the secondary cooling of the continuous casting machine. The method is based on the heat transfer model, which simulates the temperature distribution of the cast, and on the use of direct digital control (DDC). Off-line simulation results considering the control method is presented.

1. INTRODUCTION

The control of the strand temperature field and the shell thickness along the casting machine are of central importance in continuous casting operation. Both have a considerable influence on defects which can be formed in cast material. When achieving to use direct charging/rolling, the slabs with maximum heat content should be got out at the machine exit. Accurate knowledge of the liquid pool end location is important especially when using soft reduction near the pool end.

Many mathematical models for simulating the strand temperature field and the shell thickness profile have been developed and applied in recent years [1,3,5,13]. Numerical calculations are mostly performed by using the finite difference method (FDM) or the finite element method (FEM). These methods permit the use of temperature dependent material properties, complex boundary conditions and moving phase boundaries at the solidification front. However, most of the developed models are suitable only for steady-state casting conditions.

This work is a continuation of the works of authors [6,8] and those of [7,12] where is presented the method for defining the optimal secondary cooling for the steady-state continuous casting process. In this paper we present a nonlinear parabolic-type simulation model which simulates the multiphase heat transfer during solidification in unsteady-state continuous casting process. The simulation model is applied for the real-time secondary cooling control of the continuous casting process. The aim of the control system is to keep the surface temperature distribution of the strand constant with respect to time, in spite of the casting speed variations. The simulator calculates the actual internal and surface temperature of the strand using a mathematical model based on a heat conduction formula with phase changes. The optimal spray water flow rate for each spray cooling zone will be calculated by minimizing the deviation between the target surface temperature and the calculated actual surface temperature. The DDC-algorithm then directly guides the flow valve to maintain the optimal spray water flow rates, without need for an external valve positioners. Off-line simulation results considering the control method will be presented.

[1]University of Jyväskylä, Department of Mathematics, Seminaarinkatu 15, SF-40100 JYVÄSKYLÄ, FINLAND

J. Manley et al. (eds.), Proceedings of the Third European Conference on Mathematics in Industry, 401–408.

2. FORMULATION OF THE REAL-TIME TEMPERATURE SIMULATION MODEL FOR THE UNSTEADY-STATE CONTINUOUS CASTING PROCESS

Since in dynamic, real-time simulation a high calculation speed is required, only a one-dimensional approximation in space is applied in the mathematical model. Moreover, the following assumptions considering the continuous casting characteristics are assumed in the model:

(i) Because the model is one-dimensional in space, heat transfer through the thickness of the wide side alone is considered. It means that the model can be applied accurately only to the cross-sections which have a sufficiently large wide side/narrow side ratio. According to [11], the ratio should be greater than 3. Heat conduction to the withdrawal direction can be neglected, due to the relatively high withdrawal rate and low thermal conductivity of the steel.

(ii) The solid fraction f_s is assumed to vary piecewise linearly between solidus and liquidus temperatures. The latent heat emission is proportional to the change of the solid fraction.

(iii) Specific heat, density and thermal conductivity of steel are defined separately for several disjoint temperature intervals, inside which they are assumed to be constants.

(iv) Convection inside the liquid pool was taken into account by using an effective thermal conductivity. It was assumed to be five times greater than the nominal value of thermal conductivity in liquid steel.

The nonlinear heat conduction equation reads:

$$(1) \qquad \rho(T)c(T)\frac{\partial}{\partial t}T = \mathrm{div}(k(T)\nabla T) + q,$$

Here $T = T(x,t)$ is the temperature in a point $x \in \Omega \subset \mathbf{R}$ at a time t; $q = q(T)$ describes the latent heat of the material. Furthermore, $c(T)$ denotes the specific heat, $\rho(T)$ the density and $k(T)$ the thermal conductivity.

By applying an expression $f_s = f_s(T)$ which describes the solid fraction in the mushy zone, q can be expressed as follows:

$$(2) \qquad q(T) = \rho L \frac{\partial f_s}{\partial T}\frac{\partial T}{\partial t} \qquad T_s < T < T_l,$$

where T_s and T_l are the solidus and liquidus temperatures of steel.

By defining a smooth enthalpy function:

$$(3) \qquad H(T(x,t)) = \int_0^{T(x,t)} \left[\rho(\xi)c(\xi) - \rho(\xi)L\frac{\partial f_s}{\partial T} \right] d\xi$$

which takes into account both the latent heat and the specific heat, and by applying the Kirchhoff's transformation:

$$(4) \qquad K(T(x,t)) = \int_0^{T(x,t)} k(\xi)d\xi$$

we get the following enthalpy formulation for the equation (1):

$$(5) \qquad \frac{\partial}{\partial t}H(T(x,t)) = \Delta K(T(x,t)),$$

In order to solve the strand temperature distribution from the equation (5), the boundary and initial conditions must be specified. The following assumptions considering the boundary and initial conditions are assumed in deriving the model:

(v) Heat transfer from the strand surface to the mold cooling water is mainly governed by the air gap formation between the solid shell and the mold face. Many investigators have attempted to predict the heat transfer coefficient profiles, $h_{mold}(x,t)$, in the mold [2,11].
Knowing the total heat transfer, Q_{av}, the heat transfer coefficient in the mold, h_{mold}, can be approximated by

$$(6) \qquad h_{mold} = \frac{Q_{av}}{A \cdot (T - T_{mold})}$$

where
   – A is the area of mold faces
   – T is the surface temperature of the strand
   – T_{mold} is the ambient mold temperature

In this model the total heat transfer, Q_{av}, through the mold wall (kW) is approximated as:

$$(7) \qquad Q_{av} = \rho_w \cdot c_w \cdot W \cdot \Delta T$$

where
   – ρ_w is the density of cooling water
   – c_w is the specific heat of cooling water
   – ΔT is the temperature difference between input and output temperatures of the cooling water circuit
   – W is cooling water flow rate (l/min)

Above ΔT and W are the actual measured values (input parameters). Hence, the heat flux through the mold wall can be expressed as:

$$(8) \qquad -k(T)\frac{\partial}{\partial n}T = h_{mold}(T - T_{mold})$$

(vi) In the secondary cooling region the heat extraction is characterized by:

(9)
$$-k(T)\frac{\partial}{\partial n}T = h(T - T_{H_2O}) + \sigma\varepsilon(T^4 - T_{ext}^4)$$

where $h = h(x,t)$ is the heat transfer coefficient in the secondary cooling region. T_{H_2O} is the spray water temperature and T_{ext} is the air temperature. The relationship between the heat transfer coefficients h and the spray conditions like spray water flow rates, spray pressure, nozzle type and steel surface temperature must be determined experimentally. In the developed model, for slab geometry and carbon steels, an experimental correlation by Nozaki et al. [9] was used to describe this relationship in the secondary cooling zones:

(10)
$$h = \frac{1.57 \cdot W^{0.55}(1. - 0.0075 \cdot T_{H_2O})}{\alpha}$$

where α is an empirical factor to take account of the presence of support rolls. By using α, the starting up conditions when the support rolls are cooler, can also be taken into account.

(vii) In the air cooling region it is assumed that heat is extracted by radiation only. Hence, the heat flux across the boundary can be expressed as:

(11)
$$-k(T)\frac{\partial}{\partial n}T = \sigma\varepsilon(T^4 - T_{ext}^4)$$

The emissivity value of 0.8–0.9 is normally used for the oxidized strand surface. In cases, where insulating covers are used to protect the strand against heat loss, the value of emissivity must be adjusted respectively.

(viii) Initial conditions (at meniscus level): the melt is assumed to have a temperature equal to the incoming metal temperature at distance $z=0$:

(12)
$$T(x,t) = T_0(x,t) \qquad \text{in } \Omega$$

Let $z \in \mathbf{R}$ be the longitudinal coordinate along the casting machine and τ the 'age' (residence time) of a one-dimensional slice created at the distance $z = 0$. The 'age', τ, of a slice at a distance z is defined as follows:

$$z = \int_{t-\tau}^{t} v(\xi)d\xi$$

where $v(\xi)$ is the withdrawal rate and t the real-time.

Then at a present time $t \in \mathbf{R}$, the temperature distribution $T(z,x;t)$ at a distance z from the meniscus can be calculated by solving the system:

$$\begin{cases} v_{av}\dfrac{\partial}{\partial z}H(T(z,x;t)) = \Delta K(T(z,x;t)) & x \in]0,l[; z \in]0,L[\quad (13) \\[2mm] -k(T)\dfrac{\partial}{\partial x}T(z,x;t) = \begin{cases} h_{mold}(T - T_{mold}) & x = 0, l; z \in]0,z_1] \\ h(T - T_{H_2O}) + \sigma\epsilon(T^4 - T_{ext}^4) & x = 0, l; z \in]z_1, z_2] \\ \sigma\epsilon(T^4 - T_{ext}^4) & x = 0, l; z \in]z_2, L[\end{cases} (14) \\[2mm] T(z_0,x;t_0) = T(z - v_{av}\Delta t, x; t - \Delta t) & x \in]0,l[\quad (15) \end{cases}$$

where $T(z_0, x; t_0)$ is the temperature distribution at previous time event $t_0 = t - \Delta t$ at distance $z_0 = z - v_{av}\Delta t$ and $v_{av} = \frac{v(t)+v(t+\Delta t)}{2}$. Moreover, z_1, z_2 and L are the positions along casting machine where the strand exits from the mold, from the secondary cooling region and from the machine, respectively.

Now, the calculation procedure for solving the temperature distribution of the strand can be described as follows:

- The strand is divided into tracking planes, which positions to the withdrawal directions remain constants.
- For each tracking plane the actual temperature field $T(z, x; t)$ is calculated by solving the model (13)–(15) using the actual measured input data such as casting speed, mold heat flux, liquid steel temperature, cooling water flow rates, etc.
- The temperature field of the initial tracking plane (at meniscus level) is set equal to the incoming liquid steel temperature.

The discretization of the equations (13)–(15) is performed by using the FE-method in space and the implicit Euler method in time. The discretization and solving algorithm for the discretized equations are presented in [5].

3. Control system functioning

Figure 1 shows the control system in diagram form. A microcomputer (IBM-AT) is used to solve the real time temperature model which defines the actual temperature distribution of the slab on the basis of the actual casting speed, steel grade, cooling water temperature, steel pouring temperature and cooling water flow rates. A cathode ray tube (CRT) displays the real-time process information. Casting operators communicate with the microcomputer by means of the function keyboard which is used to enter manual data such as cast number, desired surface temperature (temperature setpoint) and to operate the spray control system. The microcomputer compares the calculated surface temperature with the temperature setpoint and, if needed, uses a control algorithm to adjust the actual water flow values to bring the surface temperature to the target temperature.

The spray control system is based on the direct digital control (DDC) where the microcomputer also determines the deviation between the water-flow setpoint and the flow. The microcomputer then directly guides the flow valve to maintain the flow setpoint

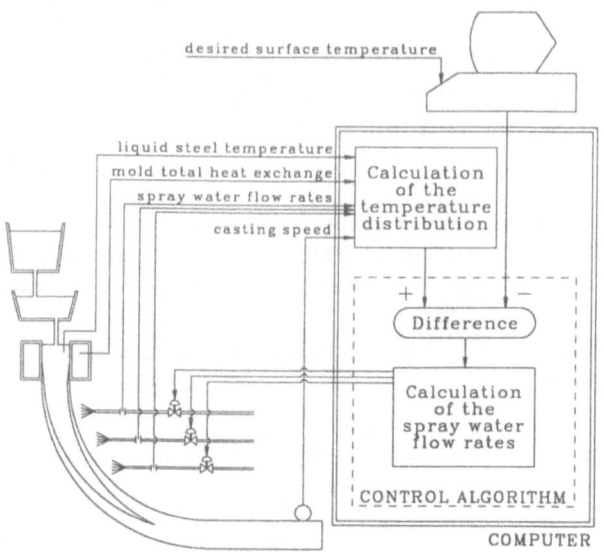

Figure 1 Schematic diagram of the dynamic spray control system.

without need for an external valve positioner. The water flow rates W^n are calculated in the following manner:

ALGORITHM 3.1:

- Give the actual casting speed v^n and the desired surface temperature Θ_A^n. Compute average casting speed v_{av} between time events t^{n-1} and t^n.
- Estimated casting temperature Θ_i^n at each control point i is calculated using the real time model (13)–(15).
- The difference, ε_i, between the estimated casting temperature, Θ_i^n, and the target temperature, Θ_A, is calculated for each control point and an average difference, ε, is calculated for each cooling zone.
- The water flow rate, W^n, to minimize the difference ε is calculated from equation

(16)
$$W^n = W^{n-1} + K_P \left[\varepsilon + \frac{1}{T_I} \int_{t^{n-1}}^{t^n} \varepsilon \, dt + T_D \frac{d\varepsilon}{dt} \right]$$

which corresponds to the principle of the analogy PID- controller. Here K_P, T_I and T_D are tuning parameters.

4. EXAMPLE

The effect of variation in casting speed to the strand surface temperatures have been studied with help of the off-line simulation runs. Simulation results for a slab caster are shown in Figures 2 (a)–(b) when casting speed varies as shown in Figure 2 (a). The material and casting data used in the simulation run were: slab geometry 1030×210 mm^2, liquid temperature $1525°C$, solid temperature $1460°C$, total mold heat flux $3080kW$. In the Figure 2 (a) the secondary cooling and the mold heat flux were kept constant during the simulation. The simulation results show that the surface temperatures fluctuate considerably, especially at the upper part of the machine.

It is evident that in order to prevent the surface temperature fluctuation during changes in casting parameters, the water flow rates in the secondary cooling zones must be controlled. In Figure 2 (b) are shown the simulation results for the same slab caster when casting speed has varied as previously but the secondary cooling has changed according to the ALGORITHM 3.1. In the Figure 2 (b) the dashed line represents the variation of the secondary cooling water flow rates and the solid lines the variations of surface temperatures. In this case the surface temperature fluctuation is not noticeably.

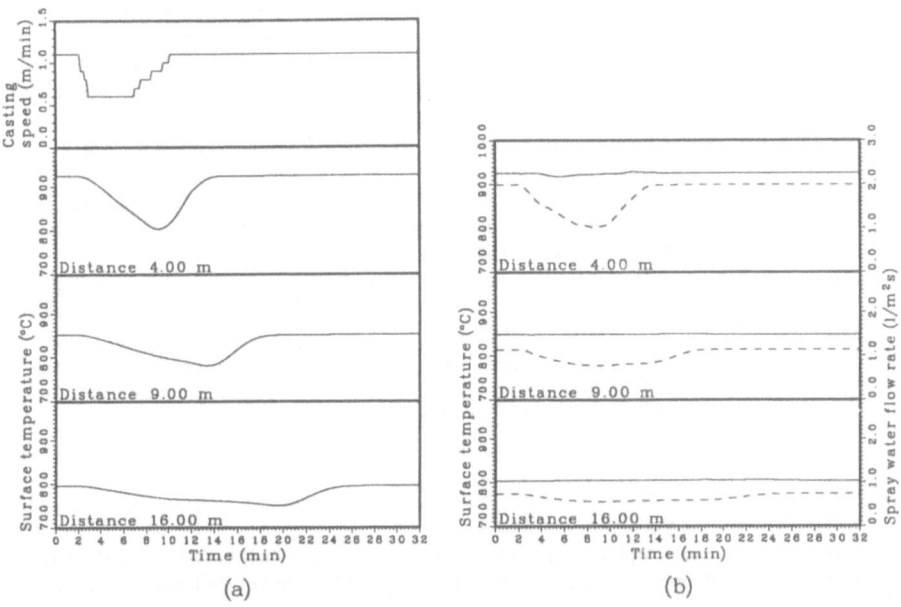

Figure 2 Fluctuation in the strand surface temperatures (a) without the secondary cooling control and (b) with the PID-control when the casting speed has varied.

5. CONCLUSION

The optimal strategy for the real-time secondary cooling control of the continuous casting process is developed. The control system is implemented on IBM AT-type microcomputers. The presented method is going to industrial tests on steel works in Finland. It is evident that the temperature simulation -based control systems become general

408

on steel works due to their flexibility and better accuracy in proportion to the conventional method (water flow proportional to speed). Especially in the continuous casting – direct rolling (CC–DR) process, conditions of greater severity are imposed on the CC secondary cooling than in the conventional continuous casting process which has the reheating furnace downstream. The requirements for the manufacture of slabs of the highest possible temperature poses various difficult problems in the method of slab cooling for the CC–DR process. The successful manufacture of high temperature defect-free slabs is one of the most important technical conditions that should be satisfied for the realization of the CC–DR process.

REFERENCES

[1]   Baptista, L.A. (1979), *Control of spray cooling in the continuous casting of steel*, Theses of University of British Columbia.

[2]   Brimacombe, J.K., Samarasekera, I.V. and Lait, J.E. (1984), Vol 2, Book Crafters, Ins., Chelsea, MI (1984), "Continuous Casting,".

[3]   Etienne, A. (1983), *Metallurgical control of a slab continuous casting machine*, Report EUR 7405 EN, Comission of the European Communities.

[4]   Kawasaki, S., Kikunaga, M., Kajita, Z., Arita, H., Chida, Y. (1984), *On the Secondary Cooling Control Technology for the Continuous Casting – Direct Rolling Process*, Nippon Steel technical report No: 23, UDC621.746.047:621.78.08.

[5]   Laitinen, E., Neittaanmäki P. (1988), *On numerical simulation of continuous casting process*, Journal of Engineering Mathematics 22, 335-354, Kluwer Academic Publishers, Dordrecht.

[6]   Laitinen, E., Neittaanmäki P. (1988), *On numerical solution of the problem connected with the control of the secondary cooling in the continuous casting process*, C-TAT, Control Theory and Advanced Technology 4 No 3, Mita Press, Tokyo, Japan.

[7]   Larrecq, M., Birat, J.P., Saquez, C. and Henry, J. (1983), *Optimization of casting and cooling conditions on steel continuous casters; Implementation of optimal strategies on slap and bloom casters*, IRSID, ACI 83 RE 1004, Jullet 1983.

[8]   Neittaanmäki, P. (1986), *On the control of cooling during continuous casting*, in Proc. of the 4th Int. Conf. on Numerical Methods in Thermal Problems, July 15–18, 1985 in Swansea (ed. Lewis, R.W., Morgan K.), 240–249, Vol I and II, Pineridge Press.

[9]   Nozaki T. et al, *A Secondary Cooling Pattern for Preventing Surface Cracks of Continuous Casting Slabs*, TISIJ 18(1986)6.

[10]   Okuno, K., Naruwa, H., Kuribayashi, T., Takamoto, T. (1987), *Dynamic spray cooling control system*, Iron and Steel Engineer 64, No 4.

[11]   Rogberg, B. (1982), *High temperature properties of steel and their influence on the formation of defects in continuous casting*, Dissertation, The Royal Institute of Technology, Department of Casting of Metals.

[12]   Saguez, C. (1986), *Optimal control of free-boundary problems*, in "Proceedings of 12th IFIP-conference on System Modelling and Optimization, Budapest September 2–6, 1985, to appear in LN in Control and Information Sciences," Springer Verlag.

[13]   Wang, Z.G. and Inoue, T. (1986), *Analysis of temperature and elastoviscoplastic stresses during continuous casting*, in "Proc. of ICCM-86, Tokyo, VIII-103 – VIII-108," Springer-Verlag.

[14]   Yamasaki, J., Miyakara, S., Kodama, M., Matsukawa, T., Matsumo, J. and Suzuki, K. (1981), *The control of the surface temperature in the continuous slab casting machine by on-line digital computer*, in "Proc. of 8th Triennial World Congress, Kyoto, Japan," IFAC Control Science and Technology, 2639–2644, 1981.

SIMULATION OF VLSI CIRCUITS: RELAXATION TECHNIQUES

Peter Lory, München (Germany)

1. Introduction

Circuit simulation programs have proven to be most important CAD tools for the analysis of the electrical performance of integrated circuits. Depending on the number of modeled transistors, these simulators require the numerical solution of initial value problems for very large, sparse systems of differential (or even differential-algebraic) equations. For a survey see Refs. 1 and 2.

Waveform relaxation (WR) has been proposed in Ref. 5 for the numerical solution of these initial value problems. It applies the well-known Gauss-Seidel and Jacobi principles for the numerical solution of systems of algebraic equations on the function space level (see also Refs. 8 and 11). Each differential equation of the system, which models an integrated circuit, corresponds to one node in this network. In its simplest form, WR solves these equations as single differential equations in one unknown, and these solutions are iterated until convergence. If this kind of node-by-node decomposition strategy is used for circuits with even just a few tightly coupled nodes, the WR algorithm will converge very slowly. Its efficiency can be greatly improved by lumping together tightly coupled nodes and partitioning the system correspondingly. The relaxation principle is then applied in a blockwise manner.

In practical applications, the groups of differential equations in the partitioned system have to be solved numerically in each cycle of the WR iteration. As stiffness is a characteristic feature in VLSI applications, most circuit simulation programs (and the present note) use the stiffly stable backward differentiation formulas for discretization (see Ref. 3). Other approaches, like Runge-Kutta related methods, can be applied in principle, too (see Ref. 9).

The present note shows that in the linear case the discretized WR algorithm is equivalent to the (algebraic) block relaxation method if the latter is applied to a properly defined linear system of algebraic equations (Section 2). Naturally the dimension of this system is greatly enlarged. However, the eigenvalue problem of the corresponding iteration matrix can be reduced to the original dimension. Consequently, estimates of the convex hull of its spectrum can be computed efficiently by an adaptive technique. This suggests a method for accelerating the rate of convergence for WR (Section 3).

J. Manley et al. (eds.), Proceedings of the Third European Conference on Mathematics in Industry, 409–414.
© 1990 *Kluwer Academic Publishers and B. G. Teubner Stuttgart.*

2. The Discretized Waveform Relaxation in the Linear Case

Valuable insight into the WR method can be gained by an investigation of the linear case

(2.1) $\quad C\dot{x} = Fx + f, \quad x(0) = x_0, \quad 0 \le t \le T.$

Here, C ("capacitance matrix") and F are constant $(n \times n)$-matrices, and f is a constant vector. Let Equ. (2.1) be partitioned as motivated in Section 1:

$$(2.2) \quad C = \begin{bmatrix} C_{11} & \cdots & C_{1p} \\ \vdots & & \vdots \\ C_{p1} & \cdots & C_{pp} \end{bmatrix}, \qquad F = \begin{bmatrix} F_{11} & \cdots & F_{1p} \\ \vdots & & \vdots \\ F_{p1} & \cdots & F_{pp} \end{bmatrix},$$

where C_{ii} and F_{ii} are $(n_i \times n_i)$ -matrices (i = 1,...,p) and $n_1+...+n_p = n$. The dimensions of the other submatrices and the partition of the vectors f and x are defined accordingly. Let the linear multistep formula

$$(2.3) \quad \sum_{k=0}^{s} \alpha_k x_{\ell-k} = h \cdot \sum_{k=0}^{s} \beta_k \dot{x}_{\ell-k}$$

be used, and let the interval $0 \le t \le T$ be subdivided by equidistant points

(2.4) $\quad t_\ell = \ell \cdot h \quad (\ell = 0,...,N), \quad h = T/N.$

For the sake of simplicity the same (uniform) stepsize h and the same formula (2.3) is used for the discretization of all the differential equations. In reality, one of the advantages of the WR algorithm is to let different subsystems be integrated according to their individual dynamical properties. Hence, the last assumption is somewhat artificial. However, this simplification allows a concise description of the basic ideas and valuable first insights. Let Equ. (2.1) be discretized by (2.3):

$$\sum_{k=0}^{s} \alpha_k C x_{\ell-k} = h \cdot \sum_{k=0}^{s} \beta_k (F x_{\ell-k} + f),$$

and let $x_{i\ell}$ denote the approximation of $x_i(t_\ell)$. Define

$$A := \begin{bmatrix} \alpha_0 & & & & 0 \\ \alpha_1 & \alpha_0 & & & \\ \vdots & \ddots & \alpha_0 & & \\ \alpha_s & \cdots & \alpha_1 & \alpha_0 & \\ & \ddots & & \ddots & \ddots \\ 0 & & \alpha_s & \cdots & \alpha_1 & \alpha_0 \end{bmatrix} \varepsilon \mathbf{R}^{N,N}, \quad B := \begin{bmatrix} \beta_0 & & & & 0 \\ \beta_1 & \beta_0 & & & \\ \vdots & \ddots & \beta_0 & & \\ \beta_s & \cdots & \beta_1 & \beta_0 & \\ & \ddots & & \ddots & \ddots \\ 0 & & \beta_s & \cdots & \beta_1 & \beta_0 \end{bmatrix} \varepsilon \mathbf{R}^{N,N},$$

$$M_\pi(h,N) := C * A - h \cdot F * B \qquad (*: \text{ direct product}),$$

$$z := (x_{11}, \ldots, x_{1N}; \ldots ; x_{n1}, \ldots, x_{nN})^T \; \varepsilon \; \mathbf{R}^{N \cdot n}.$$

Let the discretized WR Gauss-Seidel (Jacobi) method be applied to system (2.1), which is partitioned according to (2.2). It is easy to see that this algorithm is equivalent to the standard algebraic block Gauss-Seidel (Jacobi) method if the latter is applied to the following system of linear algebraic equations:

(2.5) $M_\pi(h,N) \cdot z = $ right hand side.

For the definition of block (or group) iterative methods see e.g. Refs. 10 and 12. Throughout this paper, the index π refers to the following partition of the matrices into p^2 blocks: The diagonal blocks have sizes $n_1 N \times n_1 N$, ..., $n_p N \times n_p N$; the dimensions of the off-diagonal blocks are chosen accordingly. Please note that the enlargement of the size of the matrices from n to n$\cdot$N is for theoretical purposes only. In actual computations, the explicit formation of the matrix M_π is unnecessary. Further, the eigenvalue problem for the corresponding iteration matrix can be reduced to dimension n. This fact will be proven (and exploited) in the following section.

3.   Accelerating the Rate of Convergence

Let

$$M_\pi = D_\pi + L_\pi + U_\pi \, ,$$

where D_π is the block diagonal part, L_π the strictly lower block triangular part, U_π the strictly upper block triangular part. Here, the blocking refers to partition π. The *iteration matrix of the block Gauss-Seidel method* is

(3.1) $G_\pi := -(D_\pi + L_\pi)^{-1} \cdot U_\pi .$

G_π is a ($nN \times nN$)-matrix. Let the ($n \times n$)-matrices C and F of Equ. (2.1) be split into

$$C = D_C + L_C + U_C \, , \qquad F = D_F + L_F + U_F \, ,$$

where D_C and D_F are the block diagonal parts, L_C and L_F the strictly lower block triangular parts, U_C and U_F the strictly upper block triangular parts. Here, the blocking refers to partition (2.2). Then

$$D_\pi = D_C * A - h \cdot D_F * B \, ,$$
$$L_\pi = L_C * A - h \cdot L_F * B \, ,$$

$$U_\pi = U_C * A - h \cdot U_F * B .$$

With P as the permutation matrix of Prop. 3 of Chap. 12 in Ref. 4

$$P^T U_\pi P = A * U_C - h \cdot B * U_F .$$

This matrix is lower block triangular with diagonal blocks

$$\alpha_0 U_C - h\beta_0 U_F .$$

Similarly,

$$P^T (D_\pi + L_\pi) P = A * (D_C + L_C) - h \cdot B * (D_F + L_F) ,$$

which is a lower block triangular matrix with diagonal blocks

$$(\alpha_0 D_C - h\beta_0 D_F) + (\alpha_0 L_C - h\beta_0 L_F) .$$

Consequently,

$$P^T G_\pi P = -\{P^T (D_\pi + L_\pi) P\}^{-1} (P^T U_\pi P)$$

is a lower block triangular matrix with diagonal blocks G, where

$$(3.2) \qquad G := -\{(\alpha_0 D_C - h\beta_0 D_F) + (\alpha_0 L_C - h\beta_0 L_F)\}^{-1} (\alpha_0 U_C - h\beta_0 U_F) .$$

This $(n \times n)$-matrix is the *iteration matrix for the block Gauss-Seidel method* if applied to $\alpha_0 C - h\beta_0 F$. The blocking refers to partition (2.2). So the following theorem is proven:

3.1 **Theorem:** Let $\lambda_1, ..., \lambda_n$ be the eigenvalues of G (see Equ. (3.2)). Then $\lambda_1, ..., \lambda_1$ (N-fold); . . . ; $\lambda_n, ..., \lambda_n$ (N-fold) are the eigenvalues of G_π, where G_π is the iteration matrix of the WR Gauss-Seidel algorithm (see Equ. (3.1)).

A corresponding result for the WR Jacobi algorithm can be proven in a completely analogous manner. Theorem 3.1 extends Theorem 6.1 in Ref. 11 and gives a complete explanation of the background of these results.

Chebyshev acceleration is a powerful tool for speeding up the convergence of an iterative method (see Refs. 6 and 7). It depends on two parameters which can be chosen from knowledge of the convex hull of the spectrum of the iteration matrix. Ref. 7 gives an adaptive procedure for estimating this convex hull. This suggests the following algorithm.

3.2 Algorithm (Outline):

Step 1) (Estimating the convex hull of the spectrum of G_π) Perform block Gauss-Seidel iterations (with Chebyshev acceleration) on the $(n \times n)$-matrix $\alpha_0 C - h\beta_0 F$. Here, the blocking refers to partition (2.2). Estimate the convex hull of the spectrum of G (see Equ. (3.2)). Theorem 3.1 immediately gives an estimate for the convex hull of the spectrum of the ($nN \times nN$)-matrix G_π .

Step 2) Start WR Gauss-Seidel (with Chebyshev acceleration) based upon this knowledge of the convex hull of the spectrum of the corresponding iteration matrix G_π and continue by iteratively improving the above estimates of the convex hull of this spectrum.

Please note that Step 1) is relatively cheap. The basic assumption for Chebyshev acceleration is that the real part for any eigenvalue λ of the iteration matrix satisfies $Re(\lambda) < 1$. In view of Theorem 3.1 this can be guaranteed if the spectral radius $\rho(G) < 1$. This is true if the stepsize h is sufficiently small and if the spectral radius of the Gauss-Seidel iteration matrix for C is less than 1. Note that C is strictly diagonally dominant in many practical applications.

Experiments

The following first numerical experiments give an impression of the potential of the method. In the ring oscillator with n = 201 nodes all transistors were replaced by resistors with $R_T = 10^{12}\Omega$. For the arising RC-network the matrices C and F in (2.1) are symmetric. Because of this symmetry, the WR SSOR method was used instead of the WR Gauss-Seidel method. Then the eigenvalues of the corresponding iteration matrix are real. The following table compares the asymptotic rates of convergence R_∞ for the WR SSOR method without and with Chebyshev acceleration (cf. Ref. 12). In (2.3) the backward differentiation formula with s = 4 , $\alpha_0 = 25/12$, $\beta_0 = 1$ was used; stepsize h = 0.02 nsec.

C_p (pF)	R_∞ without Ch. acc.	R_∞ with Ch. acc.	asymptotic acceleration factor
0.005	6.62	8.01	1.21
0.05	2.68	4.03	1.50
0.5	0.59	1.62	2.75
5.0	0.08	0.57	7.13

The table indicates that for tightly coupled circuits the asymptotic acceleration factor is quite high.

Acknowledgement: The author is indebted to Mr. R. Gollreiter for performing the numerical experiments.

References

1 Bulirsch, R.; Gilg, A.: Effiziente numerische Verfahren für die Simulation elektrischer Schaltungen. In Schwärtzel, H. (Ed.): Informatik in der Praxis. Berlin:

414

Springer 1986, pp 3-12

2 Bulirsch, R.; Merten, K.; Gilg, A.; Steger, K.: Numerische Simulation für VLSI-Entwurf und Technologie. Schwerpunkprogramm der Deutschen Forschungsgemeinschaft "Anwendungsbezogene Optimierung und Steuerung", Report No. 103, Mathematisches Institut, Technische Universität München, 1988

3 Gear, C. W.: Numerical Initial Value Problems in Ordinary Differential Equations. Englewood Cliffs: Prentice-Hall 1971

4 Lancaster, P.; Tismenetsky, M.: The Theory of Matrices. New York: Academic Press 1985

5 Lelarasmee, E.: The waveform relaxation method for the time domain analysis of large scale integrated circuits: theory and applications. Ph.D. dissertation, University of California, Berkeley; also Memo UCB/ERL M82/40, 1982

6 Manteuffel, T. A.: The Tchebychev iteration for nonsymmetric linear systems. Numer. Math. 28 (1977) 307-327

7 Manteuffel, T. A.: Adaptive procedure for estimating parameters for the nonsymmetric Tchebychev iteration. Numer. Math. 31 (1978) 183-208

8 Miekkala, U.; Nevanlinna, O.: Sets of convergence and stability regions. BIT 27 (1987) 554-584

9 Rentrop, P.; Roche, M.; Steinebach, G.: The application of Rosenbrock-Wanner type methods with stepsize control in differential-algebraic equations. Report TUM-M8804, Mathematisches Institut, Technische Universität München, 1988

10 Varga, R. S.: Matrix Iterative Analysis. Englewood Cliffs: Prentice-Hall 1962

11 White, J. K.; Sangiovanni-Vincentelli, A.: Relaxation Techniques for the Simulation of VLSI Circuits. Boston: Kluwer Academic Publishers 1987

12 Young, D. M.: Iterative Solution of Large Linear Systems. New York: Academic Press 1971

Author's address:

Peter Lory, Mathematisches Institut, Technische Universität München, Postfach 20 24 20, D-8000 München 2, Germany (West)

DEVELOPMENT OF MODELS FOR FLASHING TWO-PHASE JET RELEASES FROM PRESSURISED CONTAINMENT

K. McFarlane

Shell Research Ltd., Thornton Research Centre, P.O. Box 1, Chester CH1 3SH.

§1: Introduction:

Reliable assessment of the hazards associated with accidental releases of flammable or toxic gases is essential for the design and operation of chemical plant and storage facilities. Models are needed to predict the rate of formation and initial physical state of gas clouds (source models) and their subsequent transport and dilution (dispersion models). This paper summarises progress in modelling the accidental release from storage of a pressurized liquid-gas. The model need be no more complex than is required in context, should be of comparable accuracy to subsequent dispersion models, and should possess similar mathematical structure.

Previous work (Ooms 1972, Birch et al. 1984, van den Akker et al. 1983, Botterill et al. 1984, Wheatley 1987, Raj and Morris 1988) on dense and buoyant plumes, pressurized gas jets, and liquid-gas jets, supported the view that predictions accurate in context could be obtained by means of an essentially simple, one-dimensional, sectionally-averaged model.

§2: Jet Development: A Control-Volume Analysis:

We regard the jet and ambient atmosphere as a single fluid (of variable composition) occupying the upper half-plane above a horizontal ground surface. The jet and ambient merge infinitesimally so that no jet "boundary" exists at finite distance from the jet-axis; entrainment occurs therefore "at infinity". The release and early jet development is presumed essentially horizontal, aligned with the ambient (horizontal) wind. No slip occurs amongst the constituent phases of the developing jet. Mean-flow within both ambient and jet is everywhere steady.

We begin by introducing a set of control-volumes $\tau(x)$, $x>0$, an analysis of which results in an integral-averaged description of jet development independent of detailed assumptions regarding induced and ambient turbulence.

J. Manley et al. (eds.), Proceedings of the Third European Conference on Mathematics in Industry, 415–428.
© 1990 Kluwer Academic Publishers and B. G. Teubner Stuttgart.

The Control-Volume $\tau(x)$: First construct a vertical surface at such distance upwind of the release point that ambient flow is negligibly perturbed. Second, at arbitrary distance $x>0$ downwind of the release-point, construct a vertical "cross-section" $\Sigma(x)$ through the developing jet. Third link these (semi-infinite) planes by skirting the

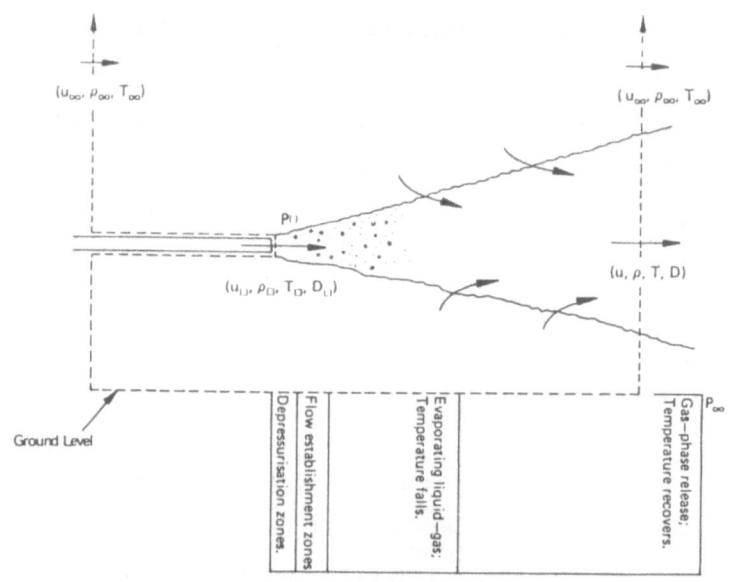

FIG. 1 — Cross-section through a typical control-volume $\tau(x)$.

Table 1

Steady-state, Full-bore, Wind alligned, Liquid Propane Releases

Trial name	Reservoir Pressure	Reservoir Temperature	Choke Pressure	Choke Temperature
P25	1.2 MPa	22°C	2.6 bar	-18°C
P26	1.6 MPa	25°C	2.8 bar	-17°C
P29	1.1 MPa	15°C	2.9 bar	-16°C
P30	1.7 MPa	18°C	3.5 bar	-10°C

(Pressure are absolute not gauge)

Trial name	Mass-Flux	Ambient Wind-speed	Ambient Temperature and Humidity
P25	5.7 kg/s	10 m/s	20°C, 50%
P26	10.5 kg/s	10 m/s	21°C, 57%
P29	6.6 kg/s	7.5 m/s	17°C, 70%
P30	9.7 kg/s	7.5 m/s	17°C, 70%

(Pipe Diameter: 5.0 cm; Pipe Length: 100 m; Pipe Exit Height: 1.0 m)

Note:

The enthalpy at the release plane is calculated assuming, as may be justified, negligible frictional or heat-transfer losses between reservoir and release orifice.

ground and pipework surfaces and passing through the jet at the plane of release. Finally construct a fourth bounding surface Σ_∞ at great (notionally infinite) distance from the jet-axis such as to enclose the (infinite) volume $\tau(x)$. [See Figure 1.]

Integration of the basic equations of motion over such a control-volume results, given an assumed "thin" jet (Hinze 1959), in the integral forms:

$$\iint_\Sigma c\mathbf{u}_\bullet dA = (c_\square/\rho_\square)dm/dt|_\square;$$

$$\iint_\Sigma (\rho\mathbf{u}-\rho_\infty\mathbf{u}_\infty)_\bullet dA = dm/dt|_\square \; [1-(\rho_\infty u_\infty)/(\rho u)] - \iint_{\Sigma_\infty} \rho\mathbf{u}_\bullet dA;$$

$$\iint_\Sigma [\rho\mathbf{u}(u_x-u_\infty)+(p-p_\infty)\mathbf{e}_x]_\bullet dA = (u_x^\square-u_\infty)dm/dt|+(p_\square-p_\infty)A_\square;$$

$$\iint_\Sigma [\rho\mathbf{u}(h+\tfrac{1}{2}u^2-h_\infty-\tfrac{1}{2}u_\infty^2)]_\bullet dA = (h_\square+\tfrac{1}{2}u_\square^2-h_\infty^\square-\tfrac{1}{2}u_\infty^{\square 2})dm/dt|_\square.$$

[Notation: $dm/dt|_\square$, mass flow rate issuing from the release point; $(\rho,c,\mathbf{u},p,h)$, density, pollutant mass-concentration, velocity, absolute pressure, and specific enthalpy of the ensemble-averaged flow. The affix "$\square$", identifies conditions at the release-plane; the suffix "∞" conditions within the unperturbed atmosphere.]

These equations express in integral form conservation of pollutant mass-flux, air entrainment, and excess above ambient of (horizontal) momentum and energy. They enable "matching" of conditions at the release orifice to conditions immediately following the external "flash".

§3: Sectionally Averaged Jet Motion:

The above integral equations may be further simplified in the zone of approximately self-similar or established flow in order to provide a "top-hat" or sectionally averaged description of jet development. We replace the detailed but unknown variation within each "cross-section" of the jet velocity and density by a single value representative of the entire cross-section; and assign to the jet a notional "jet-width" beyond which the conditions revert to ambient. Substitution then yields a "top-hat" description comprising a single entrainment relationship in the form of a quadrature, together with the three first integrals:

$$d/dx[\pi/4D^2(\rho u-\rho_\infty u_\infty)] = Entr_{shear}(D,u,\rho,c) ;$$

$$dm/dt|_\square = \pi/4D^2\rho u ; \; Entr_{shear} = d/dx \iint_{\Sigma_\infty} \rho_\infty\mathbf{u}_\bullet dA ;$$

$$[u-u_\infty]/[u_\square-u_\infty+(p_\square-p_\infty)/\rho_\square/u_\square] = (c/\rho)/(c_\square/\rho_\square) ;$$

$$(h+\tfrac{1}{2}u^2-h_\infty-\tfrac{1}{2}u_\infty^2)/(h_\square+\tfrac{1}{2}u_\square^2-h_\infty-\tfrac{1}{2}u_\infty^2) = (c/\rho)/(c_\square/\rho_\square) .$$

[Notation: D jet diameter, ρ section-mean density, c section-mean pollutant mass-concentration, u section-mean velocity, $dm/dt|_0$ release mass flow-rate.]

These equations are closed by specifying the entrainment function $Entr_{shear}$. The jet pressure is essentially ambient throughout the developed region. Local thermodynamic equilibrium is assumed.

This description of the development of a liquified gas jet is of limited scope: it enables a broad description of liquid evaporation, temperature fall and recovery, and the reduction in flow-speed that accompanies air entrainment.

§4: The Jet Cross-Section; Relationship to the "top-Hat" Description:

No information regarding the cross-sectional variation of such key parameters as contaminant concentration or jet temperature is provided, at any rate directly, by the simple "top-hat" description of jet development. However, the known point-local thermodynamic behaviour may be combined with empirically determined "self-similarity" profiles in order to deduce the sectional variation in, say, temperature consistent with the integral-averaged fluxes of released liquid-gas, momentum-excess, entrained air, and total-energy.

Introduce similarity profiles within the zone of established flow of the general form;

$$c/c_* = \phi_c(y/D_*, z/D_*; D_*; \lambda^2); \quad (u-u_\infty)/(u_*-u_\infty) = \phi_u(y/D_*, z/D_*; D_*);$$

in which $c_*(x)$ and $u_*(x)$ are the section-local and centre-line values of contaminant mass-concentration and flow-speed within the jet, $D_*(x)$ a representative "jet-width", and in which finally λ^2 is the turbulent Schmidt number measuring the relative diffusivities within the jet of momentum and concentration; ϕ_u and ϕ_c are universal functions independent of the downwind displacement $x>0$.

Consistency then requires that the several physically meaningful fluxes be invariable amongst "top-hat" and each "cross-sectional" description; it results in the linking of jet "centre-line" (c_*, u_*, D_*) and "sectional-mean" (c,u,D) variables via the non-linear integral equations:

$$\iint_{\Sigma(x)} cu \ d\Sigma = \pi/4 D^2 cu;$$

$$\iint_{\Sigma(x)}(\rho u-\rho_\infty u_\infty)~d\Sigma= \pi/4D^2(\rho u-\rho_\infty u_\infty);$$

$$\iint_{\Sigma(x)}\rho u(u-u_\infty)~d\Sigma= \pi/4D^2\rho u(u-u_\infty).$$

The notation is that the symbols of the integrands refer to turbulent-mean point-values within the chosen cross-section $\Sigma(x)$. The identical symbols of the right-hand sides refer to sectionally averaged values resulting from a "top-hat" model evaluation.

§5: Jet Asymptotics; "Top-Hat" and "Gaussian" Descriptions:

It remains to determine an appropriate form for the "top-hat" entrainment function and for the self similarity functions that describe established (self-similar) flow. In the far-field the profile shapes and entrainment law are well established: profile shapes are (neglect ground effects) broadly Gaussian (List 1982, Abramovich 1963), so that

$$\phi_u(r/D_*)= \exp(-4r^2/D_*^2); ~~\phi_c(r/D_*)= \exp(-4r^2/\lambda^2/D_*^2); ~~0\le r<\infty;$$

r being the radial displacement "off-axis" in the plane of the cross-section $\Sigma(x)$. It follows from the consistency equations that "top-hat" and "centre-line Gaussian" parameters are asymptotically related via the explicit forms (Compare Henderson-Sellers 1981, Davidson 1986).

$$(T_*-T_\infty)/(T-T_\infty)= (\rho_*-\rho_\infty)/(\rho-\rho_\infty)= c_*/c= 2/\lambda^2;$$

$$(u_*-u_\infty)/(u-u_\infty)= (D_*/D)^2= 2; ~~u_\infty\ne 0;$$

results deduced for a trace (ideal) gas contaminant in possibly humid but not saturated air. Following temperature recovery therefore jet centre-line concentrations are deduced for much diluted jets by a simple scaling applied to the associated "top-hat" results: solution of a set of non-linear integral equations for the centre-line values is not required.

§6: Jet Development: The "Top-Hat" Entrainment Function:

We return to the "classical" theory of jets and the theoretical analysis of Albertson (1948), Hinze (1959), or Abramovich (1963), all of which suggest that a simple "mixing-length" closure of the kinetic energy equation might provide an acceptable entrainment relationship. Consider the kinetic-energy equation when integrated over the standard control volume $\tau(x)$:

$$d/dx \iint_\Sigma \rho u(\tfrac{1}{2}u^2-\tfrac{1}{2}u_\infty^2)~d\Sigma= \iint_{\Sigma(x)}\Sigma_{xr}^{trb}\partial u/\partial r~d\Sigma.$$

Σ_{rx}^{trb} is a Reynolds' stress. The Prandtl closure for neutrally buoyant jets has the simple form (Hinze 1959)

$$\Sigma_{xr}^{trb}= -\mu^2\rho_\infty D_*^2|\partial u/\partial r|\partial u/\partial r;$$

in which μ^2 is a constant. Presumption of a Gaussian form for the velocity profile and the auxiliary condition $\rho= \rho_\infty$ yields an entrainment function:

$$Entr_{shear}= 6\mu^2\rho_\infty D_*(u_*-u_\infty)^3/u_*/(u_*+2u_\infty).$$

The equivalent "top-hat" representation is, in the far-field, simply

$$Entr_{shear}= e\pi\rho_\infty D(u-u_\infty)^3/(u+\tfrac{1}{2}u_\infty)/(u-\tfrac{1}{2}u_\infty); \quad \mu^2= 3e/\pi\ 2^{\frac{1}{2}};$$

The entrainment coefficient e has for the (neutral) far-field the well established value e= 0.08 (Morton, Taylor, Turner 1956, Ooms 1972, Petersen 1978,1987).

Consider next the influence of density upon the "top-hat" entrainment function. Evidence indicates that the influence of (positive) density differences between gas jet and ambient is somewhat to reduce the degree of entrainment that would otherwise occur (Yoshinobu and Saima 1977, Resplandy 1969), and certainly this seems intuitively satisfactory, though at variance with the marginally dense jet results of Ricou and Spalding (1961), in which a slight increase in entrainment above neutral buoyancy was recorded. We assume that the effect of density differences upon the dominant Reynolds' stress is described by a multiplicative factor of the general form

$$\Sigma_{xr}^{trb}= -\rho\psi(\rho/\rho_\infty)\mu^2 D_*^2|\partial u/\partial r|\partial u/\partial r; \quad \psi(\rho/\rho_\infty)\to 1 \text{ as } \rho/\rho_\infty\to 1.$$

Consider further the asymptotic behaviour of the "Gaussian"-jet entrainment function under the assumption that velocity differences between jet and ambient wind are large. Reduction in entrainment with increasing density occurs only if the leading order expression of the densimetric correction to the dominant Reynolds' stress has the form $\psi(\rho/\rho_\infty)= \rho_\infty/\rho$. It follows that the (Gaussian) entrainment function has the asymptotic form

$$Entr_{shear}= 2^{-\frac{1}{2}}e\pi D_* u_*\eta(\rho_*/\rho_\infty); \quad \Delta\rho_*= \rho_*-\rho_\infty; \quad u_*/u_\infty,\rho_*/\rho_\infty\gg 1;$$

$$\eta(\rho_*/\rho_\infty)= [1+2\lambda^2/(2\lambda^2+1)\Delta\rho_*/\rho_\infty]/[1+2\lambda^2(3\lambda^2+2)/(2\lambda^2+1)/(3\lambda^2+1)\Delta\rho_*/\rho_\infty];$$

valid for isothermal gas jets (contaminant concentration proportional to the density difference between jet and ambient air). Further simplification requires a value for the turbulent Schmidt number which in the far field is approximately 1.35 (Rouse, Yih, and Hyumphreys 1952) but which in the dense jet has a value near unity (Birch and Brown 1987). Setting therefore $\lambda^2=1$ we propose an equivalent "top-hat" entrainment function of the form

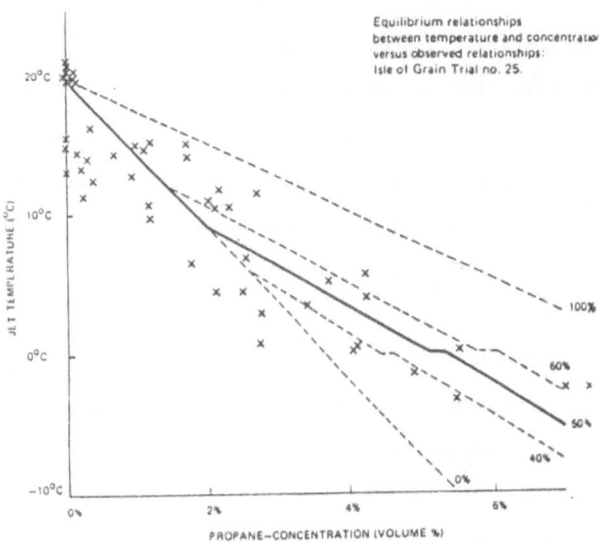

FIG. 2(a) — P25: Shifted temperatures.

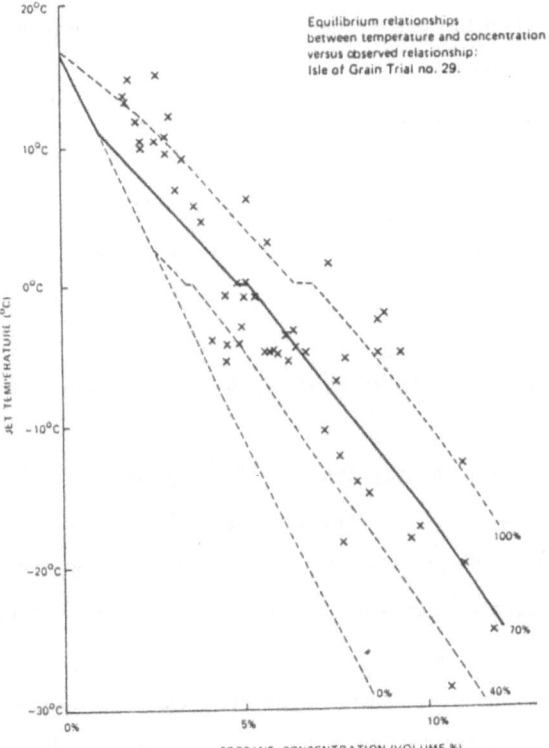

FIG. 2(b) — P29: Shifted temperatures.

$$\text{Entr}_{shear}= e\pi D\eta(\rho/\rho_\infty)(u-u_\infty)^3/(u+\tfrac{1}{2}u_\infty)/(u-\tfrac{1}{2}u_\infty); \quad u>u_\infty;$$

$$\eta(\rho/\rho_\infty)= [1+4/3/(\rho/\rho_\infty-1)]/[1+5/3/(\rho/\rho_\infty-1)].$$

§7: Jet Development: Theory versus Experiment:

We compare the above developed "top-hat" model with a programme of large-scale, pressurized, liquid-propane releases conducted jointly by Shell Research Ltd and by BP Research International (Cowley and Tam 1988). The experimental site was the (former) BP refinery at the Isle of Grain. We consider a subset (Table 1) of the eighty-four trials comprising steady, unignited, wind-aligned, "full-bore" (full pipe diameter) releases of liquid propane. These trials involved the discharge of tonne quantities of liquid propane via a 50mm diameter 100m long pipe placed parallel to and 1m above levelled ground. The data here summarised is this author's interpretation of very extensive, "instantaneous", sensor data: it is recognised that differing instrumental averaging times may affect somewhat the details of concentration/temperature correlations.

Considerable flashing was evident within the release pipe, with two-phase choked flow at the orifice. The steady release-rate and release (orifice) pressure were experimentally determined. Release-plane void-fraction is calculated upon the basis of assumed thermodynamic equilibrium and negligible inter-phase slip: temperatures recorded at the orifice are broadly consistent with thermodynamic equilibrium at the choke-front. Energy losses within the release pipe due either to heat transfer to atmosphere or to wall friction are assumed negligible. Validation is little affected by uncertainties in the prediction of two-phase choked flow within pipes (van den Akker and Bond 1984).

Consider first the assumption of thermodynamic equilibrium (van den Akker, Spoelstra, and Snoey 1983). Figure 2 illustrates the degree of departure between local thermodynamic equilibrium and a set of (instantaneous) pairs of propane concentration and temperature measurements conducted both on- and off-axis at distances of 9 and 14 metres downwind of the release plane. Generally good agreement exists between the theoretical prediction of an essentially linear relationship between volumetric concentration (%) and temperature ($^\circ$C), and the mean trend of the experimental

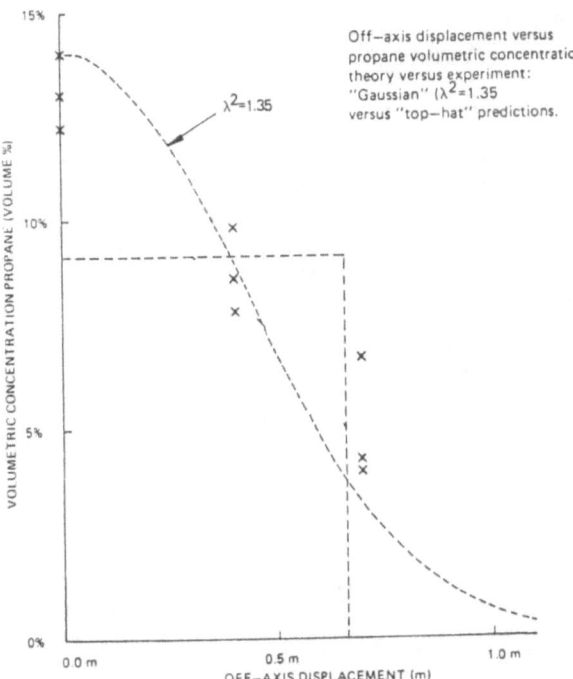

FIG. 3 — Isle of Grain P26: Downwind displacement 9.0 m (theory 9.1 m)

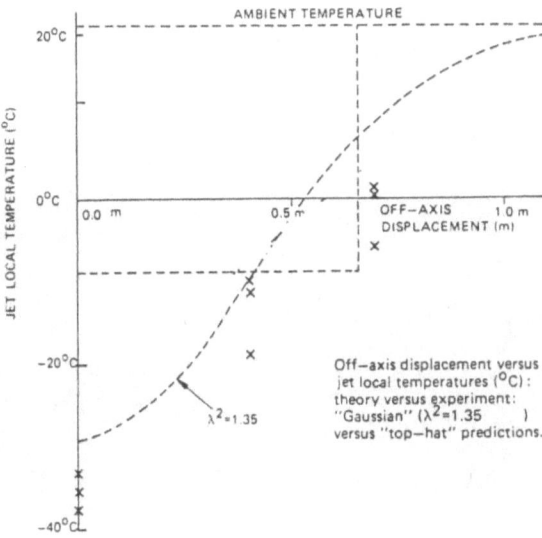

FIG. 4 — Isle of Grain P26: Downwind displacement 9.0 m (theory at 9.1 m).

424

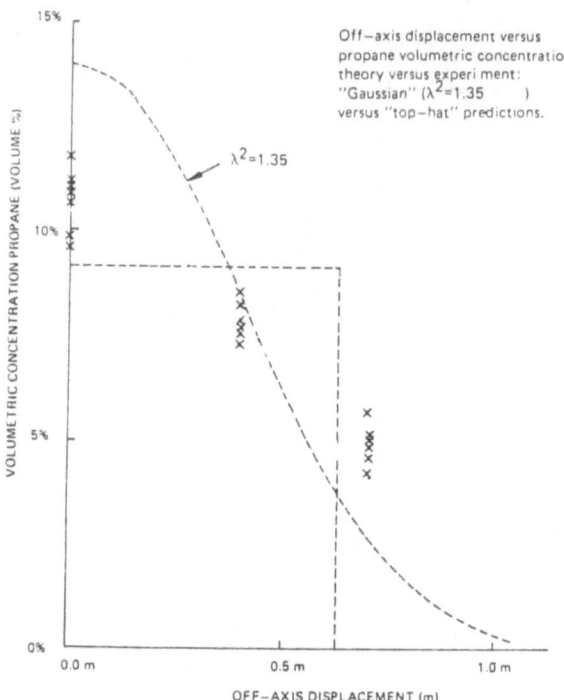

FIG 5 — Isle of the Grain P29: Downwind displacement 9.0 m (theory 9.1 m).

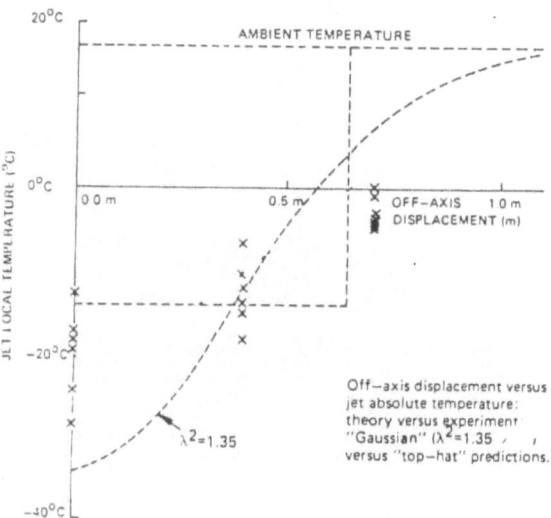

FIG. 6 — Isle of Grain P29: Downwind displacement 9.0 m (theory at 9.1 m).

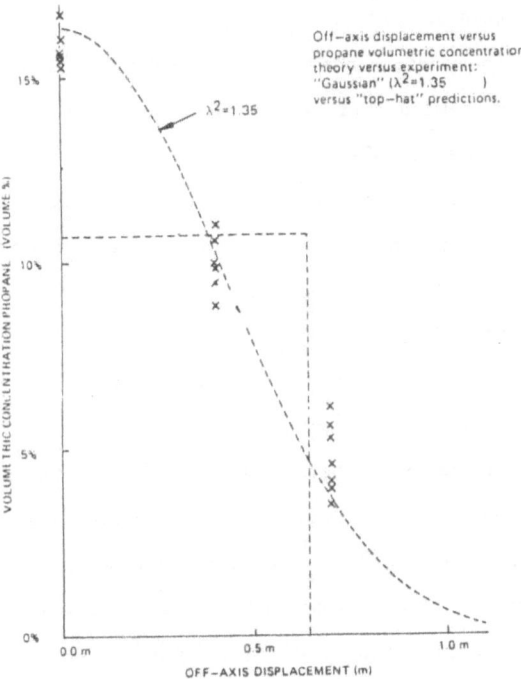

FIG. 7. — Isle of Grain P30: Downwind displacement 9.0 m (theory at 9.1 m).

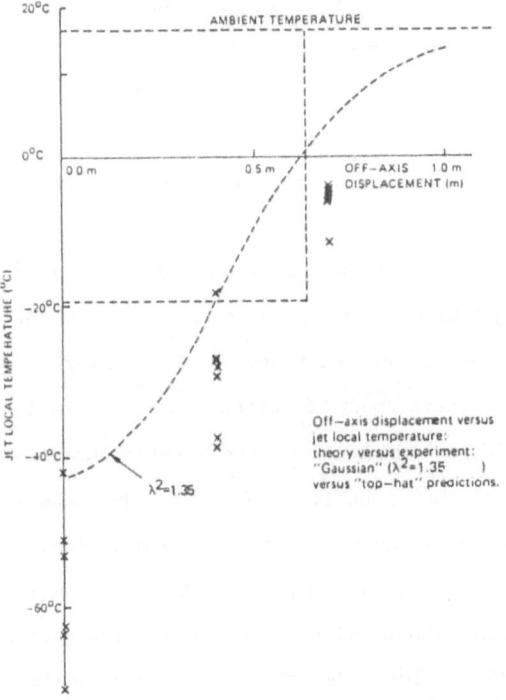

FIG. 8. — Isle of Grain P30: Downwind displacement 9.0 m (theory at 9.1 m).

results. A large degree of scatter is however evident.

Consider next the performance of the "top-hat" entrainment function in reproducing the down-wind variation of temperature, and "volumetric" concentration of propane. The chosen value of λ^2 was precisely unity. Figures 3 through 8 illustrate the degree of agreement between theory and momentary point-local values plotted as a function of downwind distance and distance off-axis. As can be seen the general agreement between the simple "top-hat" model and data is reasonable as regards jet "width" and mean "subcooling" below ambient; whereas (clearly) the detailed prediction of centre-line concentration or temperature, or the displacement off-axis of (say) the flammability limits is beyond the model scope, there being generally large differences between sectional-mean and centre-line values of either temperature or concentration. Note, however, the sensitivity of predictions at distance off-axis to fluctuations in the ambient wind-direction (resulting in greater scatter and inaccuracy at the jet "edge").

Finally contrast the predictions of the point-local concentration and temperature profiles deduced from the sectionally averaged values given assumed Gaussian velocity and (mass) concentration profiles with the experimental data of Figures 3 through 8 ($\lambda^2 = 1.35$). Generally the centre-line variation is well captured, though some systematic underprediction of values significantly off-axis is evident.

§8: **Conclusions:**

A simple but effective model has been developed describing the steady release to coflowing ambient of a pressurised liquid-gas. Several features of the analysis appear new or at least new in this context: in particular the derivation of shear entrainment function for a neutrally buoyant jet via (Prandtl) "mixing-length" closure of the kinetic-energy equation; and the associated derivation and functional form of the densimetric correction $\eta(\rho/\rho_\infty)$ (§6). Focus upon shear entrainment permits (§6) the derivation of point-local from sectional-mean predictions given only self-similarity and an assumed "profile" form, and in addition allows the straightforward analysis of jet "asymptotics" (§5). The model stucture allows sucessive estimation of "sectional-mean" or "top-hat" properties and of point-local predictions conditional upon an assumed self-similar "profile" form; and has merit in permitting step-by-step tuning to experimental data.

Novel experimental data on the large-scale release of liquidified propane have been presented. These show good agreement with theoretical prediction.

References:

[1]: Abramovich G N: The Theory of Turbulent Jets, MIT Press, Cambridge, Massachusetts, 1963; §1.7 Heat Transfer in a Submerged Jet,pp 17-22; §1.8 Diffusion of Constituents in a Submerged Jet, pp 23-25.

[2]: Albertson M L, Dai Y B, Jensen R A, and Hunter Rouse M: Diffusion of Submerged Jets, Transactions American Society of Civil Engineers, Paper No 2409, Proceedings December 1948, pp 639-665.

[3]: Birch A D, Brown D R, Dodson M G, and Swaffield F: The Structure and Concentration Decay of High Pressure Jets of Natural Gas, Combustion Science and Technology 36(1984), 249-261.

[4]: Birch A D and Brown D R: The Use of Integral Models for predicting Jet Flows, in Mathematics in Major Accident Risk Assessment, Cox R A (editor), Proceedings of the IMA Conference on Mathematics in Major Accident Risk Assessment, July 1986, in preparation 1988.

[5]: Botterill J A, Williams I, and Woodhead T J: The Transient Release of Pressurized LPG from a Small-Diameter Pipe, in The Protection of Exothermic Reactors and Pressurised Storage Vessels, Proceedings of the Institution of Chemical Engineers Symposium, Chester, England, April 1984, Published by Pergamon Press, 1984, pp 265-278.

[6]: Cowley L T and Tam V H Y: Consequences of Pressurized LPG Releases: The Isle of Grain Full Scale Experiments, submitted 13th Int. Conf. LNG/LPG Conference and Exhibition, Kuala Lumpur, Malaysia, October 18-21, 1988.

[7]: Davidson G A: Gaussian versus Top-Hat Profile Assumptions in Integral Plume Models, Atmospheric Environment 20(1986), 471-478

[8]: Hanna S R: Turbulent Diffusion: Chimneys and Cooling Towers, in Plate E J: Engineering meteorology, Elsevier Scientific Publishing Company, Amsterdam, 1982; Chapter 10, pp429-451, especially §10.2.3, pp438-439.

[9]: Henderson-Sellers B: Shape Constants for Plume Models, Boundary Layer Meteorology 21(1981), 105-114.

[10]: Hinze J O: Turbulence An Introduction to its Mechanisms and Theory, McGraw-Hill Book Company, London, 1959, Chapter 6, §6.7: pp 404-447.

[11]: List E J: Turbulent Jets and Plumes, Ann. Rev. Fluid Mech. 14(1982), 189-212.

[12]: Morton B R, Taylor G I, and Turner J S: Turbulent Gravitional Convection from Maintained and Instantaneous Sources, Procs. Roy,. Soc. Series A, 234(1956), 1-23.

[13]: Ooms G: A New Method for the Calculation of the Plume Path of Gases emitted by a Stack, Atmospheric Environment 6(1972), 899-909.

[14]: Petersen R L: Plume Rise and Dispersion for varying ambient turbulence, thermal stratification, and stack exit conditions, PhD thesis, Colorado State University, Fall, 1978. Published by University Microfilms International, London, WC1R 4EJ, 1986.

428

[15]: Petersen R L: Performance Evaluation of Integral and Analytical Plume Rise Algorithms, JAPCA 37(1987), 1314-1319.

[16]: Raj P K, and Morris J A: "Source Characterization and Heavy Gas Dispersion Models for Reactive Chemicals", Technology & Management Systems INc., Burlington, Massachusetts 01803-5128, AFGL-TR-88-0003(I), December 1987.

[17]: Resplandy A: Etude experimentale des proprietes de l'ammoniac (conditionant les mesures a prendre pour la securite du voisinage des stockages industriels), Chimie et Industrie Genie Chimique, 102(1969), 691-702.

[18]: Ricou F P and Spalding D B: Measurements of Entrainment by axisymmetrical turbulent jets, J. Fluid Mech. 11(1961), 21-32.

[19]: Rouse H, Yih C S, and Hyumphreys H W: Gravitational convection from a Boundary Source, Tellus 4(1952), 201-210.

[20]: Schatzmann M: The Integral Equations for Round Buoyant Jets in Stratified Flows, J Applied Mathematics and Physics (ZAMP) 29(1978), 608-630.

[21]: Schatzmann M: An Integral Model of Plume Rise, Atmospheric Environment 13(1979), 721-731.

[22]: van den Akker H E A, Spoelstra H, and Snoey H, Discharges of Pressurized Liquified Gases through Apertures and Pipes, 4th International Symposium on Loss Prevention and Safety Promotion in the Process Industries (EFCE Event no 290), Harrowgate, North Yorkshire, England, September 1983, Proccedings distributed by Pergamon Press, Oxford, pp E23-E35.

[23]: van den Akker H E A, and Bond W M: Discharge of saturated and superheated liquids from pressure vessels. Prediction of homogeneous choked two-phase flow through pipes. in The Protection of Exothermic Reactors and Pressurised Storage Vessels, Proceedings of the Institution of Chemical Engineers Symposium, Chester, England, April 1984, Published by Pergamon Press, 1984, pp 91-108.

[24]: Wheatley C J: A Theoretical Study of NH^3 Concentrations in Moist Air arising from Accidental Releases of Liquified NH^3, using the Computer Code TRAUMA, Safety and Reliability Directorate (SRD), United Kingdom Atomic Energy Authority (UKAEA), Wigshaw Lane, Culcheth, Warrington, Cheshire, England, WA3 4NE, SRD/HSE/R 393, February 1987.

[25]: Wheatley C J: Discharge of Liquid Ammonia to Moist Atmospheres - Survey of Experimental Data and Model for estimating initial conditions for dispersion calculations, Safety and Reliability Directorate (SRD), United Kingdom Atomic Energy Authority (UKAEA), Wigshaw Lane, Culcheth, Warrington, Cheshire, England, WA3 4NE, SRD/HSE/R 410, April 1987.

[26]: Yoshinobu E R A and Saima A: Turbulent mixing of gases with different densities, JSME 20(1977), 63-70.

Author's Address:

Shell Research Limited,

Thorton Research Centre,

Post Office Box 1,

Chester, CH1 3SH.

Robust Recursive Estimation: The L_p Approach

D.W. McMichael

(The Control Systems Centre, UMIST, PO Box 88, Manchester, M60 1QD)

Outlier-contaminated normal errors in regression problems are modelled by *exponential power distributions* and the resulting maximum likelihood estimators involve L_p minimisations ($1 < p \leq 2$). L_1, estimation is minimax outlier-robust and minimax covariance-robust over the neighbourhood of exponential power distributions. Efficiency loss is negligible. Convergent and consistent recursive gradient-type L_p estimators are derived.

Key Words: recursive estimation, robust regression, L_p optimisation, minimax robustness, outliers, exponential power distribution.

1 Introduction

The aim of this paper is to present a family of recursive estimators for regression coefficients, which are insensitive to the presence of outliers in the data, and yet are not significantly less efficient than conventional least squares. Underlying the discussion is the practical concern that the minimisation problems implied by a robust estimation criterion have to be solved by recursive gradient-type methods. This is a significant limitation.

2 An Outlier Distribution Model

Models for outlier-contaminated error distributions generally assume that the errors arise from two sources: most cases are affected by errors that are the sum of many small random disturbances, but some are affected by large errors which are sums of comparitively few large disturbances. Clearly, only the former type are likely to be distributed in close approximation to normality. This line of reasoning has frequently led to modelling with mixtures of normal and some long-tailed distribution (typically Cauchy, or a large variance normal distribution). The family of *exponential power distributions*, $E_p(p,a)$ proposed here are not of the mixture variety. They enable different degrees of outlier contamination to be modelled by choice of the parameter p ($1 < p \leq 2$) [McMichael, 1987]. The net result is similar — they lie in a perturbation neighbourhood that includes the normal distribution, and with that exception are all leptokurtic. The exponential power distribution's probability density function (pdf) is

$$p_\xi(\xi) = \frac{p\, e^{-|\xi/a|^p}}{2a\, \Gamma(1/p)}. \tag{2.1}$$

a is a scale (dispersion) parameter > 0, and p is a dimensionless parameter affecting the kurtosis of the distribution. When p=2 the distribution is normal with variance $1/\sqrt{2}$, and when p=1 the distribution is Laplacian. Its variance is

$$var(\xi) = a^2 \frac{\Gamma(3/p)}{\Gamma(1/p)}. \tag{2.2}$$

J. Manley et al. (eds.), Proceedings of the Third European Conference on Mathematics in Industry, 429–435.
© *1990 Kluwer Academic Publishers and B. G. Teubner Stuttgart.*

As p decreases from 2 to 1 the distribution develops longer thicker tails; and as a consequence, outlying errors are more probable. Underlying the concept of the outlier is rarity, they are 'surprising' and improbable data points: by extending the tails of the model distribution we are admitting that they are more likely.

3 MLEs Based on Exponential Power Distributions

The maximum likelihood estimator for a distribution parameter chooses estimates which maximise the probability (likelihood) of the data. We shall be concerned wht the problemof estimating θ_0 in the linear-in-the-parameters model

$$y_i = x_i^T \theta_0 + \xi_i. \tag{3.1}$$

where y_i is the physical measurement corresponding to the ith case, ξ_i is a $n \times 1$ vector of regressors, θ_0 is a $n \times 1$ vector of parameters and xi_i is a random error term. The pdf of ξ_i is $p_i(\xi_i)$, and since the error is additive,

$$p_{y_i}(y_i|\theta_0) = p_{\xi_i}\left(y_i - x_i^T \theta_0\right). \tag{3.2}$$

If the ξ_i are mutually independent and identically distributed the joint probability density of the observations, y_i is,

$$p(y_1, ..., y_N|\theta_0) = \prod_{i=1}^{N} p_{\xi_i}\left(y_i - x_i^T \theta_0\right). \tag{3.3}$$

If the l.h.s. of (3.3) is regarded as a function of θ then it is interpreted as a *likelihood function*:

$$L(y_1, ..., y_N|\theta) = \prod_{i=1}^{N} p_{\xi_i}\left(y_i - x_i^T \theta\right). \tag{3.4}$$

The MLE for θ maximises the likelihood function over θ giving

$$\hat{\theta} = arg\ \max_{\theta}\ L(y_1, ..., y_N|\theta). \tag{3.5}$$

Aside from the intuitive appeal of MLEs they possess a number of optimum properties. However, for exponential power distributions, when $p \neq 2$, $\hat{\theta}$ is not a sufficient statistic for θ_0, but the asymptotic properties of the MLE criterion remain. As $N \to \infty$ the estimator is consistent ($\theta \to \theta_0$ w.p. 1), sufficient, normal and attains the Cramer-Rao minimum variance bound (MVB) [Kendall and Stuart, 1979].

Off-line Solutions of the MLE Estimation Problem

It is usually easier to maximise $ln(L)$ rather than L itself — particularly if the distribution is exponential. This approach results in the following minimisation problem,

$$\hat{\theta} = arg\ \min_{\theta}\ \sum_{i=1}^{N} \phi\left(y_i - x_i^T \theta\right); \tag{3.6a}$$

$$\phi(x) \propto -ln\, p_\xi(x) + const. \tag{3.6b}$$

$\phi(.)$ is known as the *case cost function* (ccf). For exponential power distributitons the MLE for θ_0 is

$$\hat{\theta} = arg\ min_\theta \sum_{i=1}^{N} |y_i - x_i^T\theta|^p. \tag{3.7}$$

Notice that this is an L_p normed minimisation problem, and also that $\hat{\theta}$ is independent of the scale parameter a.

To prove the asymptotic results using the methods of Kendall and Stuart [1979] requires the existence of the first derivative of the ccf and the existence of the expectation of the second derivative. Unfortunately, when $p \neq 2$ $\phi(0)'$ does not formally exist, but this difficulty is overcome by augmenting $\phi(x)'$ by the point at zero such that $\phi(0)' = 0$. The second derivative of $\phi(x)$ is infinite at $x = 0$, but its expectation does exist and the conventional proofs of the optimal asymptotic properties of MLEs apply.

When p=2 equation (3.7) has the least squares analytical solution — the *normal equations*:

$$\begin{aligned}
\hat{\theta} &= [X^T X]^{-1} X^T Y; \\
X^T &= [x_1, ..., x_N]^T, \\
Y^T &= [y_1, ..., y_N]^T.
\end{aligned} \tag{3.8a}$$

However, when $p \neq 2$ numerical methods are required (see Press *et al.* [1986]). Insight into the process of robustisation can be obtained from the numerically weak method of *iteratively reweighted least squares* (IRLS) mentioned by Box and Draper [1987] and equivalent to the recursive algorithm described by Dutter [1975]. Dutter [1975] has shown that at each iteration the method consistently reduces the estimator cost in (3.6a) provided that $\phi(x)$ is convex and symmetric, $\phi'(x)/x$ is bounded and monotone decreasing for $x < 0$, and the minimum has not already been reached. The iterative scheme is as follows:

$$\hat{\theta}_{(k)} = [X^T W_{(k-1)} X]^{-1} X^T W_{(k-1)} Y; \tag{3.9a}$$

$$w_{(k)ij} = \begin{cases} \phi'\left(\varepsilon_{(k)i}\right) / \varepsilon_{(k)i} & i = j \\ 0 & i <> j \end{cases}$$

$$\varepsilon_{(k)i} = y_i - x_i^T \hat{\theta}_{(k)}. \tag{3.9c}$$

$W_{(k)}$ is a diagonal weighting matrix which, when $\phi(x)$ is weaker than x^2, progressively penalises those cases with large residuals ($\varepsilon_{(k)i}$). To achieve consistency it may be necessary to modify $\phi(x)'/x$ near the origin to prevent unboundedness. It is revealing to compare (3.9a) with the Markov estimator for uncorrelated data in which the diagonal matrix W is the inverse of covariance matrix of the error vector

$$[\xi_1, ..., \xi_N]^T.$$

The Markov estimator weights cases in inverse proportion the corresponding error variance. Robust regression can be interpreted as Markov estimation with iterative estimation of var(ξ_i). We shall see later that this approach to solution of (3.7) is instrumental in estimating the influence of single outlying cases on the estimates.

4 Robustness Properties of Lp Estimation

It has been shown that in comparison to least squares L_p estimators are more robust to single large errors [McMicchael, 1988].

A second approach to the robustness properties of estimators is to examine their asymptotic behaviour, — principally bias and covariance. Provided the score function is odd and the actual error distribution is symmetric, the estimator (3.6) will be unbiassed if the expectation

$$E\left[\frac{\partial ln\tilde{L}}{\partial \theta}\right]_{\theta=\theta_0}$$

(4.1)

exists. $\tilde{L}$ is the likelihood function corresponding to the assumed error distribution. In order to emphasize the fact that p can be viewed both as a parameter of the error distribution, *and* as a design parameter, we shall denote the parameter of the error distribution $E_p(p,a)$ by "p", and the design parameter in the minimisation problem (3.7) by "p̃". In $L_{\tilde{p}}$ estimation the condition (4.1) corresponds to a requirement for the existence of the $(\tilde{p}-1)^{th}$ moment of the error distribution $(1<\tilde{p}\leq 2)$ — always satisfied by real data.

If, by lucky chance p̃=p, the asymptotic covariance reaches the minimum bound. A little integration shows that this is

$$\underset{N \to \infty}{Lim}\left\{-\left[E\left(\frac{\partial^2 lnL}{\partial \theta^2}\right)\right]^{-1}\right\} =$$

$$\frac{a^2\Gamma(1/p)}{p^2\Gamma(2-1/p)}\underset{N \to \infty}{Lim}[X^TX]^{-1}.$$

(4.2)

This is the lowest asymptotic covariance obtainable when the errors are distributed as $E_p(p,a)$. If, however p̃ is not equal to p then

$$\underset{N \to \infty}{Lim}\left\{Cov\left(\hat{\theta}\right)\right\} =$$

$$\frac{a^2\Gamma(1/p)\Gamma((2\tilde{p}-1)/p)}{p^2\Gamma((\tilde{p}-1)/p+1)^2}\underset{N \to \infty}{Lim}[X^TX]^{-1},$$

(4.3)

[McMichael,1988]. The scalar factor of this expression is the *asymptotic covariance factor*.

For asymptotically normal estimators the *asymptotic efficiency* is simply the ratio of the minimum variance bound to the estimator's variance. Since the scale and matrix factors of both are equal, the appropriate multivariate extension is to define the efficiency as the ratio of the scalar factor of (4.8) to that of (4.9):

$$\eta = \frac{\Gamma((\tilde{p}-1)/p+1)^2}{\Gamma(2-1/p)\Gamma((2\tilde{p}-1)/p)}.$$

(4.4)

Design: Choosing p̃

In the absence of definite information about the error distribution a sensible approach to design is to optimise the procedure for the 'worst case' within Ω, a nominated neighbourhood of possible perturbations. This argument is formalised by the minimax design procedure. If $C(Y,\tilde{p},\Omega)$ is a measure of estimator degradation due to perturbation of the data from the null model (normality) then the minimax optimal choice of p̃ is

$$\tilde{p} = arg \ \min_{x} \left\{ \sup_{\Omega} C\left(Y, x, \Omega\right) \right\}. \tag{4.5}$$

The minimax criterion is conservative and may result in safeguarding the estimator from unlikely worst cases. To obtain sensible estimators it may be necessary to choose Ω rather carefully. Minimax robust designs may themselves be unrobust to this choice. Happily, neither of these criticisms applies to the designs below.

Minimax SIC Design. It has been shown that minimax SIC design, taking Ω as $\{1 < p \leq 2\}$ is p̃=1⁺ [McMichael, 1988].

Minimax Asymptotic Covariance Design. $L_{\tilde{p}}$ estimation is unbiassed under the assumption that the error distribution is symmetric, so the asymptotic property of interest is covariance. For the reasons stated we choose the perturbation neighbourhood Ω as $\{1 < p \leq 2\}$ and choose $C(Y,\tilde{p},\Omega)$ to be the *asymptotic covariance factor*

$$C\left(Y, \tilde{p}, \Omega\right) = a^2 \ \frac{\Gamma\left(1/p\right) \Gamma\left(\left(2\tilde{p} - 1\right)/p\right)}{p^2 \ \Gamma\left(\left(\tilde{p} - 1\right)/p + 1\right)^2}. \tag{4.6}$$

It has been shown that regardless of p̃ its highest value occurs uniformly when p=1. Choosing p̃=1 minimises this worst case variance. Thus, the minimax realizable covariance design also chooses p̃=1⁺.

As mentioned previously, there is little loss in efficacy incurred by using L_{1+} estimation when the noise is Gaussian, and the general criticism of minimax methods for being too conservative is not justified in this case. Furthermore, the lower boundary of Ω is not arbitrary.

5 Recursive Lp Estimation

By recursive estimation we mean numerical procedures that enable sequential cases to be included into the estimator one at a time, in such a way as the computational effort required is less than recalculating the estimates with off-line techniques each time a new case arrives. The net result is usually that the cost minimisation is less accurate (recursive least squares is a notable exception). Recursive techniques are appropriate when the case data arrives serially and the estimates are needed in real-time; for example in self-tuning control. When there is a substantial cost associated with gathering each new case sequential estimation procedures become attractive. Recursive procedures are particularly suited to parameter tracking problems where adaptivity is required [Ljung and Soderstrom, 1983].

Two sorts of gradient-type methods are given here: a stochastic Newton method [Kushner and Clark, 1978], and a slower but computationally less demanding group of steepest descent stochastic approximation methods [Robbins and Monro, 1951; Wasan, 1969]. Stochastic approximation finds the solution of the general equation

$$\underset{y}{E}\left[g\left(\theta, y\right)\right] = 0 \tag{5.1}$$

with the iterative scheme:

$$\hat{\theta}\left(t\right) = \hat{\theta}\left(t - 1\right) + K\left(t\right) g\left(\hat{\theta}\left(t - 1\right), y\left(t\right)\right). \tag{5.2}$$

If the noise distribution and the case cost function $\phi(x)$ are symmetric, minimising

$$V(\theta) = \underset{y_i}{E}\left[\phi\left(y_i - x_i^T\theta\right)\right] \tag{5.3}$$

is asymptotically equivalent to the minimisation (3.6a). If $V(\theta)$ is convex then we seek solutions to

$$\frac{\partial V(\theta)}{\partial \theta} = E\left[\phi'\left(\varepsilon_i(\theta)\right)x_i\right] = 0. \tag{5.4}$$

For the stochastic Newton method we also require an estimate of the Hessian of $V(\theta)$. Defining

$$\frac{\partial^2 V(\theta)}{\partial \theta^2} := R(\theta) = E\left[\phi''\left(\varepsilon_i(\theta)\right)x_i x_i^T\right]; \tag{5.5}$$

we can construct a problem analogous to (5.1):

$$E\left[\phi''\left(\varepsilon_i(\theta)\right)x_i x_i^T - R(\theta)\right] = 0. \tag{5.6}$$

Equations (5.4) and (5.6) may be solved in parallel by the following recursive predction error estimation algorithm

$$\hat{\theta}(t) = \hat{\theta}(t-1) + \frac{\widetilde{R}(t)^{-1}}{t} x(t) \phi'(\varepsilon(t)); \tag{5.7}$$

$$\widetilde{R}(t) = \widetilde{R}(t-1) +$$

$$\frac{1}{t}\left[\phi''(\varepsilon(t)) x(t) x^T(t) - \widetilde{R}(t-1)\right], \tag{5.8}$$

$$\varepsilon(t) = y(t) - x(t)^T\hat{\theta}(t-1). \tag{5.9}$$

Note that $\tilde{R}(t)$ only an approximation for $R(\hat{\theta}(t-1))$. It is of some practical importance that this scheme can be realised by a slight modification of the recursive least squares algorithm:

$$\hat{\theta}(t) = \hat{\theta}(t-1) + P(t) x(t) \phi'(\varepsilon(t-1)) \tag{5.10}$$

$$P(t) = \frac{P(t-1)}{\lambda(t)}[I-$$

$$\frac{x(t) x(t)^T P(t-1)}{\lambda(t)/\phi''(\varepsilon(t)) + x(t)^T P(t-1) x(t)}]. \tag{5.11}$$

In this realization $P(t) = [\tilde{R}(t)]^{-1}$, and $\lambda(t)$ is a *forgetting factor* $(0 < \lambda(t) \leq 1)$. If $\lambda(t)<1$ then the estimates will be able to track slowly varying parameters. The update (5.11) can be implemented accurately, efficiently and stably using Thornton and Bierman's UDU^T factorisation algorithm [see Bierman, 1977]. Notice that $\phi''(\varepsilon(t))$ must be bounded. This is not true for the L_p case cost function at $\varepsilon(t) = 0$, so the L_p case cost (or at least its second derivative) must be modified near the origin. An modification reminiscent of Huber's M-estimator [Huber, 1964] is to introduce a quadratic section near the origin

$$\phi_{p^m}(x) = \begin{cases} \left(|x| - c\left(1 - \frac{p}{2}\right)\right)^p & |x| \geq c \\ x^2 \left(\frac{p}{2}\right)^p c^{p-2} & |x| < c \end{cases} \tag{5.12}$$

which is continuous and its first derivative exists everywhere. c should be chosen to be the same order of magnitude as the noise variance. Its value is not critical, and becomes less so as p increases.

The time-update requires $n(n+1)/2$ single precision storage locations and performs $(n-1)(n-2)$ divisions per update. If computational capacity is limiting scalar gain steepest-decent methods are appropriate. Sensible choices for $K(t)$ in

$$\hat{\theta}(t) = \hat{\theta}(t-1) + K(t) x(t) \phi'(\varepsilon(t)) \tag{5.13}$$

$$K\left(t\right) = \frac{\alpha}{t}$$

$$K\left(t\right) = \frac{\alpha}{t}\left[\beta + \phi''x\left(t\right)^T x\left(t\right)\right]^{-1}$$

$$K\left(t\right) = \alpha\left[\beta + \sum_{i=1}^{t} \phi''x\left(i\right)^T x\left(i\right)\right]^{-1}$$

(5.14)

with $\alpha,\beta>0$. These gains can be modified to include the forgetting factor, $\lambda(t)$ by implementing a recursion for $K_1(t)$ calculated from

$$K_1\left(t\right) = \frac{K_1\left(t-1\right)}{\lambda\left(t\right)} - \left[K\left(t-1\right) - K\left(t\right)\right].$$

(5.17)

It has been demonstrated that the modified L_{p^m} estimator with case cost defined by (5.12) is θ_0 w.p. 1 if the errors are uncorrelated [McMichael, 1988].

6 Conclusion

This paper has presented a range of consistent recursive implementations of the L_p estimator using algorithms that are closely related to recursive least squares and linear recursive steepest descent. The robustness and efficiency properties of this estimator are excellent as p becomes close to 1. Of all consistent gradient-type recursive estimators L_{1^+} estimation is both minimax SIC, and asymptotic covariance robust. The estimator does not require any scale parameters to be known accurately or to be estimated on-line. The recursive implementations can be made adaptive so as to track time-varying parameters.

Bibliography

Bierman, G.J., 1977, "Factorization methods for discrete system estimation", Academic Press.

Box, G.E.P. and Draper, N.R., 1987, "Empirical model building and response surfaces", John Wiley, New York.

Dutter, R., 1975, "Robust regression: different approaches to numerical solutions and algorithms", Report no. 6, Fachgruppe für Statistik, Eidgen, Techniche Hochschule, Zurich.

Hoaglin, D.C. and Welsch, R., 1978, "The hat matrix in regression and ANOVA", Amer. Statitian, vol. 32, pp 108–115.

Huber, P., 1964, "Robust estimation of a location parameter", Ann. Math. Stat., vol. 35, pp 73–101.

Kendall, M.G. and Stuart, A., 1979 (4th ed.) "The advanced theory of statistics", vol. II, Griffin, London.

Kushner, H.J. and Clark, D.S., 1978, "Stochastic approximation methods for constrained and unconstrained systems", Springer-Verlag, New York.

Ljung, L. and Söderstrom, T., 1983, "Theory and practice of recursive identification", MIT Press.

McMichael, D.W., 1987, "On-line fault detection: a system non-specific approach", D.Phil. Thesis, Univ. of Oxford: OUEL Report no. 1729/88.

McMichael, D.W., 1988, "Robust recursive L_p estimation", IEE Procs part D, to be published.

Pregibon, D., 1981, "Data analytic methods for generalized linear models", Ph.D. Thesis, University of Toronto.

Press, W.H, Flannery, B.P., Teukolsky, S.A. and Vetterling, W.T., 1986, "Numerical recipes", Cambridge U.P.

Robbins, H. and Monro, S., 1951, "A stochastic approximation method", Ann. Math. Stat., vol. 22., pp 400–407.

Wasan, M.T, 1969, "Stochastic approximation", Cambridge U.P.

NUMERICAL APPROXIMATION OF FREE BOUNDARY PROBLEMS IN POLYMER CRYSTALLIZATION

Salvatore Mazzullo, Maurizio Paolini, Claudio Verdi

Abstract. The skin-core morphology of an injection moulded polymer product is due to the melting-crystallization phase transition. It is thus described in terms of thermal history and overall (nucleation and crystal growth) crystallization kinetic. Two Stefan problems have been analyzed to study the moving boundary appearing during polymer crystallization. Their difference is due to the constitutive equation for the heat flux. An improved Avrami equation is adopted for the overall crystallization kinetic. This accounts for the formation of a mushy region besides liquid and crystal phases. Both models have been efficiently discretized by a stable finite element method based on a semi-explicit finite difference approximation in time. The skin-core structure is well predicted by the Fourier-Stefan model.

Key words: partial differential equations, Stefan problems, numerical approximation

1. Formulation of the mathematical models. Physical properties of polymeric materials mainly depend both on the chemical identity of monomers and the amorphous or semicrystalline nature of the polymer. The properties can also be influenced by the method of processing. In fact, due to melting-crystallization phase transition, large variations in the microstructure occur in different parts of an injection moulded product. The best known aspect of these heterogeneities is the so-called skin-core structure: in the case of polypropylene, starting from the surface of a microtomed slice, it is possible to observe up to four layers, more or less distinguishable (see Fig. 1 and Fig. 2).

In order to get an insight into the phenomena responsible for the skin-core morphology, we concentrated on both packing and cooling stages, ignoring the flow dynamics of the mould filling stage. Morphology is thought to be described in terms of thermal history and crystallization kinetic. Thus, the mathematical models consist of the kinetic equation and the energy transport equation.

To our ends it would be appropriate to separate the kinetic of crystallization into two steps: nucleation and crystal growth, [10]. But, as a first attempt, we dealth only with an overall kinetic of crystallization. Denoting by θ the absolute *temperature* and by

437

J. Manley et al. (eds.), Proceedings of the Third European Conference on Mathematics in Industry, 437–443.
© 1990 *Kluwer Academic Publishers and B. G. Teubner Stuttgart.*

438

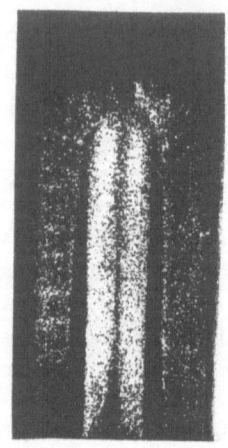

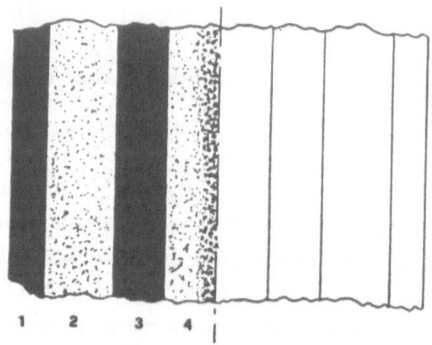

Figure 2. Scheme of the morphology
Zone 1 (skin): no visible crystalline structure
Zone 2: bright microspherulitic structure
Zone 3: dark band
Zone 4 (core): bright spherulitic, size
increasing toward the centre

Figure 1. Example of skin-core morphology
(microtomed slice from injection moulded
polypropylene; magnification 40x)

w the *crystallinity* of the injected polymer, we considered the kinetic equation

$$(1.1) \qquad \frac{\partial w}{\partial t} + H^{-1}(w) \ni f(\theta, w),$$

where H^{-1} denotes the inverse of the Heaviside graph and forces the constraint $0 \leq w \leq 1$.
Thus, (1.1) is equivalent to

$$\frac{\partial w}{\partial t} = f(\theta, w)^{+} \ \text{ if } w = 0, \quad = f(\theta, w) \ \text{ if } 0 < w < 1, \quad = -f(\theta, w)^{-} \ \text{ if } w = 1.$$

Consolidated experimental observations show that f is a bell-shaped function of the
temperature, which vanishes outside the range (Θ_g, Θ_m), [10]. Here Θ_m stands for the
equilibrium melting temperature and Θ_g for the *glass transition temperature*. The kinetic
equation (1.1) thus account for the formation of a mushy region, which proceeds as
follows. As soon as $\theta < \Theta_m$, by (1.1) we have $\frac{dw}{dt} > 0$ and release of latent heat, namely
crystallization, starts. If the cooling is fast enough, θ can reach Θ_g before $w = 1$; in that
case $\frac{dw}{dt} \ll 1$ and the uncrystallized material forms an amorphous glassy phase with glass
concentration equal to $1 - w$; the remaining fraction w is occupied by crystal phase. We
adopted for the rate of crystallization f the function proposed by Malkin et al. [4]:

$$(1.2) \qquad f(\theta, w) := K_0(\theta)[w_{eq} - w][1 + C_0 w] \quad \text{for } 0 \leq w \leq w_{eq}(\theta) < 1,$$

being: $K_0(\theta) = K e^{-\frac{E + \psi R}{k\delta} - \frac{\phi}{\Theta_m - \theta}}$ if $\theta < \Theta_m$, $= 0$ if $\theta \geq \Theta_m$ the *nucleation kinetic
constant* (see Fig. 4) and $0 \leq w_{eq} = a\sqrt{\Theta_m - \theta} \leq 1$ the *equilibrium crystallinity*. Another
law well fitting the experimental crystallinity, as a function of time, can be found in [1].

The effects of the pressure P on the kinetic of crystallization may be quantitatively evaluated. From preliminary experimental results (see Fig. 3) it may be envisaged that pressure induces a shift of the equilibrium melting temperature, toward higher values, in such a way that the Clausius-Clapeyron equation may be applied also to molten polymers: $\frac{dP}{d\theta} = \frac{\lambda}{\theta\,\delta v}$, being λ the *latent heat of crystallization* and δv the variation of specific volume from liquid to solid phase. Such information will be used during the packing stage.

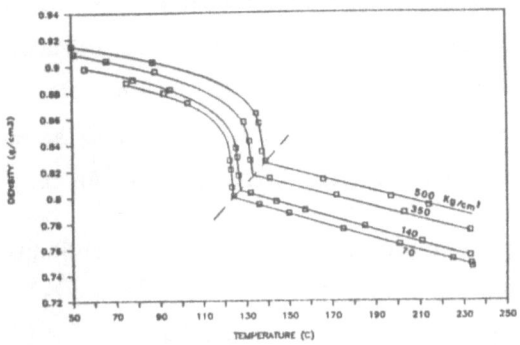

Figure 3. P-V-T diagram

Denoting by $\Omega \subset \mathbf{R}^d$ $(d \leq 3)$ the domain occupied by the quiescient molten polymer and by $[0, T]$ the time interval in which crystallization occurs, we consider the *energy balance equation*, in $\Omega \times (0, T)$:

(1.3)
$$\rho C_p \frac{\partial\theta}{\partial t} + \rho\lambda\frac{\partial w}{\partial t} + \mathrm{div}\,\mathbf{q} = 0.$$

Equation (1.3) is coupled with the following constitutive equation for the *heat flux* $\mathbf{q}$:

(1.4)
$$\alpha\frac{\partial\mathbf{q}}{\partial t} + \mathbf{q} = -k\,\mathrm{grad}\,\theta.$$

The case $\alpha = 0$ corresponds to the Fourier *conduction law*, whereas a positive *relaxation time* $\alpha > 0$ was introduce by Cattaneo [3], with the following motivation. The classical Fourier conduction law has an essentially empirical base and has been generated by a *steady* experiment of heat conduction through an ideal slab of infinite extent, [2]. A natural question is to ascertain whether such a relation obtained under steady conditions may be considered valid also under nonsteady conditions. Cattaneo overcame this criticism in that equation (1.4) is asymptotically coincident with the classical Fourier conduction law, for $\alpha \to 0$. We refer to [6] for a detailed analysis of the use of Cattaneo constitutive equation in polymer crystallization models.

The system of equations (1.1)÷(1.4) must obviously be completed by initial and boundary conditions, more precisely:

(1.5) $\qquad \theta(x,0) = \theta_0, \quad w(x,0) = w_0, \quad q(x,0) = q_0 \quad \text{in } \Omega,$

(1.6) $\qquad \theta = g_D \quad \text{on } \Gamma_D \times (0,T), \quad q \cdot \nu = p\theta + g_N \quad \text{on } \Gamma_N \times (0,T),$

(here $\Gamma_D \cup \Gamma_N = \partial\Omega$ and ν denotes the unit outward normal vector on Γ_N).

The problem has to be understood in variational sense, [9]. Under suitable regularity assumptions on the data, existence and uniqueness of the solution are known for the case of Fourier conduction law, $\alpha = 0$, [9]. The case $\alpha > 0$ needs further investigations; more precisely, in [9] existence, but not uniqueness, is proved for the simple crystallization kinetic $f(\theta, \chi) = c\theta$, whereas no existence results are known for the general case.

2. Numerical approximation of the continuous problems. We give a unified treatment of both Fourier ($\alpha = 0$) and relaxed Fourier-Cattaneo ($\alpha > 0$) models. For any integer number N ($\tau := \frac{T}{N}$ stands for the *time step*), the semi-explicit time discrete algorithm can be stated as follows:

Set $\theta^0 := \theta_0$, $w^0 := w_0$, $q^0 := q_0$, and, for any $1 \le n \le N$, find θ^n, w^n and q^n such that the following equations hold in Ω:

(2.1) $\qquad \rho C_p \dfrac{\theta^n - \theta^{n-1}}{\tau} + \rho\lambda \dfrac{w^n - w^{n-1}}{\tau} + \text{div} q^n = 0,$

(2.2) $\qquad \alpha \dfrac{q^n - q^{n-1}}{\tau} + q^n = -k\nabla\theta^n,$

(2.3) $\qquad \dfrac{w^n - w^{n-1}}{\tau} + H^{-1}(w^n) \ni f(\theta^{n-1}, w^{n-1})$

and $\theta^n = g_D(n\tau)$ on Γ_D, $q^n \cdot \nu = p\theta^n + g_N(n\tau)$ on Γ_N.

Let us comment on the *existence* and *uniqueness* of the solution $\theta^n \in V(g_D(n\tau))$, $w^n \in L^\infty(\Omega)$, $q^n \in L^2(\Omega)^d$ of the variational formulation of the problem (here $V(g) := \{v \in H^1(\Omega) : v = g \text{ on } \Gamma_D\}$). Noting that the inverse of the maximal monotone graph $(I + H^{-1})$ is a monotone function, (2.3) determines uniquely $w^n \in L^\infty(\Omega)$:

(2.4) $\qquad w^n = (I + H^{-1})^{-1}(w^{n-1} + \tau f(\theta^{n-1}, w^{n-1})) \quad \text{a.e. in } \Omega.$

Then, rearranging (2.2) as follows:

(2.5) $\qquad q^n = -k\dfrac{\tau}{\alpha + \tau}\nabla\theta^n + \dfrac{\alpha}{\alpha + \tau}q^{n-1} \quad \text{in } L^2(\Omega)^d,$

(2.1) can be viewed as a linear elliptic equation in the unknown $\theta^n \in V(g_D(n\tau))$:

(2.6)
$$< \rho C_p \theta^n, v > + k \frac{\tau^2}{\alpha + \tau} < \nabla \theta^n, \nabla v > + \tau \ll p\theta^n, v \gg = -\tau \ll g_N(n\tau), v \gg +$$
$$+ < \rho C_p \theta^{n-1} - \rho\lambda(w^n - w^{n-1}), v > + \frac{\alpha\tau}{\alpha + \tau} < q^{n-1}, \nabla v > \quad \forall\, v \in V(0)$$

(here $< v, w >:= \int_\Omega vw\, dx$ and $\ll v, w \gg := \int_{\Gamma_N} vw\, d\sigma$). Thus the discrete time problem has a unique solution (note that $p = p(x) \geq 0$).

By using standard variational techniques, it is very easy to prove that the scheme is *stable in energy spaces*, [5, 8]. For the case of Fourier conduction law, $\alpha = 0$, it is well known that stability entails convergence of the discrete solutions to the continuous one (see [9], where compactness arguments were used). Due to the lack of a uniqueness proof, for the case $\alpha > 0$, $f(\theta, \chi) = c\theta$, one can only prove that the discrete solutions converge, possibly taking a subsequence, to a solution of the continuous problem, [9].

For the spatial discretization of the time discrete scheme (2.1)÷(2.3), we use continuous linear finite element for the temperature variable and piecewise constant element for the crystallinity and the heat flux. The use of the vertex quadrature rule to evaluate the integrals makes the fully discrete scheme very easily implemented on a computer, [5, 8]. In fact, the algorithm consists of three steps per time level: (2.4) just corresponds to an element-by-element evaluation of a given nonlinear function; the linear system related to (2.6) can be solved by using either a S.O.R. method or a preconditioned conjugate gradient method; finally, in (2.5), the discrete heat flux is easy computed, as the discrete temperature is linear on each element. For carrying out the numerical experiments we made use of a finite element package specially designed for free boundary evolution problems, in general two dimensional space domains, [7].

The numerical tests show that, even for the case $\alpha > 0$, the discrete solutions seem to converge, despite the lack of a uniqueness proof for the continuous problem.

3. Numerical experiments. In order to present some numerical simulations, we refer to a $2 - D$ rectangular trasversal section of a standard sample . The walls of the sample are assumed to guarantee an efficient heat exchange, and simmetry is used to define the domain Ω. Initially, the molten polymer is at a temperature $\theta_0 = 587\ °K$, $\Theta_m = 501\ °K$ being the melting and $\Theta_g = 333\ °K$ the glass transition temperatures. The physical constants of the material (Nylon 6) was basically found in [4, 5]. For simplicity we assume that the density of both the molten and crystallized polymer remains unchanged during the phase change. Finally, to simulate the packing stage, we apply a sudden pressure of 100 MPa to the sample, a given time (7.5 sec) after the filling of the mould. Due to the constancy of the density, the only effect of the packing stage is on the crystallization kinetic, through the Clausius-Clapeyron equation (see section 1). In these numerical experiments we used the experimental value: $\frac{dP}{dT} = 0.4 \ ^\circ K/MPa$.

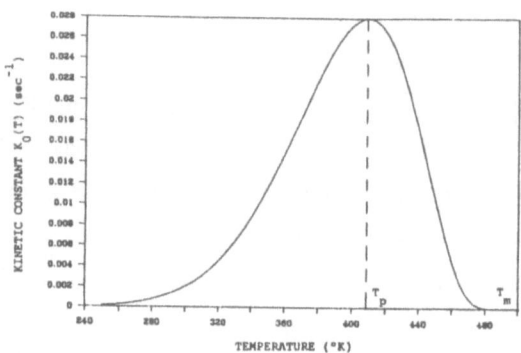

Figure 4. Nucleation kinetic constant as a function of temperature

Crystallization occurs in a time interval $[0, T]$ where $T = 45$ seconds. Fig. 5 shows the crystallinity at final time T for the relaxed Fourier-Stefan model, by using $N = 90$ time steps and a uniform mesh size $h = 100 \ \mu m$. No packing is simulated, so that crystallization is stress-free. Crystallinity varies smoothly from the outer surfaces to the inner core. A low level of crystallinity appears near the surfaces (skin).

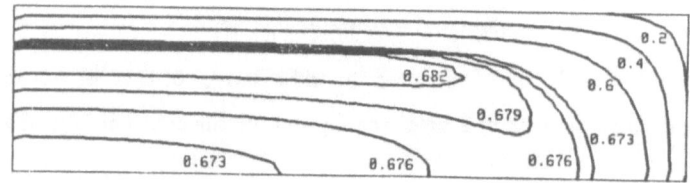

Figure 5. Crystallinity level curves (Fourier-Stefan model; no packing)

The effect of packing, 7.5 seconds after the end of the filling of the mould, is shown in Fig. 6. Crystallinity has now a non-monotonic variation within the sample, and it reaches an internal maximum before reaching the maximum on the core. A low level of crystallinity near the surfaces (skin) is followed by a first maximum, then a minimum and finally a second maximum on the core. With reference to Fig. 2, the internal maximum may be associated to the bright microspherulitic, zone 2, the minimum to the dark band, zone 3, and finally the core maximum to the macrospherulitic, zone 4. In such a way, it seems that a description of the skin-core morphology has been obtained.

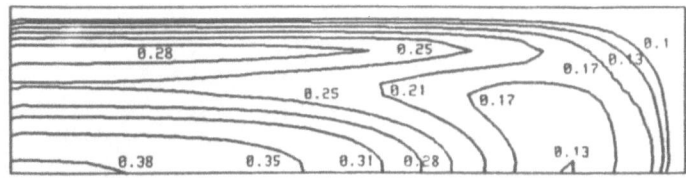

Figure 6. Crystallinity level curves (Fourier-Stefan model; packing after 7.5 seconds)

The numerical results obtained by solving the Stefan-Cattaneo model are strictly similar to those illustrated above for the case of Fourier conduction law. A motivation for this similarity is that the order of magnitude of the relaxation time α for heat transport in polymer is comparable with that of the time step τ. Thus, the effects of discontinuity in heat transport mix up with the truncation error of the numerical scheme.

References

[1] J. BERGER, W. SCHNEIDER, A zone model of rate controlled solidification, Plastics and Rubber Processing and Applications, 6 (1986), 127-133.

[2] H. S. CARSLAW, J. C. JAEGER, Conduction of heat in solids, Oxford University Press, Oxford (1959).

[3] C. CATTANEO, Sulla conduzione del calore, Atti Semin. Fisico-Matem. Univ. Modena, 3 (1948), 83-101.

[4] A. Ya. MALKIN, V. P. BEGHISHEV, I. A. KEAPIN, Z. S. ANDRIANOVA, General treatment of polymer crystallization kinetics, part 2: the kinetics of nonisothermal crystallization, Polym. Eng. Sci., 24 (1984), 1402-1408.

[5] S. MAZZULLO, M. PAOLINI, C. VERDI, Polymer crystallization and processing: free boundary problems and their numerical approximation, preprint 626 I.A.N. of C.N.R., Pavia (1988).

[6] R. OCONE, G. ASTARITA, Continuous and discontinuous models for transport phenomena in polymers, A. I. Ch. E. J., 33 (1987), 423-435.

[7] M. PAOLINI, G. SACCHI, C. VERDI, Finite element approximations of singular parabolic problems, Int. J. Numer. Meth. Eng., 26 (1988), 1989-2007.

[8] C. VERDI, A. VISINTIN, Numerical analysis of the multidimensional Stefan problem with supercooling and superheating, Boll. Unione Mat. Ital., (7) I - B (1987), 795-814.

[9] A. VISINTIN, Stefan problem with phase relaxation, IMA J. Appl. Math., 34 (1985), 225-245.

[10] B. WUNDERLICH, Macromolecular physics, Vol. 2, Academic Press, New York (1976).

S. Mazzullo, HIMONT ITALIA, Centro Ricerche "G. Natta", 44100 Ferrara, Italy

M. Paolini, Istituto di Analisi Numerica del C.N.R., 27100 Pavia, Italy

C. Verdi, Dipartimento di Meccanica Strutturale, Università di Pavia, 27100 Pavia, Italy

Ignition/Extinction Phenomena: An Investigation

of

Parametric Sensitivity for a Strongly Nonlinear

Reaction-Diffusion System

Dirk Meinköhn

Abstract: Combustion of a porous fuel agglomerate or exothermic catalytic conversion involving a porous catalyst pellet represents a chemical process which
is irreversible due to an intermediate adsorption step. For stationary operations, two different temperature regimes of the process exist depending on
whether there is chemical (i.e., kinetic) or diffusive control. Within the set
of all stationary states of the process, jump-like transitions from one regime
into the other are termed "ignition" or "extinction" depending on the direction
of the temperature change. The mathematical treatment of ignition and extinction presented in this paper centers on an investigation of the stability boundary on the state surface which represents the set of all stationary states in
a suitable thermodynamic state space. Projecting the state surface into the
subspace of (independent) control parameters is shown to generate contours by
means of a visibility condition, with the contours representing the projection
image of the stability boundary. The contours and, in particular, their singularities are investigated with the help of lower bounds, the main advantage of
this approach resting in the fact that qualitative and quantitative information can be gained by employing linear methods.

1   Physical Problem and Mathematical Model

For a highly porous catalyst particle or fuel pellet residing in an ambient atmosphere of gaseous reactants, conditions for ignition and extinction
can be derived from the following boundary value problem of a single differential equation [1], [2], if a single reaction step and suitable boundary conditions are assumed:

(1) $\Delta y + \lambda w(y;\xi,\beta,n) = 0$ in D, $y = 0$ on ∂D

Here, Δ designates the Laplacian, D the region occupied by the pellet, ∂D its
surface, y the temperature distribution in D , ξ, β, λ, n the control parameters. The function w is of the following form:

(2) $w(y;\xi,\beta,n) = (1 - \xi y)^n \exp(y/(1+\beta y))$

From physical considerations [1], [4], the ranges of the parameters are as
follows :

(3) $\lambda \geq 0$, $\beta \geq 0$, $n \gtrless 0$, $\xi \gtrless 0$

Negative values of ξ are admitted in order to include "generalized" reaction
rate laws [5]. Only exothermic chemical reactions are considered and therefore
function w must only be positive :

445

J. Manley et al. (eds.), Proceedings of the Third European Conference on Mathematics in Industry, 445–452.
© 1990 Kluwer Academic Publishers and B. G. Teubner Stuttgart.

$$(4) \quad w(y;\xi,\beta,n) \; > \; 0$$

Thus, the following differential inequality arises from Eq.(1) :

$$(5) \quad \Delta y \; \leq \; 0 \; \text{ in } D \; , \; y \; = \; 0 \; \text{ on } \partial D$$

On account of the boundary condition of Eq.(5), application of the minimum principle [6] furnishes :

$$(6) \quad 0 \; < \; y \; \leq \; y_m \quad \text{ in } D \; .$$

In order to ensure that w is nonvanishing, bounded, and positive for ξ positive, the range of w is restricted by stipulating for the maximum y_m of y :

$$(7) \quad \xi \; > \; 0 \; \rightarrow \; y_m \; < \; 1/\xi$$

Assumption of a suitable configuration for D (D has to be starshaped with a center of symmetry - e.g.: a sphere, a finite or infinite slab, a finite or infinite right circular cylinder) results in a solution y of Eq.(1) with a unique maximum y_m located at the center of symmetry of D. Therefore, y_m may formally be regarded as an independent parameter, whereby the solutions of Eq.(1) result in parametric representation :

$$(8) \quad y \; = \; y(x;y_m,\beta,\xi,n) \quad , \quad \lambda \; = \; \lambda(y_m,\beta,\xi,n)$$

The catalyst pellet represents an open thermodynamic system. Therefore, $\lambda(y_m,\beta,\xi,n)$ represents the state surface in a thermodynamic state space spanned by y_m and λ, β, ξ, n . Here, y_m characterizes the internal thermodynamic state of the system whereas λ, β, ξ, n represent control parameters, i.e., the independent parameters. Due to the properties of w (cf.Eq.(2)), the boundary value problem of Eq.(1) is strongly nonlinear. The advantage of introducing the state surface $\lambda(y_m,\beta,\xi,n)$ consists in that linear methods exist by which it is possible to characterize the solutions of Eq.(1) as to their stability and, in particular, as to the stability boundary on the state surface which separates the regions of stable and unstable solutions of Eq.(1) ([7], [8]). For the curve $\lambda(y_m)$ resulting from the intersection of the state surface with a plane defined by β = const, ξ = const, n = const, the following properties have been shown to hold of necessity [7] : for stable solutions of Eq.(1), $\lambda(y_m)$ is strictly increasing and for critical solutions (i.e., solutions on the statbility boundary), $\lambda(y_m)$ has a stationary point : $\partial\lambda/\partial y_m = 0$. Consequently, the projection of the state surface along the y_m-axis into the subspace of control parameters λ, β, ξ, n will furnish "visible" contours which are the projection image of the stability boundary. Information on these contours is of particular interest, because the control parameters represent the independent variables of Eq.(1).

2 Disappearance of Criticality in Terms of Lower Bounds

If $w(y,\xi,\beta,n)$ is increasing with respect to y for $0 \le y \le y_0$, then a lower bound $\lambda_M(y_m)$ of the intersection $\lambda(y_m)$ can be derived with the help of linear methods [8]. In particular, for $0 \le y_m \le y_0$:

(9) $\lambda_M(y_m) \le \lambda(y_m)$

with

(10) $\lambda_M = c \dfrac{y_m}{w(y_m)}$

Here, c represents a constant which depends on the geometry [8] (e.g., c=2 for the infinite slab of unit thickness).
In order to determine the interval $0 \le y \le y_0$ for which $w(y)$ is increasing, $w'=\partial w/\partial y$ has to be investigated :

(11) $w' = \dfrac{w}{(1+\beta y)^2 (1-\xi y)} (1-\xi y-n\xi(1+\beta y)^2)$

Because of Eqs.(4),(6),(7), $w > 0$, $(1-\xi y) > 0$, and thus, two cases arise:
Case 1: $\text{sign}(n)= -\text{sign}(\xi) \leftrightarrow n\xi < 0$, whereby $w' > 0$ and w strictly increasing in y for $0 \le y < 1/\xi$ if $\xi > 0$ or for $0 \le y$ if $\xi < 0$.
Case 2: $\text{sign}(n)=\text{sign}(\xi) \leftrightarrow n\xi > 0$. Under the provision of

(12) $1 - n\xi > 0 \leftrightarrow \xi < 1/n$ for $\text{sign}(n)=+1$ and $\xi > 1/n$ for $\text{sign}(n)= -1$

there exists an interval $0 \le y \le y_0$, for which w is strictly increasing with respect to y . As $n\xi > 0$ for case 2, an application of Descartes' rule of signs shows that the factor $(1-\xi y-n\xi(1+\beta y)^2)$ of w' possesses a positive and a negative root. Therefore, due to $w'(0)=1-n\xi > 0$, w can have no minimum for $0 \le y < 1/\xi$ if $\xi > 0$ or for $0 \le y$ if $\xi < 0$.
The stationary points of $\lambda_M(y_m)$ (cf.Eq.(10)) result for values of y_m , which are the roots of the following equation :

(13) $w(y_m) - y_m w'(y_m) = 0$.

In order to establish a relationship between the stationary points of $\lambda_M(y_m)$ and of $\lambda(y_m)$, the positive roots y_{mi} of Eq.(13) have to satisfy: $0 \le y_{mi} \le y_0$. Case 1 poses no problem as w is strictly increasing in y . For case 2 and under the provision of Eq.(12) , only a single maximum of w may occur for $0 \le y < 1/\xi$ if $\xi > 0$ or $0 \le y$ if $\xi < 0$, with the location of this maximum given by y_o. Because the roots y_{mi} of Eq.(13) designate those tangents of $w(y)$ which intercept at the origin, geometrical considerations lead to the conclusion that for w defined according to Eqs.(2),(4), the positive roots y_{mi} of Eq.(13) must satisfy $0 \leq y_{mi} < y_o$. Thus, under the provision of Eq.(12), the stationary points of $\lambda_M(y_m)$ on the positive y-axis remain within the interval for which Eq.(9) holds. A function $w(y)$ for which

$$(14) \quad w(y) - yw'(y) \geq 0$$

is termed concave (in a generalized sense). It has been shown [7], that for functions w that are concave on the interval $0 \leq y \leq \bar{y}_m$, the corresponding intersection $\lambda(y_m)$ is strictly increasing in y_m for $0 \leq y_m \leq \bar{y}_m$. Because of Eq.(2), $w(o) > 0$. Thus, for a function of the class defined by Eq.(2), there always exists a nontrivial interval $0 \leq y \leq \bar{y}_m$ where w is concave. If $\lambda(y_m)$ possesses an extremum, then there exists a first maximum λ^* which will be attained at a value $y_m > \bar{y}_m$. Because of Eqs.(13),(14) being identical, $\bar{y}_m$ is the lowest positive root of Eq.(13) where $\lambda_M(y_m)$ (cf.Eq.(9)) reaches a maximum λ_M^* . Therefore, the following condition holds of necessity :

$$(15) \quad \lambda_M^* < \lambda^*$$

Further, if w is such that there exist at most two positive roots of Eq.(11), then there exists at most a minimum λ_M^{**} of $\lambda_M(y_m)$ beyond the maximum λ_M^* , such that on account of Eq.(9) the lowest minimum λ^{**} of $\lambda(y_m)$ must of necessity satisfy the following inequality:

$$(16) \quad \lambda_M^{**} \leq \lambda^{**}$$

It has been shown in [8], that for some convex functions w in the class defined by Eq.(2), continuous changes in the control parameters β, ξ, n will deform the function w in such a way that it becomes strictly concave for all of the positive y-axis, i.e.: $w(y) - yw'(y) > 0$. Such a change-over in w from convex to strictly concave entails the disappearance of criticality in the sense that a maximum/minimum pair (λ^*,λ^{**}) of $\lambda(y_m)$ must disappear in a "collision" process, leaving the function $\lambda(y_m)$ to be strictly increasing. Because of Eqs.(13),(14) being identical, this process of disappearance of criticality is replicated in terms of lower bounds, with a maximum/minimum pair $(\lambda_M^*,\lambda_M^{**})$ of $\lambda_M(y_m)$ disappearing in an analogous collision process.

Upon projecting $\lambda(y_m,\beta,\xi,n)$ or $\lambda_M(y_m,\beta,\xi,n)$ along the direction of the y_m-axis into the subspace of control parameters λ,β,ξ,n, a contour is generated for those points, for which the visibility condition $\partial\lambda/\partial y_m=0$ or $\partial\lambda_M/\partial y_m=0$ holds. Singularities in the contours occur if the projection is along an asymptotic direction, i.e., in the direction of a tangent which is of an order of tangency larger than 1 [9]. The merging of a maximum/minimum pair of an intersection $\lambda(y_m)$ or $\lambda_M(y_m)$ results in a stationary point of inflection, for which the projection is in the direction of a tangent of tangency order 2. Thus, the merging of a maximum/minimum pair of such an intersection leads to a singularity in the resulting contour in control parameter space which is of lowest order in the hierarchy of singularities: a cusp.

For the class of functions w defined by Eq.(2), the condition of Eq.(13) is equivalent to the vanishing of the following polynomial of third order in y:

$$(17) \quad P(y) = (n-1)\xi\beta^2 y^3 + (\beta^2+\xi+2(n-1)\xi\beta)y^2 + ((n-1)\xi+2\beta-1)y + 1 = 0$$

Thus, at most three roots of Eq.(13) and, therefore, at most three stationary points of $\lambda_M(y_m)$ exist for a functiom w defined by Eq.(2). The merging of a maximum/minimum pair of $\lambda_M(y_m)$ is associated with a double root of the polynomial $P(y)$ (cf.Eq.(17)), which is to occur if the discriminant of $P(y)$ vanishes. In consequence of Eq.(6), only those mergers are of interest which take place on the positive y_m-axis. Therefore, two cases have to be investigated separately. Abbreviating the polynomial coefficients of $P(y)$ by a_i , we find :

$$(18) \quad P(y) = a_3 y^3 + a_2 y^2 + a_1 y + 1$$

Case (a): $a_3 > 0 \leftrightarrow (n-1)\xi > 0$. According to Descartes' rule of signs, there will be no positive root for $sign(a_1)=sign(a_2)=+1$, whereas for the other sign combinations, a pair of positive roots or a positive double root occurs. Continuous changes in the control parameters β,ξ,n may thus lead to the "collision" of two simple roots of Eq.(17) which coalesce in a double root and disappear subsequently. The projection of $\lambda_M(y_m,\beta,\xi,n)$ into the subspace of control parameters λ,β,ξ,n will thus produce a contour containing cusp ridges which represent the set of positive double roots of Eq.(17).

Case (b): $a_3 < 0 \leftrightarrow (n-1)\xi < 0$. According to Descartes' rule of signs, a single positive root of Eq.(17) occurs in case (b), except for $sign(a_2)=+1$, $sign(a_1)=-1$, for which a positive treble root of Eq.(17) is possible which is tantamount to a third order of tangency in the projection of the state surface. Therefore, continuous changes in β,ξ,n exist such that a pair of double roots of Eq.(17) is made to disappear in a "collision" process with the subsequent appearance of a simple root. In control parameter space, this collision process is associated with a higher order singularity in the contours which is of the "swallow-tail" type (cf. [9] for the nomenclature).

450

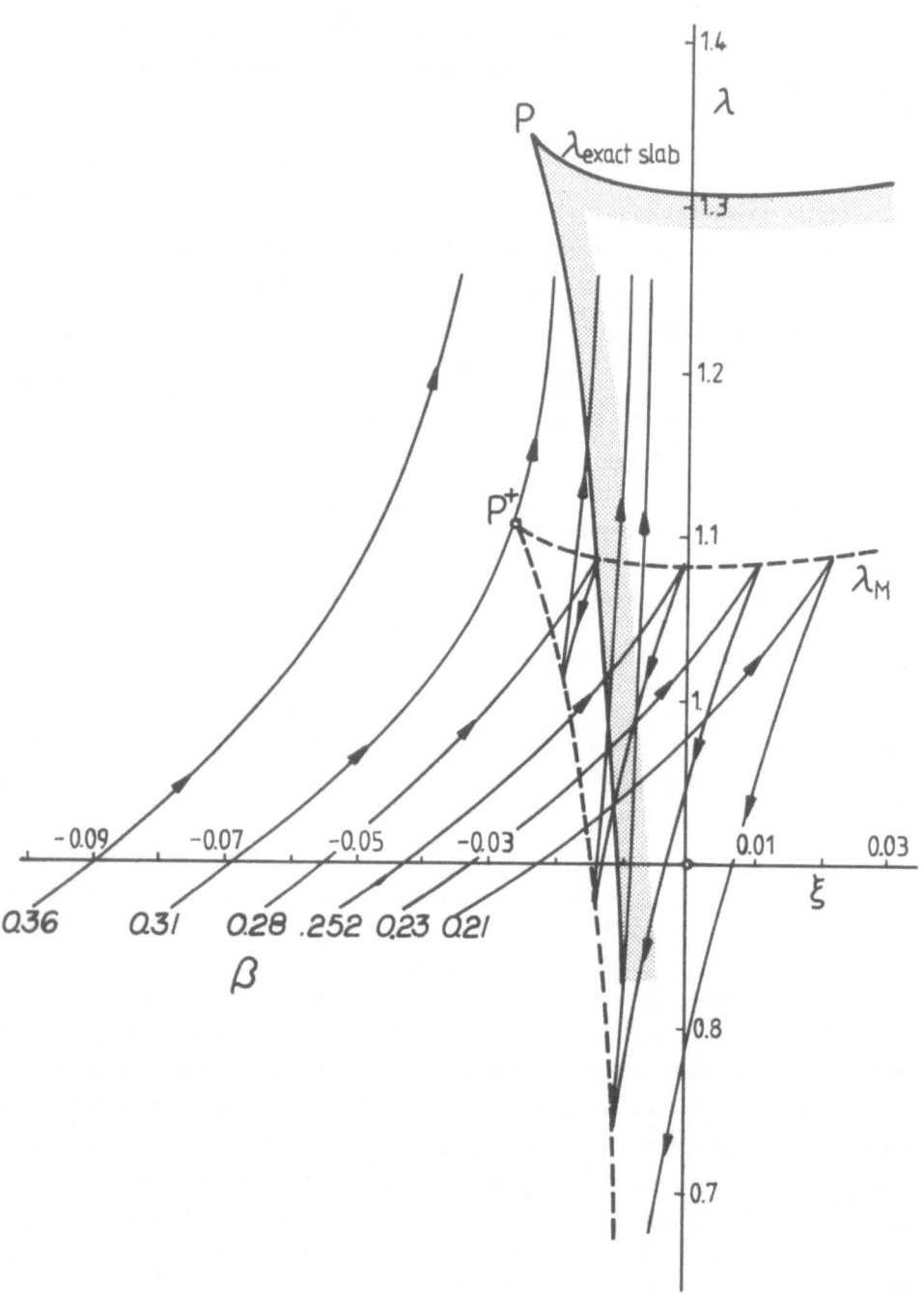

Fig.1 Swallow-Tail Singularity in the Bifurcation Set for n=2.o

3   A Singularity of Swallow Tail Type

Selecting n=2.0 entails $\xi > 0$ according to case (a) and $\xi < 0$ according to case (b). A visible contour results from the projection image of the set of stationary points of the intersections $\lambda_M(y_m)$ in state space which are defined by Eqs.(10),(17). Solving for ξ in $P(y_m)=0$ (cf.Eq.(17)) furnishes:

$$\lambda_M(y_m) = c \frac{y_m}{(1-\xi y_m)^n} \exp(-y_m/(1+\beta y_m))$$

(19)

$$\xi(y_m) = \frac{y_m-(1+\beta y_m)^2}{y_m(y_m+(n-1)(1+\beta y_m)^2)}$$

$(\lambda_M(y_m),\xi(y_m),y_m,\beta,n)$ represents the set of stationary states on the state surface $\lambda_M(y_m,\beta,\xi,n)$. The contours of this surface are obtained by projecting along the y_m-axis :

(20) $(\lambda_M(y_m),\xi(y_m),y_m,\beta,n) \rightarrow (\lambda_M(y_m),\xi(y_m),\beta,n)$

$(\lambda_M(y_m),\xi(y_m))$ furnishes a parametric representation of the visible contours for given values of the parameters β,n . Fig.1 displays the resulting contours for n=2.0 and values of β ranging from 0.36 to 0.21 .

For n=2.0, case (a) amounts to positive values of ξ , whereas case (b) is equivalent to negative values of ξ . Therefore, two stationary points of $\lambda_M(y_m)$ exist at most for positive values of ξ , whereas for negative values of ξ , at least one and at most three stationary points of $\lambda_M(y_m)$ exist. Therefore, only contour singularities of the cusp type appear for positive values of ξ , whereas for negative values of ξ , a higher order singularity of the swallow-tail type is possible beyond the cusp singularities. In the example of Fig.1, a swallow-tail P^+ appears for $\beta=0.31$. The pair of cusp ridges which issue from the swallow-tail P^+ are indicated by broken lines. While a cusp of the contour is equivalent to the disappearance of a maximum/minimum pair of the intersection $\lambda_M(y_m)$, inspection of Fig.1 reveals that a swallow-tail gives rise to the disappearance of a pair of cusps, with a regular stationary point of $\lambda_M(y_m)$ appearing subsequently.

For the infinite slab, a solution of Eq.(1) can be derived by quadrature [3]. With the help of numerical calculations it can be shown that a swallow-tail singularity P occurs in the contours of the exact state surface. In Fig.1, the cusp ridges issuing from P have been marked $\lambda_{exact\ slab}$, with the resulting solid curves underlined by shading. Inspection of Fig.1 reveals that the broken lines of cusp ridges of λ_M provide a continuous lower bound for the

shaded curve representing the exact locations of the cusp ridges in the case
of the infinite slab.

4   Conclusion

With the help of lower bounds, a novel feature in the class of reaction-
diffusion systems defined by Eqs.(1),(2) has been derived, namely a higher
order singularity of the swallow-tail variety in the projection of the state
surface into the subspace of control parameters. This has been shown to agree
with what can be derived for the only generally solvable model of Eqs.(1),(2)
which results for the infinite slab geometry.

5   References

[1]   Aris,R.: The Mathematical Theory of Diffusion and Reaction in Permeable
       Catalysts, Oxford: Clarendon Press 1975

[2]   Stakgold,I.: Reaction-Diffusion Problems in Chemical Engineering,
       Lecture Notes in Mathematics No. 1224, Berlin: Springer 1986

[3]   Gavalas,G.R.: Nonlinear Differential Equations of Chemically Reacting
       Systems, Berlin: Springer 1968

[4]   Glassman,I.: Combustion, New York: Academic Press 1987

[5]   Boddington,T.; Feng, C.-G.; Gray, P.: Thermal Explosions, Criticality,
       and the Disappearance of Criticality in Systems with Distributed Tem-
       peratures, Proc.R.Soc.Lond. A 390 (1983) 247-264

[6]   Protter,M.H.; Weinberger,H.F.: Maximum Principles in Differential
       Equations, Englewood Cliffs: Prentice-Hall 1967

[7]   Meinköhn, D.: Stability and Criticality of the Stationary Solutions of
       the Nonlinear Fourier Equation in Frank-Kamenetzki's Theory,
       J.Chem.Phys. 74 (1981) 3603-3608

[8]   Meinköhn, D.: Disappearance of Criticality in Nonuniform Systems I,
       SIAM-J.Appl.Math. 48 (1988) 536-548

[9]   Arnold, V.I.: Catastrophe Theory, Berlin: Springer 1984

Dirk Meinköhn
Fakultät für Physik
Universität Bielefeld
D-4800 Bielefeld 1 , FRG

ON THE METHODS FOR OPTIMAL SHAPE DESIGN

Pekka Neittaanmäki

Abstract. A short survey of the numerical methods for solving optimal shape design problems is given.

1. Setting of the optimal shape design problem

The design of the geometry is the primary problem of designers of structural systems. In spite of graphical work stations and modern software for analyzing the stucture, finding the best geometry for the stucture by "trial and error" is still a very tedious and timeconsuming task. The goal in optimal shape design (structural optimization, or redesign) is to computerize the design process and therefore shorten the time it takes to design or improve the existing design. Structural optimization is already used in certain applications in the automobile, marine, aerospace industries and in designing truss and shell structures (with minimum weights). In general, however the structural optimization is just beginning to penetrate the industrial community. The integrated FEM and CAD (Computer Aided Design) technologies within optimization loop will (hopefully quite soon) fully computerize the design loop.

From the mathematical point of view optimal shape design (or optimum design, optimization of the domain, structural optimization) is a relatively new branch of calculus of variations and especially of optimal control where the study is devoted to the problem of finding the optimal shape of an object. In an optimal shape design process one wishes to optimize a criteria involving a solution of a partial differential equation with respect to its domain of definition.

Let $\Omega \in \mathcal{O}$ (= set of admissible domains) be a domain (a geometrical layout of the structure) for which we want to find an optimal design. In order to know state $y(\Omega)$ of the system we solve in Ω so called state problem (typically partial differential equation with appropriate boundary conditions). Finally, we define a criterion functional (a cost functional) by which we measure the goodness of our design. Summarizing the situation we have the mappings

$$\Omega \mapsto y(\Omega) \mapsto J(\Omega; y(\Omega)) \tag{1}$$

and we settle the problem: Find $\Omega^* \in \mathcal{O}$ such that

$$J(\Omega^*; y(\Omega^*)) \leq J(\Omega; y(\Omega)) \qquad \text{for all } \Omega \in \mathcal{O}. \tag{2}$$

In many cases domain Ω can be parametrized: We can define a mapping $u \mapsto \Omega(u)$ by which $\Omega(u)$ is defined and $u \in U_{\mathrm{ad}}$ (= set of admissible controls). Consequently, we have mappings

$$u \mapsto \Omega(u) \mapsto y(u) \mapsto J(u; y(u)) \tag{3}$$

453

J. Manley et al. (eds.), Proceedings of the Third European Conference on Mathematics in Industry, 453–459.

and we settle the problem: Find $u^* \in U_{\mathrm{ad}}$ such that

$$(4) \qquad J(u^*; y(u^*)) \leq J(u; y(u)) \qquad \text{for all } u \in U_{\mathrm{ad}} \, .$$

A large range of important optimal shape design problems which arise in structural mechanics, acoustics, electric fields, fluid flow and other areas of engineering and applied sciences can be formulated as (2) or (4). Typically,

$$(5) \qquad J(\Omega; y(\Omega)) = \int_\Omega dx \qquad \text{(minimization of the weight)}$$

$$(6) \qquad J(\Omega; y(\Omega)) = \int_\Omega (y(x))^2 \, dx \qquad \text{(minimization of displacements)}$$

$$(7) \qquad J(\Omega; y(\Omega)) = \int_\Omega (\nabla y(x))^2 \, dx \qquad \text{(minimization of stresses)}$$

$$(8) \qquad J(u; y(u)) = \int_{\Gamma(u)} \frac{\partial}{\partial n} y(x) \, dx \qquad \begin{array}{l}\text{(minimization of contact stresses} \\ \text{or boundary flux).}\end{array}$$

In the proceedings [2, 4, 9, 17, 24] and in books [5, 14, 15, 16, 21, 22, 28, 31] one can find tens of examples of optimal shape design problems with industrial applications.

2. On the numerical methods for solving optimal shape design problems

One solves the problems (2) and (4) iteratively by using the techniques of calculus of variations and mathematical programming. When the discretization of the state problem has been done by FEM, FD or BEM, the backbone in numerical realization of optimal shape design problems are nonlinear programming algorithms (or multicriterion optimization). This means that the detailed study of this subject is at the interface of no less than four fields:

– optimal control,
– partial differential equations,
– numerical analysis
– optimization.

During the past 20 years, optimization theory has been developed widely for the optimal design of many structural and mechanical systems. The optimal shape design problems have been studied extensively by the engineers when the state equations are governed by partial differential equations with suitable boundary conditions. In the last 15 years there has been increased mathematical interest to study the question of existence of an optimal shape, numerical approximations, convergence and sensitivity analysis and related area. A detailed reference related to the mathematical questions and the engineering applications can be found in the monograph [14].

The design sensitivity analysis is the most tedious step in numerical realization of the solution procedure for finding the optimal shape. A substantial literature has been developed in the field of shape design sensitivity analysis. Contributions to this field has been made using two fundamentally different approaches. The first approach uses the discretized model, based on finite element analysis, and proceeds to carry out shape design sensitivity analysis by controlling finite element node movement and differentiating the algebraic finite element equations (called here algebraic approach). The second approach to sensitivity analysis uses the material derivative method of continuum mechanics to account for changes in the shape of the domain into consideration. This method is called in the literature material derivative (or shape derivative) method [36]. In this method the sensitivity information is expressed as integrals (domain or boundary integrals) of the solution of the state and adjoint state. The numerical evaluation of these integrals may cause inaccuracies.

In algebraic approach the difficulty lies in calculating the sensitivities for stiffness matrix and mass matrix and for force vector. We shall outline here a simple method for doing this technically tedious step. As we have shown in [14], when the material derivative approach is applied to the discrete model obtained by finite element method (linear elements or in general isoparametric elements), the algebraic approach and material derivative approach give the same gradient.

Numerical realization of optimal shape design yields optimization problems, where typically

 – the evaluation of the objective function is costly
 – objective function may not be convex (see Example 1)
 – objective function may not be smooth (see Example 2)
 – objective function may be vector valued (multicriterion optimization).

Concerning general algorithms and software which are useful as well in shape optimization we mention the survey article [29] and the book [10].

In the case when the objective function is nonconvex, one could apply the method of global optimization [27, 35]. Because the evaluation of the objective function is costly, the methods of global optimization are very expensive. In many applications the resulting finite dimensional problem leads without regularization to nonsmooth optimization problem. For the method of nonsmooth optimization (subgradient algorithm, bundle methods) we refer to [7, 20].

In the cases where several conflicting design criteria appear it is natural to apply multicriterion (multicriteria, multiobjective, multiple criteria, vector) optimization. Instead of one scalar objective function several criteria are minimized simultaneously in the feasible set. The objective function is a vector which includes the different criteria as components. The main advantange in multicriterion approach is its flexibility to generate good solutions for the designer. It can be interpreted as a systematic sensitivity analysis of the most important value judgements. The literature in this field is scattered into many different application fields and especially in engineering design the number of publications is increasing rapidly. Especially we mentioned wide possibilities to use multicriterion optimization in structural optimization. For the methods of multicriterion optimization we refer to [9, 32, 34]. On fuzzy optimum design of structures see [11].

During recent years knowledge based systems of optimal design have been tested and developed. This field is still in very beginning but shows already now interesting advances, see [1, 19, 25]. We further mention the works of Hartmann [12, 13] on application of artificial intelligence tools for reanalysis with structural optimization, and articles of Papalambros [25, 26] on the use of artificial intelligence in design optimization. One can find in [9] articles for example on the use of interactive multicriterion approach in structural design and on knowledge-based approach in multicriterion design.

There are several general purpose CAD and FEM software packages which already include first modules to join shape optimization to the design procedure, see [3] and [8]. In the article [18] the features of about 30 internationally used program systems for structural optimization are listed.

3. Two examples

Concerning the nonconvexity of the criterion function in (4) we shall give an example due to Cea [6].

EXAMPLE 1. The mapping $u \mapsto J(u)$ is not necessarily convex. Let $\Omega(u)$ be an interval of the real line

$$\Omega(u) = \{x \in \mathbf{R} \mid 0 < x < u\}.$$

Let the govering state equations be

(9)
$$\begin{cases} -y(u;x) = 2, & x \in (0, u) \\ y'(u;0) = 0, \\ y(u;u) = 0. \end{cases}$$

The cost function to be minimized is

(10)
$$J(u) = \int_{\Omega(u)} (y(u) - 1)^2 \, dx = \int_0^u (y(u;x) - 1)^2 \, dx.$$

As the solution of (9) is

$$y(u;x) = u^2 - x^2,$$

the cost function reads by (10)

$$J(u) = \frac{8}{15} u^5 - \frac{4}{3} u^3 + u.$$

Thus the mapping $u \mapsto J(u)$ is not convex and usual optimization methods are not sure to converge to the optimal solution. If one puts box constraints of the type $u_1 < u < u_2$ in an appropriate way, one gets a convex problem.

In practice we meet problems whose behaviour is described by variational inequalities (contact between deformed bodies, obstacle problems, fluid through porous media, phase transition in the solidification of liquids, etc.). The field of applications of variational inequalities are steadily growing. The most charasteristic property of variational inequalities is the fact that their solution does not depend smoothly, in general, on the control. We present an example due to Mignot and Puel, [23], where the system is described by variational inequality and the control is on the right hand side. We shall see that the corresponding optimal control problem is nonsmooth.

EXAMPLE 2. Consider the optimal control problem in $\mathbf{R}^1$:

$$(11) \qquad \min_{u \in \mathbf{R}^1} \{ J(u) = (y(u) - 1)^2 + u^2 \}$$

where $y = y(u) \in K$ is the solution of the variational inequality

$$(12) \qquad y(\varphi - y) \geq u^+(\varphi - y) \qquad \text{for all } \varphi \in K ,$$

with $K = [0, \infty)$. We find that

$$y(u) = (-1 + u)^+$$

is the solution of (12). Consequently,

$$J(u) = \begin{cases} 2u^2 - 4u + 4 & \text{for } u \geq 1 \\ u^2 + 1 & \text{for } u \leq 1 . \end{cases}$$

We find that $J(u)$ is not only nonsmooth but also nonconvex.

Concerning more details of methods outlined here we refer to [14], where among others optimal shape design for contact problems in elasticity, for elastic – perfectly plastic bodies, state constrained (constraints for displacements or stresses, etc.) optimal shape design are handled as well from the theoretical as the numerical point of view.

5. References

[1] Arora, J.S.; Baenziger, G.: Uses of artificial intelligence in design optimization. Comp. Meth. in Appl. Mech. and Eng. <u>54</u> (1986) 303–323

[2] Atrek, E.; Gallagher, R.H.; Ragsdell, K.M.; Zienkiewicz, O.C. (eds.): New Directions in Optimum Structural Design. Chichester: John Wiley & Sons 1984

[3] Beckers, P.; Braibant, V.; Fleury, C.: Shape optimal design – an approach matching CAD and optimization consepts. In J.S. Gero (ed): Optimization in Computer-Aided Design. Elsevier Science Publishers (North-Holland) 1985

[4] Bennett, J.A.; Botkin, M.E. (eds.): The Optimum Shape: Automated Structural Design. Plenum Press 1986

[5] Carmichael, D.G.: Structural Modelling and Optimization (A General Methodology for Engineering and Control). Chichester: Ellis Horwood Ltd. 1981

[6] Cea, J.: Problems of shape optimal design. In [17], pp. 1005–1048

[7] Clarke, F.H.: Optimization and nonsmooth analysis. New York: John Wiley & Sons 1983

[8] Encarnacao, J.; Schlechtendahl, E.G.: Computer Aided Design, Fundamentals and System Architectures. Heidelberg: Springer-Verlag 1983

[9] Eschenauer, H.; Koski, J.; Osyczka, A. (eds.): Multicriterion Design Optimization – Procedures and Applications. Berlin: Springer Verlag 1989, to appear

[10] Gill, P.E.; Murray, W.; Wright, M.H.: Practical Optimization. London: Academic Press 1981

[11] Guang-Yan, W.; Wen-Quan, W.: Fuzzy optimum design of structures. Eng. Opt. 8 (1985) 291–300

[12] Hartmann, O.: Application of AI tools for re-analysis within structural optimization. Eng. Opt. (1987) 355–367

[13] Hartmann, D.: Selection and evaluation of structural optimization strategies by means of expert systems. (1988)

[14] Haslinger, J.; Neittaanmäki, P.: Optimal Shape Design Problems. Theory and Applications. Chichester: John Wiley & Sons 1988

[15] Haug, E.J.; Arora, J.S.: Applied Optimal Design, Mechanical and Structural Systems. New York: J. Wiley & Sons 1979

[16] Haug, E.J.; Choi, K.K.; Komkov, V.: Design Sensitivity Analysis of Structural Systems. Orlando: Academic Press 1986

[17] Haug, E.J.; Cea, J. (eds.): Optimization of Distributed Parameter Structures. Parts I & II. Alphen aan den Rijn: Sijthoff & Noordhoff 1981. Nato Advances Study Institute Series, Series E

[18] Hörnlein, H.R.E.H.: Take-off in optimal structural design. In [24], pp. 901–919

[19] Jozwiak, S.F.: Application of artificial intelligence notions in structural optimization programs. Comput. Struct. 24 (1986) 1009–1013

[20] Kiwiel, K.C.: Methods of Descent for Nondifferentiable Optimization. New York: Springer-Verlag 1985. Lecture Notes in Mathematics 1133

[21] Komkov, V.: Variational Principles of Continuum Mechanics with Engineering Applications. Vol 2. Introduction to Optimal Design Theory. Dordrecht: D. Reidel Publ. Company 1988

[22] Lawo, M.: Optimum Structural Design. Braunschweig: Vielveg 1986. (in German)

[23] Mignot, F.; Puel, J.P.: Optimal control of some variational inequalities. SIAM J. Control and Optimiz. 22 (1984) 466–478

[24] Mota Soares, C.A. (ed.): Computer Aided Optimal Design: Structural and Mechanical Systems. Berlin: Springer-Verlag 1987. NATO ASI Series F, Vol 27

[25] Papalambros, P.: Knowledge based systems in optimal design. In [24], pp. 759–804

[26] Papalambros, P.: Integration of knowledge forms in design optimization. In: Proc. of NSF Workshop on the Design Process. 1987 pp. 217–436

[27] Pardalos, P.M.; Rosen, J.B.: Constrained Global Optimization Methods: Algorithms and Applications. Berlin: Springer-Verlag 1987. Lecture Notes in Computer Science 268

[28] Pironneau, O.: Optimal Shape Design for Elliptic Systems. New York: Springer-Verlag 1984 Springer Series in Computational Physics

[29] Schittkowski, K.: Software for mathematical programming. In [30], pp. 383–451

[30] Schittkowski, K. (ed.): Computational Mathematical Programming. Berlin: Springer-Verlag 1985 NATO ASI Series F, Vol 15

[31] Sokolowski, J.; Zolésio, J.P.: Introduction to Shape Optimization, Shape Sensitivity Analysis. A book in preparation

[32] Stadler, W.: Natural structural shapes (a unified optimal design philosophy). In [33], pp. 355–390

[33] Stadler, W. (ed.): Multicriteria Optimization in Engineering and in the Sciences. New York: Plenum Press 1988. Mathematical Concepts and Methods in Sciences and Engineering 37

[34] Steuer, R.E.: Multiple Criteria Optimization: Theory, Computation and Application. New York: J. Wiley & Sons 1986

[35] Törn, A.; Žilinskas, A.: Global Optimization. (to appear)

[36] Zolésio, J.P.: The material derivative (or speed) method for shape optimization. In [17], pp. 1089–1151

P. Neittaanmäki, University of Jyväskylä, Department of Mathematics, Seminaarinkatu 15, SF-40100 Jyväskylä, Finland

Steady-State Optimization of
Large Gas Networks
A.J. Osiadacz and D.J. Bell, Manchester.

Summary: An algorithm is described for steady-state optimization of large
gas networks. The problem of optimization is treated as a nonlinear
problem with nonlinear constraints. It is assumed that the structure of the
network is known.

1. Introduction

The objective of a gas distribution network is to supply gas to

customers. To achieve this, gas is first transported through the high

pressure part of the network incorporating all its compressors, valves and

regulators. Compressors supply energy to the network in order to

compensate for a loss of pressure along the pipes as a result of friction

between the gas and the pipe wall. The problem is to know what is the

minimum that the compressors have to supply whilst maintaining the

demand. A solution to this problem can be found by treating it as an

optimization problem, in which objective function is chosen as the cost of

running the compressors. This work is concerned with the minimization of

operating costs for high pressure gas networks under steady-state

conditions. The steady-state in a gas network is described by system of

algebraic non-linear equations. In steady-state problems, since loads and

supplies are not functions of time, an algorithm for optimization determines,

once and for all, the structure of the network (i.e. the number of sources,

compressors, valves and regulators called-units which must be on). In

addition, the algorithm must determine the optimal parameters of the

operation, namely nodal pressures and flows through branches (pipes). For

these reasons, the problem of optimization is formulated as a mixed integer

problem (Wilson et al. 1988). By linearizing the flow equation, the nonlinear

constraints, and the objective function, the problem of steady-state

optimization has been expressed as a mixed-integer linear programming

problem (Wilson et al. 1988). In the investigation described in this paper,

461

J. Manley et al. (eds.), Proceedings of the Third European Conference on Mathematics in Industry, 461–467.
© 1990 Kluwer Academic Publishers and B. G. Teubner Stuttgart.

it was assumed that the structure is known, and the problem of optimization is treated without simplification, i.e. as a nonlinear problem with nonlinear constraints.

2. Problem formulation

Generally, the cost of running the compressors in the system is a function of the variables (flows through branches, flows through units and nodal pressures) and also how many units (compressors, valves, regulators and sources) are switched on.

The running cost of a compressor is expressed in terms of the fuel necessary for the engine to produce a power required for the compression of gas by that compressor. The power required from the prime - mover is expressed by the following equation

$$(2.1) \qquad\qquad N = kM/\Psi(p_d/p_s)\beta{-}1 \qquad [W]$$

where k is the coefficient dependent on the properties of gas and the inlet
   temperature,

   M is the mass flow of gas through compressor (kg/s),

   Ψ is the efficiency of the compressor,

p_d, p_s is the discharge and suction pressure respectively (Pa),

$\beta = (\gamma{-}1)/\gamma$, where γ is the ratio of specific heats.

It has been assumed that the flow of gas through a pipe is described by an algorithm due to Wilson et al. (1988). In this case, the problem of optimization is formulated as follows:

$$(2.2) \qquad\qquad J = \min \sum_{i=1}^{k} N_i$$

where k is the number of operating compressors subject to the constraints. Two kinds of constraints have to be considered (see Wilson et al. 1988). One relates to equality conditions arising from Kirchhoff's Laws and the flow equation (Osiadacz, 1987). The other corresponds to the operational restrictions on the units appearing in the network, which assume the form of inequality constraints.

3. Problem solution

The above problem has been formulated as a nonlinear programming one with nonlinear constraints in the form

(3.1) $$\min f(\underline{x})$$

subject to equality and inequality constraints

(3.2) $$c_i(\underline{x}) = 0, \ i = 1,2,...,m'$$

$$c_i(\underline{x}) \geqslant 0, \ i = m'+1,...,m$$

and the functions $f(\underline{x})$ and $c_i(\underline{x})$; $i = 1,2,...,m$ are real and differentiable. To solve this problem, an iterative method is used at each iteration which minimizes a quadratic approximation to the Lagrangian function subject to sequentially linearized approximations to the constraints (Powell, 1978, 1982). A line search procedure utilising the "watchdog technique" (Chamberlain, 1980) is used to force convergence when the initial values of the variables are far from the solution. To begin the calculation a starting point $\underline{x}_0$ has to be chosen together with an n×n positive definite matrix $\underline{B}_0$. The iterations generate a sequence of points $\underline{x}_k$ (k=1,2,...) that usually converges to the required vector of variables and also they generate a sequence of positive definite matrices $\underline{B}_k$ (k=1,2,3,...).

At the beginning of the k-th iteration both $\underline{x}_k$ and $\underline{B}_k$ are known. The vector $\underline{d} = \underline{d}_k$ is obtained by minimizing the quadratic function

(3.3) $$Q(\underline{d}) = f(\underline{x}_k) + \underline{d}^T \nabla f(\underline{x}_k) + 1/2\underline{d}^T \underline{B}_k \underline{d}$$

subject to linear conditions:

$$c_i(\underline{x}_k) + \underline{d}^T \nabla c_i(\underline{x}_k) = 0, \ i = 1,2,...,m'$$

$$c_i(\underline{x}_k) + \underline{d}^T \nabla c_i(\underline{x}_k) \geqslant 0, \ i = m'+1,...,m$$

The vector $\underline{x}_{k+1}$ has the form:

$$\underline{x}_{k+1} = \underline{x}_k + \alpha_k \underline{d}_k$$

where α_k is a positive step-length.

The calculation of $\underline{d}_k$ is a quadratic programming problem having the property that, if all constraints are linear and if $f(\underline{x})$ is a quadratic function whose second derivative matrix is $\underline{B}_k$, then $\underline{x}_k + \alpha_k \underline{d}_k$ is the required vector of variables. Given that $\underline{\lambda}_k$ is the m–component vector of Lagrange multipliers at the solution of the quadratic programming problem which defined $\underline{d}_k$, the definition of $\underline{B}_{k+1}$ must depend on $\underline{\lambda}_k$ in order to take account of any constraint curvature.

4. Results of investigations

To prove the correctness of the stated algorithm a non–trivial example, involving part of real network, has been solved. We present the results for the gas network shown in Fig. 1. which comprises 38 pipes, 33 nodes, 2 compressors, 2 sources and 1 valve. All the pipes have the same diameter, which is equal to 0.9144m. (36 inch). The following constraints have been imposed on nodal pressures:

$$p_9 \leqslant 56.0 \text{ bars, } p_{20}, p_{23} \text{ and } p_{26} \geqslant 52.0, \; p_{12} \geqslant 55.0$$

Maximum discharge pressure and maximum compressor ratio for both compressors (Comp-1 and Comp-2) cannot exceed 60.0 bars and 1.5 respectively. For sources (S1 and S2) it was assumed that $p_{max} \leqslant 60$ bars, and maximum flow (m^3/h) $\leqslant$ 1420000.

Selected results of the optimization are given in Table 1 and Table 2.

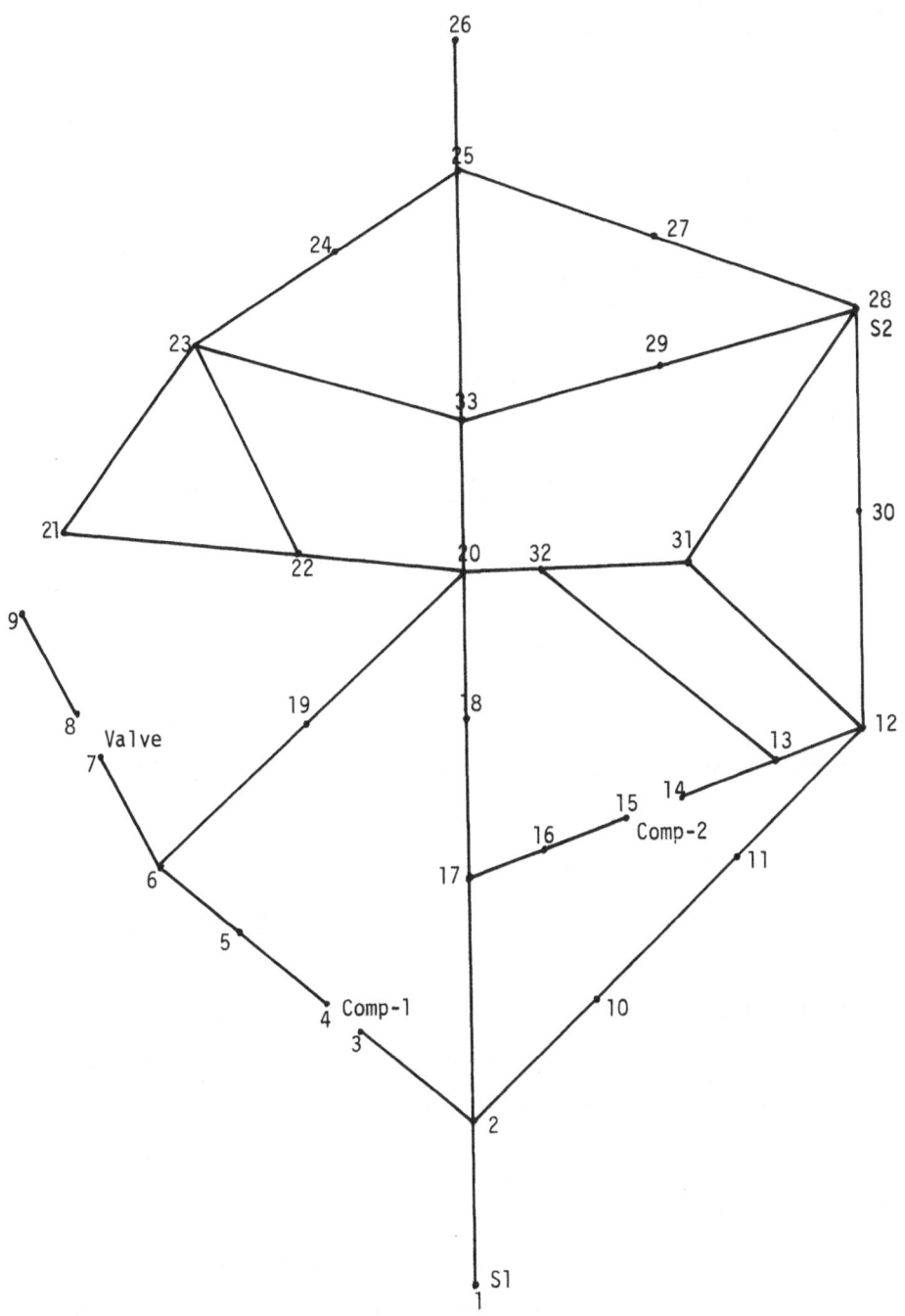

Fig.1 Structure of the network

Tab.1 Optimal parameters of work for compressors

Compressor number	Suction pressure(bar)	Discharge pressure(bar)	Flow (m^3/h)	Power (kw)
Comp-1	53.6	59.3	618605.8	3385
Comp-2	50.9	58.0	465222.7	3144

Tab.2 Optimal parameters of work for sources

Source number	Output pressure(bar)	Flow (m^3/h)
S1	60.0	943897
S2	57.2	1415844

Investigations here illustrated the correctness of the proposed algorithm. Future comparison between two versions should show which algorithm is better in the sense of accuracy, reliability and time consumption.

5. References

[1] Chamberlain, R.M., et al.: The watchdog technique for forcing convergence in algorithms for constrained optimization. Report DAMTP 80/NA9, University of Cambridge (1980).

[2] Osiadacz, A.J.: Simulation and analysis of gas networks, E&FN Spon Ltd. London (1987).

[3] Powell, M.J.D.: A fast algorithm for nonlinearly constrained optimization calculations. In Numerical Analysis, Dundee 1977, Lecture Notes in Mathematics 630, ed. G.A. Watson, Springer-Verlag (1978).

[4] Powell, M.J.D.: Extensions to subroutine VFO2AD. In System Modelling and Optimization, Lecture Notes in Control and Information Sciences 38, eds. R.F. Drenick and F. Kozin, Springer-Verlag (1982).

[5] Wilson, J.G., Wallace, J. and Furey, B.P.: Steady-state optimization of large gas transmission systems. In Simulation and Optimization of Large Systems, ed. A.J.Osiadacz, Oxford 1988.

Department of Mathematics, UMIST, Manchester, M60 1QD, U.K.

Acknowledgement

The authors wish to express their thanks to British Gas plc for providing data for the given example. Much help received from Mr. A.E. Fincham of the London Research Station of British Gas has been particularly valuable. This work was carried out under an SERC/British Gas collaborative grant No.GR/C56696.

Modelling and Control of Acid–Base Blending Systems

Ulf Palmquist
Department of Mathematics and Mathematical Statistics
Chalmers University of Technology
Gothenburg, Sweden

Abstract

In this paper mathematical models for the dynamic and chemical properties of systems for continuous blending of acids and bases are studied. The basic ideas behind the general models are briefly indicated by studying a neutralization system.

A method developed for pH control is described. This method is based on the estimation of the concentrations of acids and bases from pH measurements and it aims at minimizing the deviation in pH from the desired value as well as the consumption of acid and base used to control the pH.

Background

The properties of many chemical and biological processes in solution depend on the concentration of hydrogen ions (protons), $[H^+]$. Changing the pH value of the solution $(pH = - \log[H^+])$ by adding acid or base is, in these cases, a method for affecting the process.

In industry several of these processes, e.g. purification of chemical and biological waste water, are performed in accordance with Figure 1. The solution is continuously fed into a stirred tank and, based on pH measurements, pH_e, in the effluent, a regulator controls valves for addition of acid and base.

Today the regulator used for pH control is often of traditional PID type. This means that the volumetric flow rate of reagents, Q_c, is a linear functional of the deviation in measured pH from the reference value, pH_r, as shown in the formula

J. Manley et al. (eds.), Proceedings of the Third European Conference on Mathematics in Industry, 469–477.

470

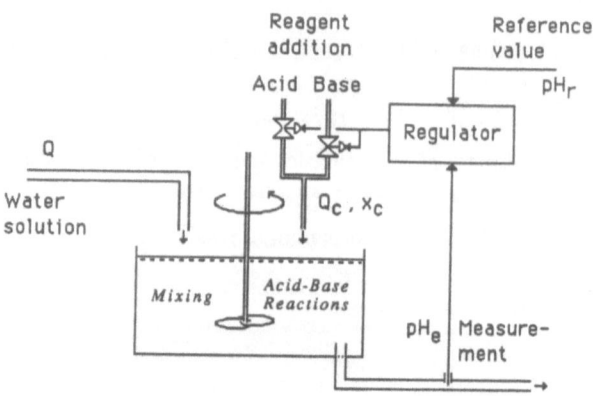

Figure 1   System for Blending of Acids and Bases

$$Q_c(t) = K_P\{pH_e(t) - pH_r\} + K_I \int_0^t \{pH_e(s) - pH_r\}ds + K_D \frac{d}{dt} \{pH_e(t)-pH_r\}.$$

In some implementations the parameters K_P, K_I and K_D are changed between prechosen values as pH_e passes fixed levels, see e.g. {4}. Lately, adaptive pH regulators have also been considered, see e.g. {1}.

For pH control, these types of regulators are, however, rarely successful. The problems are caused by mainly two reasons:

i.   The pH value is a strongly non-linear function of the concentrations of acids and bases in the solution.

ii.   The concentrations of acids and bases in the feed are not constant. Changes in the processes up-stream of the blending tank cause variations, often in an unpredictable fashion.

pH as a Function of Acid and Base Concentrations

By an example we will demonstrate the non-linearity of pH and also introduce a function for determining the pH in a (fairly) general water solution of acids and bases.

Consider a dilute water solution of hydrochloric acid, HCl, sodium hydroxide,

NaOH, and carbonic acid, H_2CO_3. The acid–base reactions and their equilibrium relations are

$$H_2CO_3 \rightleftarrows H^+ + HCO_3^- \qquad K_1 = \frac{[H^+][HCO_3^-]}{[H_2CO_3]} \qquad (1)$$

$$HCO_3^- \rightleftarrows H^+ + CO_3^{2-} \qquad K_2 = \frac{[H^+][CO_3^{2-}]}{[HCO_3^-]} \qquad (2)$$

$$HCl \rightleftarrows H^+ + Cl^- \qquad K_{HCl} = \frac{[H^+][Cl^-]}{[HCl]} \qquad (3)$$

$$NaOH \rightleftarrows Na^+ + OH^- \qquad K_{NaOH} = \frac{[Na^+][OH^-]}{[NaOH]} \qquad (4)$$

$$H_2O \rightleftarrows H^+ + OH^- \qquad K_w = [H^+][OH^-]. \qquad (5)$$

Above, [M] denotes the concentration of substance M and the K's are dissociation constants. In dilute water solutions, the strong hydrochloric acid and sodium hydroxide can be treated as completely dissociated, implying that $K_{HCl} \simeq \infty$, $K_{NaOH} \simeq \infty$.

The condition of electrical neutrality (sum of signed ion charges = 0) can be written as

$$[H^+] - [OH^-] = [HCO_3^-] + 2[CO_3^{2-}] + [Cl^-] - [Na^+]. \qquad (6)$$

Given the amounts of H_2CO_3, HCl and NaOH in the solution, the concentration of protons, and hence the pH value, can be determined as that value $[H^+]$ satisfying relation (1) – (6).

When modelling dynamic acid–base systems like the one in Figure 1, the form of the acid–base relations (1) – (6) are not particularly convenient. A more suitable formulation can, however, be found. For the example above let

$$Y = [H^+]$$
$$X_1 = [H_2CO_3]_{total} = [H_2CO_3] + [HCO_3^-] + [CO_3^{2-}]$$
$$X_2 = [HCl]_{total} - [NaOH]_{total} = [Cl^-] - [Na^+].$$

Then, using the equilibrium relations (1) – (5), one finds that the condition of electrical neutrality (6) can be expressed as

$$Z(Y) = c_1(Y)X_1 + c_2(Y)X_2 \qquad (7)$$

where we have introduced the so called equilibrium functions, defined for $y > 0$ as

472

$$Z(y) = y - K_w/y, \qquad \text{(water)}$$
$$c_1(y) = (2+y/K_2)/(1+y/K_2+y^2/(K_1K_2)), \qquad (H_2CO_3)$$
$$c_2(y) = 1, \qquad \text{(HCl–NaOH)}.$$

Now, given the total concentrations X_1 and X_2 there exists one unique positive value Y satisfying (7). Hence, under the restriction $Y > 0$, (7) defines the proton concentration $Y = [H^+]$ and the pH value $pH = -\log Y$ as a function of the concentrations X_1 and X_2. For this example, the (titration) graph of $pH(X_1, X_2)$ is shown in Figure 2. Its non-linearities explain the difficulties to control the pH value.

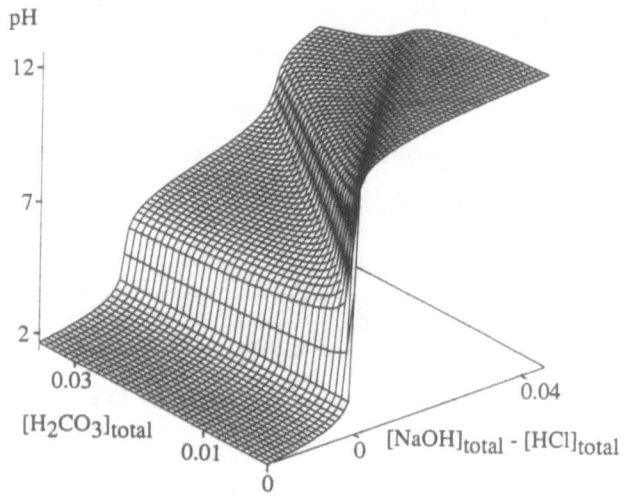

Figure 2 Titration Surface for the Equilibrium System $(H_2O, H_2CO_3, NaOH, HCl)$

In {2}, section 6, the above way of determining pH is proved valid for a large class of acid–base water solutions. More explicit, let $X_1, X_2,..., X_m$ denote the total concentrations of m different acids and bases in a dilute water solution. Then each acid and base can be associated with an equilibrium function $c_i(y)$, $y > 0$, such that

$$Z(Y) = c_1(Y)X_1 +...+ c_m(Y)X_m$$
$$Y > 0 \qquad (8)$$
$$pH(X) = -\log Y$$

defines pH as an implicit function of the concentrations $X = [X_1 \ldots X_m]^T$. The equilibrium function for any acid (base) is rational and its parameters depend only on the dissociation constants of that acid (base).

Models of Acid–Base Blending Systems

In order to derive pH control methods it is neccesary to model the dynamics and the chemical properties of the system.

Assume that the feed of the system in Figure 1 is a dilute water solution of m different acids and bases. Let $X_{f,i}$, $X_{e,i}$ denote the total concentration in the feed and effluent, respectively, of the i:th acid (base). Often the feed concentrations deviate from their mean values $\bar{X}_i$ in an unpredictable way due to disturbances in the operation of processes up–stream of the blending tank. This can be expressed by modelling the deviation $\tilde{X}_{f,i} = X_{f,i} - \bar{X}_i$ as an ARMA process,

$$\tilde{X}_{f,i}(k) + a_1 \tilde{X}_{f,i}(k-1) + \ldots + a_n \tilde{X}_{f,i}(k-n) = b_1 W_i(k) + \ldots + b_n W_i(k-n). \quad (9)$$

Here W_i is a white stochastic process and a_k, b_j are parameters (which have to be estimated).

The chemical engineering approach for relating the effluent concentration $X_{e,i}$ to the feed concentration $X_{f,i}$ is to assume that the solution in the tank is well mixed and then make a massbalance wrt. the i:th acid (base). This yields

$$\dot{X}_{e,i} = \lambda(X_{f,i} - X_{e,i}) + \lambda u_i \quad (10)$$

where λ^{-1} is the time constant of the tank and u_i is the controlled reagent addition of the i:th acid (base) scaled as a feed concentration.

Remark 1. Since $X_{f,i}$ and $X_{e,i}$ are defined as total concentrations of an acid (base) they are not affected by the acid–base reactions in the solution.

Remark 2. In {2}, section 2 and 3, relations between feed and effluent concentrations of an inert molecular matter are derived by modelling the molecular mixing in the tank as random. The analysis there show that if the tank solution is well mixed then (10) holds almost as a deterministic relation (possibly with a feed transport lag).

The dynamic relations (9) − (10) for the m acids and bases can now be collected into a state space model (note that no interaction between acids and bases exist since $X_{f,i}$ and $X_{e,i}$ are similar to the concentration of an inert species). In discrete time (sampled system) the structure of the model is

$$X(k+1) - \bar{X} = A(X(k) - \bar{X}) + Bu(k) + W(k) \qquad (11)$$

where X is the state vector containing $X_{f,i}$, $X_{e,i}$, $i = 1,...,$ m and all other dynamic states, $\bar{X}$ is the expectation of X, u is the control vector, W is a white stochastic vector, A and B are parameter matrices.

Acid−base reactions are in general fast and may therefore be modelled as finding their equilibria instantly. Thus, referring to the previous section, in particular (8), it follows that if Y and pH denote the concentration of protons and the pH value in the effluent, then

$$Z(Y) = c_1(Y)X_{e,1} +...+ c_m(Y)X_{e,m}$$
$$Y > 0 \qquad (12)$$
$$pH = - \log Y$$

defines Y and pH as functions of the concentration state vector X. In (12), c_i, $i = 1,...,$ m are the equilibrium functions for the m acids and bases in the solution.

Measurement of the effluent pH is not exact and is therefore modelled as

$$pH_e = pH + e \qquad (13)$$

where e is a random measurement error, independent of the concentrations.

In summary, the model of the dynamic and chemical properties of the acid−base blending system is given by (11) − (13).

Concentration Estimator

For determination of the exact amount of reagent addition required to obtain the desired effluent pH value, it is necessary to know the concentration state vector, (simply controlling directly on the measured pH value, like the PID regulator, is not sufficient if more than one acid is present, c.f. Figure 2). Real−time measurement of the acid and base concentrations are, in general, costly and in most cases impossible. A method (an estimator), which based on the measured effluent pH, recursively in time yields statistical estimates of the acid–base concentrations, has therefore been derived. The idea behind the estimator is as follows (complete derivations are reported in {2} and {3}).

Let $\mathcal{F}_j = \{pH_e(i); i \leq j\}$ and assume that the processes W and e are independent, white and Gaussian. Denote the estimate of X(k) given $\mathcal{F}_j$ by $\hat{X}(k|j)$ and assume that the distribution of X(k) given $\mathcal{F}_{k-1}$ is $N(\hat{X}(k|k-1), \Gamma_{k|k-1})$. Then the estimate of X(k) given $\mathcal{F}_k$ is determined as the Maximum Likelihood estimate corresponding to the conditional distribution of X(k) given $\mathcal{F}_k$. Some (rather lengthy) manipulations show that

$$\hat{X}(k|k) = \arg\min\{(pH_e(k) - pH(x))^2/\sigma_e^2 + \|x - \hat{X}(k|k-1)\|^2_{\Gamma_{k|k-1}}\}, \quad (14)$$

where σ_e^2 is the variance of e(k). The time up–dating $(k \to k+1)$ is achieved by approximating the distribution of X(k) given $\mathcal{F}_k$ with the Gaussian distribution $N(\hat{X}(k|k), \Gamma_{k|k})$ and then up–dating according to the dynamic relation (11). (The covariance $\Gamma_{k|k}$ follows from the minimization in (14)).

In words, the estimator is a "one–step" Maximum Likehihood estimator with Gaussian approximation after each measurement conditioning. In {5}, the estimator has been evaluated, several appealing properties have been found.

pH Control

Two performance criterion have been considered when deriving pH regulators. The first one measures the error in effluent pH and also the consumption of reagents. It is of the form

$$J_1[u] = \int_0^T \{(pH(X(\tau)) - pH_r)^2 + \alpha u^2(\tau)\} d\tau$$

where α is a cost parameter. The optimal control is that for which the expectation of J_1 attains its minimal value. Since pH is a non–convex function of the stochastic state

X, the optimization is far from trivial (in particular in real-time). An algorithm yielding (crude) approximations has however been derived.

Let $Y_r = 10^{-pH_r}$. From the pH function formula (12) it follows that $|Z(Y_r) - c_1(Y_r)X_{e,1} - ... - c_m(Y_r)X_{e,m}|$ increases whenever $|pH(X)-pH_r|$ does. This fact motivates considering the (discrete time) criteria

$$J_2[u] = \sum_{k=1}^{N} \{(Z(Y_r)-c_1(Y_r)X_{e,1}(k)-...-c_m(Y_r)X_{e,m}(k))^2+\alpha u^2(k-1)\}.$$

Using the method of dynamic programming one finds that the control minimizing the expectation of $\dot{J}_2$ is of the feedback form ($\mathcal{L}_i$ = gain vector)

$$u(i) = -\mathcal{L}_i^T \hat{X}(i|0).$$

Figure 3 shows a block diagram of the model and also how the concentration

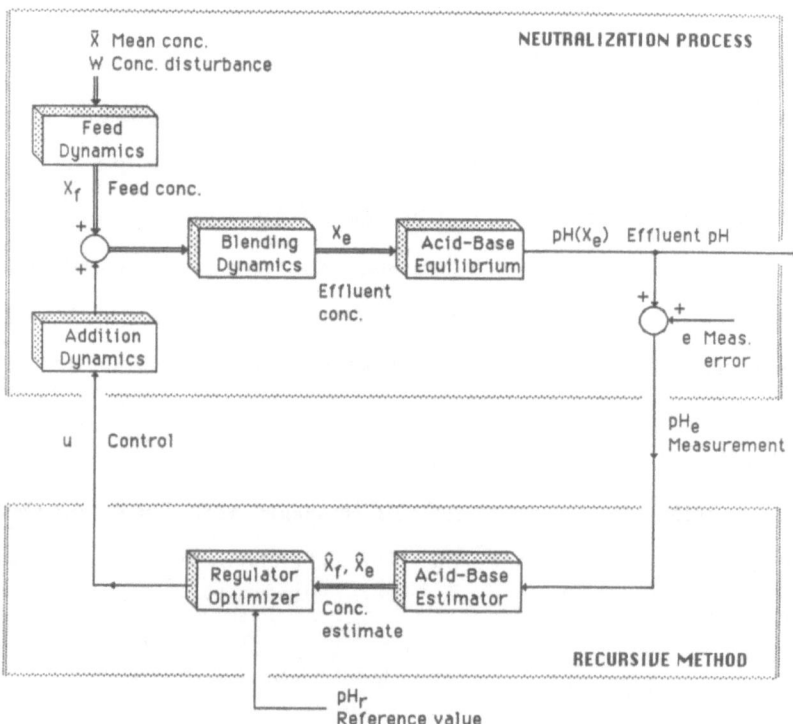

Figure 3   pH Control based on Concentration Estimation and Economical Reagent Addition

estimator and any of the two derived regulators are constituents of a pH control system. In {3}, the experience of this pH control system, implemented at a plant for treatment of industrial waste water, is reported. Several advantages in comparison with traditional PID control have been observed.

Acknowledgement   The support of this work given by Chalmers University of Technology, EKA Nobel AB, The Swedish Institute of Applied Mathematics and The Swedish Board for Technical Development is gratefully acknowledged.

References

{1}   Pajunen, G.A. Comparison of Linear and Non-Linear Adaptive Control of a pH-Process. IEEE Control Systems Magazine Vol. 7, No. 1, Feb. 1987.

{2}   Palmquist, U. State Estimation for Acid-Base Blending Systems. Department of Mathematics, Chalmers Univ. of Tech., Göteborg, 1985.

{3}   Palmquist, U. Estimation and Control for Acid-Base Blending Systems. Department of Mathematics, Chalmers Univ. of Tech., Göteborg, 1987.

{4}   Shinskey, F.G., A Self-Adjusting System for Effluent pH Control. Control Engineering, Feb. 1974.

{5}   Sjöström, J. Aspects on Concentration Estimators for Acid-Base Blending Systems. Department of Mathematics, Chalmers Univ. of Tech., Göteborg, 1987.

Authors address: Dr. Ulf Palmquist
                          Department of Mathematics
                          Chalmers University of Technology
                          S-412 96 Göteborg
                          Sweden

Key words: Acid-base equilibrium processes, Neutralization processes, pH systems, pH regulators, pH control, Non-linear systems, Non-linear state estimation.

THE PREDICTION OF CYCLIC PLASTIC STRAIN GROWTH BEHAVIOUR FOR SEVERE THERMAL LOADING PROBLEMS IN STRUCTURAL ENGINEERING USING UPPER BOUND METHODS AND LINEAR PROGRAMMING

A.R.S. Ponter and K.F. Carter

Introduction

In structural engineering, severe thermal loading problems are known to occur in liquid metal cooled fast reactors as well as many other applications. There is therefore a need to be able to predict, at the design stage, which combinations of thermal and mechanical load will induce deformation of a given structure by cyclic plastic strain growth. For structures already in service, knowledge of the amount of accumulated plastic strain, and thus assessment of residual lifetime in terms of the number of loading cycles is also desirable.

The upper bound method is capable of satisfying both these requirements and of giving a general understanding of the different types of structural behaviour that may be encountered. The result is to produce a diagram of thermal load plotted against mechanical load for a particular thermal loading cycle, known as an interaction diagram, which shows the areas corresponding to the regimes of structural behaviour as well as the deformation mechanisms along shakedown/ratchetting boundary. The technique is independent of the number of loading cycles and leads to simple and reliable computer software together with easily understandable graphical output. The upper bound method has a number of advantages over the alternative full inelastic calculation which can be computationally very slow, and is only applicable to a single loading point. As the full inelastic calculation only allows a few loading cycles to be analysed it is not always possible to establish whether shakedown or ratchetting is taking place, and is unable to give any information on the behaviour of the system in changed conditions.

J. Manley et al. (eds.), Proceedings of the Third European Conference on Mathematics in Industry, 479–485.

Extended Upper-Bound Method

Upper-bound theory [1] requires that the energy dissipated by the deformation mechanism must be greater than or equal to the work done by the
mechanical and thermal loads. At high temperatures, the thermo-elastic
stresses $\hat{\sigma}_{ij}(\underline{x},t)$ can become sufficiently large at some point in the structure that they cannot be contained within the yield surface. Theory requires
that the thermo-elastic solution can be translated by a residual stress
field $\underline{\rho}(\underline{x})$ so that it is contained within yield. The material is assumed to
be perfectly plastic with complete cyclic hardening (which can be shown to
be a conservative assumption) and constant mechanical load throughout the
thermal loading cycle (t=0 to t=τ).

$$(1) \quad \int_V \int_0^\tau \sigma_{ij}^c(\underline{x},t)\ \dot{\varepsilon}_{ij}^c(\underline{x},t)\ dt\ dV \geq \psi \int_S \underline{P}.\underline{\Delta U}^c\ ds$$

$$+ \int_V \int_0^\tau (\hat{\sigma}_{ij}(\underline{x},t) + \underline{\rho}(\underline{x}))\ \dot{\varepsilon}_{ij}^c(\underline{x},t)\ dt\ dV$$

V denotes the volume and S the surface area of the body and the superscript
c denotes that the stresses σ_{ij} and strain rates $\dot{\varepsilon}_{ij}$ are plastic in origin.
$\underline{\Delta U}^c$ are the plastic displacements of the deformation mechanism. The equation
(1) may now be rearranged to give a form suitable for minimization.

$$(2) \quad \int_V \int_0^\tau [\sigma_{ij}^c(\underline{x},t) - (\hat{\sigma}_{ij}(\underline{x},t) + \underline{\rho}(\underline{x}))]\ \dot{\varepsilon}(\underline{x},t)\ dt\ dV \geq \psi \int_S \underline{P}.\underline{\Delta U}^c\ ds$$

Now the structure can be divided into two regions; V_s in which the total
stress is contained within the yield surface, and V_f where the total stress
exceeds yield. By the virtual work condition the contributions from the two
regions can be separated and equated.

$$(3) \quad \int_{V_s} \underline{\rho}(\underline{x})\ \Delta\varepsilon\ dV = - \int_{V_f} \underline{\rho}(\underline{x})\ \Delta\varepsilon\ dV$$

As yield has been exceeded in the region V_f, then the cost function associated with this region must be zero. Use is made of the virtual work condition enabling the upper-bound form to be written as

(4) $\displaystyle\int_{V_s}\int_0^\tau [\sigma_{ij}^c(\underline{x},t) - \hat{\sigma}_{ij}(\underline{x},t)]\ \dot{\varepsilon}_{ij}^c(\underline{x},t)\ dt\ dV$

$\displaystyle\qquad + \int_{V_f} \rho(\underline{x})\ \Delta\varepsilon_{ij}^c\ dV \qquad\qquad\qquad \geq \psi \int_S \underline{P}.\underline{\Delta U}^c\ ds$

The yield criteria used is based on the limited interaction model of Drucker and Shield [2], which consists of a Tresca yield surface in the axial and hoop stress directions. The plastic strain ε_{ij}^c can be expressed as a linear combination of plastic multipliers defined by the Tresca yield surface. As the upper bound method finds the closest solution to the true solution from above, the values of $\underline{\Delta U}^c$ and ε_{ij}^c may not be the correct solution values. However by requiring

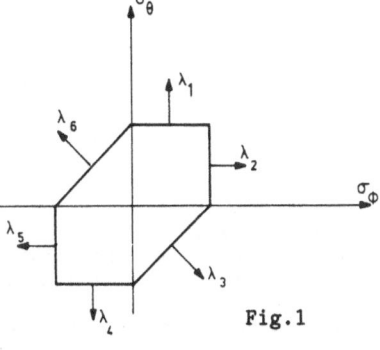

(5) $\displaystyle\int_S \underline{P}.\underline{\Delta U}^c\ ds = 1$

the problem can be reduced to the minimization over the volume of the body throughout the loading cycle.

Fig.1

Numerically the technique is to divide the structure into a series of finite elements connected with plastic hinges at the nodes which accomodate the bending moments in the axial direction. This has been applied to axisymmetric thin shell structures, which are the most commonly found type of component in fast reactor design. The displacements calculated from the appropriate strain/displacement relationships must also be continuous throughout the structure and any plastic hinge angle can be expressed in terms of the displacements within the elements adjoining that particular node. This gives a constraint at each node relating hinge angle to the normal displacements at the node. The constant mechanical load can be separated into an axial load component and a pressure component acting normal to the surface. This allows a variety of different loading situations to be studied.

Minimization of the Upper Bound by Linear Programming Techniques

The upper bound method is then the straightforward translation of the strain/displacement relations into the energy dissipation and work done terms, giving in linear programming terms a cost function and a general constraint respectively. The minimization takes place to find the mechanism of smallest cost for a given amount of mechanical work done, subject to the boundary constraints, mechanical work done constraint and hinge angle constraints. The cost function is evaluated by calculating the left hand side of equation (4). The essential assumption of this procedure is that the plastic strain rate field $\dot{\varepsilon}^c_{ij}(\underline{x},t)$ can only be nonzero where there is plastic deformation taking place, and this can only happen when the function $\Delta\sigma_{ij}(\underline{x},t) = [\sigma^c_{ij}(\underline{x},t) - \hat{\sigma}_{ij}(\underline{x},t)]$ is at a minimum. The procedure for evaluating the cost function may be summarized as follows:

For each of the plastic multipliers, corresponding to a face of the yield surface, find the minimum of the stress difference through the entire loading cycle, for a given point in space. This gives a set of times in the loading cycle at which the minima occur. Integrate through the volume of each element using the stress difference minima determined above.

The procedure for calculating the cost function for the plastic hinges is similar. The relationship between the hinge angle and the strain/displacement equations are used to form a system of constraints at each node throughout the structure. The mechanical work done can now be expressed as the sum of contributions from the plastic multipliers within each element and forms a general constraint to the linear programming method determining the size of the mechanism of deformation. The boundary constraints are usually of the type :- Displacements either normal and/or tangential to the axisymmetric axis (or shell midsurface depending on mechanical loading type) at one or both ends of the structure - set to zero.

The minimum is found within the system of constraints by a sparse matrix linear programming package known as XMP [3]. The resultant minimum cost is then the mechanical load at the shakedown/ratchetting boundary for a given thermal loading cycle and from the plastic multipliers and hinges active in the solution, the deformation of the structure is calculated. The solution process is repeated for varying values of the thermal load, by linear scaling of the temperature history, until an interaction diagram is calculated containing sufficient points.

Interaction Diagrams

Interaction diagrams for a wide variety of cases have been calculated by the upper bound method of Ponter, Karadeniz and Carter [4]. The computer program EECS3 which generates the results is capable of finding interaction diagrams for generalised axisymmetric thin shells with arbitrary linearly varying temperature histories. Four regimes of behaviour can be found within an interaction diagram, namely

Elastic (E): In this region any combination of mechanical load and thermo-elastic stress is always within the plastic yield surface throughout the structure and thus deformations are purely elastic.

Shakedown (S): Here the total loading initially exceeds the plastic yield surface somewhere in the structure, but the thermo-elastic stress itself is less than twice the yield stress of the material. The effect is that for the first few loading cycles plastic strain growth occurs as a residual stress field develops due to these strains, so that after a number of cycles the behaviour of the plastically deformed structure becomes fully elastic. Inside the shakedown region there is no low cycle fatigue.

Plastic Shakedown (F): In this region the thermo-elastic stresses exceed twice yield, and plastic strains occur at two or more instants in the thermal loading cycle, but accumulate to zero strain over the complete cycle. Typically this can occur when a region of material, whose thermo-elastic stresses cannot be accommodated within the plastic yield surface, is prevented from deforming plastically by surrounding material in the elastic region. In the plastic shakedown region the lifetime of the structure is determined by fatigue.

Ratchetting (R): Here the total loading exceeds the plastic yield surface and the structure exhibits incremental cyclic plastic strain growth which settles to a constant rate after a few cycles. The usual design criteria is that the structure shall not ratchet, however if this is unavoidable, the lifetime of the structure is then determined by the number of loading cycles for which the accumulated plastic strains remain within a given tolerance.

That interaction diagrams are unique for each structure and cyclic loading history, although there is information within the diagram on the likely behaviour of the structure when the loading conditions are changed. Typical shapes of exact interaction diagrams for different thermal loading

484

cases in axisymmetric thin cylinders are shown in Figs 2 to 4. In Fig.2 the through thickness temperature gradient is dominant, thus the ratchetting boundary is close to the Bree line [5] and the lifetime of the structure at high thermal and low mechanical loads is determined by fatigue. In Fig.3 the axial temperature gradient is dominant resulting in ratchetting at zero mechanical load for moderate or high levels of thermal load where the plastic shakedown region may be small or absent. Fig.4 shows the effects of movement of a predominantly axial temperature gradient over a significant length of tube, causing ratchetting of the structure at low levels of the thermal load.

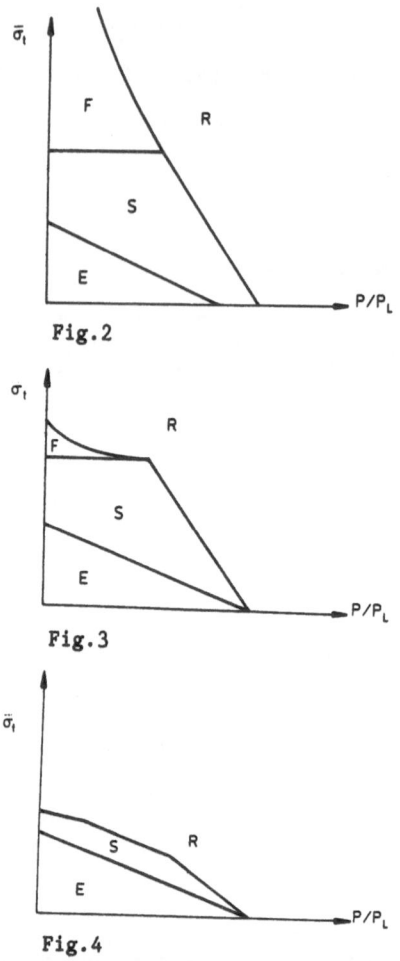

Fig.2

Fig.3

Fig.4

Vertical Axes – Thermal Load normalised by the plastic yield stress at the initial temperature. Horizontal Axes – Mechanical Load normalised by the collapse load at the initial temperature.

Conclusion

The interaction diagram method based on upper-bound theory gives a theoretical understanding of structural behaviour in a wide range of circumstances, which is confirmed by the available experimental evidence, enabling the design and assessment of structures to be carried out with confidence which is the primary requirement of industry.

Links with Industry

The construction of design codes based on the interaction diagram method is taking place in collaboration with the UKAEA at Risley. This work is being supervised by a joint working group between the UKAEA, NNC (National Nuclear Corporation) and Leicester University. It is hoped that the final draft of these design codes will be included as a United Kingdom contribution to the European Fast Reactor project, for incorporation into the RCCMR set of design rules.

In support of the CEGB (UK Central Electricity Generating Board), extension of these techniques to cracked bodies, including creep, fatigue and creep-fatigue phenomena, will be undertaken. This work is to help assess the residual lifetime of structural components in existing reactors, in conjunction with the CEGB's own well-known R5 procedure.

Acknowledgments
The authors gratefully acknowledge the support of the United Kingdom Atomic Energy Authority, the Science and Engineering Research Council, and the European Economic Communities Fast Reactor Coordinating Committee. The work reported in this paper forms part of a study of the effects of severe thermo-mechanical loading in both the plastic and creep ranges.

References

[1] Karadeniz S.; Ponter A.R.S.: A Linear Programming Upper Bound Approach to the Shakedown Limit of Thin Shells subjected to Variable Thermal Loading. J.Strain Anal. 19 (1984) 221

[2] Drucker D.C.; Shield R.T.: Limit Analysis of Symmetrically Loaded Thin Shells of Revolution. ASME J.Appl.Mech. 26 (1959) 61

[3] Marsten R.E.: The Design of the XMP Linear Programming Library. Assc.Comp.Mach.- Trans.Math.Soft.(ACM TOMS) 7 (1981) 481

[4] Karadeniz S.; Ponter A.R.S.; Carter K.F.: The Plastic Ratchetting of Thin Cylindrical Shells subjected to Axisymmetric Thermal and Mechanical Loading. ASME J.Press. Vessel Tech. 109 (1987) 387

[5] Bree J.: Elastic-Plastic behaviour of Thin Tubes subjected to Internal Pressure and Intermittent High Heat Fluxes with application to Fast Nuclear Reactor Fuel Elements. J.Strain Analysis 2 (1967) 226

Prof. A.R.S. Ponter and Dr. K.F. Carter,

Engineering Department, Leicester University,

University Road, Leicester LE1 7RH. U.K.

Three Problems in the Integration of Electric Circuits by ROW-Type Methods

Peter Rentrop, TU München

1.   Introduction

The modified nodal analysis for an electric circuit leads
to a quasilinear implicit ordinary differential equation system

$$C \; U'(t) = F(t,U(t)) \tag{1.1}$$

for given consistent initial values $U(t_o) = U_o$. The n-vector $U(t)$
includes the unknown voltages at the nodes. The matrix C con=
sists of the (possibly voltage depending) capacities, the right
hand side $F(t,U)$ stands for the resistancies and the characteri=
stics of the transistors.

If C is a regular matrix, the system (1.1) can be solved with
modified standard integrators, see Bulirsch, Gilg [1], Stoer,
Bulirsch [10]. It is not advisable to invert C explicitly, be=
cause a possible sparse structureof C is not preserved in C^{-1}.
Furthermore, this is not necessary - due to the desired steep
courses of the voltages, the system (1.1) behaves very stiff.
It must be treated with stiff integration software, like ROW-
methods, which are described in Kaps, Rentrop [5] and Kaps, Wan=
ner [6]. The stiff integrators can be easily adapted to system
(1.1) for a regular, constant matrix C.

If C is a singular matrix, and this happens in electric circuit
simulation, the standard software fails. The convergence assump=
tions of a ROW-method are violated and the order reduces to one.
In practical computations, a stepsize control, based essentially
on two methods of different order, fails. In the case of a sin=

J. Manley et al. (eds.), Proceedings of the Third European Conference on Mathematics in Industry, 487–493.
© 1990 Kluwer Academic Publishers and B. G. Teubner Stuttgart.

gular C, (1.1) describes a differential-algebraic system.

$$U(t) = (y(t), z(t)), \quad y: n_1\text{-vector}, \quad z: n_2\text{-vector} \qquad (1.2)$$

$$y'(t) = f(t, y, z),$$
$$0 = g(t, y, z) \qquad n_1 + n_2 = n$$

The Implicit Function Theorem guarantees locally an unique solu= tion, provided the matrix $(\frac{\partial g}{\partial z})$ is regular - the so-called index -1 assumption. Obviously a singular matrix C creates at least three main difficulties:

(I)   How should a ROW-method be modified, to handle a singular matrix C?

(II)   What are the consistent initial values? It is not possib= le to prescribe n initial values independently.

(III) During the integration, the index-1 assumption must be checked.

In the following the three questions will be discussed for two circuits. Although they are less complicated than real-life cir= cuits, they offer some insight into the so-called reliability problem of standard integration software - questions of their efficiency will not be treated. We will show, that the numeri= cal techniques are already available to handle most of the cir= cuit simulation problems in a reliable way.

2.   Linear Case: Operational Amplifier

Applying Kirchhoff laws to the amplifier circuit (2.1) we obtain the linear implicit ordinary differential equation

$$C \, U'(t) + B \, U(t) = 0 \qquad (2.2)$$

with:

$$U = (U_1, U_2, U_3, U_4, i), \qquad G_j = 1/R_j, \quad j=1,\ldots,4 \qquad (2.3)$$

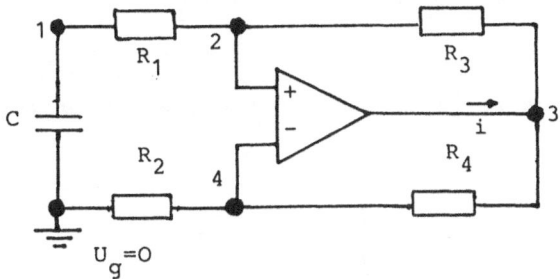

Circuit (2.1): Operational Amplifier

$$B = \begin{bmatrix} G_1 & -G_1 & 0 & 0 & 0 \\ -G_1 & G_1+G_2 & -G_2 & 0 & 0 \\ 0 & -G_2 & G_2+G_4 & -G_4 & 1 \\ 0 & 0 & -G_4 & G_3+G_4 & 0 \\ 0 & 1 & 1/A & -1 & 0 \end{bmatrix}, \quad C = \begin{bmatrix} C & 0 & 0 & 0 & 0 \\ 0 & 0 & 0 & 0 & 0 \\ 0 & 0 & 0 & 0 & 0 \\ 0 & 0 & 0 & 0 & 0 \\ 0 & 0 & 0 & 0 & 0 \end{bmatrix}$$

An unique solution of (2.2), (2.3) exists, if the matrix pencil

$(\lambda C + B)$ is regular for all λ, excluding the eigenvalues, see

Wilkinson [11] and the further literatur cited there. We obtain:

$$\det(\lambda C + B) = \lambda C \left\{ (G_1G_4-G_2G_3) - \frac{1}{A} (G_1+G_2)(G_3+G_4) \right\} \qquad (2.4)$$
$$- G_1G_2G_3 - \frac{1}{A} G_1G_2(G_3+G_4)$$

For arbitrary values of λ det$(\lambda C + B)$ is not equal 0. We have the

solvability of (2.2), (2.3) - see problem (III). The numerical

treatment of linear systems is discussed in [11] and can be also

performed with the ROW-methods from part 3. However, question

(II) is not answered yet. For the initial values we get:

if $\{...\}$ = 0: all initial values are prescribed by the structure of the system.

if $\{...\}$ ≠ 0: one initial value is free.

Circuit (2.1) shows, that special values of the resistancies and

of the amplification factor A fix the number of independently

choosable initial values - and not the structure of the circuit.

490

3.   Nonlinear Case: Two Transistor Amplifier

The nodal voltage analysis of a two transistor amplifier, due to Glashoff, Oberle [4], leads to the descriptor form (3.2).

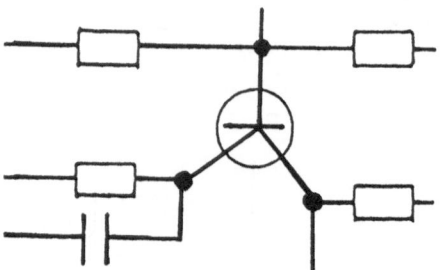

Circuit (3.1): Typical part of a two transistor amplifier

$$C\ U'(t) = F(U(t)), \qquad U = (U_1, \ldots, U_8) \tag{3.2}$$

where the capacity matrix C is given by

$$C = \begin{pmatrix} -c_1 & c_1 & & & & & & \\ c_1 & -c_1 & & & & & 0 & \\ & & -c_2 & & & & & \\ & & & -c_3 & c_3 & & & \\ & & & c_3 & -c_3 & & & \\ & & & & & -c_4 & & \\ & 0 & & & & & -c_5 & c_5 \\ & & & & & & c_5 & -c_5 \end{pmatrix}$$

Obviously C has not full rank. Each of the blocks $\begin{bmatrix} -c_i & c_i \\ c_i & -c_i \end{bmatrix}$ is rank deficient, therefore rank(C) = 5.

There are two possibilities to treat the implicit system (3.2) numerically. By applying a singular value decomposition of C, see [4], (3.2) is transformed into a differential-algebraic sy= stem of type (1.2). There exists several approved codes to hand= le (1.2); e.g. DASSL by Petzold [7] and the modification of Füh= rer [3], LIMEX by Deuflhard, Hairer, Zugck [2] and modified ROW- methods by Roche [8] and [9].

It is also possible to apply the above mentioned methods direct=
ly to the implicit system (1.1). In [8] it is shown, that the
usual 'iteration-matrix' of a ROW method:

$$U^h(t_o+h) = U(t_o) + \sum_{i=1}^{s} c_i k_i \qquad (3.3)$$

$$(I - h\boldsymbol{\gamma}DF(U_o)) k_i = h F(U_o + \sum_{j=1}^{i-1} \alpha_{ij} k_j) + h DF(U_o) \sum_{j=1}^{i-1} \gamma_{ij} k_j$$

can be modified to

$E = C - h\boldsymbol{\gamma}DF(U_o)$, also for a singular matrix C.

However, in order to achieve the same order of the method as in
the regular case, the number of increments k_i has to be increa=
sed, since additional equations of conditions have to be satis=
fied. A code of type (3.3) is able to solve the nodal equations
(2.2), (2.3) - this answers question (I).

The methods in [2] and in [8] are also equipped with an index-
monitor. For a ROW-type method it is sufficient to control the
behaviour of the first increment k_1 for two different stepsizes
h and $\tilde{h} \angle h$ - usually h and $\tilde{h}$ are obtained from a rejected step
in the stepsize control. The index-monitor - this answers que=
stion (III).

It is still an open question, as how consistent initial values
should be computed. There are three possibilities to answer the
question (II):

(i) It would be preferable to determine the initial values by
an analytic method, but this is seldom done in practice. Here
both circuits would allow this.

(ii) One can apply Newton-type techniques combined with homoto=
py methods to obtain consistent initial values, see [10].

492

(iii) The stiffness of the system (3.2) can be used in the sen=
se, that perturbations in the initial values are damped out. We
are interested in the exit voltage $U_8(t)$ for the initial signal
$U_e(t) = 0.1 \sin(2000\,\pi t)$. To achieve appropriate initial values
the integration starts with $\tilde{U}_e(t) = 0.1$, assuming wrong initial
conditions $U_j(0)$, $j=1,\ldots,8$. Due to the inherent stability of
the system the integration ends in an asymptotic phase, which
can be used as consistent initial values. A typical exit signal
is presented in Figure (3.4).

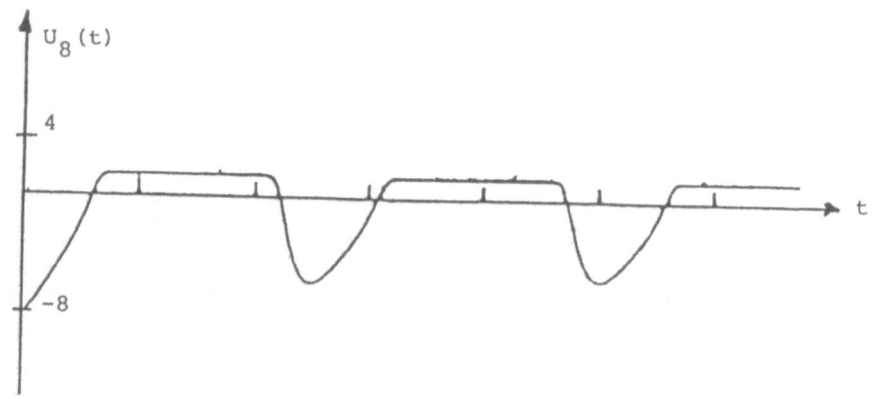

Figure (3.4):   Exit signal $U_8(t)$

References

1 Bulirsch, R., Gilg, A.: Effiziente numerische Verfahren für
   die Simulation elektrischer Schaltungen. In H. Schwärtzel
   (ed.): Informatik in der Praxis. Springer Verlag, Berlin
   1986, p. 3-12

2 Deuflhard, P., Hairer, E., Zugck, J.: One step and extrapola=
   tion methods for differential-algebraic systems. Numer. Math.

$\underline{51}$ (1987) 501-516

3   Führer, C.: Differential-algebraische Gleichungssysteme in mechanischen Mehrkörpersystemen. Thesis, TU München, Rep. TUM-M8807, August 1988

4   Glashoff, K., Oberle, H.J.: Private Communication, 1987 Universität Hamburg, FB Mathematik

5   Kaps, P., Rentrop, P.: Generalized Runge-Kutta methods of order four with stepsize control for stiff ordinary diffe= rential equations. Numer. Math. $\underline{33}$ (1979) 55-68

6   Kaps, P., Wanner, G.: A study of Rosenbrock-type methods of high order. Numer. Math. $\underline{38}$ (1981) 279-298

7   Petzold, L.R.: A description of DASSL - a differential-al= gebraic system solver. IMACS Trans. Sci. Comp. $\underline{1}$, ed. R.S. Stepleman, 1982

8   Rentrop, P., Roche, M., Steinebach, G.: The application of Rosenbrock-Wanner type methods with stepsize control in differential-algebraic equations. TU München, Rep. TUM-M8804, June 1988

9   Roche, M.: Rosenbrock methods for differential-algebraic systems. Numer. Math. $\underline{52}$ (1988) 45-63

10   Stoer, J., Bulirsch, R.: Introduction to Numerical Analysis. Springer Verlag: 1980

11   Wilkinson, J.H.: Note on the practical significance of the Drazin Inverse. In L.S. Campbell (ed.): Recent applications of generalized inverses. Pitman Publ. No. $\underline{66}$ (1982) 82-99

Address: Prof. Dr. P. Rentrop, Mathematisches Institut,
         TU München, Arcisstr. 21,
         D - 8000 München 2

CONTROLLER DESIGN FOR INDUSTRIAL MULTIPASS PROCESSES

E. ROGERS and D.H. OWENS

ABSTRACT

The disturbance decoupling with stability problem for a
differential non-unit memory linear multipass process   is formulated
from practical considerations and some initial results are presented.
In particular, the concept of an error actuated controller with memory
is used to produce a solution in one special case.

1. INTRODUCTION

The essential characteristic of a multipass process is the presence
of a repetitive, or recursive, action with interaction between
successive passes. Industrial examples include long-wall coal cutting
and the rolling of metal strip.

In the most general case, a multipass process has nonlinear
dynamics and a pass length which varies from pass to pass. Suppose,
however, that the pass length is constant and the dynamics are linear.
Then a large number of practical examples are known, Rogers [2], to
satisfy these conditions. Further, an abstract representation and
stability theory for this class has been formulated, Owens [1], which
includes differential non-unit memory linear multipass processes as a
special case.

To date, no work has been reported on the design of control schemes
for this sub-class, which can be used to model certain simplified modes
of operation of bench mining systems, and hence the subject area of this
paper. In particular, the so-called multipass disturbance decoupling
with stability problem for this case is formulated from practical
considerations. Further, the possibility of obtaining a 'geometric
style' solution, in the spirit of Wonham [3], is discussed. Finally,

495

J. Manley et al. (eds.), Proceedings of the Third European Conference on Mathematics in Industry, 495–502.
© 1990 *Kluwer Academic Publishers and B. G. Teubner Stuttgart.*

the concept, Rogers [2], of an error actuated controller with memory is used to solve this problem in one special case.

2. PROBLEM FORMULATION

This paper considers the case of differential non-unit memory linear multipass processes whose state-space model has the form

$$\dot{X}_{k+1}(t) = AX_{k+1}(t) + BU_{k+1}(t) + \sum_{j=1}^{M} B_{j-1}Y_{k+1-j}(t)$$

$$Y_{k+1}(t) = CX_{k+1}(t), \quad X_{k+1}(t) \in R^n, \quad Y_{k+1}(t) \in R^m, \quad U_{k+1}(t) \in R^{\ell}$$

$$0 \le t \le a, \quad X_{k+1}(0) = d_{k+1}, \quad k \ge 0 \tag{1}$$

This model has practical applications since it can, Rogers [2], be used to describe certain simplified modes of operation of bench mining systems where, typically, M, the so-called memory length, is in the range ten to seventeen. Finally, the details of the stability theory for the general class which includes (1) as a special case can be found in Owens [1].

Necessary and sufficient conditions for so-called, Owens (1977), stability along the pass of (1) take the following form.

Theorem 2.1: Under certain well defined controllability and observability assumptions, (1) is stable along the pass if, and only if,

(a) all eigenvalues of the matrix A have strictly negative real parts, and

(b) there exists real numbers $\epsilon > 0$, $r_\infty < \lambda < 1$ such that

$$|sI_n - A - \sum_{j=1}^{M} B_{j-1}z^{-j}C| \ne 0 \tag{2}$$

for all complex numbers z, s satisfying $|z| \ge \lambda$ and $Re\{s\} \ge -\epsilon$.

Stability along the pass is an obvious necessary requirement of any practically feasible control policy for (1). Further, a detailed study,

Rogers [2], of practical examples has produced cases where desired behaviour also requires the following:-

(i) The pass profile $Y_k(t)$, $0 \leq t \leq \alpha$, should be indpendent of pass profiles $Y_{k-j}(t)$, $0 \leq t \leq \alpha$, $1 \leq j \leq M$, for all passes $k \geq k^* \geq 1$ with an optimum choice of $k^* = 1$.

(ii) Suppose that (i) above holds. Then for all passes $k \geq k^* \geq 1$ the dynamics are described by the so-called memoryless differential multipass process obtained from (1) by deleting the previous pass terms $\sum\limits_{j=1}^{M} B_{j-1} Y_{k+1-j}(t)$. Further, the triple A,B,C is also easily shown to characterise the state-space description of the interpretation of the so-called, Owens [1], limit profile, denoted Y_∞, for this case.

Hence, in effect, the limit profile is reached on pass k^* and has the transfer-function matrix description.

$$Y_\infty(s) = G^P(s) U_\infty(s)$$
$$G^P(s) = C(sI_n - A)^{-1}B \tag{3}$$

Additionally, dynamics of (3) are required to be 'acceptable', where the precise interpretation of this term is a matter for judgement based on knowledge of the particular application or example under consideration. For the purposes of this paper, the interpretation taken is the essential necessary requirement of stability in the sense that all eigenvalues of the matrix A have strictly negative real parts.

The requirements of (i) and (ii) above constitute the so-called multipass disturbance decoupling with stability problem for (1).

3 ANALYSIS

As a preliminary, it is instructive to briefly review the well known disturbance decoupling with stability problem from conventional differential linear systems theory. Consider, therefore, the system of (4) below where q(t) represents a disturbance which is assumed not to be directly measurable by the controller

$$\dot{X}(t) = AX(t) + BU(t) + Dq(t)$$

$$Y(t) = CX(t), \ X(t)\epsilon R^n, \ Y(t)\epsilon R^m, \ U(t)\epsilon R^\ell, \ q(t)\epsilon R^\nu \qquad (4)$$

Further, suppose that the linear state feedback law $U(t) = FX(t)$ is applied to (4). Then the disturbance decoupling problem for the resulting closed-loop system is to find a suitable F such that q(t) has no influence on the controlled output Y(t). Equivalently, this closed-loop system is said to be disturbance decoupled relative to the pair Y(t),q(t) if, for each initial condition $X(0)\epsilon R^n$, the output Y(t), $t \geq 0$, is the same for all $q(t) \in R^\nu$.

The above problem has been the subject of much research effort. One element of which has been the use of such geometric concepts as (A,B) - invariant subspaces to obtain necessary and sufficient conditions for solvability. Note, however, that these do not guarantee that the resulting closed-loop system is stable. In particular, they do not guarantee that all eigenvalues of A+BF have strictly negative real parts which is obviously essential for applications. This has led to the so-called disturbance decoupling with stability problem for which necessary and sufficient conditions for solvability take the form of theorem 5.8 in Wonham [3].

Return now to the multipass case and, since the following discussion generalises in a natural manner, consider the special case of a unit memory process and hence M = 1 in (1). Further, interpret $Y_k(t)$, $0 \leq t \leq a$, $k \geq 0$, as a disturbance which is not directly measured by the controller on pass k+1. Then a clear structural similarity exists with (4) if the so-called, Rogers [2], current pass state feedback law

$$U_{k+1}(t) = FX_{k+1}(t), \quad 0 \leq t \leq a, \quad k \geq 0 \tag{5}$$

is applied. To see this, consider again requirement (i) of the subject problem and interpret it in terms of the resulting closed-loop system. In which case it follows immediately that this requirement holds relative to the pair $Y_{k-1}(t)$, $Y_k(t)$, $0 \leq t \leq a$, $k \geq k^* \geq 1$, if, for each initial condition $X_k(0) = d_k \in R^n$, the output $Y_k(t)$ is the same for all $Y_{k-1}(t) \in R^m$. Hence multipass disturbance decoupling simply means that the contribution of the previous pass profile to the current one is zero, $0 \leq t \leq a$, $k \geq k^* \geq 1$.

The second requirement of the subject problem in this case requires that F be selected such that all eigenvalues of A+BF have strictly negative real parts. Hence this multipass problem has a clear structural similarity to its conventional linear systems counterpart which was briefly reviewed earlier in this section. Consequently it can be conjectured that a solution can be developed using geometric concepts such as (A,B) - invariant subspaces with the subsequent possibility of a natural generalisation to the non-unit memory case. Work on this approach is currently progressing and will be reported in due course.

Previous work, Rogers [2], has developed controller structures with memory for (1), i.e. controllers which explicitly use information from the previous passes. To conclude this section the application of one of these controller structures to the subject problem is considered. This leads to a solution in one special case.

Define the current pass error vector for (1) as

$$e_{k+1}(t) = R_{k+1}(t) - Y_{k+1}(t), \quad 0 \leq t \leq a, \quad k \geq 0 \tag{6}$$

where $R_{k+1}(t) \in R^m$ is a new external vector taken to represent desired behaviour on this pass. Then a so-called memoryless dynamic unity-negative feedback controller with proportional multipass minor loop compensation constructs the input as

$$U_{k+1}(t) = Y_{k+1}^C(t) - \sum_{j=1}^{M} K_j Y_{k+1-j}(t), \quad 0 \le t \le \alpha, \ k \ge 0 \quad (7)$$

where K_j, $1 \le j \le M$, are constant $\ell \times m$ matrices and $Y_{k+1}^C(t)$ is the output from a memoryless differential process with input $e_{k+1}(t)$ of the form

$$\dot{X}_{k+1}^C(t) = A^C X_{k+1}^C(t) + B^C e_{k+1}(t) \quad (8)$$

$$U_{k+1}(t) = C^C X_{k+1}^C(t) + D^C e_{k+1}(t) \quad 0 \le t \le \alpha, \ k \ge 0$$

where $X_{k+1}^C(t) \in R^{n_1}$ denotes the internal state of (7). Further, combining (1), (6), (7) and (8) yields

$$\dot{\tilde{X}}_{k+1}(t) = (\tilde{A} - \tilde{B}\tilde{C})\tilde{X}_{k+1}(t) + \tilde{B}R_{k+1}(t) + \sum_{j=1}^{M} \tilde{B}_{j-1} Y_{k+1-j}(t)$$

$$Y_{k+1}(t) = \tilde{C}\tilde{X}_{k+1}(t), \quad 0 \le t \le \alpha, \ k \ge 0 \quad (9)$$

where $\tilde{X}_{k+1}(t) = [X_{k+1}(t)^T, X_{k+1}^C(t)^T]^T \in R^N$, $N = n + n_1$ and

$$\tilde{A} = \begin{bmatrix} A & BC^C \\ 0 & A^C \end{bmatrix}, \quad \tilde{B} = \begin{bmatrix} BD^C \\ B^C \end{bmatrix}, \quad \tilde{B}_{j-1} = \begin{bmatrix} B_{j-1} - BK_j \\ 0 \end{bmatrix}, \quad 1 \le j \le M$$

$$\tilde{C} = [C \ 0] \quad (10)$$

This state-space model has an identical structure to its open-loop counterpart (1). Hence theorem 2.1 can be applied to provide conditions for closed-loop stability along the pass. Further, a multipass disturbance decoupling with stability problem can be formulated in an identical manner to that described above.

Consider, therefore, the multipass disturbance decoupling with stability problem for (9)-(10) in the special case when $n = m = \ell$ and $|B| \ne 0$; a not uncommon situation in bench mining systems. Further, let the controller matrices K_j, $1 \le j \le M$, be selected according to the following rule.

$$K_j = B^{-1} B_{j-1}, \quad 1 \le j \le M \quad (11)$$

Then $\mathring{B}_{j-1} = 0$, $1 \leq j \leq M$, in (10) and hence in (9) the pass profile
$Y_k(t)$, $0 \leq t \leq \alpha$, is independent of the pass profiles $Y_{k-j}(t)$,
$1 \leq j \leq M$, for all passes $k \geq 1$. Equivalently, multipass disturbance
decoupling is achieved in this case with the optimum choice of $k^* = 1$.

Suppose, therefore, that (11) holds and apply the Laplace transform
to obtain, after some manipulation,

$$Y_\infty(s) = (I_m + G^P(s)G^C(s))^{-1}G^P(s)G^C(s)R_\infty(s) \tag{12}$$

where

$$G^C(s) = C^C(sI_{n_1} - A^C)^{-1}B^C + D^C \tag{13}$$

The design problem can now be completed by choosing $G^C(s)$ to ensure
desired behaviour. This can be undertaken using an appropriate
frequency domain technique from differential conventional linear systems
theory

4. CONCLUSIONS

This paper has considered the design of control schemes for
differential non-unit memory linear multipass processes which can be
used to model certain simplified modes of operation of bench mining
systems. In particular, the multipass disturbance decoupling with
stability problem for these processes has been formulated from practical
considerations. Further, the clear structural similarity between this
problem and the well known disturbance decoupling with stability problem
from differential conventional linear systems theory has been noted.
This has led to a discussion of the possibility of obtaining a
'geometric style' solution as per the conventional problem. Finally,
the concept of an error actuated controller with memory has been used to
solve this problem in one special case.

REFERENCES

[1] OWENS, D.H., 1977, Stability of linear multipass processes. Proc.
     I.E.E., 124, (11), pp.1079-1082.

[2] ROGERS, E., 1987, Feedback and stability theory for linear
     multipass processes. Ph.D. thesis, University of Sheffield.

[3] WONHAM, W.M., 1974, Linear multivariable control: A geometric
     approach, New York: Springer Verlag.

E. ROGERS AND D.H. OWENS

Centre for Industrial Modelling and Control,

University of Strathclyde,

Glasgow.

G1 1XJ

Scotland.

COMPUTER-INTEGRATED PRODUCTION-PLANNING & INVENTORY-CONTROL

AT AN AUTOMOBILE-ENGINE PRODUCER

Walter Schneider
Institut für Informatik
A-4040 Universität Linz

1.The production

On 2 conveyor-belts with auxiliary transfer groups the following engine families are produced:
- 4 cylinder Otto-engine with Bosch fuel injection
- 6 cylinder Otto-engine with Bosch fuel injection
- 6 cylinder Otto-engine optimized for Economy, designated "ETA" after the greek letter used for efficiency by engineers
- 6 cylinder Diesel-engine turbocharged
- 6 cylinder Diesel-engine naturally aspirated.

The above mentioned engine families include a great manifold of different engine-types (at about 200 variants). Whereby all 4 cylinder engines are mounted on one *assembly line* and all 6 cylinder on the other one.

Special parts which spread up the combinatoric manifold, such as camshafts, craftshafts, cylinder heads, valves e.g. are produced on 29 *production groups* - some of them joined together to a *transfer-group* - which are input to the above mentioned assembly lines.

2. State of the art

Nowadays production planning is a multi-criteria optimization where minimal waste time on the assembly lines has to be combined with the "optimal" lot size for the 29 production groups. Further goals which also are regarded are:
- inventory costs
- term of delivery
- production smoothing.

Todays greatest problems are:
- Lot sizes of the production groups dont fit to each other.
- Waste-time minimization of the assembly lines is done without regard to the inventory list. Thus infeasible solutions occur.
- The production plan - if feasible - is not smooth enough (this means that working time should be about 8 hours the day without great changes).
- Great set ups from one engine family to the other should be done outside the normal working time by a small staff. If I can't influence this 30 ore more workers are waiting untill 3 workers have changed production tools.

J. Manley et al. (eds.), Proceedings of the Third European Conference on Mathematics in Industry, 503–511.
© 1990 *Kluwer Academic Publishers and B. G. Teubner Stuttgart.*

3.Further policy

Caused by the market situation *JIT (Just In Time) production* has had to be introduced. This means that no more minimal waste time on the assembly lines and lot-size optimization on the production groups, but time accuracy and low inventory costs are the most important goals. Further on it was a wish of the logistic staff that production planning enables a much more interactive process including *JIT knowledge* of the foreman.

4.Solution

The above mentioned production-planning policy is realized in 3steps - each one producing a more & more detailed production plan.

4.1. 1st step: Global optimization regarding the demand, inventory-, production- & mean set up costs.

This step is done by dynamic optimization. The following abbreviations and tables are used:

Abbreviations

T	...	period
ISB(i)	...	inventory stock at the beginning of period i
ISE(i)	...	inventory stock at the end of period i
ISU(i)	...	inventory stock-up during period i
ISL(i)	...	lowest point of inv. stock during period i
ISM(i)	...	mean point of inventory stock during period i
ISX(i)	...	maximal point of inventory stock during period i
ISS(i)	...	safety point of inventory stock during period i
ISB	...	inv.stock at the beginning of the specified period
.	...	
.	...	
.	...	
ISS	...	safety point of inv.stock during specified period
D(j,i)	...	demand of product j in period i
P(j,i)	...	production of product j in period i
D	...	demand of specefied product and period
P	...	analogous
mPC	...	mean production costs without set up costs
mSUC	...	mean set up costs
mSUT	...	mean set up time
cTC	...	cumulated total costs until now
IC	...	inventory costs

Tables

product	ISX(i) i=1,6
1	700
2	800
3	1000
4	600
5	400

Tab.1:max.point of inv.can
differ over the periods

product	ISB(1)
1	300
2	500
3	600
4	200
5	200

Tab.2:inv. stock at beginning
of the 1.period

product	IC i=1,6
1	0,50
2	0,10
3	0,30
4	0,80
5	0,60

Tab.3:inventory costs
over all periods

product	ISS(i)
1	100
2	200
3	100
4	50
5	100

Tab.4:safety point of inv.stock
over all periods

product	mPC(i) i=1,6	parts per hour
1	10,0	100
2	5,0	100
3	15,0	50
4	10,0	50
5	10,0	50

Tab.5:prod.costs & -capacity
Beware: prod.costs can differ
from period to period

from \ to	1	2	3	4	5
1	0,0	1,0	1,0	3,0	2,0
2	2,0	0,0	2,0	2,0	1,5
3	3,0	1,0	0,0	2,0	3,0
4	4,0	2,0	1,0	0,0	1,0
5	1,0	1,0	1,0	3,0	0,0
mSUT	2,0	1,0	1,0	2,0	1,5

Tab.6: matrix of set-up times

prod. \ T	1	2	3	4	5	6
1	50	50	50	50	50	50
2	200	200	100	200	200	100
3	25	25	50	50	25	50
4	100	100	100	50	150	100
5	50	100	50	100	50	100

Tab.7:demand D(i,j)

product	ISL(6)	ISX(6)
1	200	400
2	250	350
3	300	500
4	200	300
5	200	300

Tab.8:range of inv.stock at the
end of planning horizon
(= 6-th period)

By means of dynamic programming we compute the optimal production and inventory policy for each product and get the following suboptimal and optimal solutions:

Production & inventory table of product 1 & period 1

ISS	...	safety stock of inventory	= 100 parts
inventory costs per part of product 1			= 0,5 money units
mPC	...	mean prod.costs per part of prod.1	= 10 money units
mSUC	...	mean set up costs	= 200 money units

T	ISB	D	P	ISL	ISE	cTC	mSUC	mPC	ISM	IC	cTC
1	300	-50	0	250	250	0,00	0	0,00	275	137,50	137,50
1	300	-50	100	250	350	0,00	200	1000,00	275	137,50	1337,50
1	300	-50	200	250	450	0,00	200	2000,00	275	137,50	2337,50
1	300	-50	300	250	550	0,00	200	3000,00	275	137,50	3337,5
1	300	-50	400	250	650	0,00	200	4000,00	275	137,50	4337,50

Tab.9:dynamic computations of product 1 and period 1

The computations ofthe following periods 2,...,5 run the same way and last not least we get:

Production & inventory table of product 1 & period 6

ISS	...	safety stock of inventory	= 100 parts
inventory costs per part of product 1			= 0,5 money units
mPC	...	mean prod.costs per part of prod.1	= 10 money units
mSUC ...		mean set up costs	= 200 money units

T	ISB	D	P	ISL	ISE	cTC	mSUC	mPC	ISM	IC	cTC
1	150	-50	100	100	200	1687,50	200	1000,00	125	62,50	2950,00
1	150	-50	200	100	300	1687,50	200	2000,00	125	62,50	3950,00
1	150	-50	300	100	400	1687,50	200	3000,00	125	62,50	4950,00
1	250	-50	0	200	200	2737,50	0	0,00	225	112,50	2850,00 *
1	250	-50	100	200	300	2737,50	200	1000,00	225	112,50	4050,00
1	250	-50	200	200	400	2737,50	200	2000,00	225	112,50	5050,00
1	350	-50	0	300	300	3787,50	0	0,00	325	162,50	3950,00
1	350	-50	100	300	400	3787,50	200	1000,00	325	162,50	5150,00
1	450	-50	0	400	400	4837,50	0	0,00	425	212,50	5050,00

Tab.10:dynamic computations of product 1 & period 6
with "*" ... optimal solution of product 1

Tracing the opt.solution of all products from period 6 (P=6) to period 1 we get:

T	ISB	D	P	ISL	ISE	cTC	mSUC	mPC	ISM	IC	cTC
6	250	-50	0	200	200	2737,50	0	0,00	225	112,50	2850,00
5	300	-50	0	250	250	2600,00	0	0,00	275	137,50	2737,50
4	150	-50	200	100	300	337,50	200	2000,00	125	62,50	2600,00
3	200	-50	0	150	150	250,00	0	0,00	175	87,50	337,50
2	250	-50	0	200	200	137,50	0	0,00	225	112,50	250,00
1	300	-50	0	250	250	0,00	0	0,00	275	137,50	137,50

Tab.11:optimal solution of product 1 over the 6 periods

T	ISB	D	P	ISL	ISE	cTC	mSUC	mPC	ISM	IC	cTC
6	300	-100	100	200	300	3905,00	100	500,00	250	25,00	4530,00
5	500	-200	0	300	300	3865,00	0	0,00	400	40,00	3905,00
4	700	-200	0	500	500	3805	0	0,00	600	60,00	3865,00
3	300	-100	500	700	700	1180,00	100	2500,00	250	25,00	3805,00
2	500	-200	0	300	300	1140,00	0	0,00	400	40,00	1180,00
1	500	-200	200	300	500	0,00	100	1000,00	400	40,00	1140,00

Tab.12:optimal solution of product 2 over the 6 periods

T	ISB	D	P	ISL	ISE	cTC	mSUC	mPC	ISM	IC	cTC
6	400	-25	0	375	375	772,50	0	0,00	388	116,25	888,75
5	450	-50	0	400	400	645,00	0	0,00	425	127,50	772,50
4	500	-50	0	450	450	502,50	0	0,00	475	142,50	645,00
3	550	-50	0	500	500	345,00	0	0,00	525	157,50	502,50
2	575	-25	0	550	550	176,25	0	0,00	563	168,75	345,00
1	600	-25	0	575	575	0,00	0	0,00	588	176,25	176,25

Tab.13:optimal solution of product 3 over the 6 periods

T	ISB	D	P	ISL	ISE	cTC	mSUC	mPC	ISM	IC	cTC
6	300	-100	0	200	200	7200,00	0	0,00	250	200,00	7400,00
5	450	-150	0	300	300	6900,00	0	0,00	375	300,00	7200,00
4	100	-50	400	50	450	2640,00	200	4000,00	75	60,00	6900,00
3	200	-100	0	100	100	2520,00	0	0,00	150	120,00	2640,00
2	300	-100	0	200	200	2320,00	0	0,00	250	200,00	2520,00
1	200	-100	200	100	300	0,00	200	2000,00	150	120,00	2320,00

Tab.14:optimal solution of product 4 over the 6 periods

T	ISB	D	P	ISL	ISE	cTC	mSUC	mPC	ISM	IC	cTC
6	350	-100	0	250	250	6005,00	0	0	300	180,00	6185,00
5	200	-50	200	150	350	3750,00	150	2000,00	175	105,00	6005,00
4	300	-100	0	200	200	3600,00	0	0,00	250	150,00	3750,00
3	150	-50	200	100	300	1375,00	150	2000,00	125	75,00	3600,00
2	250	-100	0	150	150	1255,00	0	0,00	200	120	1375,00
1	200	-50	100	150	250	0,00	150	1000,00	175	105,00	1255,00

Tab.15:optimal solution of product 5 over the 6 periods

Summing up the needed production time we get

period	product	ISE in parts	production in parts	time demand in hours set up	prod.	total.
	1	250	0	0.0	0.0	0.0
	2	500	200	1.0	2.0	3.0
1	3	575	0	0.0	0.0	0.0
	4	300	200	2.0	4.0	6.0
	5	250	100	1.5	2.0	3.5
						12.5
	1	200	0	0.0	0.0	0.0
	2	300	0	0.0	0.0	0.0
2	3	550	0	0.0	0.0	0.0
	4	200	0	0.0	0.0	0.0
	5	150	0	0.0	0.0	0.0
						0.0
	1	150	0	0.0	0.0	0.0
	2	700	500	1.0	5.0	6.0
3	3	500	0	0.0	0.0	0.0
	4	100	0	0.0	0.0	0.0
	5	300	200	1.5	4.0	5.5
						11.5
	1	300	200	2.0	2.0	4.0
	2	500	0	0.0	0.0	0.0
4	3	450	0	0.0	0.0	0.0
	4	450	400	2.0	8.0	0.0
	5	200	0	0.0	0.0	0.0
						14.0
	1	250	0	0.0	0.0	0.0
	2	300	0	0.0	0.0	0.0
5	3	400	0	0.0	0.0	0.0
	4	300	0	0.0	0.0	0.0
	5	350	200	1.5	4.0	5.5
						5.5
	1	200	0	0.0	0.0	0.0
	2	300	100	1.0	1.0	2.0
6	3	375	0	0.0	0.0	0.0
	4	200	0	0.0	0.0	0.0
	5	250	0	0.0	0.0	0.0
						2.0.

Tab.16:time demand over the planning-horizon of 6 days

Summing up the capacity demand per product we get the following table:

product	production time in hours
1	4
2	11
3	0
4	16
5	14.5

The time demand per product and day shows the following figure

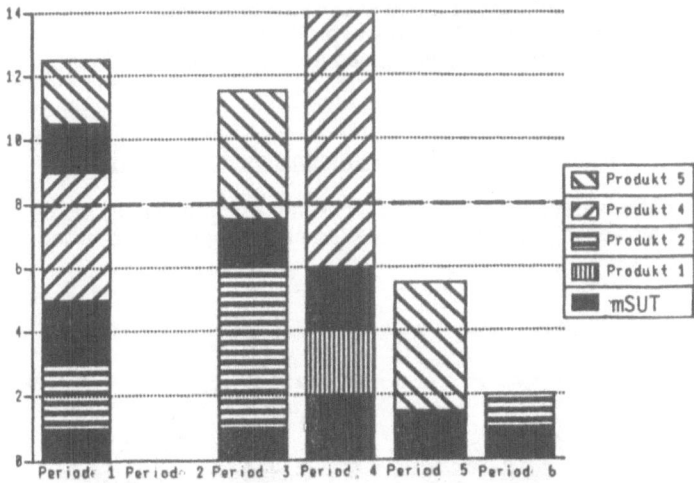

Fig.1:capacity demand per product and day (period)

4.2. 2-nd step:Capacity smoothing over the planning horizon

Corresponding to the capacity-demand of the 1-st step we compute individual production-costs per day. So the production costs of the 1.day are 12,5/8 = 1,56 times higher than before, analogousthe costs of the 2.day are 0/8 = 0 times higher and so on. Enlarging and deminishing the production costs as mentioned before.we run a second - and if necessary a third - dynamic optimization.

In the most cases this procedure computes quite good solutions.Over all the decision-maker also can rearrange the production plan by hand using the information of the preceding computations.In our case we get the following solution:

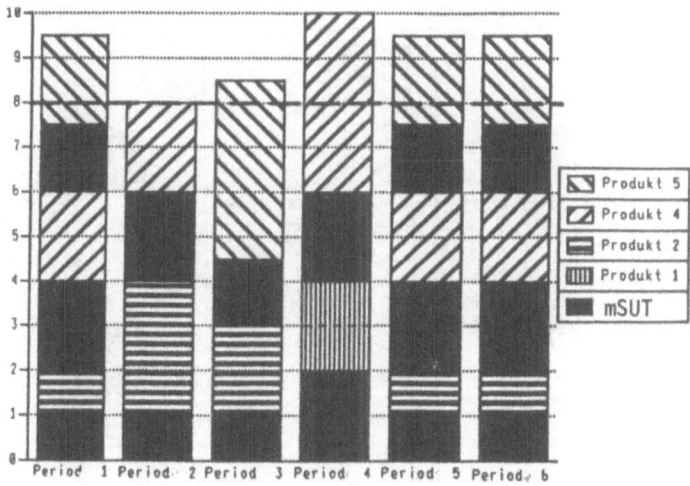

Fig.2:Capacity demand after the smoothing of the 2nd step

510

This step, although heuristic,tourned out to be 3-times effective:

1. The foreman running this program on his terminal has much more information about side-effects like personal situation as the staff of logicians.
2. A much higher degree of flexibility in production planning.
   If necessary the user can replan immediately without loss of information.
3. Psychological reasons like self-confidence, responsibility & motivation of the staff.

4.3. 3rd step: Minimizing the set-up time

By means of a bounded combinatoric algorithm we optimize the set-up times. This algorithm also enables a quick computation of all other significant solutions. We get:

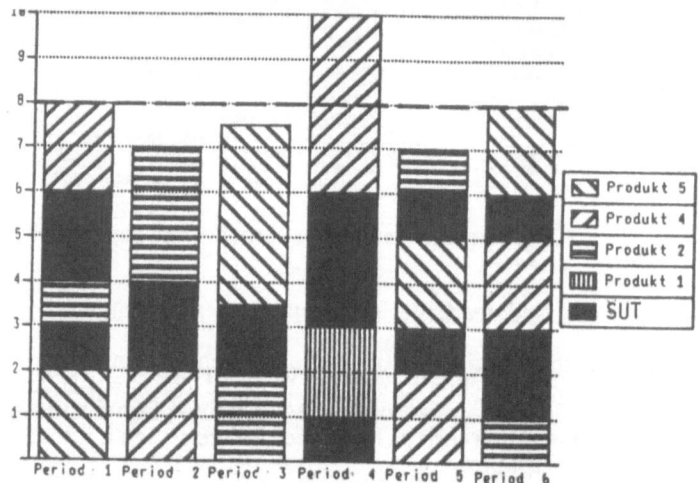

Fig.3:Sequence of production after minimizing the set-up times

Interpreting the above production plan as a continuous time tie be intersected every 8 hours we get:

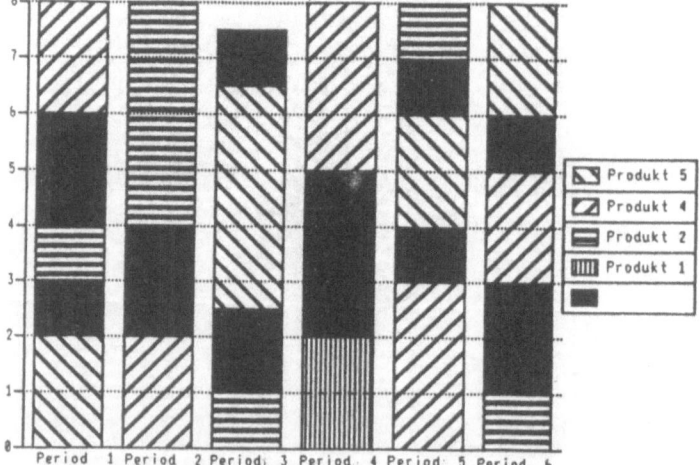

* Fig.4:The final production plan *

As in step 2 also in this step a final validity check takes place.

Comparing the needed time amount at the end of each step we get:

1st step ... 45,5 hours
2nd step ... 55,0 hours
3rd step ... 47,5 hours

5. Conclusion

Over all the quality of the solutions - feasibility, smootheness & practicability - as well as flexibility, interactivity & transparency of production planning have been enlarged in a highly significant way.

References:

1. Axsäter, Schneeweiß, Silver (edit.):Multi-Stage production planning and inventory control. Springer-Verlag, Berlin- Heidelberg- N.Y. (1986)

2. Biggs J.R.:Priority rules for shop floor control in a material requirements planning system under various levels of capacity. Int.J.Prod.Res. 23,33-46 (1985)

3. Blackburn J.D., Millen R.A.:Simultaneous lot-sizing and cap. planning in multi-stage assembly processes.EJOR 16,84-93 (1984)

4. F.Glover et al.:A Primal Simplex Variant for the Maximal-Flow Problem. Naval Res.Log.Quart. 31,41-61 (1984)

5. Knolmayer G.:A simulat. study of some simplification strategies in the development of product-mix models.EJOR 12,339-347 (1983)

6. Schneider:Dyn.Programmierung bei Einkauf-& Lagerbewirtschaftung THEXIS 3,78-82 (1987)

7. Schneider:Trim-loss minimizat. in a crepe-rubber mill opt.sol. vers. heuristic in the 2(3)-dimensional case.EJOR 34,273-281(1988)

Numerical Solution of a Liquid Crystal Problem by Optimization

Seymour Singer

1. Introduction

The deformation of a nematic liquid crystal layer sandwiched between parallel metallically coated glass plates and subjected to an external electric field perpendicular to the plate, is of interest in the design of optical devices. Schadt and Helfrich [5] showed how such layers can be switched from light to dark and dark to light. Leslie [3] obtained equations describing the deformation of a twisted orientation pattern in a magnetic field. Deuling [1] considered the deformation of a nematic layer contained between metallized glass plates in an electric field when one plate is twisted through an angle ω_m around an axis perpendicular to the plates. The nematic layer exhibits a helical orientation pattern even for small electrical fields. In practical applications, it is important to describe these twisted orientation patterns in terms of twist and tilt angles of the liquid crystal molecules as a function of depth in the layer. We have developed an algorithm to solve Deuling's equations by a nonlinear optimization technique applied to an integral formulation of the governing equations. The effective implementation of this technique is facilitated by the availability of high quality optimization software found in the IMSL and NAG Libraries.

2. The Governing Equations

The model is based on the continuum description of the liquid crystal as developed by Oseen [4] and Frank [2]. The mathematical theory has been developed extensively by Leslie [3]. We shall consider a twisted nematic cell of thickness L and filled with positive dielectric anisotropic liquid crystal i.e. $\varepsilon_{11} > \varepsilon_{\perp}$. The algorithm to be described calculates the

513

J. Manley et al. (eds.), Proceedings of the Third European Conference on Mathematics in Industry, 513–522.
© 1990 Kluwer Academic Publishers and B. G. Teubner Stuttgart.

deformation pattern of the cell when subjected to a switching electric voltage U/U_o. The natural voltage unit U_o is

$$U_o = \pi \left[\frac{k_{11}}{\varepsilon_o (\varepsilon_{11} - \varepsilon_{\perp})} \right]^{1/2}$$

(2.1)

where ε_{11}, $\varepsilon_{\perp}$ are dielectric constants of the nematic liquid parallel (respectively perpendicular) to the molecular axis and ε_o is the vacuum permittivity.

The liquid crystal molecules are prealigned parallel to the electrode plates, one of which is subjected to a twist angle ω_m. The perpendicular electric field distorts the helical orientation pattern and causes the less tightly anchored molecules to realign parallel to the applied field. This realignment is characterized by the tilt angle φ and twist angle ω of the director vector $\underline{n}$. The threshold voltage for realignment is given by

$$U_T = U_o \left[1 + \left(\frac{\omega_m}{\pi} \right)^2 \left(\frac{k_{33}}{k_{11}} - \frac{2k_{22}}{k_{11}} \right) \right]^{1/2}$$

(2.2)

The free energy per unit area of the liquid crystal is given by

$$F = \frac{1}{2} \int_0^L [k_{11}(\text{div } \underline{n})^2 + k_{22}(\underline{n}.\text{curl } \underline{n})^2 + k_{33}(\underline{n} \times \text{curl } \underline{n})^2 - \underline{D}.\underline{E}]dz$$

(2.3)

where k_{11}, k_{22}, k_{33} are, respectively, splay, twist and bend elastic constants and $\underline{n}$ is the director vector of the liquid crystal molecules.

Maxwell's equations

$$\underline{\nabla} \times \underline{E} = \underline{0}, \quad \underline{\nabla} \cdot \underline{D} = 0$$

(2.4)

together with the boundary conditions

$$\omega(0) = 0, \; \omega(L) = \omega_m, \; \omega\left(\frac{L}{2}\right) = \frac{\omega_m}{2}, \; \varphi(0) = \varphi(L) = 0, \; \varphi\left(\frac{L}{2}\right) = \varphi_m$$

(2.5)

determine the tilt angle φ and twist angle ω of molecules along the cell by minimizing the free energy (2.3). Deuling [1] has shown that this procedure leads to a coupled set of nonlinear equations viz.

$$\omega_m = 2\beta \int_0^{\varphi_m} \frac{\sqrt{1+k\,\sin^2\varphi}}{g(\varphi)\cos^2\varphi(1+\alpha\,\sin^2\varphi)}\, d\varphi \equiv f_1(\varphi_m,\beta) \qquad (2.6a)$$

$$\overline{U} \equiv U/U_o = \frac{2}{\pi} \int_0^{\varphi_m} \frac{\sqrt{1+k\,\sin^2\varphi}}{g(\varphi)(1+\gamma\,\sin^2\varphi)}\, d\varphi \equiv f_2(\varphi_m,\beta) \qquad (2.6b)$$

which must be solved simultaneously to determine the maximum tilt angle φ_m and integration constant β. In the foregoing equations we have introduced the function

$$g(\varphi) = \left\{ \frac{\sin^2\varphi_m - \sin^2\varphi}{(1+\gamma\,\sin^2\varphi)(1+\gamma\,\sin^2\varphi_m)} + \beta^2\,\frac{1+k}{1+\alpha}\left[\frac{1}{(1+\alpha\,\sin^2\varphi_m)\cos^2\varphi_m} \right.\right.$$

$$\left.\left. - \frac{1}{(1+\alpha\,\sin^2\varphi)\cos^2\varphi} \right]\right\}^{1/2} \qquad (2.6c)$$

and the normalized parameters

$$k = (k_{33} - k_{11})/k_{11}, \quad \gamma = (\epsilon_{11} - \epsilon_\perp)/\epsilon_\perp, \quad \alpha = (k_{33} - k_{22})/k_{22} \qquad (2.7)$$

A computer program has been written to solve equations (2.6a), (2.6b) for values of φ_m, β given an applied voltage $\overline{U}$ exceeding threshold and a maximum twist angle ω_m. The first phase of the program determines φ_m, β by a Levenberg-Marquardt optimization algorithm, ZXSSQ, in the IMSL Library. The maximum tilt angle φ_m will occur at the middle of the cell where boundary anchoring forces are least effective.

The second phase of the program uses the values of φ_m, β obtained in phase 1 to determine the local tilt angle $\varphi(Z/L)$ and local twist angle $\omega(Z/L)$ of the liquid crystal molecules. As shown by Deuling, the following relation determines $\varphi(Z/L)$, implicitly as a function of depth in the layer.

$$Z/L = \frac{1}{2} \int_0^{\varphi(Z/L)} \frac{\sqrt{1+k\,\sin^2\varphi}}{g(\varphi)}\,d\varphi \bigg/ \int_0^{\varphi_m} \frac{\sqrt{1+k\,\sin^2\varphi}}{g(\varphi)}\,d\varphi \qquad (2.8)$$

Using this local tilt angle, the corresponding local twist angle is obtained from the formula

$$\omega(Z/L) = \beta \int_0^{\varphi(Z/L)} \frac{\sqrt{1+k\,\sin^2\varphi}}{g(\varphi)\cos^2\varphi(1+\alpha\,\sin^2\varphi)}\,d\varphi \qquad (2.9)$$

3. Numerical Aspects

Examination of the function (2.6c) reveals that both integrals in (2.6a), (2.6b) are singular at the upper limit. Indeed, both integrals are of order $O(|\varphi-\varphi_m|^{-1/2})$ as $\varphi \to \varphi_m$ and thus are convergent. This behavior suggests the introduction of Gauss – Jacobi quadrature formulas. The latter are designed to yield optimal estimates for integrals of the form $\int_{-1}^{1} f(x)\omega(x)dx$ where f is sufficiently smooth and $\omega(x) = (1-x)^\mu(1+x)^\nu$ with $\mu > -1$, $\nu > -1$, is a nonnegative integrable weight function. Information on Gauss – Jacobi quadrature is available in Krylov [6]. A linear transformation $\varphi = \varphi_m(s+1)/2$ is used to map the integrals (2.6a), (2.6b) onto the range $-1 \leq s \leq 1$. In terms of the mapped domain, equations (2.6a), (2.6b) become

$$\omega_m = \beta\sqrt{2\varphi_m} \int_{-1}^{1} \sqrt{\frac{\frac{\varphi_m}{2}(1-s)}{\sin\frac{\varphi_m}{2}(1-s)}} \frac{\sqrt{1+k\,\sin^2\frac{\varphi_m}{2}(s+1)}}{\sqrt{\sin\frac{\varphi_m}{2}(s+3)}}$$

$$\cdot \frac{1}{\left[1+\alpha\,\sin^2\frac{\varphi_m(s+1)}{2}\right]\cos^2\frac{\varphi_m(s+1)}{2}\,h(s)} \frac{ds}{\sqrt{1-s}} \qquad (3.1a)$$

$$\overline{U} = \frac{\sqrt{2\varphi_m}}{\pi} \int_{-1}^{1} \sqrt{\frac{\frac{\varphi_m}{2}(1-s)}{\sin\frac{\varphi_m}{2}(1-s)}} \frac{\sqrt{1+k\,\sin^2\frac{\varphi_m(s+1)}{2}}}{\sqrt{\sin\frac{\varphi_m}{2}(s+3)}}$$

$$\cdot \frac{1}{\left[1+\gamma\,\sin^2\frac{\varphi_m(s+1)}{2}\right]h(s)} \frac{ds}{\sqrt{1-s}} \qquad (3.1b)$$

with the corresponding expression to (2.6c)

$$h(s) = \left\{ \frac{1}{\left[1+\gamma \sin^2 \frac{\varphi_m}{2}(s+1)\right](1+\gamma \sin^2 \varphi_m)} \right.$$

$$\left. + \beta^2 \left(\frac{1+k}{1+\alpha}\right) \frac{1+\alpha\left[\sin^2 \frac{\varphi_m(s+1)}{2} - \cos^2 \varphi_m\right]}{\left[1+\alpha \sin^2 \frac{\varphi_m(s+1)}{2}\right]\left[1+\alpha \sin^2 \varphi_m\right]} \right.$$

$$\left. \cdot \frac{1}{\cos^2 \frac{\varphi_m(s+1)}{2}} \cdot \frac{1}{\cos^2 \varphi_m} \right\}^{1/2} \qquad (3.1c)$$

The system of integral equations (3.1a), (3.1b) is approximated by a system of nonlinear transcendental equations obtained by applying Gauss - Jacobi quadrature formulas to the integrals occurring therein. In practice we have used both six- and 12- point quadrature formulas with similar results. Values of φ_m, β are obtained by applying nonlinear optimization routines in the IMSL and NAG libraries. We have used the following routines in numerical work

1) E04HBF, E04JBF: A quasi-newton routine requiring only function values for determining a minimum of a function subject to fixed bounds on the independent variables.

2) ZXSSQ: A Levenberg - Marquardt nonlinear least squares routine in the IMSL Library.

The basic idea is to start with initial estimates of φ_m, β and refine them iteratively until the least squares error

$$\|\underline{R}\|^2 = |R_1|^2 + |R_2|^2 \qquad (3.2)$$

is minimized. The expressions

$$R_1 = f_1(\varphi_m, \beta) - \omega_m, \quad R_2 = f_2(\varphi_m, \beta) - \overline{U}$$

are the residuals corresponding to the current estimates of φ_m, β. The integrals occurring in the numerators of (2.8), (2.9) are regular for

$0 \leq Z/L \leq 1/2$ and are evaluated using a twelve-point Gauss - Legendre formula. If X is an initial estimate of the parameter vector $\underline{X} = [\varphi_m, \beta]$, a sequence of estimates is generated by the iteration

$$\underline{X}^{(n+1)} = \underline{X}^{(n)} - [\alpha_n D_n + J_n^T J_n]^{-1} J_n^T R(\underline{X}^{(n)}) \qquad (3.4)$$

Here J_n is an estimate of the Jacobian matrix $\dfrac{\partial(f_1, f_2)}{\partial(\varphi_m, \beta)}$ at $\underline{X}^{(n)}$, D_n is a diagonal matrix matching the diagonal of $J_n^T J_n$ and α_n is a positive scalar called the Marquardt parameter. In IMSL routine ZXSSQ, strict descent towards the minimum is enforced by increasing α_n until a smaller least squares error is obtained.

4. Discussion of Results

Equations (2.6a), (2.6b) were solved for various sets of values of ω_m and $\bar{U}$. As a test we repeated Deuling's calculations with parameter values $\alpha = 0.6$, $\varkappa = 0.8$, $\gamma = 0$. The results of the computer runs were identical to those displayed in Deuling. We also used a parameter range $1 \leq \bar{U} \leq 6$, $0 \leq \omega_m \leq \pi/2$ typical of those used in display devices. The following physical parameters were used

$$k_{11} = 9.5 \times 10^{-12}, \quad k_{22} = 8 \times 10^{-12}, \quad k_{33} = 18 \times 10^{-12} \text{ M.K.S.}$$

$$\varepsilon_{11} = 16, \quad \varepsilon_\perp = 5, \quad \varkappa = 0.89, \quad \alpha = 1.25, \quad \gamma = 2.2, \quad U_0 = 0.98 \text{ volts.}$$

These correspond to the E7 liquid crystal mixture. With reasonable initial parameters, the algorithm typically converged to machine accuracy within 10-15 iterations.

The IMSL routine ZXSSQ supplies values of the residual least squares error, residuals for the final parameter estimates, the L_2 norm of the gradient vector $\underline{\nabla} \|R\|^2$, and an estimate of the number of significant figures in the parameter estimates. In addition, plots of φ, ω as functions of depth Z/L are produced for each case, and the gradual steepening one expects with

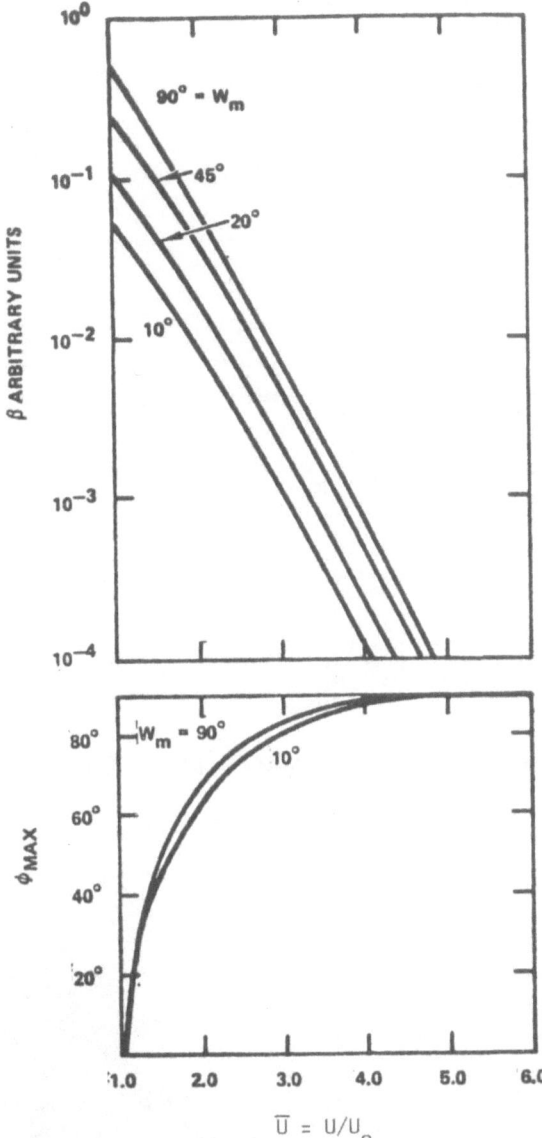

Fig. 1   Maximum tilt angle φ_m and integration parameter β
vs. reduced voltage U/U_o for various values of
maximum twist angle ω_m

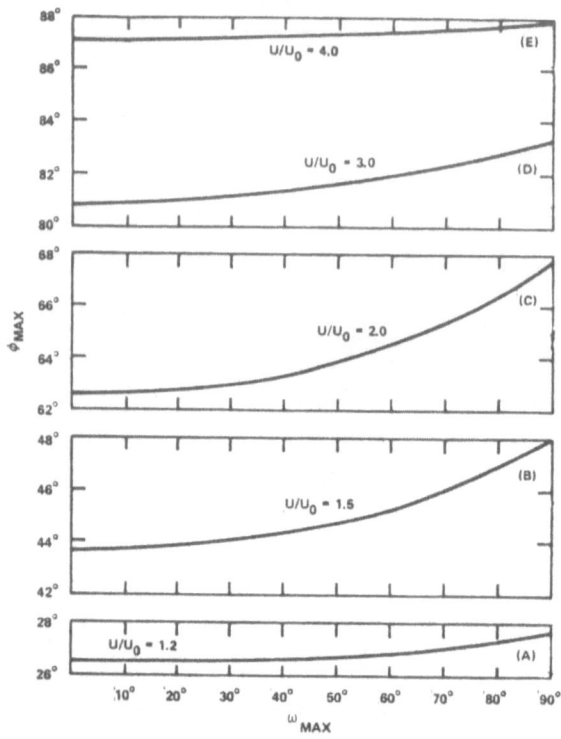

Fig. 2   Maximum tilt angle ω_m vs. maximum twist angle ω_m for various values of reduced voltage U/Uo.

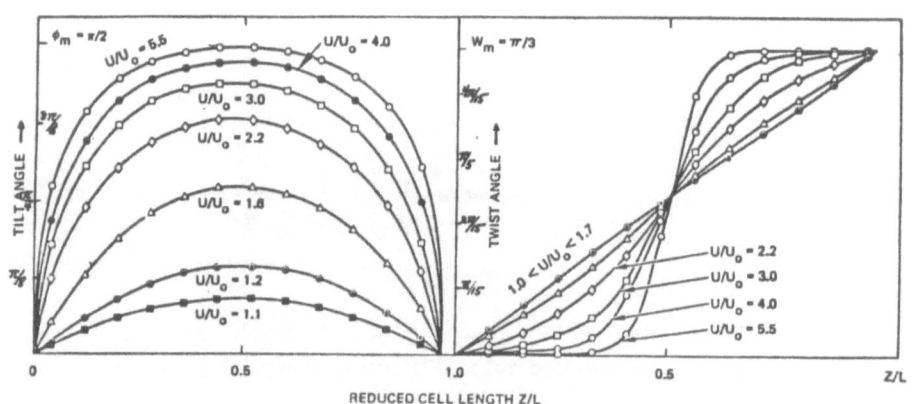

Fig. 3   Tilt angle and twist angle vs. reduced cell length z/L for maximum twist angle $\pi/3$ radians.

increasing voltage is apparent. Values of the residual sum of squares ranged from 10^{-27} to 10^{-32}. Residuals for final parameter estimates were of order 10^{-14}. All of these measures yield confidence in the results.

In Figure 1 we have plotted the variation of β and φ_m with applied voltage $\bar{U}$ for the E7 liquid crystal mixture. The maximum tilt angle φ_m reaches a saturation at $\bar{U} > 4.0$. It depends only weakly on ω_m. However, β changes exponentially with $\bar{U}$ and is highly sensitive to maximum twist ω_m.

In Figures 2(A) through 2(E), we show the dependence of φ_m on ω_m for various applied voltages $\bar{U} = 1.2, 1.5, 2.0, 3.0, 4.0$. At voltages near threshold, $1 \leq \bar{U} \leq 1.2$, or near saturation, φ_m increases slightly with ω_m. At intermediate voltages, there is a ten percent increase in φ_m with increasing ω_m.

Using the previously tabulated values of φ_m, β the program generates twist and tilt angles of the liquid crystal molecules along the cell. An example with $\omega_m = 60^\circ$ is shown in Figure 3. The electric field tilts the molecules at the center of the cell and the twist spiral starts to break down. At intermediate voltages, the twist angle is split into three domains. Near each electrode, the molecules tend to achieve parallel alignment because of the surface anchoring treatment. At the central portion of the cell, the molecules transfer the twist. This regime contains highly tilted molecules whose orientation becomes perpendicular as the voltage is increased. At higher voltages, the high tilt regime extends along most of the cell.

5. Conclusion

A numerical optimization algorithm available in the IMSL and NAG Libraries was used to calculate the deformation pattern of liquid crystal (LC) molecules in a twisted nematic cell subjected to an electric field. The numerical computation of tilt and twist angles of the (LC) molecules

522

was used to determine how the final deformation is related to the applied field and the maximum twist angle prealignment. The results replicate the computations reported by Deuling [1].

References

[1] Deuling, H.J.: Deformation pattern of twisted nematic layers in an electric field. Mol. Cryst. Liq. Cryst. $\underline{27}$ (1974) 81-93.

[2] Frank, F.C.: Disc. Faraday Soc. $\underline{25}$ (1958) 19-28.

[3] Leslie, F.M.: Distortion of twisted orientation patterns in liquid crystals by magnetic fields. Mol. Cryst. Liq. Cryst. $\underline{12}$ (1970) 57-72.

[4] Oseen, C.W.: Ark. Mat. Astron. Fys. $\underline{19}$ (1925).

[5] Schadt, M.; Helfrich, W.: Voltage-dependent optical activity of a twisted nematic liquid crystal. Appl. Phys. Lett. $\underline{18}$ (1971) 127-128.

[6] Krylov, V.I.: Approximate Calculation of Integrals, Macmillan (1962) 111-121.

1325 El Encanto Drive,

Brea,

California 92621-1918,

U.S.A.

LACTEO: A dairy management and forecasting system

D. Sprevak   R. S. Ferguson

Abstract

A forecasting procedure for the total milk yield of a dairy cow which uses the existing information on the average characteristics of a herd and the information on lactation yield, as it becomes available, is shown to give reliable estimates of the total milk yield early in the lactation.

1   Introduction

This paper describes a low cost turnkey package based on a hand held microcomputer which was designed for the storage and analysis of milk yields of dairy cows. An important facility in the package is its ability to predict the total milk yield for a cow very early in the lactation. Predictions of the total yield after 15 weeks of lactation are generally within 5 % of the actual yield produced. The paper is mainly concerned with the description of the algorithm for total milk yield prediction.

To complement the prediction algorithm, a comprehensive data acquistion system and database for herd management was developed. This complete package frees the 'farmer' from the necessity of a special recording device, and a personal computer for analysis. The whole hardare/software package is built around a PSION organiser that has the capability to obtain and store all the information necessary for the prediction algorithms for a herd of up to 200 cows.

The structure of the software is described in Figure 1 , it consists of three sections. The data acquisition section, the herd management section and the forecasting section.

J. Manley et al. (eds.), Proceedings of the Third European Conference on Mathematics in Industry, 523–530.
© 1990 Kluwer Academic Publishers and B. G. Teubner Stuttgart.

524

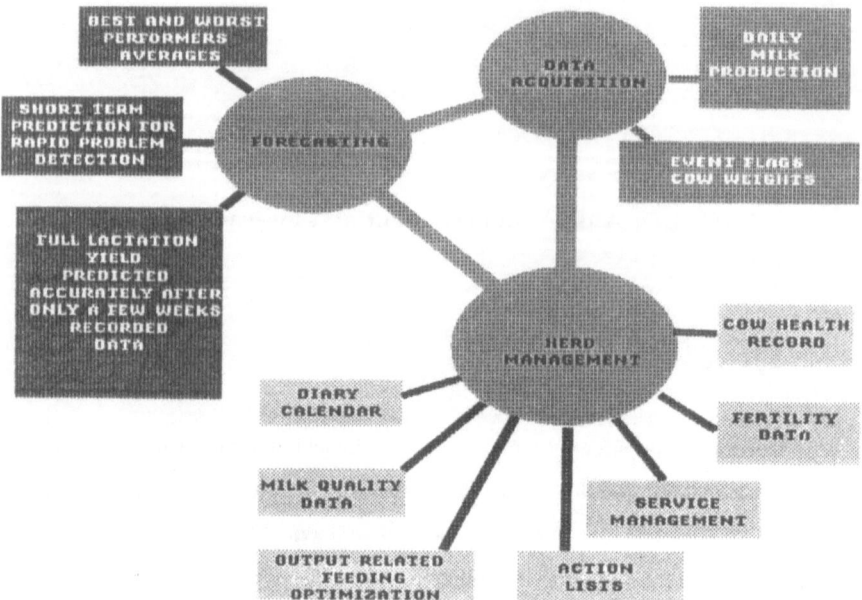

Figure 1: Diagram of the structure of the computer package LACTEO.

2   The forcasting of milk yield

A mathematical model of the lactation curve of a dairy cow was proposed by
Wood [5] [6] which relates the daily milk yield y to the lactation week t in the
form

$$y(t) = at^b \exp(-ct)(1 + \epsilon(t)) \tag{1}$$

where a, b and c are parameters which depend on the cow and $\epsilon(t)$ is a random term
which takes account of the variations about the lactation curve. It is convinient to
linearise the model by a logarithmic transformation, thus

$$\ln y(t) = \ln(a) + b\ln(t) - ct + \ln(1 + \epsilon(t)) \tag{2}$$

Goodall and Sprevak [3] have shown that the term $\ln(1 + \epsilon(t))$ forms a correlated
series and that the series can be modelled with a first order autoregressive model
of the form :

$$\ln(1 + \epsilon(t)) = \alpha \ln(1 + \epsilon(t-1)) + \epsilon'(t) \tag{3}$$

where the term $\epsilon'(t)$ is a series of independent normally distributed random variables with zero mean and variance σ^2 .

Goodall and Sprevak [4] proposed a recursive algorithm for the estimation of $\ln(a), b, c$ and $\ln(1 + \epsilon(t))$, given the observations $\ln y(t)$ up to time t . As new observations become available the estimates are upgraded recursively. The method also requires an initial guess for the values of the unknown parameters and their variance covariance matrix. For this a-priori information we used the averaged values of the parameters and their variance covariance matrix which were obtained for the whole herd on the previous year's lactation.

In this paper we extend the development of the algorithm so it can apply to Autumn calving cows. Cows calving between late September and late November are commonly referred to as Autumn calvers. These cows show a trend change in milk yield in April, when they go out to grass. Goodall [2] proposed to account for the effect of date of calving by adding an extra parameter to the model given by equation (2) to give the new equation, whose parameters need to be re-estimated,:

$$\ln y(t) = \ln a + b \ln t - ct + dD + \ln(1 + \epsilon(t)) \tag{4}$$

where D is a categorical variable which is set to zero for October to March production and to one for April to September production. To apply the estimation procedure to the parameters of the model given by equation (4) one simply uses the equations given in Goodall and Sprevak [4] with a change in the dynamic equations which define the linear model for the lactation curve, which for the Autumn calving cows are given by :

$$\theta_t = \begin{bmatrix} \ln a \\ b \\ c \\ d \\ \ln(1 + \epsilon(t)) \end{bmatrix} = \begin{bmatrix} 1 & 0 & 0 & 0 & 0 \\ 0 & 1 & 0 & 0 & 0 \\ 0 & 0 & 1 & 0 & 0 \\ 0 & 0 & 0 & 1 & 0 \\ 0 & 0 & 0 & 0 & a \end{bmatrix} \begin{bmatrix} \ln A \\ b \\ c \\ d \\ \ln(1 + \epsilon(t-1)) \end{bmatrix} + \begin{bmatrix} 0 \\ 0 \\ 0 \\ 0 \\ \epsilon'(t-1) \end{bmatrix} \tag{5}$$

$$Y_t = \begin{bmatrix} 1 & \ln t & -t & D & 1 \end{bmatrix} \theta_t \tag{6}$$

Where the variable

$$D = \begin{cases} 0 & \text{if } t < 27 \\ 1 & \text{otherwise} \end{cases} \tag{7}$$

The initial values are taken as the average characteristics of the herd, i.e.

$$
\hat{\theta}_0 = \begin{bmatrix} 2.841 \\ 0.366 \\ 0.053 \\ 0.178 \\ 0.000 \end{bmatrix}, \hat{s}_0 = \begin{bmatrix} 0.305 \\ 0.176 \\ 0.016 \\ 0.145 \\ 0.618 \end{bmatrix}, \hat{G}_0 = H_0 \begin{bmatrix} 1 & -0.8 & -0.32 & 0.28 & 0 \\ -0.8 & 1 & 0.8 & 0.5 & 0 \\ -0.32 & 0.8 & 1 & 0.2 & 0 \\ 0.28 & 0.05 & 0.2 & 1 & 0 \\ 0 & 0 & 0 & 0 & 1 \end{bmatrix} H_0
$$

(8)

where H_0 is a diagonal matrix whose diagonal elements are those of $\hat{s}_0$, and the element $\hat{G}(5,5)$ was set to its estimated asymptotic value, i.e. $\sigma^2/(1 - \alpha^2)$.

The other two parameters which completely specify the system were also set to the estimated values for the previous year's data. With $\alpha = 0.69$ and $\sigma^2 = 0.043$. Goodall and Sprevak [3] have found that the 95 % confidence interval for α is very wide and therefore a fixed value for α for the whole herd is a sensible choice. However the algorithm is very sensitive to the choice of the value for σ^2. If σ^2 is set too small the Kalman filter sometimes leads to divergence of estimating errors. This situation occurs when the estimated covariance values become smaller than the actual covariances. If on the other hand σ^2 is set too high the performance of the filter is degraded through poor estimates of the unknown parameters. It was found that by estimating σ^2 recursively, as data is received, the filter behaves in a stable manner.

To test the response of the estimation algorithm 'synthetic' lactation data, having particular statistical characteristics, was generated using a simulation technique. For the different cows in a herd, the parameters which specify a lactation curve (a, b, c, D, α) can be considered as the outcome of a multivariate normal random variable with a mean vector estimated by the vector sample mean and a variance-covariance matrix estimated with the sample variance-covariance matrix for the herd. For a herd of Autumn calving cows from the Agricultural Research Institute of Northern Ireland at Hillsborough it was found for the vectors of means:

$$
\mu^T = \begin{bmatrix} 2.841 & 0.366 & 0.053 & 0.178 & 0.69 \end{bmatrix},
$$

(9)

for the vector of standard deviations:

$$
s^T = \begin{bmatrix} 0.305 & 0.176 & 0.016 & 0.145 & 0.245 \end{bmatrix}
$$

(10)

where the superscript T denotes the transpose of the matrix.

And for the upper half of the (symmetric) correlation matrix:

$$R = \begin{bmatrix} 1 & -0.8 & -0.32 & 0.28 & 0.15 \\ & 1 & 0.8 & 0.5 & -0.10 \\ & & 1 & 0.2 & 0.20 \\ & & & 1 & 0.5 \\ & & & & 1 \end{bmatrix} \qquad (11)$$

The variance-covariance matrix Σ is obtained from

$$\Sigma = HRH \qquad (12)$$

Where H is a diagonal matrix whose diagonal elements are those of s . Thus a random sample from a multivariate normal distribution $N(\mu, \Sigma)$ would output a set of values for the parameters of a given lactation curve. To generate such sample X we use the transformation

$$X = \mu + BU \qquad (13)$$

where U is a vector of independent normally distributed, zero mean, variance one, random variables, with the same dimensions of X ; and B is the lower diagonal matrix obtained from the transformation, Chatfield and Collins [1]

$$\Sigma = BB^T \qquad (14)$$

Then a set of independent random variables are generated by standard techniques to form the components of U and using equation (13) we obtain the set X which defines the parameters of a lactation curve. The simulation of milk yields is finally accomplished by adding to the lactation curve values sampled according to the autoregressive process (3). Figure 2 shows the simulated values for a given sample together with real lactation data and a sample of simulated lactation curves.

The pictures show the ability of the simulation method to generate realistic synthetic data. The simulation technique was used to generate data corresponding to a herd of 50 cows. Two types of data were generated. One type corresponding to spring calving cows, the other corresponding to Autumn calving cows, differing from the first one only by the change in trend after the 27th week of lactation.

528

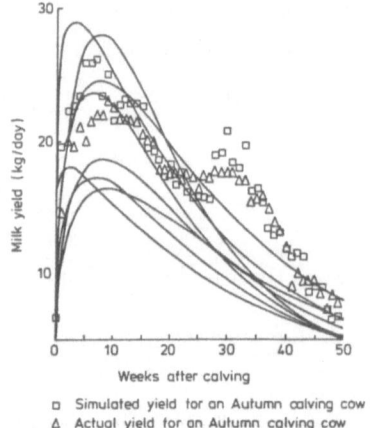

Figure 2: Simulated lactation curves. Note the change in trend for the plotted data occurring at about the 27th week of lactation which is characteristic of Autumn calving cows.

The prediction algorithm was used to estimate the the total milk yield for these data. Figure 3 shows the box-plot for the calculated relative error of prediction for the estimation of the total milk yield as a function of the number of weeks after calving. Figure 4 shows the equivalent graph for a real herd of 50 cows, in this case the date of calving varies from cow to cow and is approximately uniform over the year.

The simulated lactation data shows that the effect of the change of trend which occurs when the Autumn cows go out to grass does not significantly affect the performance of the forecasting algorithm. The results presented here indicate that the forecasting algorithm gives reliable estimates of the total milk yield early in the lactation.

3   Conclusions

The forecasting algorithm based in a form of recursive least squares with the addition of a-priori information on the average characteristics of the herd was shown to give reliable estimates of the total milk yield of a dairy cow early in the lactation. However, for the algorithm to be used in the dairy industry it needs to be presented

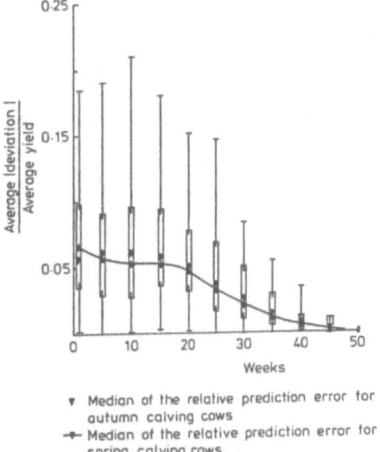

Figure 3: Results for the simulated herd. Box and whisker plot of the relative error of estimation of total milk yield using the forecasting algorithm for the autumn calving cows.

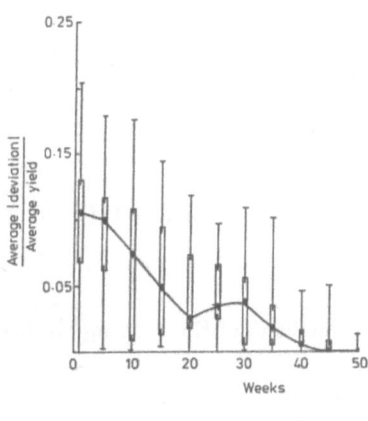

Figure 4: Results for the real herd. Box and whisker plot of the relative error of estimation of total milk yield using the forecasting algorithm for cows calving the year round.

in a system which provides the farmer with a complete self-contained low-cost package. The package should allow the user to conduct his entire herd management analysis. To that extent Lacteo has been developed and it is at present under trial with the Agricultural Research Institute of Northern Ireland at Hillsborough.

4 Acknowledgements

The authors thank Dr. S. Mayne from the Agricultural Research Institute of Northern Ireland at Hillsborough for the data on milk yields.

5 References

[1] Chatfield,C. and Collins,A.J.,1980.Introduction to multivariate analysis. Chapman and Hall, London.

[2] Goodall,A.E.,1983. An analysis of seasonality of milk yield production.Anim.Prod. 36:69-72.

[3] Goodall,A.E. and Sprevak D., 1984. A note on a stochastic model to describe the milk yield of a dairy cow. Anim.Prod. 38:133- 136.

[4] Goodall,A.E. and Sprevak D., 1985. A Bayesian estimation of the lactation curve of a dairy cow. Anim.Prod. 40:189-193.

[5] Wood,P.D.P.,1967. Algebraic model of the lactation curve in cattle. Nature, 216:164-165.

[6] Wood, P.D.P., 1969. Factors affecting the shape of the lactation curve in cattle., Animal Production, 11, 307-316 .

- D. Sprevak, is with the Department of Engineering Mathematics The Queen's University of Belfast

- R. S. Ferguson, is with the Department of Electrical and Electronic Engineering The Queen's University of Belfast

Identification of Amplitude and Phase Discontinuities in the Intensity Signal of a Nd-YAG solid state laser

Hans-Georg Stark

Department of Mathematics

University of Kaiserslautern

Abstract: The intensity signal of a Nd-YAG solid state laser can be used for calibration purposes. Some irregularities in this signal are identified based on a physically motivated description of the emitted intensity. This leads to a nonlinear parameter optimization problem.

Keywords: Signal-processing, ultrashort laser pulses, parameter optimization

1. Introduction

In calibrating optical devices on uses the signal of a Nd-YAG solid state laser, which is operated in the socalled "Q-switch mode": the corresponding intensity signal consists of a short peak, on which some oscillatory signal is superimposed (the signal analyzed here is shown in fig. 1). For calibration purposes these oscillations are undesirable and they can be compensated by a hardware set up. This compensation procedure works effectively only, if no sudden changes in amplitude and/or phase of these oscillations occur. But such discontinuities may happen due to physical circumstances. For calibrating it is then necessary, to obtain the differences in amplitude and phase before and after the discontinuity. By experience only the first two harmonics contribute to the oscillations and the following model function for the measured intensity is suggested:

$$J(t) = J_0(t) + J_1(t) \sin(\omega_1 t + \phi_1) + J_2(t) \sin(\omega_2 t + \phi_2), \quad (1.1)$$

where $J_0(t)$ denotes the basic signal, $\omega_2 \approx 2\omega_1$ and the frequencies are assumed to be constant in the region of interest (which is not fulfilled exactly, see below). Let t_J denote the (known!) time, where the discontinuities occur and the problem may then be formulated as: Determine ΔJ_i, $\Delta\phi_i (i=1,2)$, where

$$\Delta J_i = J_i(t_J - 0) - J_i(t_J + 0) \qquad (1.2)$$

J. Manley et al. (eds.), Proceedings of the Third European Conference on Mathematics in Industry, 531–537.
© 1990 Kluwer Academic Publishers and B. G. Teubner Stuttgart.

532

$$\Delta\Phi_i = (\omega_i t + \phi_i)^{before} - (\omega_i t + \phi_i)^{after} \qquad (1.3)$$

(we have written the expression for the phase discontinuity in the general form (1.3) in order to account for possible slight changes in the local frequency).

The Fourier spectrum of the signal (for the spectrum of our signal from fig. 1 see fig. 2) shows, that the ω_1 and ω_2 parts are relatively well separated, which would suggest the following procedure: Reconstruct $J_i(t)$ $\sin(\omega_i t + \phi_i)$ from the corresponding peak in the frequency spectrum only and try to approximate $J_i(t)$ before and after t_J by some polynomial (in [1] polynomials of second and fourth degree have been used) with a least squares fit. This method has two drawbacks. The reconstructed functions depend too strongly on the particular interval around the frequencies ω_i, taken for reconstruction (i.e.: the ω_i-peaks in the Fourier spectrum are not separated well enough for this purpose). Secondly it turned out, that the shape of the envelopes J_i cannot be approximated satisfactorily by polynomial functions.

Thus we tried, to approximate the full signal (1.1) by model functions, taking into account the physics behind this signal, as described below. We were able, to improve the approximation in the amplitudes in this way, but could not improve the results in phase, as compared to [1], due to the non stationarity of the frequencies.

2. The model function

A solid state laser emits electromagnetic radiation in modes with discrete frequencies $\tilde{\omega}_i = \tilde{\omega}_0 + i\Delta\omega$ ($i \in M \subset Z$, $|i\Delta\omega| \ll \tilde{\omega}_0$ for all $i \in M$). With the electromagnetic field

$$\vec{E}_i(t) = \vec{e}_i \sin(\tilde{\omega}_i t + \psi_i) \qquad (2.1)$$

one obtains for the effective intensity, averaged over one period,

$$I_i(t) = \frac{\tilde{\omega}_0}{2\pi} \int_t^{\frac{2\pi}{\omega_0}+t} \vec{E}_i^2 d\tau = \frac{\vec{e}_i^2}{2}, \qquad (2.2)$$

where $|i\Delta\omega|\ll\tilde{\omega}_0$ has been used. The resulting electromagnetic field $\vec{E} = \sum_i \vec{E}_i$ leads with $|i\Delta\omega|\ll\tilde{\omega}_0$ to a total effective intensity

$$I = \frac{1}{2}\sum_i e_i^2 + \sum_{n>k} \vec{e}_n\vec{e}_k \sin((n-k)\Delta\omega t + \phi_{nk}). \qquad (2.3)$$

This can be written as

$$I = \sum_i I_i + \sum_{n>k} \alpha_{nk}\sqrt{I_n}\sqrt{I_k} \sin((n-k)\Delta\omega t + \phi_{nk}). \qquad (2.4)$$

This explains the form (1.1): J_0 contains the sum of all partial intensities, the oscillations occur due to interference between modes with different frequencies. Of course all of this is well known and may be found in many books on laser physics [2].

Assuming, that different modes essentially retain the same shape (i.e. $I_k(t)/I_1(t) \approx$ const for $k\neq l$) and taking the empirical fact, that only the first two harmonics contribute, into consideration, (2.4) motivates the following ansatz for the total intensity (1.1):

$$I(t) = J_0(t) + J_1(t) \sin(\Delta\omega t + \phi_1) + J_2(t) \sin(2\Delta\omega t + \phi_2) \qquad (2.5)$$

The amplitudes $J_k(t)$ are hereby assumed to be proportional to the intensity I of a __single__ mode. For a solid state laser, operated in the Q-switch mode, one has [2]

$$I(t) \sim (n_a - n(t) + n_s \ln \frac{n(t)}{n_a}) \qquad (2.6)$$

Here n(t) denotes the socalled population inversion, whose time dependence during a laser pulse qualitatively is depicted in fig. 3. I reaches its maximum at $n(t) = n_s$, the end inversion is obtained as the solution of the equation

$$n_a - n_e + n_s \ln(n_e/n_a) = 0 \qquad (2.7)$$

Decay processes as for the population inversion often can be modelled by a Weibull type distribution, which in our case leads to the ansatz

$$n(t) = (n_a - n_e) \exp(-(t/\tau)^\beta) + n_e. \qquad (2.8)$$

Thus J_k in (2.5) can be parametrized by a set $(n_{a,k}; n_{s,k}; \tau_k; \beta_k)$ $(k=0,1,2)$ and reads

$$J_k(t) = (n_{a,k} - n_k(t) + n_{s,k} \ln(n_k(t)/n_{a,k})). \qquad (2.9)$$

In this formula $n_k(t)$ is defined as in (2.8) with $n_{e,k}$ obtained from $n_{a,k}$ and $n_{s,k}$ by solving eq. (2.7), which can be done very easily using a Newton iteration scheme.

If, moreover, the frequencies and phases in (2.5) are considered as parameters also, in order to admit small deviations from the ideal relation in (2.5), the model function (2.5) is described by 16 parameters. This model function has to be adjusted to the measured intensity by a least squares fit before and after the discontinuity in the signal and of course this procedure leads to a highly nonlinear parameter optimization problem in the parameters describing the model function (2.5). To this end we employ a Levenberg-Marquardt-type optimization procedure in a version modified by R. Fletcher [3], which is especially well adapted to the minimization of square sums. Since the quality of the approximation depends essentially on the choice of "good" initial values for the parameters, an algorithm was developped to determine reasonable initial values out of the measured data. Unfortunately we cannot describe it here due to lack of space. For details refer to [4]. Of course it is also possible, to start the iteration with initial values, which proved to be useful by experience.

3. Examples
In order to determine the ΔJ_i, $\Delta \Phi_i$ defined in (1.2) and (1.3) respectively, one chooses with $t_l < t_J < t_r$ two intervals $[t_l, t_J]$ and $[t_J, t_r]$, which are small enough, such that the assumption of constant frequency in (2.5) can be justified. The model

function (2.5) is then adjusted to the measured signal on these intervals and $\Delta J_i, \Delta \Phi_i$ are computed from the corresponding parts of (2.5) as described in (1.2) and (1.3). Usually there occurs no discontinuity in the smooth part of the signal described by J_o. The complexity of the algorithm is then reduced as follows: Reconstruct the smooth part from the Fourier spectrum by retaining the low frequency part only, which usually is separated quite well from the ω_i-peaks. Fit J_o to this signal on $[t_1, t_r]$ and keep the resulting parameters ($n_{a,o}$; $n_{s,o}$; τ_o; β_o) fixed in the subsequent approximations on $[t_1, t_J]$ and $[t_J, t_r]$.

This procedure has been applied to the signal shown in fig. 1 with $t_J = 70$. Figures 4 and 5 show the original signal (——) together with its approximation (–o——o–) on the respective intervals before and after the discontinuity. The plots show a reasonable agreement of the measured and fitted intensities.

In order to check the reliability of the calculated ΔJ_i-values, the procedure was also applied to a signal, which has been obtained by adding together three contributions corresponding to the basic peak and the two oscillatory parts. An artificial discontinuity was inserted, by multiplying the oscillations with 0.7 for all times greater than some t_J. The relative decrease in the fitted amplitudes J_k was 0.54 for $k = 1$ and 0.75 for $k = 2$. Unfortunately we have not been able, to get comparable results for the phase discontinuity due to non stationarity of the frequencies.

Acknowledgement:

I thank B. Claus, who had the tedious task of programming and Dr. K. Schreiner for supplying us with a programmed version of the optimization procedure in [3].

References

[1] W. Krüger, M. Schulz-Reese: Preliminary report of the technomathematics group, Fachbereich Mathematik der Universität Kaiserslautern 1986, unpublished

[2] H. Weber: Laser Physics, Lecture notes, Fachbereich Physik der Universität Kaiserslautern, 1984/85 and references quoted therein, in particular

536

S.L. Shapiro: Ultrashort Light Pulses, Topics in Applied Physics, Vol. 18, Springer Verlag, Berlin, Heidelberg, New York, 1977

A. Yariv: Quantum Electronics, 2. ed., J. Wiley, New York, 1975

[3] R. Fletcher: A Modified Marquardt Subroutine for Non-Linear Least Squares, Report R. 6799, AERE, Harwell, England (1971).

[4] H.-G. Stark: Report no. 23 of the technomathematics group, Fachbereich Mathematik der Universität Kaiserslautern, 1987

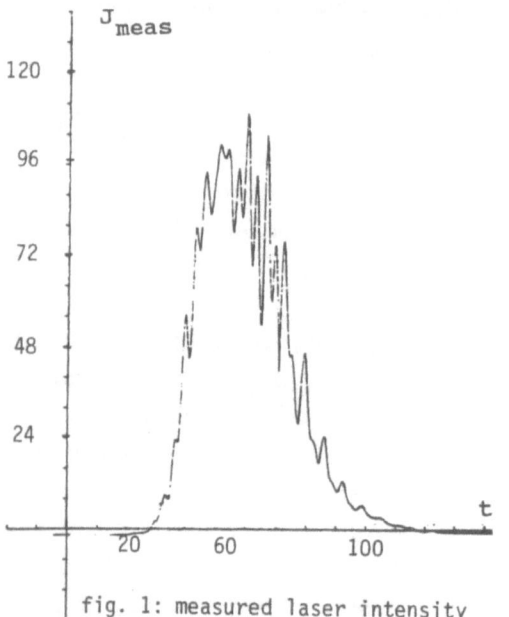

fig. 1: measured laser intensity

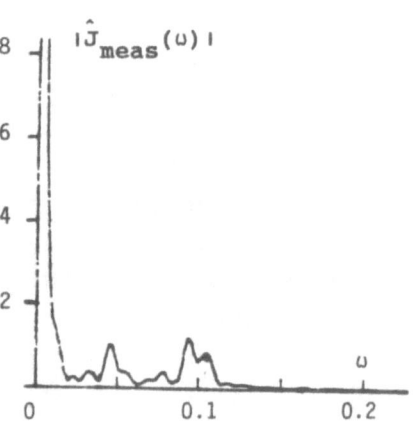

fig. 2: Fourier spectrum

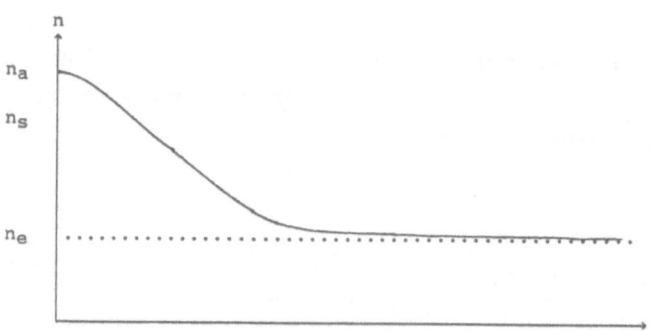

fig. 3: population inversion

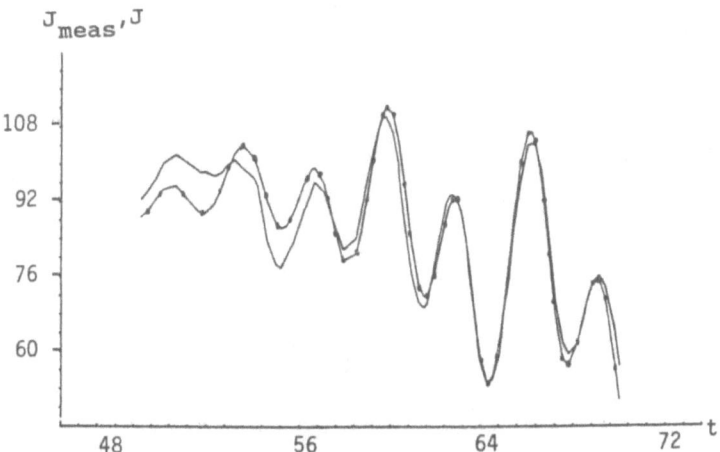

fig. 4: measured (-) and fitted (θ─θ) intensity
before discontinuity

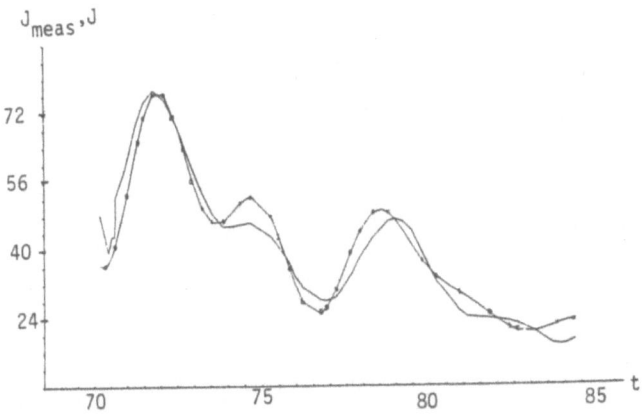

fig. '5: measured (-) and fitted (θ─θ) intensity
after discontinuity

Author's adress:

H.-G. Stark

Fachbereich Mathematik der Universität Kaiserslautern

Postfach 3049

D-6750 Kaiserslautern

FRG

GEL ELECTROPHORESIS AND GRAPH MATCHING

by

D.Swailes and S. McKee

University of Strathclyde
Department of Mathematics
Glasgow, Scotland.

1. Introduction

Two-dimensional gel electrophoresis is an effective tool for the separation
of protein macro-molecules. In particular, two-dimensional gel
electrophoresis systems which give good resolution for separating a complex
mixture of proteins combine the technique of isoelectric focusing in the
presence of a neutral detergent in the first dimension and slab gel
electrophoresis using sodium dodecyl sulfate in the second dimension [1].

Proteins have an isoelectric point which causes them to be negatively
charged in solutions where the pH is above their isoelectric point and
positively charged in solutions where the pH is below their isoelectric
point. The separation makes use of this property and movement of the
macromolecules is due to this isoelectric phenomenon, an imposed external
electric field, diffusion, and retardation by the gel. The result is a two-
dimensional map of spots or protein 'blobs' which, in principle, can be used
to characterise the particular plant or animal tissue since a protein group
in essence represents a genetic system. Thus genetically induced diseases
or deficiencies in animals may be recognised by such techniques, and genetic
differences between plant cultivators can be utilised in variety
identification and in plant breeding.

2. Existing Systems

A system for automatic analysis of gel electrophoresis separations must
include the following elements: scanning and image acquisition; filtering,
noise reduction and distortion removal; spot detection; editing spot matching;

J. Manley et al. (eds.), Proceedings of the Third European Conference on Mathematics in Industry, 539–545.
© 1990 *Kluwer Academic Publishers and B. G. Teubner Stuttgart.*

and interpretation. Scanning and image acquisition converts a continuous
(analogue) image into a digital form. Elimination of noise, caused by
protein streaking, by staining (to aid visualisation) and by noise induced
by the equipment itself, can be achieved by a technique similar to the "Top
Hat Transformation" (see [2]). The operator has the option to intervene and
perform an edit or a spot match using his knowledge of protein chemistry.
Spot matching can differ according to the application: the interest here is
matching a collection of spots (current gel) to another collection (target
gel). The final step is the interpretation of the results. Automatic
systems which analyse vast amounts of data from such gel separations and
produce spot matchings have recently been developed e.g. TYCHO [3], QUEST [4],
GELLAB [5], HERMES [6].

   More recently, a description of a system, known as GESA, of which this
work constitutes a part, has been given by Van Hoff et al [7]. Figure 1
illustrates the superposition of the current and target gel after filtering
spot identification and alignment. This contribution is concerned with the

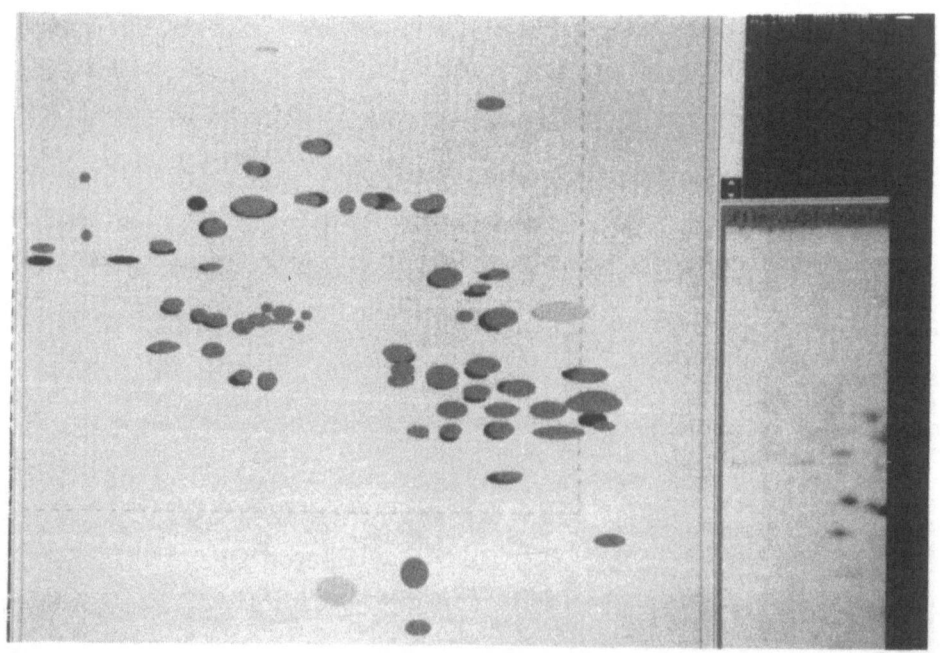

3. Spot Matching

A number of authors have been involved with spot matching, namely [3], and
[5]. Potter [8] described a technique where he used an array processor and
Skolnick ([9], [10]) proposed an automatic comparison technique. Most
recently, Hoff [11] suggested employing Skolnick's ideas with generalised
Gabriel graphics. All these methods involve a certain amount of heuristics
and as such are difficult to analyse rigorously. This paper presents the
first serious attempt at an efficient automatic algorithm which does not rely
on heuristics. The starting point is the realisation that spot matching can
be viewed as a linear assignment problem.

4. Linear Assignment Problem

The first algorithm for the linear assignment problem was due to Kuhn [12].
Other algorithms have been proposed by Barr, Glover and Klingman [13], Hung
and Rom [14], Bertsekar [15] and Balinski [16], Mack [17], Tomizaura [18] and
Dorhout [19] described, developed and improved the shortest augmenting path
algorithm. Most recently, Jonker and Volgenant [20] have derived an
augmenting path algorithm for the linear assignment problem which they claim
to be uniformly faster than those algorithms already mentioned.

5. Gel Electrophoresis Spot Matching Problem

Here we wish to determine if two sets of spots can be matched and, if so, can
we find the best match. Since the two sets of spots may not therefore be
identical and because filtering and spot identification is not exact the two
sets of spots do not contain the same number in general and from this somewhat
unsatisfactory situation we have to decide whether a match should be made.

Given two gels produced by two way gel electrophoresis, let $A = \{a_i\}_{i=1}^m$
represent the set of blobs (spots) on the target gel and $B = \{b_j\}_{j=1}^m$ the set
of blobs on the current gel. If we characterise each blob by the following
attributes: its position (relative to some x-y coordinate system), its area

and its optical density (averaged intensity of staining), then we can write

$$a_i = (\alpha_{11}, \alpha_{12}, \alpha_{13}, \alpha_{14}), \quad i = 1,\ldots,m$$

and

$$b_j = (\beta_{j1}, \beta_{j2}, \beta_{j3}, \beta_{j4}), \quad j = 1,\ldots,n$$

α_{11}, β_{j1} are the x-coordinates of a_i and b_j respectively,

α_{12}, β_{12} are the y-coordinates of a_i and b_j respectively,

α_{13}, β_{13} are the area of a_i and b_j respectively,

α_{14}, β_{14} quantify the density of a_i and b_j respectively.

Since it is reasonable to suppose that vectors corresponding to matching blobs lie 'close' to each other, a set of 'potential' b_j matches for any given a_i can be defined by setting up a window $W = (w_1, w_2, w_3, w_4)$: $b_j \in B$ is then a potential match for $a_i \in A$ if and only if

$$|\alpha_{ik} - \beta_{ik}| \leq \frac{1}{2} w_k, \quad k = 1,2,3,4. \tag{1}$$

This enables us to define an $m \times n$ bipartite graph $G = (A,B,E)$, where $(a_i, b_j) \in E$ if and only if a_i and b_j satisfy (1). We label the edge (a_i, b_j) of G with

$$L(a_i, b_j) = \left[\sum_{k=1}^{4} s_k(\alpha_{ik} - \beta_{jk})^2 \right]^{\frac{1}{2}}$$

where s_k are weighting coefficients.

Obviously the 'best' choice of window is unkn own a priori: too small a window may result in inaccurate matches or no matches at all, while too large a window could result in artificial matches. This led to the idea of defining a sequence of windows lying in a specified range. An upper window $U = (u_1, u_2, u_3, u_4)$ and a lower window $L = (\ell_1, \ell_2, \ell_3, \ell_4)$ with $\ell_i \leq u_i$ are defined, together with a sequence of windows $w^k = (w_1^k, w_2^k, w_3^k, w_4^k)$ by

$$w^0 = L$$

and

$$w^k = w^{k-1} + \frac{1}{N} (U - L) \quad k = 1,\ldots,N$$

where N is an integer determining the number of windows in the sequence.

As above a graph G^k is constructed for each window w^k and the assignment algorithm is applied to each. This results in N+1 sets of matches between the two gels which we record in the form of an $m \times n$ frequency matrix $F = (f_{ij})$, f_{ij} being the number of times a_i and b_j have been matched. Assuming that the larger the value of f_{ij} the more likely it is that b_j is a correct match for a_i the final set of matches is based on the distribution of the larger values of F. This is done by applying the assignment algorithm to find the min (m,n) entries of F, no two in the same row and column, the sum of which is maximum. If these entries are given by f_{i_k}, f_{j_k} $k = 1,\ldots,$min (m,n) we then match a_{i_k} and b_{j_k} provided f_{i_k,j_k} is non-zero, otherwise (if $f_{i_k,j_k} = 0$) we leave a_{i_k} and b_{j_k} unmatched. We also leave unmatched those a_i and b_j not among these min (m,n) entries. A probability measure can be defined and this used to decide the acceptability of a particular match (see [21]).

6. Concluding Remarks

A succinct description of spot matching in gel electrophoresis has been given. This was part of a larger project which was concerned with the development of a system known as GESA. This was an engineering project on a relatively short time scale. Since any existing technique would have required modification, we therefore used a graph matching technique developed by the authors in-house: a full description can be found in [21]. Although we do not claim that our linear assignment technique is necessarily 'best possible', it was shown to be $O(\min(n^3,m^3))$ and was therefore likely to be comparable with that of Jonker and Volgenant [20]; a proper comparison, although not germaine to this project, would nevertheless be of interest.

544

Acknowledgements

This research was carried out as a multidisciplinary project by Unilever
Research, Colworth, the Plant Breeding Institute, The Turing Institute and
the University of Strathclyde with funding under the Alvey programme of
research and development (Project IKBS/104).

References

[1] O'Farrel, P.H.: J. Biol. Chem., 1975, 250, 4007-4021.

[2] Meyer, F., In: Quantitative Analysis of Microstructures in Material
     Science, Biology and Medicine. J.L. Cheremant (Ed.) 1978, 374-380,
     Rieder-Verlag, Stuttgart.

[3] Anderson, N.L., Taylor, J., Scandora, A.E., Coulter, B.P. and
     Anderson, N.G.: Clin. Chem. 1981, 27, No.11, 1807-1820.

[4] Garrels, J.I., Farrar, J.T. and Burwell, C.B.: In: Two-Dimensional
     Gel Electrophoresis of Proteins - Methods and Applications. J.E. Celis
     and R. Bravo (Eds.) 1984, Academic Press.

[5] Lemkin, P.F. and Lipkin, L.E.: Electrophoresis, 1983, 4, 71-81.

[6] Vincens, P. and Tarroux, P.: Electrophoresis, 1987, 8, 100-107.

[7] Van Hoff, A.A., Niblett, T., Rowlands, D.G., Flook, A., Payne, P.I.
     and McKee, S.: Knowledge-Based Systems, 1988, 1, No.5, 292-300.

[8] Potter, D.J.: Comp. Biomed. Res., 1985, 18, 347-362.

[9] Skolnick, M.M.: Clin. Chem., 1982, 28, No.4, 979-986.

[10] Skolnick, M.M.: Computer Vision, Graphics and Image Processing, 1986,
     35, 306-332.

[11] Van Hoff, A.A.: 1987, Semi-Automatic Analysis of Two Dimensional
     Electrophoresis Gels. M.Sc. Thesis, University of Strathclyde, Scotland.

[12] Kuhn, H.W.: Naval Research Logistics Quart., 1955, 2, 83-97.

[13] Barr, R., Glover, F. and Klingman, D.: Math. Prog., 1977, 13, 1-13.

[14] Hung, M.S. and Rom, W.O.: Operations Res., 1980, 28, 969-982.

[15] Bertsekas, D.P.: Math. Prog., 1981, 2, 152-171.

[16] Balinski, M.L.: Operations Res., 1985, 33, 527-536.

[17] Mack, C.: New J. Stats. and Operations Res., 1969, 1, 17-29.

[18] Tomizawa, N.: Networks, 1971, 1, 173-194.

545

[19] Dorhout, B.: Report BN21/73, 1973, Mathematisch Centrum, Amsterdam.

[20] Jonker, R. and Volgenaut, A.: Computing, 1987, 38 325-340.

[21] McKee, S. and Swailes, D.: Unilever Report XCW88 0033, Unilever
      Research, Colworth, England.

University of Strathclyde,

Department of Mathematics,

Livingstone Tower,

26 Richmond Street,

Glasgow G1 1XH,

Scotland.

MATHEMATICAL MODELS OF SILICON CHIP FABRICATION

A.B. TAYLER
Mathematical Institute, 24-29 St. Giles, Oxford, U.K.

1. Introduction

It is possible to pack a large number of semi-conductor devices onto the surface of one silcon chip. Each device consists of an active region which is electrically isolated from its neighbours, and one technique for achieving this isolation is by oxidation of the silicon in the non-active region to form a silicon oxide dielectric barrier. To avoid oxidation of the active region it is masked with a silicon nitride cap, impervious to the diffusion of oxidants, which is removed by etching after the completion of the oxidation process. This dielectric oxide extends deep enough to reach a p-n junction parallel to the chip surface so that a box is formed which isolates the device.

The major design problems are to be economical with space and to avoid high stresses which could cause dislocations in the active area of the device. Since the nitride cap and silicon have very different thermal expansion coefficients they have to be separated by a thin pad of oxide which at these temperatures is visco-elastic and acts as a cushion. This will imply further oxidation under the nitride cap, thus reducing the available active region. However, the reaction rate for oxidation decreases with high stresses, inhibiting the process in the pad.

In this paper a mathematical model is proposed for the oxidation process and is analysed under some simplifying assumptions. The physical basis of the model is that the silicon oxide behaves as a slow viscous fluid through which the oxidant

J. Manley et al. (eds.), Proceedings of the Third European Conference on Mathematics in Industry, 547–556.
© 1990 Kluwer Academic Publishers and B. G. Teubner Stuttgart.

can diffuse. The silicon is a solid, impervious to the diffusion
of the oxidant, whose surface is eroded by a first order chemical
reaction in which silicon and oxygen combine to form silicon
oxide. Thus the rate of erosion determines the shape of the
surface. This rate of erosion will depend on the diffusion of the
oxidant through the oxide, which occupies a region to be
determined, and leads inevitably to a mathematical model which is
a moving boundary problem. Simpler erosion problems in which the
rate of erosion is prescribed are described by Smith, Carter and
Nobes, 1986. The nitride cap is assumed to be an elastic plate
with large stiffness so that the Euler Bernoulli beam theory
applies in a two dimensional situation.

2. One dimensional oxidation

We consider the following geometry:

$$\text{silicon} \qquad y < -f(t)$$
$$\text{silicon oxide} \quad -f(t) < y < h(t)$$
$$\text{oxidant} \qquad y > h(t)$$

where initially $f = a$ and $h = 0$.

In a one dimensional model the oxide is subject only to a constant
normal stress and everywhere moves with the same velocity $\frac{dh}{dt}$.
Thus the concentration $c(y,t)$ of oxidant diffusing through the
oxide may be modelled by

$$\frac{\partial c}{\partial t} + \frac{dh}{dt}\frac{\partial c}{\partial y} = D\frac{\partial^2 c}{\partial y^2} , \qquad (1)$$

where D is a diffusion coefficient, assumed to be constant.

At the oxidant boundary we assume there is a net flow of
oxidant into the oxide, proportional to the difference between the

equilibrium concentration c* and the local concentration c(h,t), so that

$$y = h(t) \;, \quad D\,\frac{\partial c}{\partial y} = H(c^*-c) \;, \tag{2}$$

where H is a concentration transport coefficient, also assumed to be constant.

At the silicon boundary the first order reaction is modelled by

$$y = -f(t) \;, \quad D\,\frac{\partial c}{\partial y} - (\frac{dh}{dt}+\frac{df}{dt})c = kc \;, \tag{3}$$

where k is the reaction coefficient. Note that the left hand side of equation (3) expresses the rate at which oxidant is consumed at the silicon surface, and that relative to this surface the oxide is moving with speed $\frac{dh}{dt}+\frac{df}{dt}$.

In the reaction β molecules of oxidant combine with one part of silicon to form γ parts of the oxide so that

$$\frac{1}{\beta}\,[D\,\frac{\partial c}{\partial y} - (\frac{dh}{dt}+\frac{df}{dt})c] = \frac{df}{dt} = \frac{1}{\gamma}(\frac{df}{dt}+\frac{dh}{dt}) \tag{4}$$

Suitable initial conditions must also be specified, and from (4) we note that $h = (\gamma-1)(f-a)$.

The obvious scaling to make the variables nondimensional is to scale c with c*; y, f and h with a; and t with a^2/D so that D,c* and a can be replaced by unity in equations (1)-(4). In addition there will be four nondimensional parameters

$$\frac{Ha}{D}, \; \frac{ka}{D}, \; \frac{\beta}{c^*} \text{ and } \gamma \;,$$

where at this stage we treat k as constant for simplicity.

The problem can be rewritten in a more standard form using $z = h-y$ and $s = h+f$ to obtain

$$0 < z < s(t) \ , \quad \frac{\partial c}{\partial t} = \frac{\partial^2 c}{\partial z^2} \ ;$$

$$z = 0 \ , \quad \frac{\partial c}{\partial z} = H(c-1) \ ; \quad t = 0, \ c \text{ given}, \ s = 1 \ ;$$

$$z = s(t) \ , \quad \frac{\partial c}{\partial z} + \frac{ds}{dt} c = -kc = -\frac{\beta}{\gamma} \frac{ds}{dt} \ , \tag{5}$$

where the parameters are now nondimensional. This free boundary problem has been shown to have a unique solution with a regular boundary $z = s(t)$ by Fasano and Tarzia, 1988, and is one of a class of 'penetration' problems. However, no explicit solutions are available and it is valuable to look for an asymptotic solution in terms of large values of the parameter β/γ. This is the ratio of volumes of oxygen consumed to oxide created in the reaction and can be of the order of 10. For $\beta/\gamma \gg 1$, $\frac{ds}{dt} \ll 1$ and there is no $O(1)$ change in the boundaries on this diffusion timescale. Hence we rescale time with β/γ and ignore terms of $O(\gamma/\beta)$ so that the problem reduces to:

$$0 < z < s(t) \ , \quad \frac{\partial^2 c}{\partial z^2} = 0 \ ;$$

$$z = s(t) \ , \quad \frac{\partial c}{\partial z} = -kc = -\frac{ds}{dt} \ ;$$

with the other conditions unchanged.

Thus c is linear in z and h may be determined from the positive root of $Ph^2 + Qh = t$, where P and Q are known positive constants. This is the solution obtained by Deal & Grove, 1965, using the physical argument that the oxidant diffusion is almost in equilibrium. It will not however be correct for small times when c is no longer linear in z.

3. Two dimensional oxidation

Consider now the problem in which at $t = 0$ the boundary $y = 0$, $x > 0$ consists of a nitride cap. The oxide will no longer have a uniform motion and a stream function $\psi(x,y,t)$ is introduced to describe the flow. Under the assumption of a slow viscous flow $\nabla^4 \psi = 0$ and on a time-scale large compared to the diffusion timescale $\nabla^2 c = 0$, corresponding to the one dimensional model of Deal & Grove. Four boundary conditions will be needed on the oxide/oxidant interface $y = h(x,t)$ ($x < 0$), on the nitride cap $y = g(x,t)$ ($x > 0$), and on the silicon/oxide interface $y = -f(x,t)$. This is a complicated boundary value problem discussed by Tayler and King, 1988; here we shall only consider the growth of the oxide region under the nitride cap in some special cases.

The simplest situation is for the nitride cap to be so stiff that $g \equiv 0$ and the boundary conditions on the cap are:

$$y = 0 \ , \quad \frac{\partial c}{\partial y} = 0 = \psi = \frac{\partial \psi}{\partial y} \ . \tag{6}$$

At the silicon/oxide interface the conditions are the two dimensional version of equations (3) and (4) in terms of distance s along the interface and the outward normal n to the oxide region. Ignoring terms of $O(\gamma/\beta)$ as before,

$$y = -f(x,t) \ , \quad \frac{\partial c}{\partial n} = -kc = -\frac{\partial \psi}{\partial s} \ , \quad \gamma \frac{\partial f}{\partial t} = \frac{\partial c}{\partial y} + \frac{\partial f}{\partial x} \frac{\partial c}{\partial x} \ , \tag{7}$$

together with the no-slip condition $\frac{\partial \psi}{\partial n} = 0$. Note that ψ has been made nondimensional with $\gamma \frac{(\gamma-1)}{\beta} Dc*$.

We attempt to solve this free boundary problem using the long wave or lubrication theory approximation in which length scales in the x direction are assumed to be much greater than those in the y direction. Thus if the reference length for the nondimensional time t and distance x is taken to be L, with reference length a for y and f, then $\epsilon = \frac{a}{L} \ll 1$, and conditions (7) reduce to:

$$\epsilon^2 \gamma \frac{\partial f}{\partial t} - \frac{\partial c}{\partial y} + \epsilon^2 \frac{\partial f}{\partial x} \frac{\partial c}{\partial x} = - \frac{\partial c}{\partial n} (1 + \epsilon^2 (\frac{\partial f}{\partial x})^2)^{\frac{1}{2}} .$$

But c satisfies

$$\epsilon^2 \frac{\partial^2 c}{\partial x^2} + \frac{\partial^2 c}{\partial y^2} = 0 ,$$

so that with $c = \varphi(x,t) + \epsilon^2 u(x,y,t)$,

$$\frac{\partial^2 u}{\partial y^2} = - \frac{\partial^2 \varphi}{\partial x^2} \quad \text{correct to } 0(\epsilon^2) .$$

Using (6)

$$\frac{\partial u}{\partial y} = -y \frac{\partial^2 \varphi}{\partial x^2} ,$$

and

$$\gamma \frac{\partial f}{\partial t} = f \frac{\partial^2 \varphi}{\partial x^2} + \frac{\partial f}{\partial x} \frac{\partial \varphi}{\partial x} - \frac{k}{\epsilon^2} \varphi(1 + 0(\epsilon^2)) .$$

With a new parameter $\kappa = \frac{k}{\epsilon^2 \gamma}$, assumed to be $0(1)$, in the limit as $\epsilon \to 0$

$$\gamma \kappa \frac{\partial f}{\partial t} = \frac{\partial}{\partial x} (f \frac{\partial^2 f}{\partial x \partial t}) . \tag{8}$$

Appropriate boundary conditions are: $t = 0$, $f = 1$; $x \to \infty$, $f \to 1$ exponentially; $x \to 0$, $f = f_0(t)$ where $f_0(t)$ is obtained from conditions near the edge of the nitride cap. With equilibrium concentration for $x < 0$ it may be shown by matching solutions that $f_0 = \kappa t$.

In the next section we shall obtain solutions to equation (8) but first evaluate the stream function and normal stress on the silicon surface.

Since

$$(\epsilon^2 \frac{\partial^2}{\partial x^2} + \frac{\partial^2}{\partial y^2})^2 \psi = 0 ,$$

$$\psi = a(x,t)(y^3 + \frac{3}{2} fy^2) + O(\epsilon^2) , \qquad (9)$$

and on $y = -f$,

$$\gamma \frac{\partial f}{\partial t} = - \frac{\partial \psi}{\partial x} + O(\epsilon^2) .$$

Using equations (8) and (9), and integrating once

$$\tfrac{1}{2}\kappa af^3 = -f \frac{\partial^2 f}{\partial x \partial t} + \text{constant}.$$

With no flow as $x \to \infty$, $a \to 0$ and $\frac{\partial f}{\partial x} \to 0$, so that the constant is zero and

$$a = \frac{-2}{\kappa f^2} \frac{\partial^2 f}{\partial x \partial t} .$$

In this long wave approximation the normal stress is the pressure $p(x,t)$ and is constant across the oxide layer. Hence

$$\frac{\partial p}{\partial x} = \frac{\partial^3 \psi}{\partial y^3} = - \frac{12}{\kappa f^2} \frac{\partial^2 f}{\partial x \partial t} . \qquad (10)$$

4. Penetration under the nitride cap

Equation (8) has been derived as a model for the oxide growth under the nitride cap. With a stress dependent reaction coefficient similar asymptotic arguments lead to the equation

$$\gamma \frac{\partial f}{\partial t} = \frac{\partial}{\partial x} [f \frac{\partial}{\partial x} (\frac{1}{\kappa(p)} \frac{\partial f}{\partial t})] , \qquad (11)$$

where equation (10) is replaced by

$$\frac{\partial p}{\partial x} = - \frac{12}{f^2} \frac{\partial}{\partial x} (\frac{1}{\kappa(p)} \frac{\partial f}{\partial t}) . \qquad (12)$$

The boundary conditions are: $t - 0$, $f - 1$; $x - 0$, $f - f_0(t)$, $p - p_0(t)$; $x \to \infty$, $f \to 1$ exponentially. The function $\kappa(p)$ is monotonic decreasing with p and for large values of p, $\kappa(p) \to \kappa_\infty$.

We look for a travelling wave solution of equation (11) in the form $f - f(\eta)$ where $\eta - s(t)-x$, so that

$$\gamma(f-1) - f(\frac{1}{\kappa(p)} f')' \ . \tag{13}$$

For κ constant this can be integrated to give

$$\eta - \frac{1}{(2\kappa\gamma)^{\frac{1}{2}}} \int_c^f \frac{dz}{(z-\log z-1)^{\frac{1}{2}}} \ , \tag{14}$$

where $s(t)$ is determined by conditions at $x - 0$, namely

$$s(t) - \frac{1}{(2\kappa\gamma)^{\frac{1}{2}}} \int_c^{f_0(t)} \frac{dz}{(z-\log z-1)^{\frac{1}{2}}} \ . \tag{15}$$

Hence

$$x - \frac{1}{(2\kappa\gamma)^{\frac{1}{2}}} \int_f^{f_0(t)} \frac{dz}{(z-\log z-1)^{\frac{1}{2}}} \tag{16}$$

determines f implicitly for all functions f_0. Profiles for $f(x,t)$ are in close agreement with results obtained from the process simulator TAPDANCE.

For varying κ this does not lead to a solution for p as a function of η alone except when $s - ct$. Then

$$p' - - \frac{12c}{f^2} (\frac{1}{\kappa(p)} f')' - - \frac{12c}{f^3} (f-1) \ . \tag{17}$$

For this travelling wave solution to be possible $f_0(t)$ must satisfy equation (15) with $s(t) - ct$. Equations (13) and (17) give a third order system of ordinary differential equations to analyse and solve numerically for f and p.

A further simplification is to consider a very thin initial oxide pad so that the boundary conditions on f may be replaced by

$x \to \infty$, $f \to \delta \ll 1$. Then ignoring terms $O(\delta)$ the travelling wave solution for $\eta > 0$ satisfies

$$\gamma f' = \kappa(p)\eta , \quad p' = - \frac{12c}{f^2} \qquad (18)$$

where as $\eta \to 0$, f and $f' \to 0$.

This is not a uniformly valid solution as $\eta \to 0$ since p' becomes unbounded. An inner asymptotic expansion is required with inner variables $\bar{f} = f/\delta$, $\bar{\eta} = \eta/\sqrt{\delta}$ and $\bar{p} = p\delta^{3/2}$. The problem then reduces to (13) and (17) but with $\kappa = \kappa_\infty$ so that equation (14) gives the inner solution with f and η replaced by $\bar{f}$ and $\bar{\eta}$. It will match with the outer solution determined by the solution of the pair of ordinary differential equations (18).

In applications of this model it is often the stress exerted on the silicon which is of most interest. Near the travelling wave front the pressure is $O(\delta^{-3/2})$ and there will be a moving point force, speed c, of strength $O(\delta^{-1})$ whose magnitude can be obtained from $\int_{-\infty}^\infty \bar{p}\,d\bar{\eta}$ using (14) and (17). The value of c will be determined from the conditions imposed at $x = 0$ consistent with a solution of equations (18) e.g. with κ constant $\gamma f = \frac{1}{2}(ct-x)^2$ is the solution of (18) and $f_0 = \frac{1}{2}c^2t^2$ determines c.

5. Other simple models

If the nitride cap is not rigid and is given by $y = g(x,t)$, $x > 0$ the lubrication approximation may still be applied as in Tayler and King. If in addition we ignore the erosion of the silicon surface and the oxide production the conditions on $y = -1$ are $\psi = 0 = \frac{\partial\psi}{\partial y}$. On $y = g$, $\frac{\partial\psi}{\partial y} = 0$, $\frac{\partial g}{\partial t} = - \frac{\partial\psi}{\partial x}$ and $p = \lambda \frac{\partial^4 g}{\partial x^4}$ from beam theory where λ is a stiffness parameter.

Hence

$$\psi = [\frac{1}{6} (y+1)^3 - \frac{1}{4} (y+1)^2 g] \frac{\partial p}{\partial x} ,$$

and

$$\frac{\partial g}{\partial t} = \frac{\lambda}{12} \frac{\partial}{\partial x} [(g+1)^3 \frac{\partial^5 g}{\partial x^5}] , \qquad (19)$$

where $g \to 0$ as $x \to \infty$.

A travelling wave solution satisfies

$$12cg = \lambda(g+1)^3 \frac{dg^5}{d\eta^5} , \qquad (20)$$

and the properties of solutions of (20) are not yet available.

With f comparable to g it is shown in Tayler and King that the equations are

$$\gamma\kappa \frac{\partial f}{\partial t} = \frac{\partial}{\partial x} [(g+f) \frac{\partial^2 f}{\partial x \partial t}] ,$$

$$\frac{\partial g}{\partial t} - (\gamma-1) \frac{\partial f}{\partial t} = \frac{\lambda}{12} \frac{\partial}{\partial x} [(g+f)^3 \frac{\partial^5 g}{\partial x^5}] .$$

They clearly reduce to (11) and (19) in special cases.

REFERENCES

1. B.E. Deal and A.S. Grove, General relationship for the thermal oxidation of silicon, J. App. Phys. 36 (1965) 3770.

2. A. Fasano and D.A. Tarzia, Private communication.

3. R. Smith, G. Carter and M.J. Nobes, The theory of surface erosion by ion bombardment, Proc. Roy. Soc. A 407 (1986) 405.

4. A.B. Tayler and J.R. King, Free boundaries in semiconductor fabrication, Proc. Int. Conf. on free boundaries; theory and application (Ed. Hoffmann), Irsee 1987.

ON A PARTIAL INTEGRO-DIFFERENTIAL EQUATION RELATED TO
THE DYNALISER CONCEPT FOR INDUSTRIAL RUBBER MATERIALS

by R. Van Keer and H. Serras

1.Introduction

A rubber block is strained in a stepwise manner. During a time interval $(0, T)$ it undergoes a constant relative deformation x_0. At $t = T$ the sample is suddenly released from the imposed strain. The problem is to determine the relative residual deformation $x(t)$.

During the strain period the relaxation stress is a decreasing function of time. It appears that the best fit of the relaxation signal,measured at not too small values of time,has the form,see [1],[5] ,

$$(1.1) \qquad\qquad E(t) = A + B.t^{-\alpha} \qquad (t= \text{relative time }).$$

Here A, B and α are material constants,corresponding to the elasticity, viscosity and the rate of recuperation respectively. They are provided by the Dynaliser unit. A and B are in $[N/mm^2]$, $\alpha \in]0, 1[$ is dimensionless.

In compression set tests the values of x_0 and T are standardised.

Following [3] and [5] the rubber material is idealised by the so-called generalised Maxwell model (fig.1). This consists of a linear spring and an infinite sequence of parallel simple Maxwell elements with time constant $\tau = \eta/\kappa$ varying continuously from zero to infinity. ($\kappa = $ spring constant , $\eta = $ damping constant). The spring at the left has spring constant A.

The 'distribution' function $H(\tau)$ of the element with time constant τ is found in [5] to be

$$(1.2) \qquad\qquad H(\tau) = \frac{B}{\Gamma(\alpha)}.\tau^{-\alpha-1}, \qquad \tau > 0, \quad 0 < \alpha < 1.$$

The compression set test is a common quality control test on final rubber products, mostly rubber rings and gaskets. The relevancy of readings produced by short time

J. Manley et al. (eds.), Proceedings of the Third European Conference on Mathematics in Industry, 557–564.
© 1990 Kluwer Academic Publishers and B. G. Teubner Stuttgart.

relaxation tests with digital analysis is investigated at the Instrumentation Department of Berginvest. Here we aim to correlate relaxation moduli to compression set readings in a mathematical way.

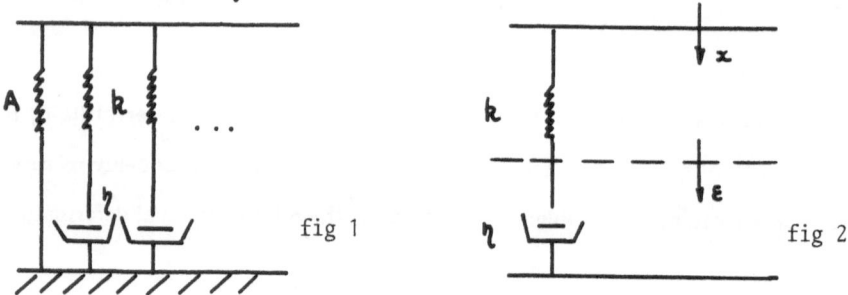

fig 1

fig 2

2.Mathematical formulation of the problem

Consider an individual Maxwell element with time constant τ (fig.2). Let $\epsilon(t,\tau)$ be the relative deformation (displacement) of the damping component — relative to the original height h_0 of the rubber,when the overall relative deformation of the Maxwell element is $x(t)$, $t > 0$. Equating the stresses in the spring (Hooke's Law) and the dashpot (Newton's Law) ,one easily arrives at

$$(2.1) \qquad x(t) = \epsilon(t,\tau) + \tau.\frac{\partial \epsilon}{\partial t}(t,\tau), \qquad t > 0, \tau > 0.$$

From this relation in the interval $]0,T[$, where $x(t) = x_0$, one has

$$(2.2) \qquad \epsilon(T,\tau) = x_0.(1 - e^{-T/\tau}), \qquad \tau > 0.$$

Finally,the balance of stresses in the generalized Maxwell model for $t > T$, i.e. when the strain has been removed,leads to

$$(2.3) \qquad A.x(t) + \int_0^\infty H(u).u.\frac{\partial \epsilon}{\partial t}(t,u). du = 0, \qquad t > T$$

(2.1) - (2.3) form the governing equations for the relative residual deformation $x(t)$, $t > T$, we are looking for. $\epsilon(t,\tau), t > T$,is seen to obey a homogeneous partial integro-differential equation of a singular Fredholm type, with (2.2) serving as an initial condition at $t = T$.

Putting $x(t) = x_0.\tilde{x}(\tilde{t})$, $t = \tilde{t} + T$, etc. and dropping the tilde ,we are left with

$$(2.4) \qquad x(t) = \epsilon(t,\tau) + \tau.\frac{\partial \epsilon}{\partial t}(t,\tau), \qquad t > 0, \tau > 0$$

$$(2.5) \quad (P) \qquad \epsilon(0,\tau) = 1 - e^{-T/\tau}, \qquad \tau > 0$$

$$(2.6) \qquad \int_0^\infty \frac{\partial \epsilon}{\partial t}(t,u).u^{-\alpha}.du + \frac{A}{B}.\Gamma(\alpha).x(t) = 0, \qquad t > 0$$

Requiring that $\epsilon(t,\tau)$ is sufficiently smooth, this constitutes the problem (P) we are dealing with in the next section.

3.A Laplace transformation method

The $\mathcal{L}$-transformation method with respect to $t > 0$ for (P) uses two hypotheses.

H_1. For each $\tau > 0$, $\frac{\partial \epsilon}{\partial t}(t,\tau)$ is $\mathcal{L}$-transformable for $s > 0$. This implies that , see [2] ,

$$(3.1) \qquad E(s,\tau) = \mathcal{L}[\epsilon(t,\tau)](s) < \infty; X(s) = \mathcal{L}[x(t)](s) < \infty, \qquad s > 0, \tau > 0$$

$$(3.2)\, \mathcal{L}[\frac{\partial \epsilon}{\partial t}(t,\tau)](s) = s.E(s,\tau) - \epsilon(0,\tau), \qquad s > 0, \tau > 0$$

H_2.

$$(3.3) \qquad \mathcal{L}[\int_0^\infty \frac{\partial \epsilon}{\partial t}(t,u).u^{-\alpha}\,du](s) = \int_0^\infty \mathcal{L}[\frac{\partial \epsilon}{\partial t}(t,u)](s).u^{-\alpha}.du < \infty, \qquad s > 0$$

Under these hypotheses the problem (P) leads to

$$(3.4) \qquad E(s,\tau) = k_1(s,\tau).\int_0^\infty k_2(u).E(s,u).du + F(s,\tau), \qquad \tau > 0 \quad (s > 0)$$

$$(3.5) \qquad F(s,\tau) = \frac{\tau}{1+\tau.s}.\epsilon(0,\tau) - \frac{1}{1+\tau.s}\int_0^\infty k_2(u).\epsilon(0,u).du.$$

$$k_1(s,\tau) = \frac{s}{1+\tau.s}, \quad k_2(u) = -\frac{B}{A.\Gamma(\alpha)}.u^{-\alpha}$$

The solution of this singular Fredholm integral equation with degenerated kernel reads

$$E(s,\tau) = F(s,\tau) + \frac{k_1(s,\tau)}{1 - \int_0^\infty k_1(s,u).k_2(u).du}.\int_0^\infty F(s,u).k_2(u).du, \qquad \tau > 0 \ (s > 0)$$

Invoking (2.5) and (3.5), as well as (2.4) and (3.2), we may arrive at

560

$$(3.6) \qquad X(s) \;=\; \frac{1}{a.s^\alpha + 1}.[a.s^{\alpha-1} - \frac{B}{A}.\int_0^\infty \frac{e^{-s.r}}{(r+T)^\alpha}.dr] \equiv \frac{D(s)}{a.s^\alpha + 1}$$

$$(3.7) \qquad E(s,\tau) \;=\; \frac{1}{1+\tau.s}.[\tau.\epsilon(0,\tau) + X(s)] \qquad \text{with} \quad a = \frac{B}{A}.\Gamma(1-\alpha).$$

$X(s)$ may be expanded as

$$(3.8) \qquad X(s) = \sum_{k=0}^{+\infty} \frac{(-1)^k}{(a.s^\alpha)^{k+1}}.D(s), \qquad s > (\frac{1}{a})^{1/\alpha}$$

Combining a convolution product theorem and a theorem on the termwise inversion of a series of image functions,see [2],we rigorously find that

$$(3.9) \qquad \mathcal{L}^{-1}[X(s)](t) \;=\; \frac{B}{A}\sum_{k=0}^{+\infty} \frac{(-1)^k}{a^{k+1}}.\frac{1}{\Gamma(\alpha(k+1))}.g_k(t) \equiv \hat{x}(t), \qquad t \geq 0$$

$$\mathcal{L}^{-1}[E(s,\tau)](t) \;=\; \epsilon(0,\tau).e^{-t/\tau} + \frac{1}{\tau}.e^{-t/\tau} * \hat{x}(t) \equiv \hat{\epsilon}(t,\tau), \qquad \tau > 0$$

$$(3.10) \qquad g_k(t) \;=\; [t^{-\alpha} - (t+\tau)^{-\alpha}] * t^{\alpha.(k+1)-1}, \qquad k \neq 0$$

$$g_0(t) \;=\; B(1-\alpha,\alpha) - (t+T)^{-\alpha} * t^{\alpha-1}, \qquad t \geq 0$$

$\hat{x}(t)$ is a continuous function of $t \geq 0$,with $\hat{x}(0) = 1$,being defined by a series of continuous functions which is uniformly convergent. Consequently $\hat{\epsilon}(t,\tau)$ is a continuous function of $t \geq 0$ for each $\tau > 0$, with $\hat{\epsilon}(0,\tau) = \epsilon(0,\tau)$. Moreover (3.9) rigorously implies that

$$(3.11) \qquad \hat{x}(t) = \hat{\epsilon}(t,\tau) + \tau.\frac{\partial\hat{\epsilon}}{\partial t}(t,\tau), \qquad t > 0, \tau > 0.$$

Verification

In order to assure that $\hat{\epsilon}(t,\tau)$ is indeed a solution of (P) it remains to show that the first step in the procedure above,i.e. from (P) to (3.4),may be reversed.

From (3.12) we may arrive at

$$(3.12) \qquad X(s) = \int_0^\infty k_2(u).\mathcal{L}[\frac{\partial\hat{\epsilon}}{\partial t}(t,u)](s)\,du, \qquad s > 0$$

If we have

(3.13) $\qquad \displaystyle\int_0^\infty k_2(u).\mathcal{L}[\frac{\partial\hat{\epsilon}}{\partial t}(t,u)](s)\,du = \mathcal{L}[\int_0^\infty k_2(u).\frac{\partial\hat{\epsilon}}{\partial t}(t,u)\,,du](s), \qquad s > 0$

then,by Lerch's theorem,[2], we obtain

(3.14) $\qquad\qquad \hat{x}(t) = \displaystyle\int_0^\infty k_2(u).\frac{\partial\hat{\epsilon}}{\partial t}(t,u).\,du \qquad$ a.e. in $]0,\infty[$.

 <u>Theorem.</u> $\hat{\epsilon}(t,\tau)$,defined by (3.9),is a solution of (P)

<u>Proof.</u> (3.14) holds if one of the iterated integrals converges absolutely (Fubini-Tonelli theorem). From (3.12) one gets that

$$Z(u,s) \equiv \mathcal{L}[|\frac{\partial\hat{\epsilon}}{\partial t}(t,u)|](s) < \infty, \qquad \forall u > 0, \quad (s > 0, \text{parameter}).$$

Moreover, $Z(u,s)$ is continuous $\forall u > 0$,as the integral converges uniformly with respect to u in each interval $[u_1,u_2], u_1 > 0$,and the integrand is continuous in $[0,+\infty[\times[u_1,u_2]$. Furthermore, (3.12) implies that $Z(u,s) = O(\frac{1}{u})$ as $u \to +\infty$. Finally we argue that $Z(u,s)$ remains bounded if $u \to 0$. Fix u and denote by $(0,a(u))$ the time interval where $\frac{\partial\hat{\epsilon}}{\partial t}(t,u) > 0$ (in fact $a(u) \to 0$ if $u \to 0$). Then we have

$$Z(u,s) = 2\int_0^{a(u)} e^{-st}.\frac{\partial\hat{\epsilon}}{\partial t}(t,u)\,dt \quad - \quad [s.X(s) - \epsilon(0,u)]\frac{1}{1+us}.$$

Clearly the second term is bounded as $u \to 0$. On account of (3.12) the integral is bounded above by $(s.u)^{-1}.e^{-T/u} \to 0$ as $u \to 0$. Hence the convergence of $\int_0^\infty k_2(u).|Z(u,s)|\,du$ follows. $\square$

<u>Numerical results</u>

As the series (3.10),defining $x(t)$,converges slowly for large values of t, the numerical inversion of $X(s)$, (3.6) is important. Here we made use of an existing routine,[4]. This requires $X(s)$ to be known in $\Omega = \{s|\Re s > 0\}$. We may resort to analytical continuation,defining suitably s^α in Ω. The results,some of which are presented below,are in good agreement with experimental values.

		Natural Rubber	Styrene Butadiene Rubber		Nitrole Butadiene Rubber	Butyl Rubber
		I	II	III	IV	V
α		0.15	0.20	0.20	0.30	0.30
B/A		0.10	0.25	0.50	0.50	1.00
$T=24h$	(1)	0.040	0.091	0.164	0.143	0.252
	(2)	0.014	0.029	0.055	0.035	0.068
$T=48h$	(1)	0.041	0.094	0.169	0.146	0.257
	(2)	0.016	0.032	0.061	0.038	0.073
$T=72h$	(1)	0.042	0.095	0.172	0.147	0.259
	(2)	0.017	0.034	0.064	0.039	0.076
$T=\infty$	(1)	0.057	0.113	0.205	0.157	0.276
	(2)	0.032	0.053	0.101	0.051	0.097

Values of $x(t)$, (1) for $t=30$, (2) for $t=1800$

4. A finite difference method

Consider a time step $\Delta t > 0$ and the time points $t_j = j.\Delta t$, $j = 0, 1, \ldots$. We define an approximation

(4.1) $\qquad \epsilon_j(\tau) \approx \epsilon(t_j, \tau)$, $j = 0, 1, \ldots, \tau > 0; x_j \approx x(t_j)$, $j = 1, 2, \ldots$

by the recurrent system (discrete problem)

(4.2) $\qquad x_j = \epsilon_j(\tau) + \tau . \dfrac{\epsilon_j(\tau) - \epsilon_{j-1}(\tau)}{\Delta t}$, $\qquad j = 1, 2, \ldots, \quad \tau > 0$

(4.3) $\qquad x_j = \displaystyle\int_0^\infty k_2(u) . \dfrac{\epsilon_j(u) - \epsilon_{j-1}(u)}{\Delta t} . du$, $\qquad j = 1, 2, \ldots$

(4.4) $\qquad \epsilon_0(\tau) = \epsilon(0, \tau)$, $\qquad \tau > 0$

The resulting equation for $\epsilon_j(\tau), j = 1, 2, \ldots$, which is similar to (3.4) may easily be solved. (4.2) then leads to

(4.5) $\qquad x_j = -\dfrac{1}{1 + a.(\Delta t)^{-\alpha}} . \displaystyle\int_0^\infty \dfrac{k_2(u)}{\Delta t + u} . \epsilon_{j-1}(u). du$, $\qquad j = 1, 2, \ldots$

Together with (4.4) and with (4.2) — solved for $\epsilon_j(\tau)$ — (4.5) constitutes a set of recurrence relations for $\epsilon_j(\tau)$ and $x_j, j = 1, 2, \ldots$. This system may be written in a more transparent form. By iteration (4.2) gives

$$(4.6) \qquad \epsilon_j(\tau) = \frac{\Delta t}{\Delta t + \tau} \sum_{l=0}^{j-1} (\frac{\tau}{\Delta t + \tau})^l . x_{j-l} + (\frac{\tau}{\Delta t + \tau})^j . \epsilon_0(\tau), \qquad j = 1, 2, \ldots$$

Substituting this expression in (4.5), we obtain

$$(4.7) \qquad x_j = \sum_{l=1}^{j-1} A_{j-1-l} . x_l + B_j, \qquad j = 2, \ldots \quad ; x_1 = B_1$$

$$(4.8) \qquad A_l = \frac{B}{A . \Gamma(\alpha)} . \frac{1}{a + (\Delta t)^\alpha} . B(1 - \alpha + l, \alpha + l), \qquad l = 0, 1, \ldots$$

$$(4.9) \qquad B_j = \frac{B}{A . \Gamma(\alpha)} . \frac{1}{a + (\Delta t)^\alpha} . \int_0^\infty \frac{r^{j-1-\alpha}}{(1+r)^j} . (1 - e^{-T/\Delta t . r}) \, dr, \qquad j = 1, 2, \ldots$$

A numerical example

For simplicity we take $T = \infty$. Then B_j, (4.9), is known exactly. We consider the case $\alpha = 0.5, a = 1$. Then $x(t) = e^t . \text{erfc}(\sqrt{t})$ is 'known'.

The table below shows a good agreement between the values of $x(t)$, obtained by the finite difference method with $\Delta t = 0.01$, and the 'exact' values.

t	0.1	0.2	0.3	0.4	0.5	1.0	2.0	3.0	4.0
(4.8)	0.727	0.646	0.594	0.555	0.524	0.428	0.337	0.288	0.256
exact	0.724	0.644	0.592	0.554	0.523	0.428	0.336	0.287	0.255

5. Concluding remarks

Some of the qualitative properties of $x(t)$, reflected by the numerical results, are:

- The (monotonous) decrease of $x(t)$ is very fast at small values of t; the smaller α, the greater the decay. This corresponds to the asymptotic expression of $x(t)$ for $t \to 0$, obtainable from the asymptotic form of $X(s)$ for $s \to \infty$, see [2]

$$(5.1) \qquad x(t) \sim 1 - \frac{1}{a}(1 + \frac{B}{A} . T^{-\alpha}) . \frac{t^\alpha}{\Gamma(\alpha + 1)}, \qquad t \to 0$$

However, $x(t)$ tends to zero only slowly.

- B/A being fixed, $x(t)$ decreases with increasing α for not too small values of t. Conversely, for given α, $x(t)$ increases with B/A. At large values of t, $x(t)$ is almost proportional to $\frac{B}{A}$. The latter agrees with

564

$$(5.2) \qquad\qquad x(t) \sim \alpha.\frac{B}{A}.T.t^{-(\alpha+1)}, \qquad t \to \infty$$

which follows from the asymptotic expression of $X(s)$ for $s \to 0$, [2].

This asymptotic form is found to be a reasonable approximation of $x(t)$ for values of t greater than $(\frac{B}{A}.10)^{1/\alpha}$.

- $x(t)$ increases with T. For small values of t , one may take $T = \infty$.

The finite difference method,resulting in (4.7) is computationally easy. However, it's applicability is restricted to values of α not smaller than 0.5, say, and for not too large values of time. The behaviour of $x(t)$ for small values of α and t can only be modelled by a very small Δt .

6.References

[1] Devis J. : 'Operational Manual of the Dynaliser Model V ' : Bergougnan Instruments,Ghent,(1987)

[2] Doetsch G. : 'Handbuch der Laplace-Transformation,I ' : Birkhäuser Verlag , Basel,(1971)

[3] Ferry J.D.: 'Visco Elastic Properties of Polymers' : John Wiley and Sons,N.Y. (1975)

[4] Honig G. , Hirdes U. : 'A Method for the Numerical Inversion of Laplace Transforms' : J.Comp.and Appl.Math.,**10** ,(1984),113-132

[5] Peters J. et al : 'The Measurement of the Elastic and Loss Moduli of Rubber and Rubberlike Materials' : Academiae Analecta,43,no.3,(1981)

R. Van Keer and H. Serras : Faculty of Applied Science , R.U.G.

St.Pietersnieuwstraat 39 , 9000 GENT , Belgium